Lecture Notes in Mechanical Engineering

Lecture Notes in Mechanical Engineering (LNME) publishes the latest developments in Mechanical Engineering—quickly, informally and with high quality. Original research reported in proceedings and post-proceedings represents the core of LNME. Volumes published in LNME embrace all aspects, subfields and new challenges of mechanical engineering. Topics in the series include:

- Engineering Design
- Machinery and Machine Elements
- Mechanical Structures and Stress Analysis
- Automotive Engineering
- Engine Technology
- Aerospace Technology and Astronautics
- Nanotechnology and Microengineering
- Control, Robotics, Mechatronics
- MEMS
- Theoretical and Applied Mechanics
- Dynamical Systems, Control
- Fluid Mechanics
- Engineering Thermodynamics, Heat and Mass Transfer
- Manufacturing
- Precision Engineering, Instrumentation, Measurement
- Materials Engineering
- Tribology and Surface Technology

To submit a proposal or request further information, please contact the Springer Editor of your location:

China: Dr. Mengchu Huang at mengchu.huang@springer.com
India: Priya Vyas at priya.vyas@springer.com
Rest of Asia, Australia, New Zealand: Swati Meherishi at swati.meherishi@springer.com
All other countries: Dr. Leontina Di Cecco at Leontina.dicecco@springer.com

To submit a proposal for a monograph, please check our Springer Tracts in Mechanical Engineering at http://www.springer.com/series/11693 or contact Leontina.dicecco@springer.com

Indexed by SCOPUS. The books of the series are submitted for indexing to Web of Science.

More information about this series at http://www.springer.com/series/11236

S. Seetharamu · Thimmarayappa Jagadish ·
Ravindra R. Malagi
Editors

Fatigue, Durability, and Fracture Mechanics

Proceedings of Fatigue Durability India 2019

Springer

Editors
S. Seetharamu
Central Power Research Institute
Bangalore, Karnataka, India

Thimmarayappa Jagadish
DHIO Research and Engineering Pvt Ltd
Bangalore, Karnataka, India

Ravindra R. Malagi
Mechanical Department
KLS Gogte Institute of Technology
Belagavi, Karnataka, India

ISSN 2195-4356 ISSN 2195-4364 (electronic)
Lecture Notes in Mechanical Engineering
ISBN 978-981-15-4778-2 ISBN 978-981-15-4779-9 (eBook)
https://doi.org/10.1007/978-981-15-4779-9

This Springer imprint is published by the registered company Springer Nature Singapore Pte Ltd.
The registered company address is: 152 Beach Road, #21-01/04 Gateway East, Singapore 189721, Singapore

Preface

This volume contains 45 technical papers from the 2019 Conference Fatigue Durability India. The theme of this 3rd International Conference was Fatigue, Durability and Fracture Mechanics. As in the past conference, the format consisted of keynote talks, invited lectures, contributed papers, exhibition and contributed posters. A total of 225 delegates participated in the conference, and 61 research papers were presented in Fatigue Durability India 2019.

Conference began with the domain experts Prof. Dattaguru, Former Professor, Indian Institute of Science, Bengaluru, India; Dr. Karisiddappa, Vice Chancellor, Visvesvaraya Technological University (VTU), Belagavi and Dr. Ravindra. R. Malagi, Technical Expert in area of fatigue, durability and fracture mechanics in nuclear energy, defence, aerospace and welded structures. This was followed by keynote talks from Prof. Dattaguru; Dr. Sam Kantimathi, President, Fatigue Concepts and Aging Aircraft Associates, California, USA; Prof. Vikram Jayaram, Professor and Chairman, Division of Mechanical Sciences, Department of Materials Engineering, Indian Institute of Science, Bengaluru, India and Shri. Prem Andrade, Senior Engineering Manager, ANSYS Inc, Pune, India.

Second day keynote presentations were made by Dr. A. Venugopal Rao, Head, Modelling and Simulation Group, Defence Metallurgical Research Laboratory, (DMRL), Hyderabad, India; Dr. Omar Ibrahim, MD Process Optimisation, USA; Shri. M. R. Saraf, Sr. Deputy Director, HOD—Technology Group, Structural Dynamics Laboratory, Environment Research Laboratory and Centre of Excellence for Fatigue and Materials, Automotive Research Association of India, (ARAI), Pune, India; Prof. Raghu Prakash, Professor, Department of Mechanical Engineering, Indian Institute of Technology, Madras, India and Dr. N. Narasaiah, Professor, Department of Metallurgical and Material Engineering, National Institute of Technology, Warangal, India.

The third day proceedings were further enriched by the following keynote speakers: Dr. Mikel Isasi Iriondo, Principal Researcher, Elastomer Material Durability, Leartiker Polymer Research Centre, Spain; Dr. R Sunder, Research Director, Bangalore Integrated System Solutions (P) Ltd. and President, Indian

Structural Integrity Society, Bangalore, India; Dr. J. M. Chandra Kishen, Professor, Department of Civil Engineering, Indian Institute of Science, Bangalore, India and Dr. Badarinarayana K., Principal Consultant—COE, Aerospace and Defence, Visvesvaraya Technological University, India.

The papers presented in the conference led to very good interactions addressing the structural integrity of critical engineering components in aerospace, nuclear, automobiles and railways, defence, power, petroleum and chemical industries.

In respect of aerospace, integrated life cycle management aspects covered structural joints, design and quality assurance requirements on the real-time warning sensors, gas turbine rotor blades, helicopter structures, launch vehicle systems subjected to vibration, system safety principles, cracked wing skin, heat exchangers in aeroengines, trainer aircraft airframes, superalloy IN718, landing gear systems' functionally graded materials and surface modification of titanium alloys.

Low cycle fatigue and fatigue crack growth behaviour of critical components and weldments including the bimetallic pipe weld joints in the nuclear industry have been brought out along with life estimation strategy for nuclear reactor pressure vessel, corrosion fatigue in the compressor blade in a heavy duty gas turbine and fatigue crack growth in Ni-base superalloys .The remaining life estimation of thermal and hydro power plant components has also been discussed.

The structural integrity aspects of automotives and railways have been looked into from cumulative damage considerations as well as durability and reliability angle along with case studies.

The deliberations have also addressed the damage index of reinforced concrete structures, building structures, irregular reinforced-concrete-based isolated structures, FRP-confined columns, spring-integrated reinforced concrete hybrid beams, crack generation in rubber components used for tyre applications, non-linear behaviour in non-masking metals and impact of thermo environmental effects on optomechanical systems.

The conference ended with the awards of presentations to

(1) Mr. Shyam Kishore of GTRE, Bengaluru; (2) Mr. Puneeth Arora of BARC Mumbai; (3) Mr. Harish Babu of Wipro, Bengaluru, and Mr. Tarang Sinde, Visvakarma University.

The conference was sponsored by Ansys; BISS Bangalore; Software Technology Park (STPI); ICAT; Dassault System; SJC Institute of Technology, Chickaballapur; CadmarC Software Pvt Ltd.; Process Optimisation (Franc3D); Endurica; True Load and Publication Partner by Springer.

This volume contains 45 papers featuring significant aspects of fatigue durability and fracture mechanics emphasizing the multi-disciplinary subjects of fracture and failure, fatigue durability and life assessment dealing with different stages in the product life cycle management encompassing technological trends like digital twins, microscopic characterization, fretting fatigue, closing the gap between lab testing and engineering practice, high temperature corrosion, thermo-mechanical fatigue, fatigue of aerospace structural joints, combined LCF and HCF, residual stress and applied load on creep relaxation and failure of cementitious materials.

We thank all the sponsors, participants and exhibitors for their contributions to FatigueDurability 2019. We specially thank the authors and reviewers of the papers in this volume. Special thanks to all organizing committee members and programme committee members of FatigueDurability 2019.

Bangalore, India	Dr. S. Seetharamu
Bangalore, India	Dr. Thimmarayappa Jagadish
Belagavi, India	Dr. Ravindra R. Malagi

Contents

About the Editors

S. Seetharamu received his Ph.D. in Mechanical Engineering from Indian Institute of Science in 1982 after obtaining M.E. in Mechanical Engineering from IISc in 1976 and B.E. in Mechanical Engineering from Bangalore University in 1974. Dr. S. Seetharamu worked in CPRI since 1985 till his retirement as Director in 2015. His areas of research interests are energy technology and materials engineering. Prior to joining CPRI, he had also worked in the industry and served as a faculty member at Toyohashi University of Technology, Japan.

Thimmarayappa Jagadish is the Technical Head of DHIO Research and Engineering Pvt Ltd., Bangalore. He completed his B.E, M.E and Ph.D in Mechanical Engineering from UVCE Bangalore University in the year 1983, 1986 and 1994 respectively, and worked as a Lecturer at RVCE, Bangalore from 1986 to 1996 and as an Assistant Professor at School of Post Graduate Studies, NTTF, Bangalore. From 1999 to 2000 he worked as Head, Competence Design Center, Federal Technologies, Bangalore, and then at GE India Technology Center (now known as JFWTC, Bangalore) as CAE Specialist from 2000 to 2002. From 2002 to 2019 worked as Professor for Post Graduate studies, Professor and Head of the Department of Mechanical Engineering in Bangalore Institute of Technology, Bangalore.

Dr. Ravindra R. Malagi is a Professor in Visvesvaraya Technological University, India. He received his Ph.D. in Mechanical Engineering & Science from Visvesvaraya Technological University (VTU) in 2010 after obtaining M.E. in Design Engineering from Karnataka University Dharwad and B.E. in Mechanical Engineering from Karnataka University Dharwad and has previously worked as a Professor in the Department of Product Design & Manufacturing in Gogte Institute of Technology, Belagavi. His research interests include finite element method, design for manufacturing, rapid prototyping, tribology, advanced theory of vibration and the theory of elasticity. He has worked as a member of B.O.E, Mechanical Engineering at VTU and Deputy Chief Coordinator for VTU Digital Valuation.

Design Accomplishment for Unforeseen Discrepancies for Better Structural Strength and Fatigue Life of Helicopter Structures

Kalinga Gulbarga and A. T. Rao

Abstract Airborne structural parts are designed for critical operating flight and landing load conditions following the certification guidelines for enough strength and rigidity. But in the development test environments they will be subjected to loads differently than expected in the operating life unexpectedly leading to premature fatigue failures. Instances like fastener loosening due to loss of positive lock in primary attachments will trigger vibratory dynamic loads and fail the structural attachments. It is essential to avoid these unsafe situations by considering such events in the design. This paper aims at dealing one such instance practically observed during flight testing of helicopter. Redesign and analysis is carried out that show simulation of critical stress levels for dynamic vibration loads for fastener loosing effects and design accomplishment to avoid such premature structural failures and maintain original damage-tolerant behavior with better strength and fatigue life levels to meet operation life requirements.

Keywords Airborne · Positive lock · Vibration · Damage tolerance · Strength

1 Introduction

Helicopter structures are designed and flight certified as per approved international Federal Aviation Regulations [1]. The design criteria is that structural parts shall have the ability to withstand severe flight and ground operation loads without permanent deformation and last for its intended design life with no fatigue failure. During the initial design and development phase, the structural parts are analyzed for critical loading conditions for the static and fatigue strength using aerodynamically estimated loads. Also certification regulation calls for design of structural parts for ultimate loads arrived from operation loads multiplied with factor of safety. The objective is that structural parts shall not fail for specified design life and if failure happens, it

K. Gulbarga (✉) · A. T. Rao
Stress Group, RWRDC, HAL, Bangalore, India
e-mail: kgulbarga@gmail.com

S. Seetharamu et al. (eds.), *Fatigue, Durability, and Fracture Mechanics*, Lecture Notes in Mechanical Engineering, https://doi.org/10.1007/978-981-15-4779-9_1

requires to survive until repair shows causing damage tolerance without diminishing strength for operation loads.

During development and flight testing phase, unforeseen instances may happen that have potential for structural damages and risk of catastrophic structural failures. The failure type, location, and size are neither genuine nor they are known from manufacturing process defects. It is important to monitor such instances, consider them in the initial design of primary structural parts of critical nature, and eliminate risk of catastrophic failures. One such practical event observed during helicopter flight testing is taken up for study as a case of design accomplishment. It is found in the inspection after helicopter flight testing that tail end horizontal control surface structural bracket near tail rotor got cracked and seen loosened bolt attachment. It could be due to loss of positive locking at the bolt nut and subsequent high "g" vibration loads at horizontal control surface could lead to structural bracket failure. Once again the structural bracket is redesigned with improved strength and attachment redundancy and verified its ability to sustain vibration fatigue loads apart from its regular operation flight load support capability.

2 Horizontal Control Surface Attachment

Horizontal control surface is an inverted aerofoil which contributes helicopter stability along the pitching axes. Its attachment is designed to withstand for maximum vertical load of push forward flight maneuver. The details of the horizontal control surface attachment are shown in Fig. 1.

The machined bracket is used to attach horizontal control surface laterally to the end of helicopter tail boom using quarter inch bolts. This horizontal control surface is assembled with vertical control surface for helicopter yawing stability. The total assembly will act as cantilever and transfers the span-wise and chord-wise aerodynamic pressure loads of horizontal and vertical surfaces to the helicopter tail boom structure through machined bracket. The vibratory load magnitudes if any due to stiffness loss will also transfer to the tail boom structure.

Originally, the machined bracket is stress analyzed for the critical push forward flight loads considering optimized attachments with quarter inch tension bolts. In the helicopter ground run flight testing, the bracket was subjected to vibratory dynamic loads (F_x, F_y, F_Z) which are entirely different from design analysis load of push forward flight load case. The load path and magnitudes got changed due to disengagement of tension bolt and resulted in crack initiation on top flange corner stressed region of machined bracket. The cracked location and the loosened bolt are shown in Fig. 2.

The cracked bracket is replaced with newly redesigned bracket which is designed and analyzed to sustain vibratory fatigue loads from unexpected instances like nut disengagement in the operation.

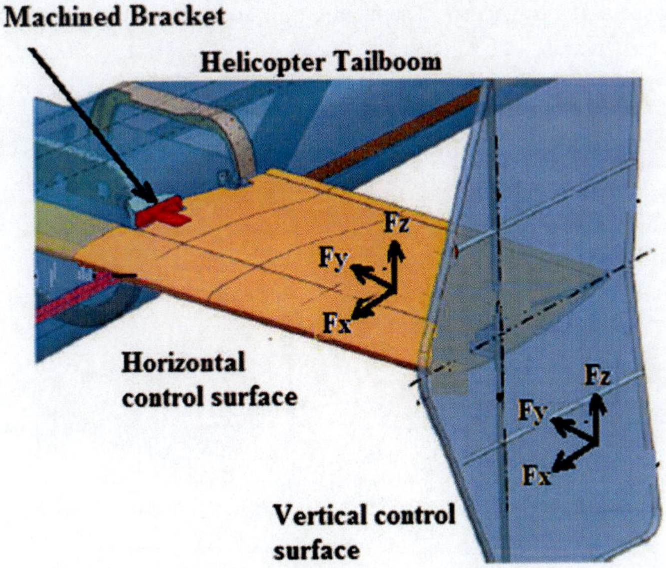

Fig. 1 Horizontal control surface attachment details

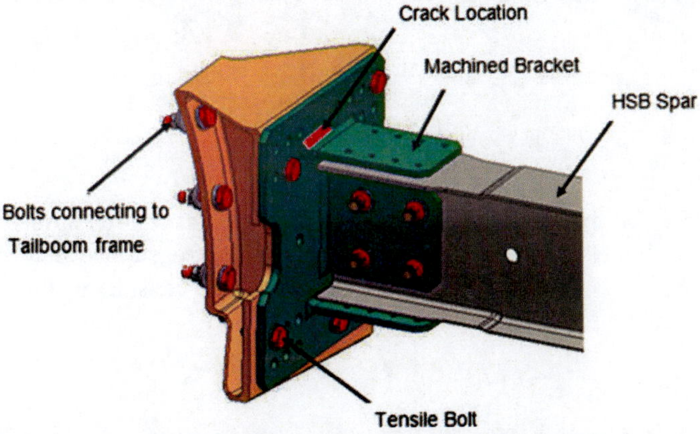

Fig. 2 Detailed view of machined bracket attachment showing failure

3 Vibratory Fatigue Loads

As per regulations, it is mandatory to monitor helicopter vibration during flight testing to track whether the vibration "g" levels are minimum. The vibration "g" magnitudes in x-, y-, and z-directions captured in the flight testing of bolt loosing instance are shown in Fig. 3. The peak vibratory load magnitudes are observed at horizontal and vertical control surfaces in all x-, y-, and z-directions.

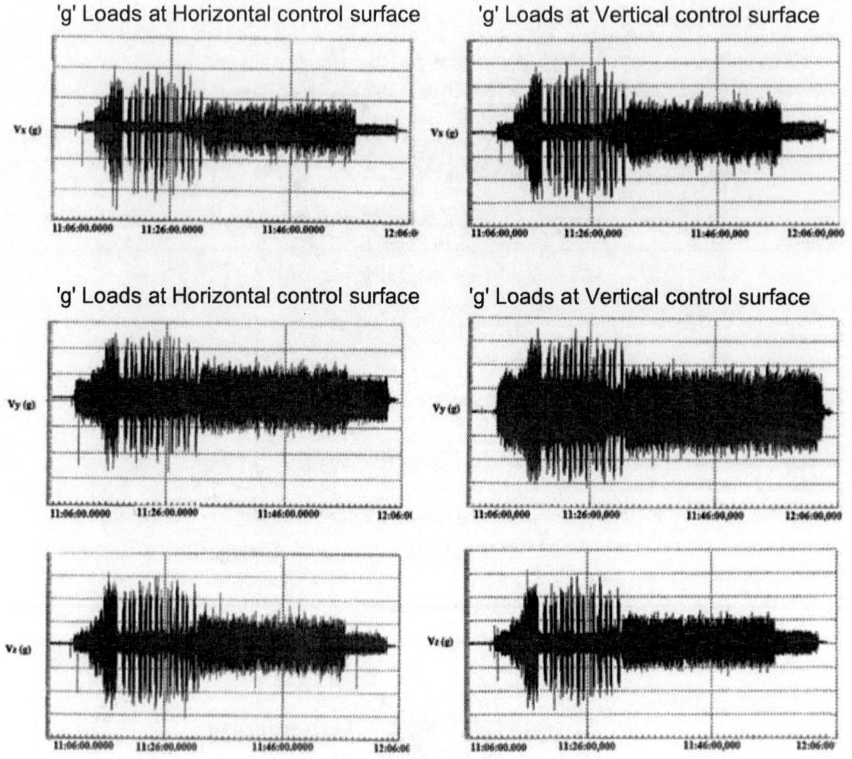

Fig. 3 Helicopter ground run vibration g loads

The estimated dynamic load magnitudes at horizontal control surface are $F_x = 230$ N, $F_y = 140$ N, and $F_z = 340$ N and at vertical control surface are $F_x = 260$ N, $F_y = 280$ N, and $F_z = 252$ N. The bracket is redesigned considering these dynamic loads and verified for fatigue life.

4 Bracket Analysis Verification

During the design phase, horizontal control surface machined bracket is stress analyzed for critical push forward flight loads with tension bolt attachments. This machined structural bracket is later stress verified for vibration load magnitudes to understand the failure with boundary conditions of original bolt connections and loosened bolt condition separately. The dynamic vibratory loads F_x, F_y, and F_Z of horizontal control surface and vertical control surface are applied and the stress variation across the bracket is obtained. These stress magnitudes are compared with the original flight push forward stress values and shown in Fig. 4.

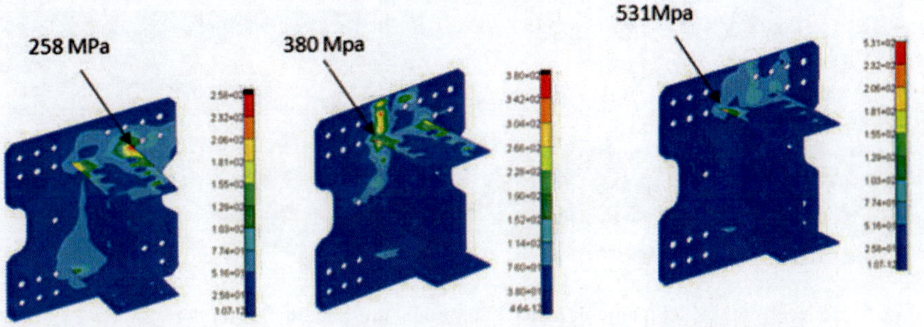

(a) Push fwd flight case (b) Dynamic load case with loosened bolt

Fig. 4 Machined structural bracket stress analysis

The original design push forward flight case design stress level is kept at 60% of bracket allowable. The bracket is made of aluminum alloy material which has satisfactory service experience and good allowable strength. The bracket is safe for flight push forward load case as stress level is within the allowable limit. For the dynamic "g" loads, stress levels raised beyond design stress level and crossed allowable strength for the condition of loosened bolt. This maximum stress region is coincided with crack location. The stress magnitude for the loosened bolt condition exceeded the bracket material static ultimate strength and hence failure of the bracket occurred.

5 Bracket Redesign and Fatigue Strength

Most effective method is to design a lower stress level to improve fatigue life. The structural bracket is redesigned to improve the redundancy, strength, and fatigue life. Stress concentration at critical location is controlled by providing ribs and increase in flange thickness. For better fail safe design, redundancy in the bolt attachments is incorporated by increasing the number of bolts. The bracket is redesigned with these features and finite element model is created which is shown in Fig. 5. Analysis has been carried out for the improved design for the dynamic vibration loads. The magnitude of bracket stress obtained is shown in Fig. 6.

The redesigned bracket stress level is well within the allowable limit of material strength and met regulation safety margins of strength for the critical loads. This bracket has ability (toughness) to sustain stress concentration. The number of cycles of the vibration peak "g" loads observed in flight case is shown in Fig. 3. From the material SN curve [2], it is seen that the number cycles the bracket can tolerate for maximum stress is nearly 6×10^7 cycles (Fig. 7). The fatigue life estimated from these cycles using the time period is meeting component design life requirements. This design life estimated assuming flight test vibration loads is conservative and

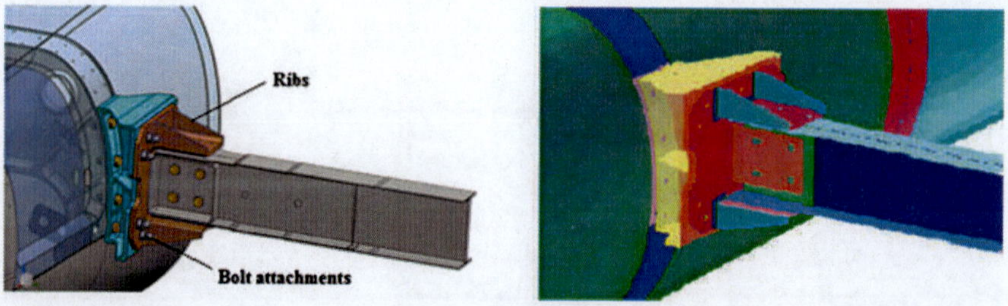

Fig. 5 Redesigned bracket with bolt attachments and corresponding FE model

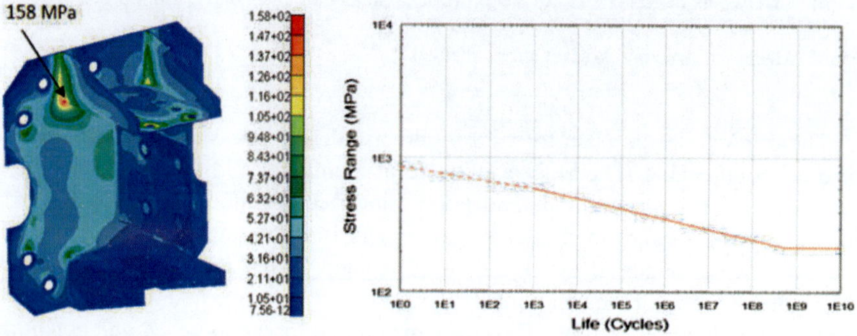

Fig. 6 Redesigned bracket stress plot for vibration dynamic loads

Fig. 7 Fatigue life cycles obtained from fatigue analysis [3]

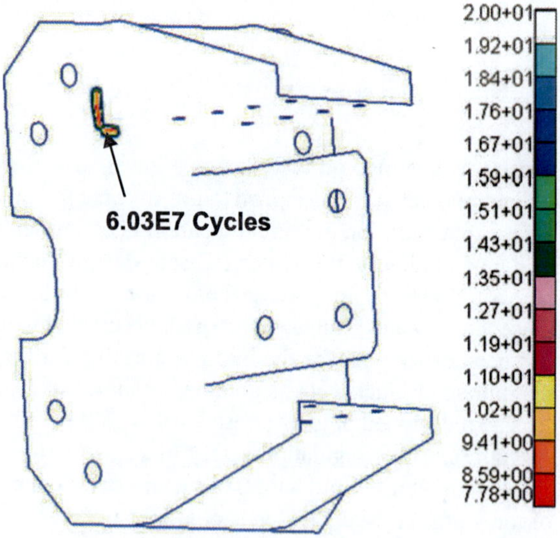

such "g" loads will not be present in the original operating conditions. Even if such load comes the redesigned bracket is able to sustain these dynamic loads and fatigue failure is highly remote and safety is ensured. It is good to explore and visualize stress critical failure possibilities in the testing phase in the critical in unsafe condition primary structures and accomplish the design such that they will not lead to unsafe conditions.

6 Conclusions

Critical structural part design and analysis shall incorporate all critical failure possibilities that may likely to occur in development other than the usual operation design flight load cases. This will accomplish the design and maintain the desired level of damage tolerance, fatigue life, and reduce the fracture-related structural failures. It is very important that manufacturing and assembly process should not reduce the damage tolerance level required by the design.

References

1. Federal Aviation Regulations (FAR 27 amdt 47–1)
2. Metallic materials Properties Development and Standardization (MMPDS), pp 3–112
3. MSC Nastran/Patran Fatigue manual

Design Optimization of Frame Rail Under Fatigue Condition Through FEA for Electric Drive Dump Truck

S. Suthakar and Y. M. Renukaraj

Abstract The objective of this study is to design optimized frame under fatigue condition through finite element analysis (FEA) for the off-highway electric drive dump truck. In today's globalization world, the off-highway sectors started adopting advance technologies to get the optimized and reliable products. Nowadays, the off-highway vehicles have to be cost-effective and must have high power-to-weight ratio for the better fuel efficiency and to increase life of the critical components such as powertrain aggregate engine, alternator, etc. In the mining sector, desirability depends on the economics of operation in terms of 'cost per ton' of material transported. The cost per ton can be achieved only through the higher payload equipment and lower cost per ton approach. In case of open cast mine, the lower cost per ton can be achieved by adopting higher capacity dump truck with optimized weight and ease of maintenance of the electric drive dump truck. Therefore, this study has made an attempt to design the optimized frame rail size under fatigue condition through FEA method for the electric dumper. It covers how and when to adopt the optimization approach while designing frame structure. Also, it has emphasized that the fatigue load is one of worst load conditions for the frame design, where the complete frame will be twisted under left and right ramp conditions. Under these conditions, the optimum frame rail was designed with optimum plate thickness to meet the design criteria. The frame rail with different sizes and plate thickness were considered and analysed under the same alternate load conditions. Based on the historical data of frame with respect to the frame life, the optimized frame rail size and required plate thickness of frame were determined at the initial stage of design. This approach results in optimized frame and meets targeted power to weight ratio. Finally, this optimized design enhances the life of powertrain aggregates and improves fuel efficiency of vehicle.

Keywords Frame · Optimization · Dumper · Electric · Powertrain · FEA (finite element analysis)

S. Suthakar (✉) · Y. M. Renukaraj
R & D, BEML Limited, Mysore, India
e-mail: gwsudha@beml.co.in

S. Seetharamu et al. (eds.), *Fatigue, Durability, and Fracture Mechanics*, Lecture Notes in Mechanical Engineering, https://doi.org/10.1007/978-981-15-4779-9_2

1 Introduction

Energy needs of the country are continuously increasing and a major share is being met by coal-based thermal power plants. In order to increase thermal power generation to meet future demands, GOI has proposed to double the current production to reach 1 billion ton by 2020. India is richly endowed with abundance of coal deposits and this provides unique opportunity for generating electricity at competitive rates which is crucial for competitiveness of manufacturing sector and overall economy. In this backdrop, manufacturing high-capacity mining machinery in India is of crucial importance to ensure self-reliance and cost competitiveness of the energy sector.

Presently, open Cast mines are predominantly operated with mechanical drive dump trucks in the range of 35 t, 60 t and 100 t for removal of over burden as well as coal. Few larger mines of M/s Coal India are also operated on 170 t DC electric drive dump trucks which have been completely imported in the past. There is a plan by the coal mines to introduce 200 Ton class dump trucks in their mega projects to enhance rate of coal production. Currently, 200 Ton class dump trucks are principally available from overseas players like Caterpillar, Komatsu, Hitachi, Belaz, etc.

In this backdrop, BEML took the initiative to design and develop BH205E AC electric drive dump truck with several design features to enhance productivity, safety, operator comfort, and reliability.

In today's competitive world, any product in the market is compelled to have reliable, optimal performance, low cost, better aesthetic and ease of maintenance. These factors have become key parameters of products in order to sustain in the competitive market.

In the mining sector, *desirability* depends on the economics of operation in terms of 'cost per ton' of material transported. The cost per ton can be achieved only through the higher payload and lower cost per ton approach. In the mining sector, especially for open cast mine, the lower cost per ton can be achieved by adopting higher capacity dump truck and easy maintenance electric-driven dump truck. Figure 1 shows the electrical-driven rear dump truck.

1.1 Problem Statement

In today's globalization, challenges faced by dump truck manufactures are that of optimized design electric drive dump truck in order to meet the mining sector demand that is the economics of operation in terms of 'cost per ton' of material transported.

Therefore, this study has made an attempt to design the optimized electric drive dump truck frame structure through FEA. To achieve the optimum design, this study has adopted FEA approach at the conceptual design itself to determine the optimum frame rail size and plate thickness in order to have longer life at low cost. The other benefits of having the optimized frame structure at initial design stage are having low weight and to meet targeted power-to-weight ratio for the better fuel efficiency.

Fig. 1 Electric drive dump truck

1.2 Comparison of Electric-Driven Dumper Over Mechanical-Driven Dumper

Figure 2 shows the comparison of electric-driven dumper over mechanical.

A few advantages of electric-driven dump truck over mechanical-driven dump truck are lower lifestyle costs such as less gear, bearing and no brake wear, easy maintenance and low cost, higher power, speed and space resulting to the economic operation.

2 Frame Design Methodology

The space claim model is initial stage of frame design and created by accommodating all the critical aggregates. It is conceptual design stage for creating space for the major aggregates such as powertrain, electrical system, hydraulic system, brake system, exhaust system fuel system, etc. In the space claim model of frame, mounting location of these aggregates was fixed. These mounting locations and its load were inputs for the designing of optimized frame under the cyclic load condition to have better fatigue life.

A few novelties of this study are, it has adopted the design optimization of frame rail size at the initial conceptual design stage itself and the required plate thickness to

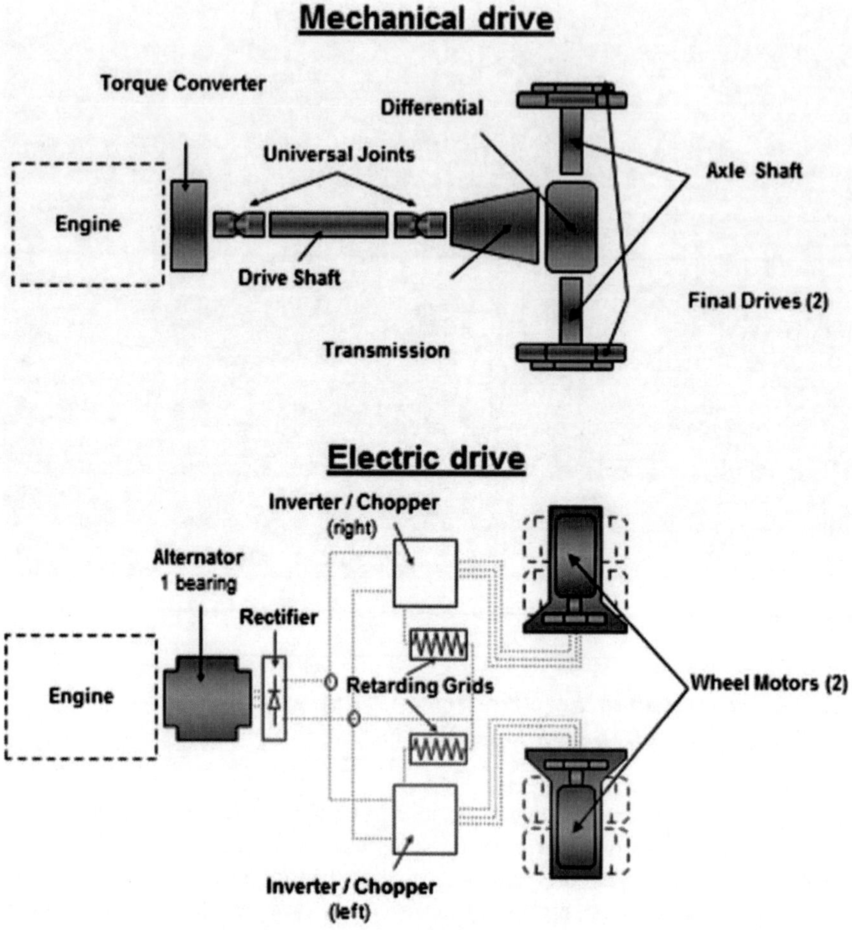

Fig. 2 Electric-driven dumper versus mechanical-driven dumper

withstand the load and meet the design criteria through the FEA and it was evaluated by theoretical method also.

2.1 *Frame Analysis Model*

The conceptual frame design model was created by using CAD software. At the initial stage of design, the major aggregate mounting locations were fixed tentatively and with these data the centre of gravity of frame was calculated. Figure 3 shows 3D model of optimized frame design.

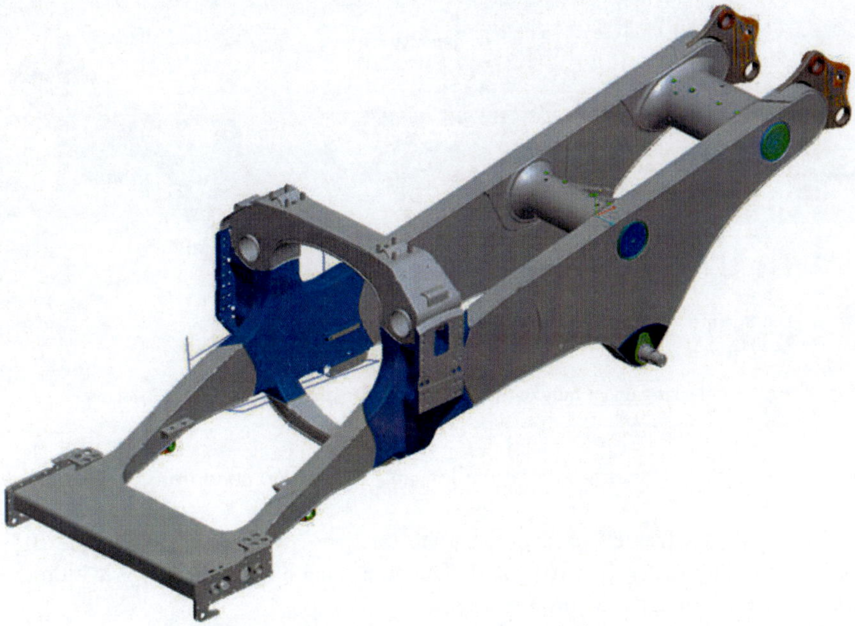

Fig. 3 3D model of frame space claim model

2.2 *Design Criteria of Frame*

Based on the historical data of dump truck and field data, the frame should be designed for 40,000 h (that is, having appro. fatigue life of ~60,000 number of cycles) and the stress value at the critical joints of frame should be within 483 MPa.

3 Results and Discussion

3.1 *Frame Rail Sizing*

The determination of optimal rail size is the most crucial stage in frame design for finding out the appropriate plate thickness of rail. Figure 4 shows the frame rail sizing under 2G condition, that is, left and right ramps through finite element analysis.

The frame structure is considered to be the load bearing structure to support and withstand the major loads such as payload, body structure weight, powertrain system and other system. It acts as back bone of vehicle and to withstand various loads such as bending, torsion, braking, etc. Among these load cases, the torsion load (also known as twisting load) is considered to be worst load for the frame structure to be fatigue condition. Under this torsional load, the complete frame structure will be

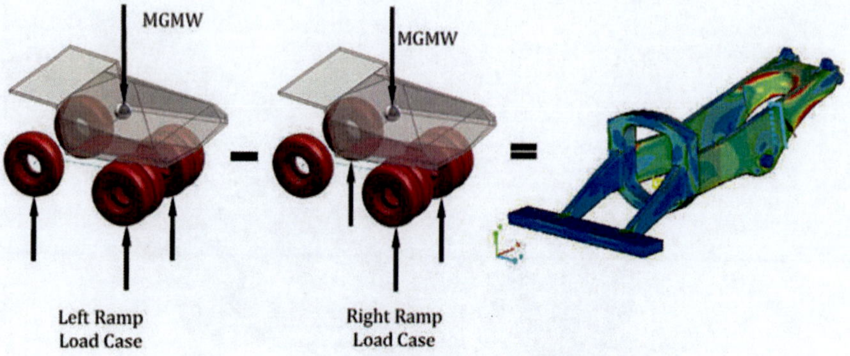

Fig. 4 Frame rail sizing under fatigue (left and right ramps) condition

twisted about its structural axis to the left and right ramp conditions (to be fatigue condition).

The optimum frame rail size and plate thickness were determined under this fatigue condition (that is, twist load). The stress value of frame rail is within the material yield strength and design criteria.

The frame rails with different sizes were modelled and analysed under the same load condition. Frame rail strength analysis was carried out under the maximum torsion load with 2G condition. These load conditions have predicated effectively the frame width and depth to withstand the maximum load under 2G conditions.

The analysis, nearly 14–16 iterations, was carried in order to find out the optimum frame rail size and required plate thickness for frame.

The plate thicknesses for bottom, side and top side of frame were determined based on the FEA analysis. The frame models of width at 200, 225, 250, 275 and 300 mm were modelled and analysed under the same load condition in order to determine the optimum frame width and depth and plate thickness to meet targeted design criteria and within the material yield strength. Figure 5 shows the FEA analysis result of two frame rail size.

From a fatigue standpoint, it is assumed that the frame must endure cyclic loading that oscillates between that left ramp and right load cases. Therefore, subtracting the results of the right ramp load case from the left ramp load case yields the full range of cyclic stress. Figure 6 shows the frame rail size FEA analysis result.

From the FEA analysis of frame, it indicates clearly that the frame rail size in range of 200 to 280 mm shows the stress value within the design criteria and material yield strength. The required plate thickness of frame rail and optimum size were determined through the FEA approach. The optimized frame design meets the international dump truck manufacture design standard (competitor models specification) and meeting the rail width and plate thickness and other design requirements such as power-to-weight ratio value and wheel base and wheel track.

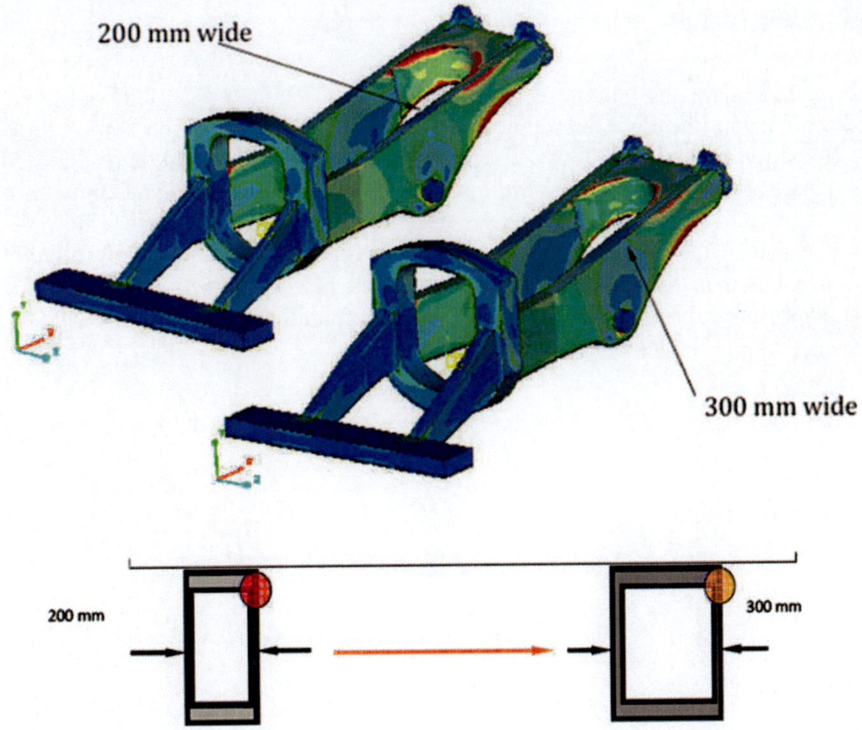

Fig. 5 FEA analysis result for the different frame rail sizes

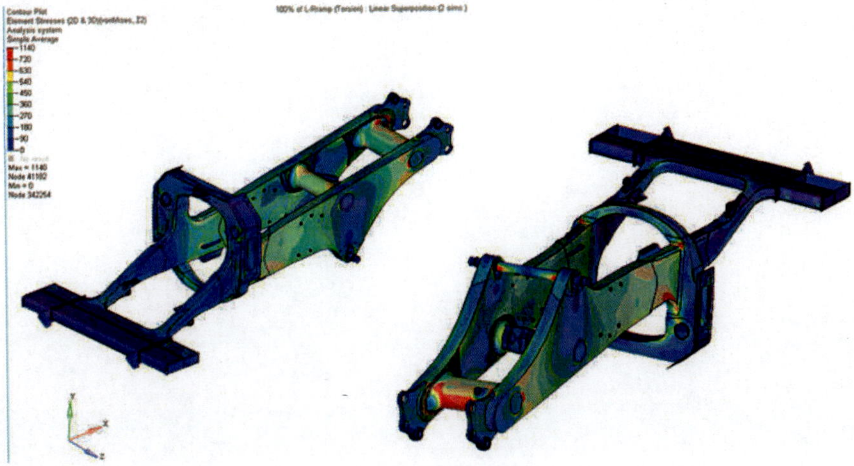

Fig. 6 Frame rail size FEA analysis result

4 Conclusion

From this study, the optimized frame rail size and required plate thickness were determined and the frame rail was designed under the torsional load condition (fatigue left and right ramp conditions). The space claim model is initial design stage and is created by accommodating all the major aggregates. The individual component location and load were considered as design inputs for frame design. Under the torsional load, the complete frame is subjected to the twist condition to left and right ramps. Based on the historical data on dump truck, newly designed optimized frame structure meets the number of cycle 60,000 cycle condition and stress value with the design criteria and material yields strength.

Total Technical Life Extension Techniques for a Trainer Aircraft Airframe

K. Manonmani, U. A. Acharya, and Appasaheb Malagaudanavar

Abstract Today, many ageing aircrafts in air forces of many countries are in service beyond their designed total technical and calendar service life. In general, typical military fighter and trainer aircraft airframe structure is designed for a service life of 20–30 years. The life extension programmes are being undertaken today for ageing aircraft considering the non-utilization of aircraft to its full potential, high acquisition cost of an aircraft limited by defence budget. This paper presents on the total technical life extension of an IAF (operator) trainer aircraft with initial Total Technical Life (TTL) of 8000 h. TTL extension programme has been carried out with operator formed task team including HAL and certifying agencies to extend the TTL of airframe beyond 8000 flight hours. Life extension studies consisted of full airframe (viz. fuselage, wing, tail plane and engine mounts, etc.) with respect to strength and fatigue life analysis margins, full scale fatigue test results, post-test inspection reports, histories of earlier overhaul data, material ageing aspects (considering environmental degradation, corrosion prevention and protective coating, wear and tear), 'g' exceedance and hard landing data, incident histories and repair scheme implemented on service aircraft. Additionally, TTL attained two lead aircrafts were strip opened, thorough inspections suggested and results were studied. After these extensive studies, the task force concluded that TTL of trainer airframe can be increased by 500 flight hours. Hence, as per recommendation, TTL of the aircraft extended to 8500 flight hours. Considering the present usage rate in fleet, the extension will ensure availability of trainer aircraft for training in another 4 years, thereby benefitting the operator for performing the intended tasks.

Keywords Total technical life (TTL) · Trainer aircraft · Full scale fatigue test (FSFT) · Fatigue modifications (MODs) · Material ageing · Inspection · Crack detection (CD) · Incident histories · Pre-survey

---

K. Manonmani · U. A. Acharya (✉) · A. Malagaudanavar
Aircraft Research and Design Centre (ARDC), HAL, Bangalore, India
e-mail: aa.udupi@hal-india.co.in

S. Seetharamu et al. (eds.), *Fatigue, Durability, and Fracture Mechanics*, Lecture Notes in Mechanical Engineering, https://doi.org/10.1007/978-981-15-4779-9_3

1 Introduction

Total Technical Life (TTL) extension programmes are being undertaken nowadays to meet the current and future mission/training requirement using existing fleet. The programme aims at ensuring the airworthiness requirements of the ageing aircraft for performance beyond initial assigned life and the assets are used to its full potential life. Extending the TTL requires various approaches for analysing all the occurrences on the airframe fleet and evaluation of study results. This will facilitate prediction of the remaining service life beyond originally awarded.

The present paper discusses on typical trainer aircraft, which forms the backbone of Indian Air force (operator) ab initio pilot training program. The trainer fleet has served operator for more than five decades and majority of them are approaching end of its awarded service life. The trainer aircraft was designed, developed and manufactured by HAL. Initially, design service life of the airframe was assigned as 5000 flight hours. The designed service life is the period of time (viz. flight cycles, flight hours or landings) established at design, during which the structure is expected to maintain its integrity when flown to the expected loads and environmental spectrum. The subsequent Full Scale Fatigue Test (FSFT) of this trainer airframe specimen had revealed that there is a potential for extending technical life of the airframe beyond this. In this case, fatigue cracks observed at various stages during testing were repaired and the test was continued till the major load-carrying member failed, wherein the test was stopped. Later Fatigue Modifications (MODs) were regularized in fleet as compliance at specified flight hours to achieve the target fatigue life of airframe. As on today, TTL of airframe was extended from originally assigned design life of 5000 flight hours to 8000 with applicable MODs at different stages.

Owing to the operational requirements, operator is requested to study feasibility of TTL extension of trainers for exploitation beyond 8000 flight hours.

Note Due to confidentiality, aircraft name is not mentioned.

2 TTL Extension Study Initial Requirement

The program requires a team of all stakeholders, viz customer, manufacturer and certifying agencies (CEMILAC & DGAQA) for studying at various aspects involved in TTL extension which ensures airworthiness and structural integrity. Two lead aircraft (i.e. operational trainer aircraft having life nearing 8000 flight hours) are required for strip examination and detailed study by the joint team. Two lead aircrafts with 7991.20 and 7999.15 h were made available for the studies. An acceptable route to airworthiness post study agreed by the stakeholders.

3 TTL Extension Methodology

3.1 Study of Historical Design Data

(a) Study of stress/fatigue analysis report of airframe components (viz. fuselage, wing, HT, VT, engine mounts, etc.) to note down the strength and life reserve margins for assigning the criticality.
(b) Review of Full Scale Fatigue Test (FSFT) results of trainer aircraft to assess the fatigue-related MODs and availability of test life margins.
(c) Study of earlier TTL extension methodologies adopted, life margins leftover, etc. of the airframe.
(d) Study of trainer airframe material ageing (environmental degradation, corrosion prevention and protective coatings, wear and tear of moving parts, etc.)

3.2 Study of Fleet Data

(a) Study of incident history (viz. hard landings, 'g' exceedances, etc.) on operational service aircraft and repair schemes implemented and fuel leaks study, etc.
(b) Study of aircraft maintenance schedules/NDT schedules, overhaul history and other abnormalities during earlier major/minor servicing, etc.
(c) Study and review of implemented various special instructions, special technical instruction and service bulletin on aircraft.
(d) Study of data on cannibalization of fuselage, wing, tail plane, fin, control surfaces, etc. during overhaul.

Strip Examination of the lead aircraft based on the inspection document generated as a result of above study.

Study of Strip examination results and feasibility for extension of life.

4 Detailed Study, Inspection and Analysis

4.1 Study of Historical Design Data

4.1.1 Study of Stress/Fatigue Analysis Reports of Airframe and Manuals

The available stress/fatigue analyses of trainer airframe documents have been studied to understand the strength and fatigue life reserve margins availability and zones requiring further inspection or special attention. From the study, the critical parts

Fig. 1 Full scale fatigue test of typical trainer airframe

were identified. Few identified critical parts are listed herein. They are removable fuselage skin stringer panel, cockpit floor, seat frames, engine mounting brackets, airbrake fittings, longerons, transport joints (front and rear fuselage), landing gear fittings, wing spars, flight control system levers, tail plane fittings, etc. Inspection zones were identified based on this study.

Initial fatigue analysis with assumed fatigue spectrum resulted in assigning the service life of 5000 h.

4.1.2 Review of Full Scale Fatigue Test (FSFT) Results and Fatigue MODs

Full Scale Fatigue Test (FSFT) of trainer airframe was carried out in Ground Test Group of ARDC, HAL during 1980s. Test results were studied by ARDC, HAL stress team. Figure 1 gives the typical trainer aircraft FSFT setup, Fig. 2 shows the typical fatigue load spectrum for FSFT and Fig. 3 shows the few pictures of observed cracks while testing. Scatter factor of 3.33 has been used on fatigue tested unfactored hours to arrive the actual factored hours. Observation and action taken during the fatigue test with respective factored hours are given in Table 1. In general, production or retrofit changes are incorporated to correct local design and manufacturing deficiencies disclosed by test.

From the study of all overhauled aircraft, it was observed that all the fatigue MODs have been implemented on trainer fleet at appropriate flight hours to meet the airworthiness and structural integrity requirements. These test results and MODs status have been studied presently to specify inspection zones.

4.1.3 Study on Earlier TTL Extension Documents

Historical data on the life enhancements done on the fleet was studied. From the study, it was seen that fatigue life of the fleet aircraft was assigned as 5000 flight

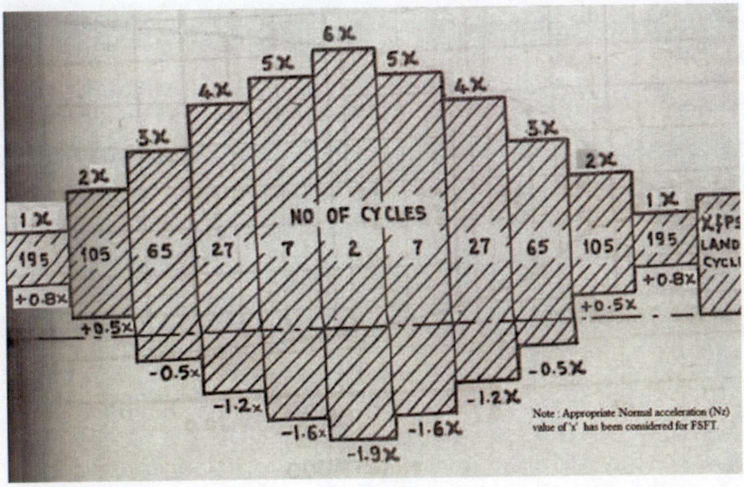

Fig. 2 FSFT fatigue load spectrum

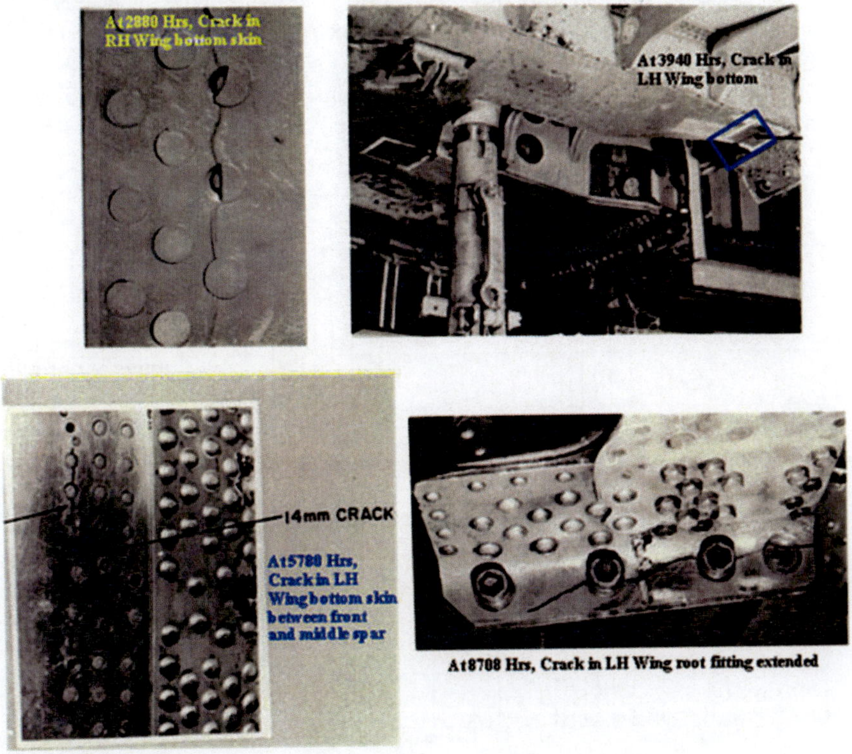

Fig. 3 Observed few cracks during full scale fatigue test of typical trainer airframe

Table 1 Observation and action taken during full scale fatigue test of trainer aircraft

Observation during fatigue test	Action taken during fatigue test after fatigue analysis	Complied MODs before flight hours (Factored)
Centre wing bottom skin crack	Skin reinforcement	2800
LH outer wing bottom skin crack near landing gear bay	Aluminium doubler replaced by a steel doubler	3900
Centre and outer wing skin and stringer crack	Centre and outer wing reinforced by doubler	5775
Centre and outer wing root fitting crack noticed	–	6330
	Test continued with crack till fracture observation	8708

hours based on analysis. Subsequent to the fatigue test based on the test observation, life of the airframe specified was 5775 h with applicable MODs initially and then extended to 6330 with compliance of centre and outer wing reinforcements. Further, life was extended to 7500 h by flight monitoring through fatigue meter (measured data) and the data were compared with design spectrum/full scale fatigue test results as given in Fig. 4.

Assimilated flight data infers that aircraft flights are within the designed fatigue spectra. Finally, with re-compliance of outer wing bottom skin steel doublers MOD,

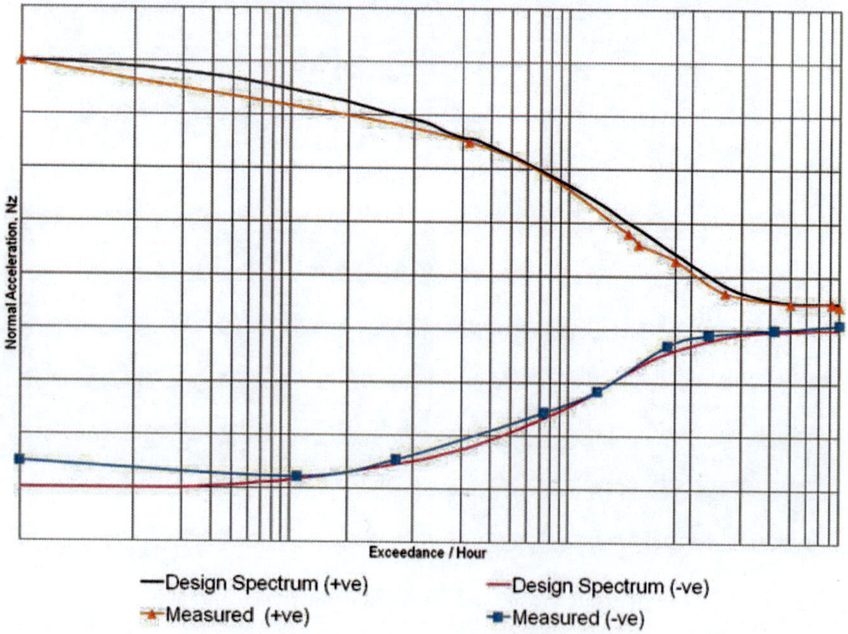

Fig. 4 Design and measured fatigue load spectrum of a typical trainer aircraft

fatigue life was assigned as 8000 flight hours. These extensions were granted with necessary modifications incorporated at appropriate life of aircraft.

4.1.4 Study on Airframe Material Ageing

Environmental degradation, corrosion prevention programmes followed, protective coatings employed, wear and tear, etc. were studied from existing service/design manuals and field data. This study was required to satisfy structural load-carrying capability. Potential corrosion problems were also considered while servicing and life extension of metallic structures of aircraft. A detailed study on damage assessment, NDT techniques and identification of common corrosion occurrences in steel, aluminium, etc. and their alloys were carried out. Inspection procedure after strip opening of aircraft and precautions while assembling back with various types of fasteners was also brought out during the activity. Further, electrical looms/cables, hoses and other fixed LRU mounting brackets were studied for environmental degradation possibilities.

4.2 Study of Fleet Data

4.2.1 Study of Incident Histories

In case of incident histories like hard landings, exceedance of higher normal accelerations beyond the design limits (i.e. higher 'g's), TTL extension studies will include issues of snags/repair schemes implemented on service aircraft, fuel leak, etc.

Analysis of the data provided by the operator showed that there was no incidence of aircraft experiencing exceedance of design limit 'g' values. Since there is no 'g' exceedance, additional checks of major fuselage attachments, distortion of panels, looseness of rivets, etc. need not be carried out. However, in one of the lead aircrafts, nose wheel indications showed 'Red' due to heavy landing. Heavy landing may result in structural distortions, fractures, skin wrinkling and failure of bolted and riveted joints. These effects are localized and are associated with heavy impact. It was observed that post heavy landing checks of these on landing gear and its attachment brackets/surrounding structures were satisfactory.

Further, customer brought out the issue of failure of rivets/fuel leaks happening frequently in air base. Detailed study on this was carried out. Later, it was suggested to carry out inspection on health of sealants (PRC) and checks for fuel tank area rivet loosening, whenever aircraft is inducted for major servicing. Further to this, the countersunk rounds to protruding rivets were recommended for compliance in the affected zones with the introduction of specially fabricated washers.

4.2.2 Study of Manuals

A detailed study of aircraft maintenance schedules/NDT schedules, overhaul history and other abnormalities during earlier major/minor servicing, etc. was carried out to arrive the inspection zones. In addition to that, study and review of various special instructions, special technical instruction and service bulletin implemented on the fleet were studied. Cannibalization details on aircraft during its service were also carried out. This is necessary to know if any structural element has put in more service life or has more life left compared to other elements. Based on this study, areas for inspection, detailed/NDT are identified for consideration of life extension.

4.3 Inspection Document Prior to Strip Opening of TTL Allotted Lead Aircraft

Multidisciplinary team studied all the relevant data of historical and fleet actual usage data. The team arrives at a document which will detail zone-wise and component-wise inspections to be done along with the type of inspections, viz visual, aided visual, micrometry, NDT techniques for crack detection, hardness detection, etc., and the document should enable to capture all possible degradations/damages that have occurred on the lead aircraft. A sample page of the inspection document is given in Fig. 5.

5 Strip Examination of Lead Aircraft

Based on the inspection documents as mentioned in Sect. 4.3, a pre-survey and complete strip opening of two TTL extension allotted aircraft were carried out. The components were checked visually as well as thorough Crack Detection (CD) checks depending on the component location and probability of occurrence of defects. Inspection results of various checks carried out were documented. The photographs of the key areas where CD checks carried out are placed in Figs. 6, 7, 8, 9.

5.1 Observation on the Pre-survey Report of TTL Extension Aircraft (Two Numbers)

The pre-survey report revealed that minor dents, scratches, etc. and fuel leaks on wing tanks were noticed on some of the secondary structures of both lead aircrafts. These minor scratches and dents were addressed by burnishing and fuel leaks were addressed with application of PRC sealants. These will not affect the structural

हिएलि HAL XXXXXXXXXX TTLE INSPECTION DOCUMENT

B.3 CENTER WING

1	Visually inspect Center Wing structure externally and check for Damage, FOD, Dents, Corrosion & Loose rivets
2	Pre survey leak check on center wing by carrying out pressurization and leak test
3	Clean and Examine center section bottom skin near booster pump
4	Clean and Examine Hinge fittings of air brake and air brake jack
5	Clean and check center wing attachment holes for any ovality
6	Visually check/Inspect front and rear spars of the wing by boroscope

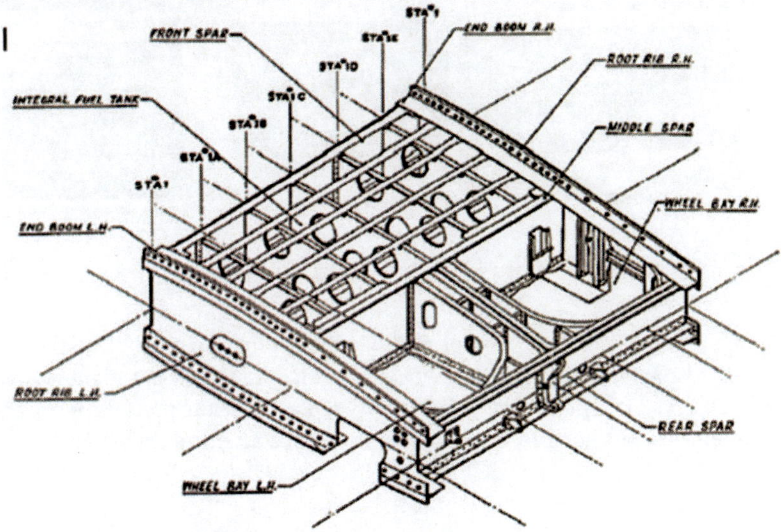

Fig. 5 Sample page of the inspection document

integrity of the aircraft. Hence, there is no implication on structural integrity if TTL extension is granted.

Further, pre-survey reports showed that some of the non-primary load-carrying members showed cracks as listed below:

- Rib supporting angle found cracked in RH outer wing.
- Stringer edge found cracked in main landing gear bay area.
- RH bottom skin near to wing root fitting leading edge found peeled-off.
- Crack found on bottom fairing front angle in centre wing assy.
- Flap assy trailing edge Aft rib crack at mould line was found.

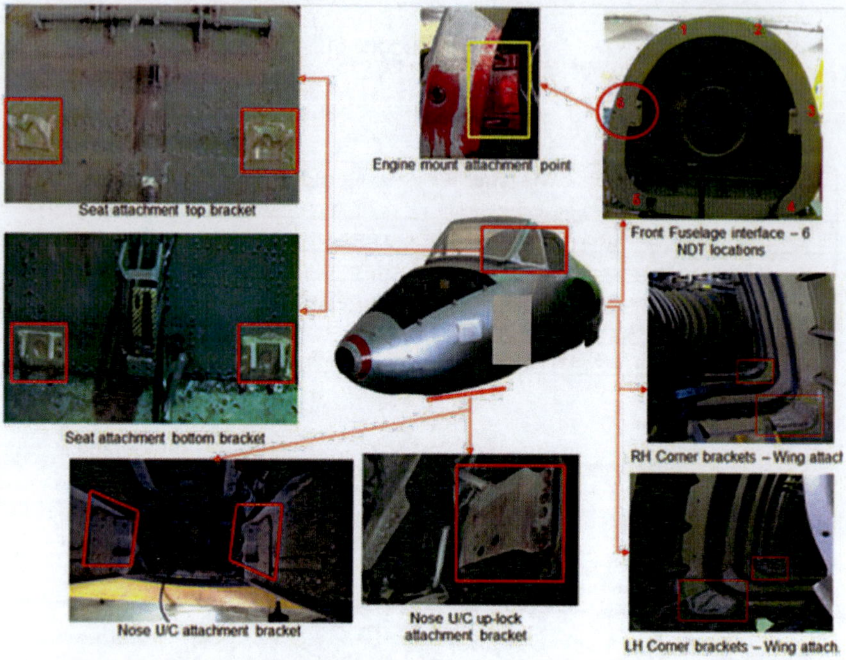

Fig. 6 CD check areas of front fuselage and nose landing gear brackets

These are minor cracks in non-primary load-carrying members. Detailed study also revealed some of the cracks are attributed to handling and not flight related. It was also observed that these parts can be easily replaced. These parts were suggested for replacement, thus not affecting the aircraft service after life extension.

5.2 *Observation on the CD Checks Report of Lead Aircraft (Two Numbers)*

Study of all the CD check reports is carried out. X-ray inspection has been carried out on fuel tank area of both aircraft LH and RH wings (centre and outer) and found satisfactory. However, a crack was noticed in RH outer wing root fitting top side during CD check in one of the lead aircrafts (refer Fig. 10). This being a primary structural member replacement of the fitting is recommended only if the crack is noticed after CD.

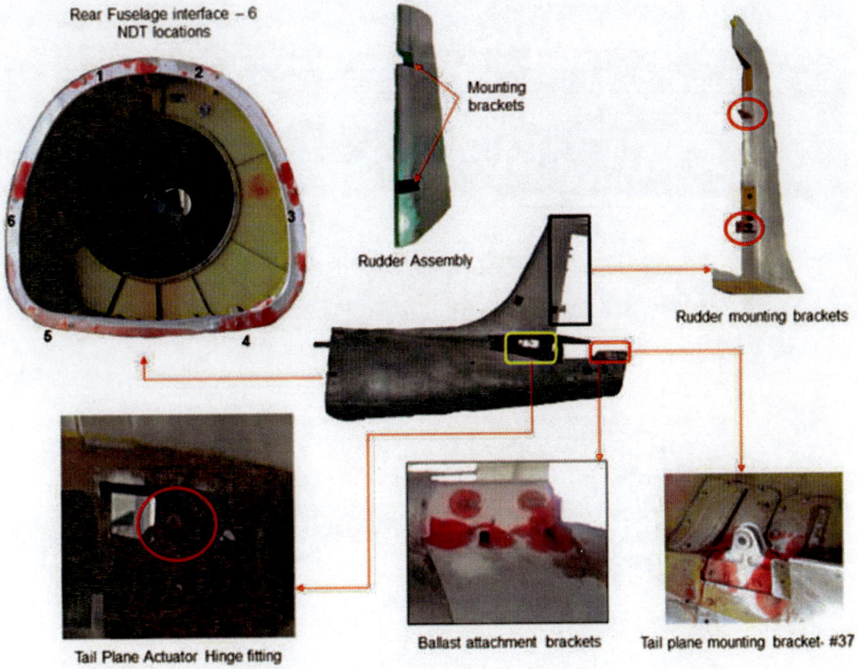

Fig. 7 CD check areas of rear fuselage

6 Scope for TTL Extension of Trainer Aircraft Fleet from Study

6.1 Observations from Inspections of TTL Extension Allotted Aircraft

From the total study carried out following observations are made:

- All fatigue modifications recommended have been implemented on time.
- The fleet data indicated no failures post MOD implementation.
- The environmental damages are very well manageable.
- The strip examination revealed cracks on secondary/non-load bearing structures which can be repaired/replaced.
- One primary structural crack observed can be detected and replacement can be effected.

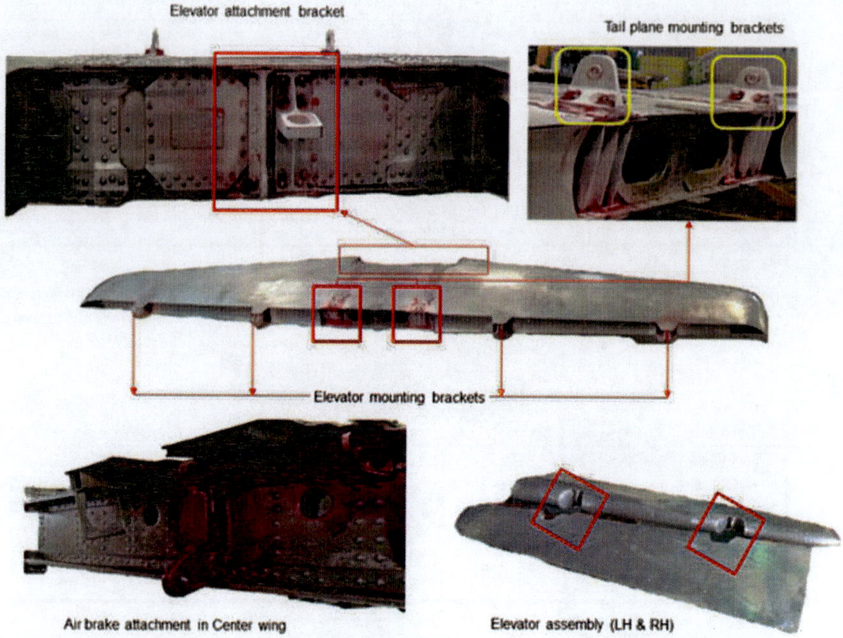

Fig. 8 CD check areas of horizontal tail, air brake and elevator attachment

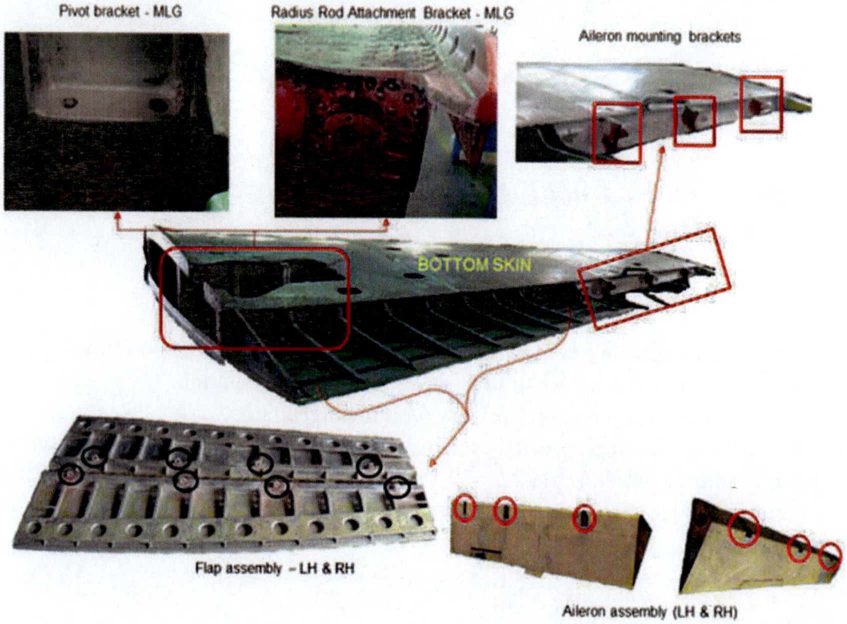

Fig. 9 CD check areas of wing, aileron and main landing gear attachment brackets

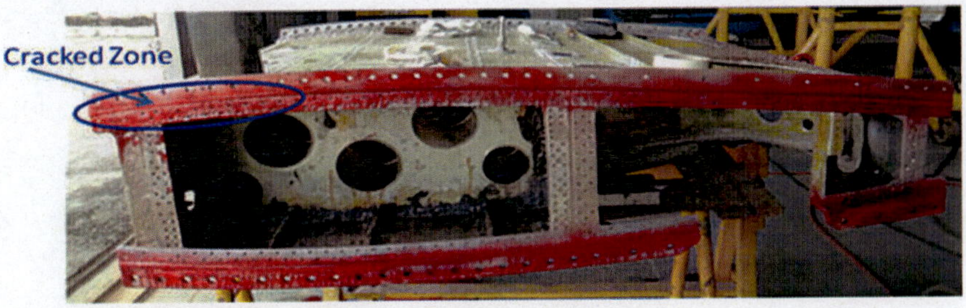

Fig. 10 CD check of RH outer wing root fitting

6.2 Feasibility of TTL Extension of Trainer Aircraft Airframe

The study of design analysis/inspection/field data and full scale fatigue test results show that margins are available for extension of life by 500 flight hours for airframe without compromising on structural integrity and airworthiness. This was arrived with the consideration of life margins of existing MODs, as well as fatigue test results till the complete failure of primary load-carrying member (wing root fitting) from initiation of crack till failure through monitoring (which happened to be 8708 flight hours). The fatigue spectrum used for testing is found to follow very closely with the actual usage. With this, an additional major servicing of the aircraft at 8000 flight hours is suggested. Further, it was proposed that whenever trainer fleet aircraft gets inducted for major servicing at 8000 h, inspection needs to be carried out specifically on all the wing root fittings. If any crack found, the root fitting needs to be replaced with a new one.

A comprehensive inspection document delineating the scope of inspection and actions that need to be carried out for granting life extension encompassing all the above studies is prepared.

7 Conclusion

i. Trainer fleet TTL can be extended by 500 flight hours beyond prevailing 8000 flight hours based on detailed data study.
ii. An additional overhaul at 8000 flight hours is recommended along with detailed areas to be inspected as per the life extension inspection document and replacement of parts that are found to be cracked after CD to ascertain the structural integrity and airworthiness.

Acknowledgements The authors are immensely grateful to the management of Hindustan Aeronautics Limited to permit publishing this paper. The authors would also like to thank design groups of ARDC, HAL, Overhaul division of HAL and certifying agencies (CEMILAC & DGAQA) as well as operators who provided the support to complete the total technical life extension task of the trainer fleet.

References

1. Regulatory article 5724—Life Extension Programme
2. Aircraft Structural Integrity program (ASIP) as per MIL-STD-1530C 01 Nov 2005
3. Stress and Fatigue analysis reports of trainer aircraft
4. Test report on Full Scale Fatigue Test of trainer aircraft
5. Aircraft Maintenance Manual (AMM) and Field Service data of a trainer aircraft

Life Estimation Strategy for a Nuclear Reactor Pressure Vessel

Upendra Singh, Vivek Shrivastav, and Rabindranath Sen

Abstract Reactor pressure vessel (RPV) of a pressurized water reactor (PWR) is an important pressure retaining equipment which decides the life of a nuclear power plant. RPV is constructed of special quality nuclear grade low-alloy steel having sufficient fracture toughness at the beginning of life and a good resistance to neutron embrittlement. However, the steel is not completely free from embrittlement phenomenon due to the presence of small amounts of impurities like copper, phosphorus, and sulfur. RPV houses the nuclear fuel core and thereby it is subjected to neutron flux which causes irradiation embrittlement of the material over a period of operation time. This results in loss of fracture toughness of the pressure boundary material leading to a possibility of brittle fracture toward the end of life. This governs the safe operating life of the RPV. In this paper, the strategy for life estimation of a RPV using LEFM approach is described and a sample calculation for a typical RPV is presented. A semi-elliptical surface crack is postulated as per ASME B & PV code, Section-III, Appendix-G, at critical locations like core belt region and nozzle-vessel junction, and the flawed structure is analyzed under the severe most loading, i.e., a rapid cooldown transient, using the finite element method. 3D FEM model of the respective portions of the RPV is made using brick elements for analyzing the behavior of the postulated cracks under the load. The model includes the 3D crack geometry as an unconstrained boundary within the structure, with quarter-point elements near the crack front for simulation of square root singularity. Fracture calculations are done to evaluate the stress intensity factors for the postulated cracks. The extent of irradiation embrittlement of the material (measured as the shift in nil-ductility-transition temperature) with respect to neutron fluence (and hence total

U. Singh (✉) · R. Sen
Advanced Light Water Reactor Division, Bhabha Atomic Research Centre, Visakhapatnam 530003, India
e-mail: upendras@barc.gov.in

R. Sen
e-mail: rabisen@barc.gov.in

V. Shrivastav
Advanced Light Water Reactor Division, Bhabha Atomic Research Centre, Trombay, Mumbai 400085, India
e-mail: viveksr@barc.gov.in

S. Seetharamu et al. (eds.), *Fatigue, Durability, and Fracture Mechanics*, Lecture Notes in Mechanical Engineering, https://doi.org/10.1007/978-981-15-4779-9_4

operation time) is assessed by using the empirical correlations as per the USNRC regulatory guide. The safe operating life of the RPV is estimated from the criticality condition of the postulated cracks.

Keywords Life prediction · Reactor pressure vessel · Fracture mechanics · Irradiation embrittlement

1 Introduction

Reactor pressure vessel of a light water nuclear power plant operates under high-temperature and high-pressure conditions, and houses the reactor core in which nuclear reactions take place and the primary heat is produced. This heat is transported from the core to the steam generator by the reactor coolant which is ordinary light water with controlled chemistry. The RPV is not only the main component of the primary pressure boundary, but also acts as a barrier to the harmful radiations from the reactor core. The structural integrity of this component is, therefore, of utmost importance to the safety of the whole plant. The fact that it is the first major barrier to the nuclear radiations makes the RPV an irreplaceable component of the power plant. The RPV is, therefore, a key component whose life governs the life of the whole plant. It is, therefore, of great importance to be able to predict the safe operating life of the RPV.

Life of any structure or component is limited because of the fact that one or more of the various forms of aging mechanisms always take place which gradually degrade the structure/component to such an extent that either its intended functional performance is severely affected or its structural integrity becomes doubtful. Some examples of aging mechanisms are creep, fatigue, thermal embrittlement, corrosion, wear, irradiation embrittlement, etc. The type(s) and severity of the aging mechanism(s) predominant for a given structure/component, and hence the safe operating life of the structure/component depend on the interaction between three basic elements: ***stress***, ***material***, and ***environment***, as described in Fig. 1. These aspects in the case of RPV are discussed in the following section.

2 Reactor Pressure Vessel

2.1 Design

A longitudinal sectional view of a typical RPV of a pressurized light water nuclear reactor plant is shown in Fig. 2.

The structure of the RPV consists of a cylindrical shell region, nozzle region, where coolant inlet and outlet nozzles are located, bottom cover welded to the shell, and a top cover bolted to the flanged portion of the nozzle region. The shell region may

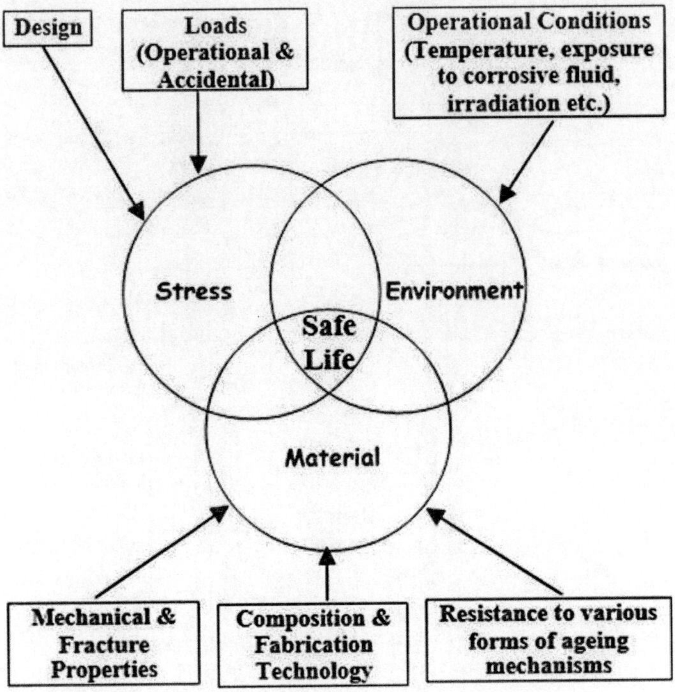

Fig. 1 Factors governing service life

or may not require having circumferential welds depending on the size of the RPV and available fabrication/transport facility. Similarly, nozzles may either be integrally machined with the vessel or may be welded to it. The aim of the designer shall be to minimize the number of welds and place them away from critical locations to increase the life of the structure. The structure of the RPV houses many vital components: reactor core containing the fuel rods, control rod drive and instrumentation and channels for guiding the coolant flow. The portion of the shell directly facing the mid-length of the reactor core is termed as the core belt region.

2.2 Loads

RPV is subjected to a variety of loads throughout its service life. The loads on a RPV can be categorized as (i) normal operating loads, (ii) operational transient loads, and (iii) accidental loads.

Normal operating loads are high internal pressure (typically 16 MPa) and high temperature (typically 300 °C).

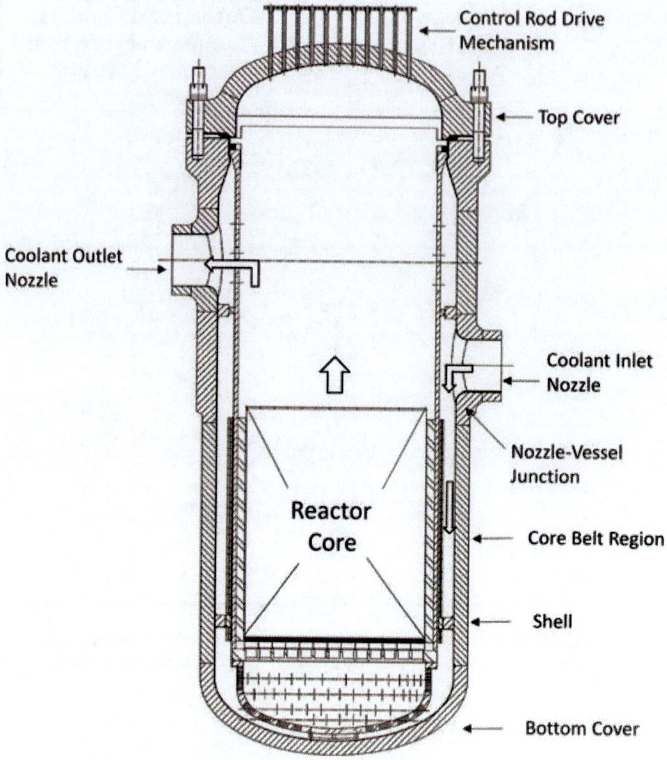

Fig. 2 Sectional view of a typical RPV

Operational transient loads arise due to normal start-up and shutdown, random fluctuations in the operating parameters, and intended power fluctuations of the reactor as per the demand.

Accidental loads on RPV may arise due to operational malfunctions of other components, leakage of coolant, emergency reactor shutdown, etc.

The most severe anticipated accidental load on a RPV is a rapid cooldown transient from normal operating temperature to ambient temperature during an emergency shutdown. Rapid cooldown is brought into effect by injecting relatively cold water into the RPV. The reason behind the severity of this transient is described in Fig. 3 which shows temperature and stress distributions across the RPV wall during a rapid cooldown transient.

Due to cold water injection during a rapid cooldown, a steep positive temperature gradient exists near the inner surface of RPV (as shown in Fig. 3) whose magnitude and profile vary with time during the transient. This causes very high tensile thermal stress in the tangential direction near the inner surface due to the self-constraint of the material against thermal contraction. The nature of the thermal stress distribution

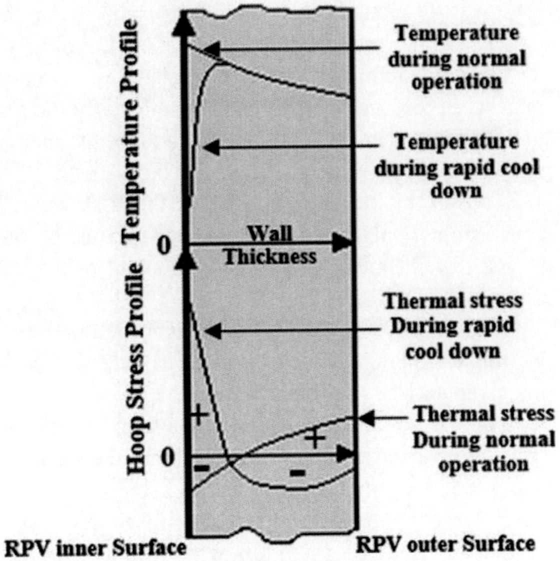

Fig. 3 Consequence of a rapid cooldown transient

across the RPV wall during a rapid cooldown transient is shown in Fig. 3. The tensile stress due to internal pressure will be acting over and above this thermal stress.

It may easily be anticipated that if a crack exists near the inner surface of the RPV, it may prove to be detrimental during such a transient load.

2.3 *Material*

RPV is fabricated from forgings of special grade low-alloy ferritic steel which has good mechanical strength (typically 450 MPa YS and 600 MPa UTS at room temperature) and simultaneously high fracture toughness.

There are two standard grades of pressure vessel steels: Mn-Mo-Ni based steel and Cr-Mo-V based steel. The steels are produced using advanced melting and casting technology to minimize the impurities, especially Cu, Ni, and P, whose presence impairs the metallurgical stability of the material under irradiation.

2.4 *Environment*

The RPV operates in adverse environment of neutron irradiation which is well known to cause embrittlement of the material. Besides, the inner surface is also exposed to high-temperature aqueous environment which can trigger various types of corrosion-assisted degradation mechanisms.

Corrosion-related mechanisms are mitigated by overlaying the inner surface of the RPV with austenitic stainless steel clad, typically 6–10 mm thick.

Although the major objective of cladding is to prevent the corrosion of the base metal, the presence of cladding is beneficial for the structural integrity of the RPV as well. Cladding acts as a buffer and reduces the thermal shock to the adjoining base metal in the event of a rapid cooldown transient.

Exposure of the inner surface of the RPV to neutron irradiation is reduced to some extent by placing a stainless steel thermal barrier between the reactor core and the RPV wall; however, the embrittlement of the RPV material is not completely eliminated.

Neutron irradiation exposure is quantified in terms of number of neutrons hitting per unit area per unit time ($n/cm^2/s$), called *flux*. The total exposure to irradiation up to a given service life (n/cm^2) is called *fluence*.

In the reactor core, the axial distribution of neutron flux is a cosine function, with a maximum at the mid-length and minimum at the ends. The core belt region, directly facing the mid-length of the core, is subjected to the maximum flux and hence the embrittlement is expected to be maximum in the core belt region of the RPV. The nozzles are usually located far away from the core, and hence the flux in the nozzle region is not usually sufficient to cause any significant embrittlement.

Embrittlement of steel due to irradiation strongly depends on the content of impurities like Cu, Ni, and P in the material [1]. Copper together with other elements is known to lead mechanism of nano-precipitation inducing matrix hardening and embrittlement. The process of Cu precipitation is enhanced in the presence of nickel. Elements like P segregate at grain boundaries embrittling the matrix. The rate and degree of embrittlement depend on the concentration of these impurities available in the steel.

As a result of irradiation exposure, ferritic steels gain strength and loose ductility. In terms of the Charpy V-notch energies, the ductile-to-brittle transition temperature (DBTT) shifts to the right (increases) and the upper shelf energy level decreases as a result of the irradiation exposure, as shown in Fig. 4.

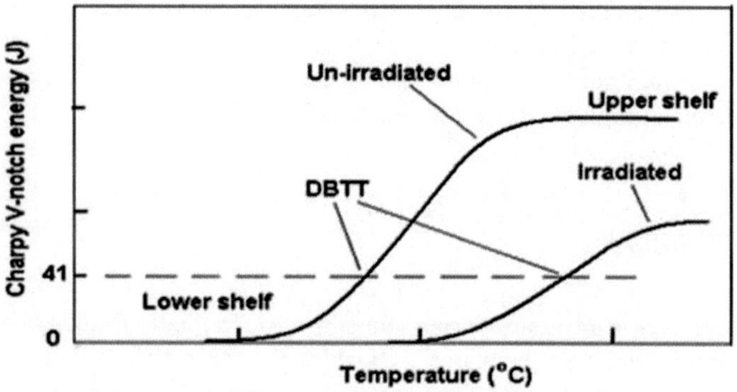

Fig. 4 Effect of irradiation on impact toughness

3 Life Prediction Strategy for RPV

3.1 Philosophy

Superposition of all factors as discussed above leads to the understanding that the life-governing aging mechanism for RPV is irradiation embrittlement. The philosophy for deciding the safe operating life of RPV is that under any circumstance, the RPV should not fail in a catastrophic manner through brittle fracture, thus ensuring safety of the public from radiation exposure. There must be sufficient margin available against failure due to brittle fracture even under the most severe and least probable anticipated load throughout the service life of the RPV.

3.2 Role of Fracture Mechanics

If a sharp crack is present in a structure made up of a brittle (or elastic-plastic with small scale yielding) material, the concepts of linear elastic fracture mechanics can be invoked to assess whether the crack will remain stable under a given load or the crack is critical and may grow in an unstable manner leading to catastrophic failure.

The parameter stress intensity factor (SIF, K_I), which is a function of geometry, load, and crack length, can be computed and compared with the critical SIF (K_{Ic}), which is a material property, to assess the state of the crack. The condition for criticality of a crack is

$$K_I \geq K_{Ic} \tag{1}$$

3.3 Assessment of Irradiation Embrittlement

The most common approach in assessment of embrittlement of RPV steel is based on shift in DBTT. The toughness-temperature transition curve (K_{Ic} vs. temperature) is experimentally determined for the unirradiated material and the curve for the irradiated material is obtained by shifting the original curve to the right by the amount of shift in DBTT. The amount of shift in DBTT for a given fluence level can be estimated by empirical correlations provided by USNRC regulatory guide (USNRG).

Reference-nil-ductility-transition temperature, RT_{NDT} as defined by ASME, which is evaluated using both Charpy impact and drop weight tests, is used as the measure of DBTT in the USNRG.

The shift in RT_{NDT} is given in USNRG-1.99 Rev. 02 for Mn–Mo–Ni type steels as follows [2]:

$$RT_{NDT}(\text{irradiated}) = RT_{NDT}(\text{initial}) + RT_{NDT}(\text{shift}) \quad (2)$$

where

RT_{NDT} (initial) = RT_{NDT} at the start of irradiation.

RT_{NDT} (shift) = (CF) × $(f)^{(0.28 - 0.10 \times \log f)}$.

where

CF = chemistry factor, given in tabular form, as function of Cu and Ni content.

f = neutron fluence in 10^{19} n/cm^2, with energy, $E > 1$ MeV.

3.4 *Procedure*

ASME Boiler and Pressure Vessel Code, section-III, Appendix-G requires that sufficient margin against non-ductile fracture of RPV must be demonstrated by postulating a crack with depth equal to 25% of wall thickness and ensuring that the postulated crack does not become critical under any of the anticipated loads. The procedure is described below:

Postulate A Surface Crack with Size

Depth, a = ¼ T (T = wall thickness),

Length = 6.a,

which is oriented in the plane normal to maximum stress.

Calculate applied stress intensity factor (K_I) either using available closed-form solutions or numerical techniques like FEM.

Material K_{Ic}

Calculate material K_{Ic} based on RT_{NDT} of the irradiated material corresponding to the fluence seen by the RPV, using Eq. [3]

$$K_{Ic} = 26.78 + 1.223 \times \exp[0.0145(T - RT_{NDT} + 160)] \quad (3)$$

(subject to the maximum value of 200 ksi.(in.)$^{0.5}$),

where T is the material temperature in °F and K_{Ic} is in ksi.(in.)$^{0.5}$.

For Prevention of Brittle Fracture

Calculated K_I should be less than material fracture toughness K_{Ic} at the crack tip. For crack to be stable, $K_I \leq K_{Ic}$. The safe operating life can therefore be conservatively predicted by postulating cracks at different critical locations of the RPV and calculating the applied SIF under the most severe anticipated load and finding out the shift in RT_{NDT} for which the material K_{Ic} will be equal to the applied SIF. The fluence required to cause this amount of shift in RT_{NDT} will give the predicted life of the RPV.

4 Life Prediction of a Model RPV

In this section, a model RPV is analyzed under a rapid cooldown transient load, and its safe operating life is conservatively predicted based on the procedure described in previous sections.

4.1 Geometry

The geometry of the model RPV is taken as

Inner radius of shell = 700 mm,
Outer radius of shell = 1000 mm,
Inner radius of nozzle = 107 mm,
Outer radius of nozzle = 165 mm,
Clad thickness = 10 mm.
No weld joint in core belt region.

It should be noted here that the dimensions of the RPV taken here for analysis are only for the demonstration purpose and the actual dimensions of the RPV of a typical pressurized light water reactor will be quite different.

The inner radius of shell (taken here as 700 mm) is deliberately taken much smaller than the usual values (of the order of 2000 mm for 1000 MW reactor) to keep the overall computational domain of FEM analysis small. However, the shell thickness (taken here as 300 mm) is taken larger than the usual values (of the order of 200 mm for 1000 MW reactor) to enhance the severity of thermal load.

4.2 Rapid Cooldown Transient Load

The load considered here is a typical severe cooldown transient where both temperature and pressure in the RPV are reducing rapidly with time. This is defined in terms of the applied temperature and pressure on the RPV inner surface during the transient as shown in Fig. 5.

4.3 Material Properties

The material properties [4] used in the present analyses are shown in Tables 1 and 2.

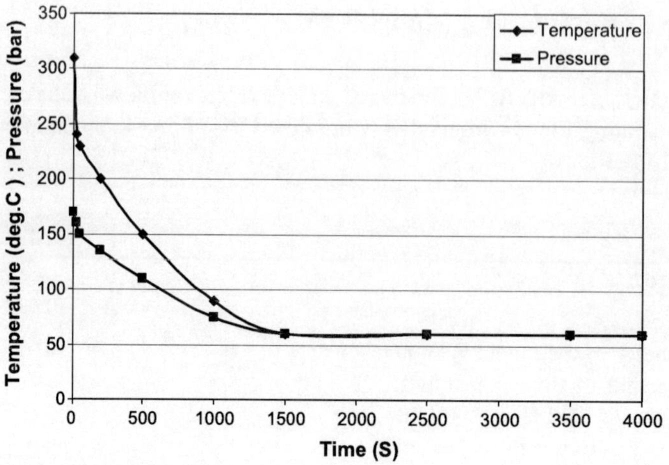

Fig. 5 Applied cooldown transient

Table 1 Mn–Mo–Ni based ferritic steel (RPV base material)

Temperature (°C)	Thermal conductivity (W/m °C)	Specific heat (J/Kg °C)	Coeff. of th. expansion	Young's modulus (GPa)	Poisson's ratio
20	37.8	449	11.6×10^{-6}	212	0.3
350	37.5	566	13.8×10^{-6}	190	0.3

Table 2 Austenitic stainless steel (clad material)

Temperature (°C)	Thermal conductivity (W/m °C)	Specific heat (J/Kg °C)	Coeff. of th. expansion	Young's modulus (GPa)	Poisson's ratio
20	15.28	497.42	16.4×10^{-6}	205	0.3
350	19.42	622.82	17.6×10^{-6}	175	0.3

4.4 Procedure

The following critical locations at inner surface of the RPV (Fig. 6) are identified and investigated with crack size taken as per the ASME B&PV code, Section-III, Appendix-G:

(i) Core belt region (due to high neutron flux and high membrane stress): Semi-elliptical crack. Depth in radial direction, 25% of shell thickness, aspect ratio, 1:6.
(ii) Nozzle-vessel junction (due to high stress concentration): Quarter circular crack with depth 25% of shell thickness.

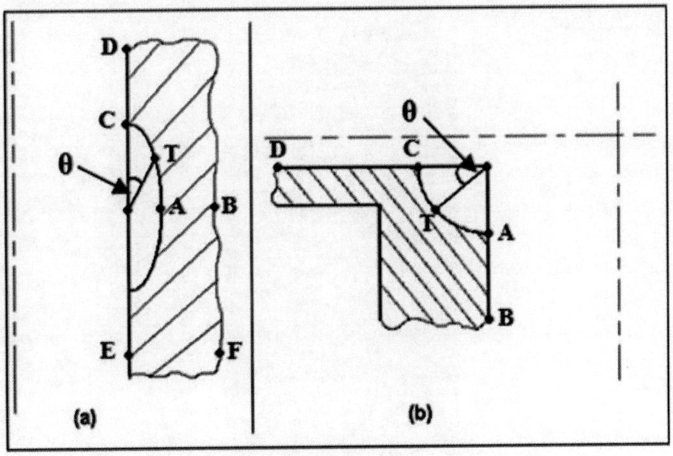

Fig. 6 Postulated cracks at **(a)** shell region and **(b)** nozzle-vessel junction

Sequential transient thermal and stress analysis of suitable 3D models of the RPV representing different critical locations along with postulated crack is performed using the finite element method, with quarter-point elements near the crack front. The displacement field solution in the vicinity of the crack is then used to calculate the stress intensity factor (SIF) values as function of time during the transient.

The applied SIF values are then compared with the material K_{Ic} in the time domain (crack severity assessment). Material K_{Ic} at a given crack tip at a given time point during the transient is calculated using the instantaneous crack tip temperature (as obtained from the thermal analysis) and invoking the ASME equation for K_{Ic} as discussed in the previous section.

The value of RT_{NDT} for which the material K_{Ic} curve becomes tangential to the applied SIF curve is taken as the limiting value of RT_{NDT} for the particular case. If the RT_{NDT} at the end of the life is more than this limiting value, then there is a possibility of non-ductile fracture of the RPV. The lowest value of limiting RT_{NDT} among all the cases investigated becomes the value of life limiting RT_{NDT} for the RPV.

4.5 *Finite Element Modeling*

Core Belt Region

The finite element (FE) model used in the analysis of the core belt region is shown in Fig. 7. Taking advantage of symmetry of the problem, only half of the geometry is modeled, the plane of symmetry being the transverse plane containing the semi-minor axis of the crack. Also, due to symmetry, only one crack face is modeled. A 45° sector of the shell is modeled. The length of the shell that is considered in the model is 1500 mm. These dimensions are conservatively chosen such that the

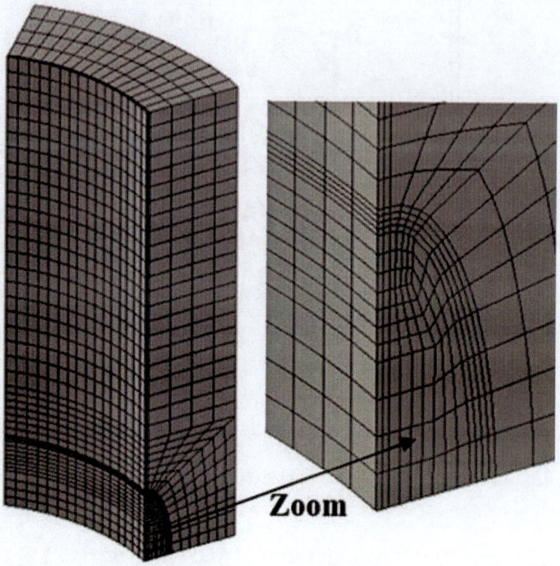

Fig. 7 FE model of core belt region

discontinuity stresses near the crack should not have an effect on the stresses on the artificial boundaries, so that proper valid boundary conditions can be applied on them.

The model is meshed with 20-noded hexahedral elements. The total number of elements is 6,240 and the total number of nodes is 27,980. Relatively, very fine mesh is taken in the neighborhood of the crack front, which is shown separately in the zoomed view in Fig. 7. The number of elements taken along the crack front is 12.

For stress analysis of the model (after thermal analysis), the crack front elements are made singular by shifting the mid-side nodes adjacent to the crack tip nodes, to the respective quarter-point locations, to simulate the square root singularity that exists near a crack. Crack is simulated in the stress analyses by leaving the crack face nodes unconstrained.

Nozzle-Vessel Junction

The FE model used in the analysis of nozzle-vessel junction is shown in Fig. 8. Quarter sectors of vessel (shell) as well as nozzle are included in the FE model. Suitable lengths of vessel (2000 mm) and nozzle (500 mm) are included. Again due to symmetry, only one crack face is modeled. The model is meshed with a combination of 20-noded hexahedral elements and 10-noded tetrahedral elements with 15-noded prism-shaped transition elements in between. The total number of elements in the model is 12,552 and the total number of nodes is 24,142. Crack is simulated in the same manner as in the case of core belt region and also the crack front elements are made singular for the stress analyses as discussed.

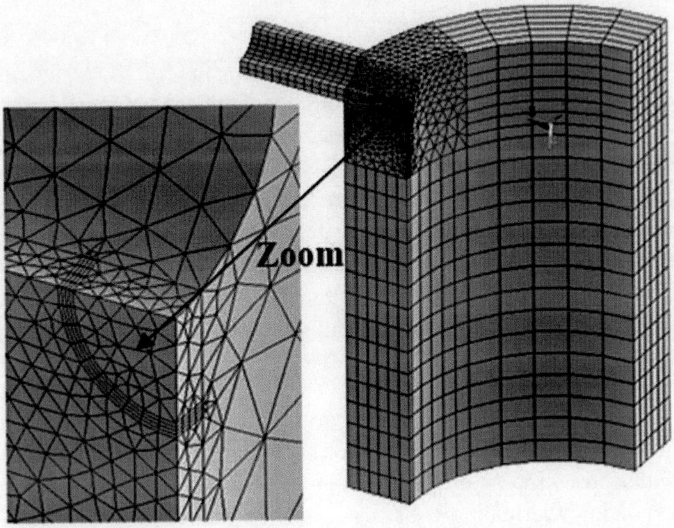

Fig. 8 FE model of nozzle-vessel junction

Finite Element Analysis

For both the models, first, a steady-state thermal analysis is done to attain the temperature distribution existing at time $t = 0$ (the time at which the transient begins). Then, a transient thermal analysis is done to determine the temperature distribution existing in the structure at different time points after the beginning of the transient.

Stress analyses are then performed at a number of time points of the transient to determine the displacement and stress fields existing at these times.

Both, stress analysis and thermal analysis, are done using the same mesh, except that, during stress analysis, the elements adjacent to the crack front are made singular by shifting the mid-side nodes to quarter-point locations.

Boundary Conditions

The loads and boundary conditions applied in the analyses of the FE models are listed below.

Thermal Analysis

(1) Convective boundary condition on the outer surfaces of the models with heat transfer coefficient (h) equal to 7.0×10^{-6} W/mm^2 K (this is estimated from Handbook correlations for heat transfer coefficient for a vertical cylinder).
(2) Bulk air temperature outside is equal to 150 °C.
(3) Temperatures are applied on the inner surfaces of the models as per the transient given in Fig. 5.
(4) All other surfaces thermally insulated.

Stress Analyses

(1) The longitudinal planes of the models (excluding the crack face) are restricted to move in the respective out-of-plane directions (this is based on the assumption that any longitudinal plane of a cylinder can displace only in the radial direction under axi-symmetric loading).
(2) The transverse plane of the models nearer to the crack location is constrained in the vertical direction.
(3) Instantaneous value of internal pressure as per the transient is applied on the inner surfaces of the models, and also on the crack face.
(4) Uniform longitudinal traction (arising due to instantaneous internal pressure) is applied to the transverse plane of the models which is away from the crack location.
(5) Longitudinal traction corresponding to the instantaneous internal pressure, at the nozzle transverse plane, in case of nozzle-vessel junction.
(6) Instantaneous temperature distribution in the models, as obtained from the corresponding thermal analysis.

Calculation of SIF

The displacement field solutions near the crack front as obtained from the finite element analyses at different time points of the transient are used for the calculation of SIF as function of time.

With reference to Fig. 9, the SIF, K_I, at the crack tip "T" in mode I case, is numerically calculated as per the following relation [5] when quarter-point elements are used:

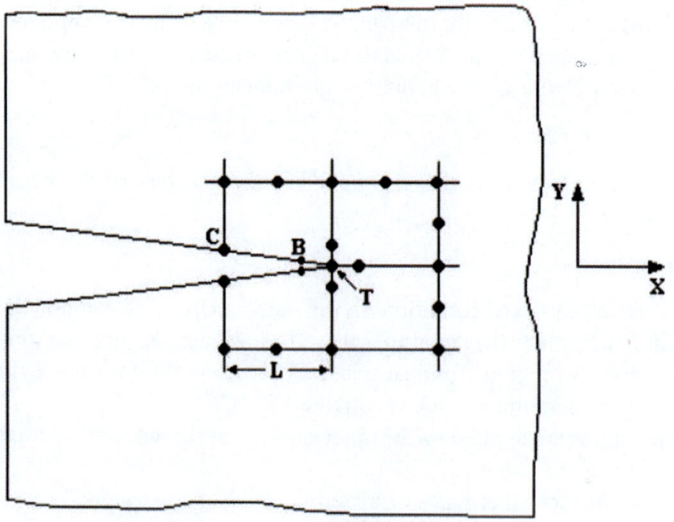

Fig. 9 Numerical calculation of SIF

$$K_I = \frac{E}{(1+\upsilon).(1+\kappa)}.\sqrt{2\pi}\left(\frac{4v_B - v_c}{\sqrt{L}}\right) \tag{4}$$

where v_B and v_C are the displacements of the nodes B and C, respectively, in the Y-direction (crack opening direction), and $\kappa = 3 - 4\upsilon$ in case of plane strain and $(3-\upsilon)/(1+\upsilon)$ in case of plane stress.

5 Results and Discussion

Figures 10 and 11 show the variations of temperature over the crack front at different time points, t, after the beginning of the transient ($t = 0$ S). The angular position of any crack tip "T" on the crack front is defined in Fig. 6.

At time $t = 0$ S, when the transient just begins, steady-state conditions corresponding to the normal operational temperature of the RPV exist, and hence the temperature variation over the crack front is small as seen in the figures. But, as the transient begins, the temperature drops more rapidly at the crack tips nearer to the inner surface of the RPV, and hence the temperature variation over the crack front increases gradually as shown in the Figs. 10 and 11.

After reaching a maximum at about 1500 s, the temperature variation starts reducing as the structure gradually approaches the steady-state condition again.

From the stress analyses of the models at various time points, it is found that the maximum crack opening displacements exist at time $t = 1500$ s, for both models. Variations of total (thermal + pressure) hoop stress, which is the crack opening stress, along paths A–B, C–D, and E–F (as defined in Fig. 6) at time $t = 1500$ s are shown in Figs. 12 and 13.

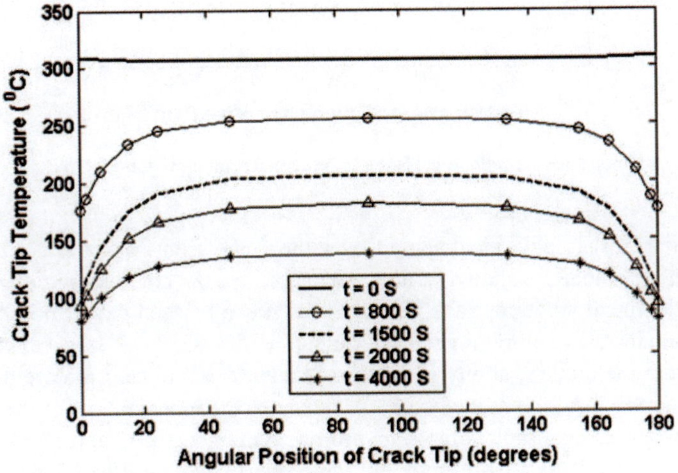

Fig. 10 Temperature distribution over crack front for core belt region crack

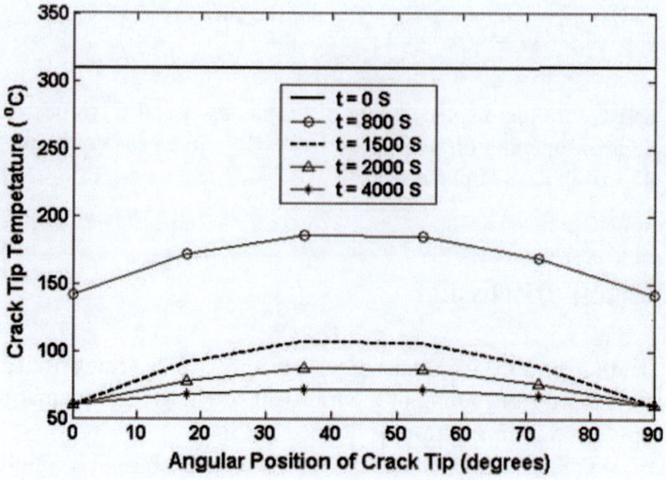

Fig. 11 Temperature distribution over crack front for nozzle-vessel junction crack

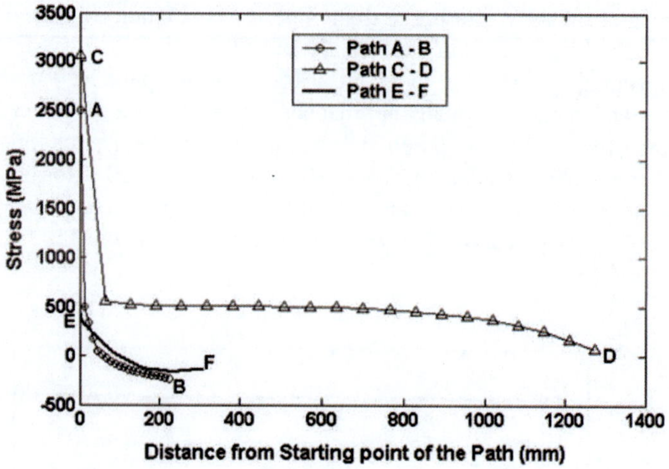

Fig. 12 Total hoop stress at time, $t = 1500$ s for core belt region model

Stress intensity factors calculated over the crack fronts at various time points during the transient are shown in Figs. 14 and 15 for the core belt region crack and the nozzle-vessel junction crack, respectively. It can be seen that in both the cases, the SIF attains the maximum values at time, $t = 1500$ s.

For both the models, at any given time (except $t = 0$ in case of core belt region crack), the SIF is higher at the crack tips nearer to the inner surface of the RPV.

Crack severity assessment plots are presented in Figs. 16, 17 and 18. For the core belt region, the results are presented for the deepest crack tip (Fig. 16) as well as the crack tip near the inner surface of the RPV (Fig. 17).

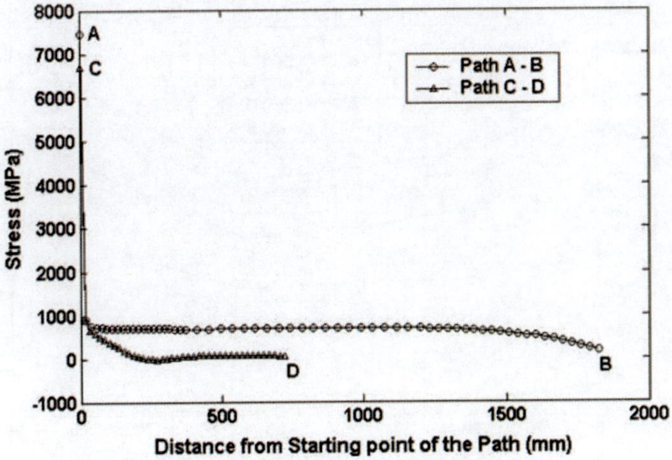

Fig. 13 Total hoop stress at time, $t = 1500$ s for nozzle-vessel junction model

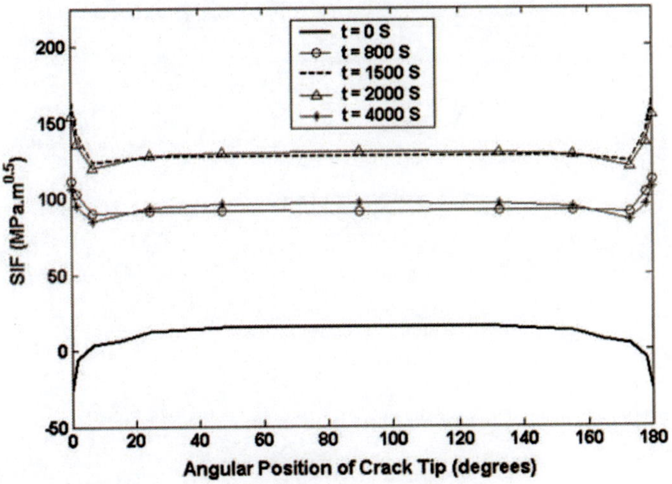

Fig. 14 SIF distribution over crack front for core belt region crack

Material K_{Ic} curves are plotted for initial RT_{NDT} (conservatively taken as 0 °C) and that value of RT_{NDT} for which the K_{Ic} curve tends to touch the applied SIF curve.

Figure 18 presents the crack severity assessment plot for the nozzle-vessel junction. Applied SIF for two extreme crack tips are plotted: crack tip at the vessel inner surface and crack tip at the nozzle inner surface. The material K_{Ic} curve in this case is plotted only for the initial RT_{NDT} value, because at this location stress intensity governs over material degradation due to fluence for integrity evaluation. The fluence at the nozzle location will generally be much less than 1×10^{18} n/cm^2 ($E > 1$ Mev),

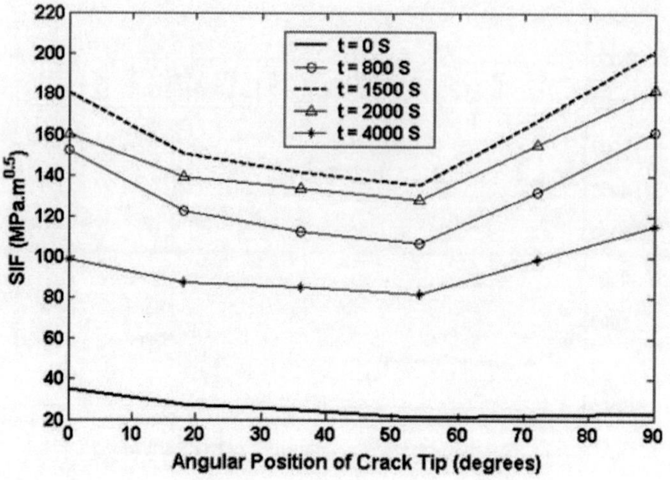

Fig. 15 SIF distribution over crack front for nozzle-vessel junction crack

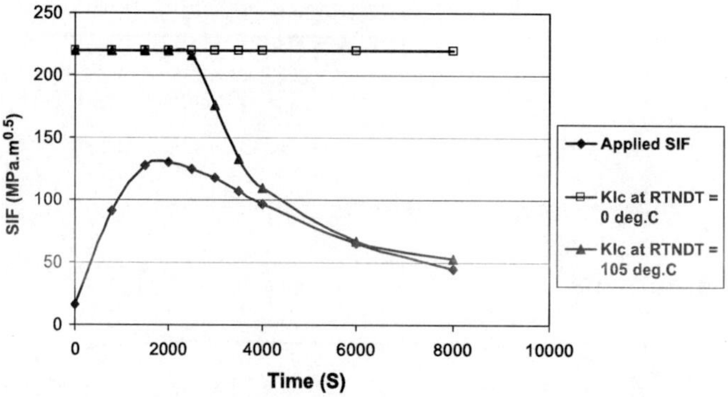

Fig. 16 Severity assessment plot for the deepest crack tip in core belt region

and hence degradation of toughness property is expected to be negligible at this location. So, the crack severity even at the end of the life will be same as depicted in Fig. 18.

From the analysis, the nozzle-vessel junction of the RPV has been proved to be less critical for structural integrity of the RPV compared to the core belt region.

From the assessment plots, the life limiting value of RT_{NDT} for the RPV is found to be 45 °C. For an unirradiated base metal with initial RT_{NDT} of 0 °C, this means that a shift in RT_{NDT} of 45 °C is tolerable till the end of the life.

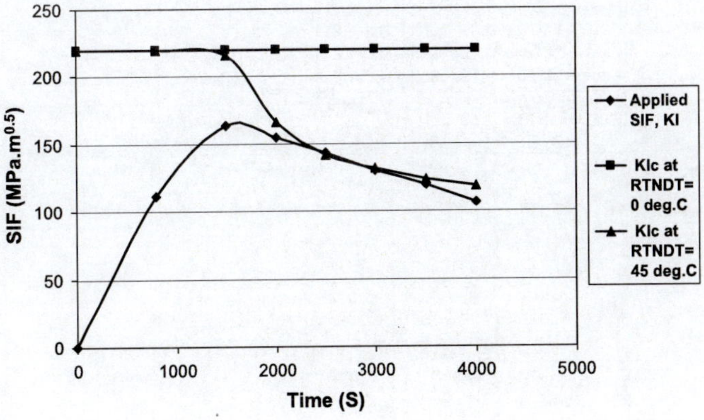

Fig. 17 Severity assessment plot for the surface crack tip in core belt region

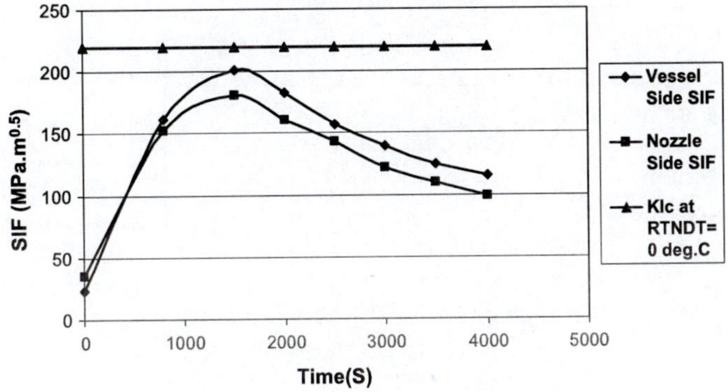

Fig. 18 Severity assessment plot for nozzle-vessel junction crack

5.1 Life Prediction

Safe operating life of the RPV can now be conservatively predicted by using the tolerable shift in RT_{NDT} up to the end of life and invoking the correlation between neutron fluence, material composition, and RT_{NDT} shift as described in previous sections. Figure 19 shows the degradation of RPV steel (increase in RT_{NDT}) as a function of neutron fluence to which it is exposed, over the period of operation, for three different compositions. The figure also shows the corresponding end-of-life fluence values for the model RPV, for which the maximum tolerable increase in RT_{NDT} is 45 °C, as found above from the fracture analyses.

Figure 20 presents the predicted life of the model RPV (in effective full power years of operation) as a function of neutron flux at the core belt region corresponding to the full reactor power, and composition of the RPV steel.

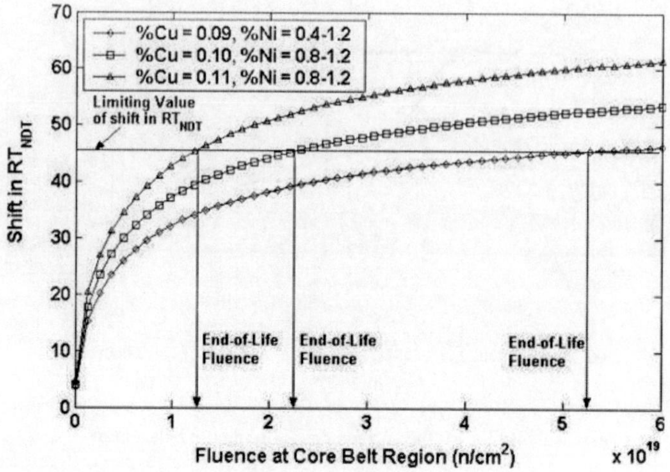

Fig. 19 Degradation of RPV steel with neutron fluence

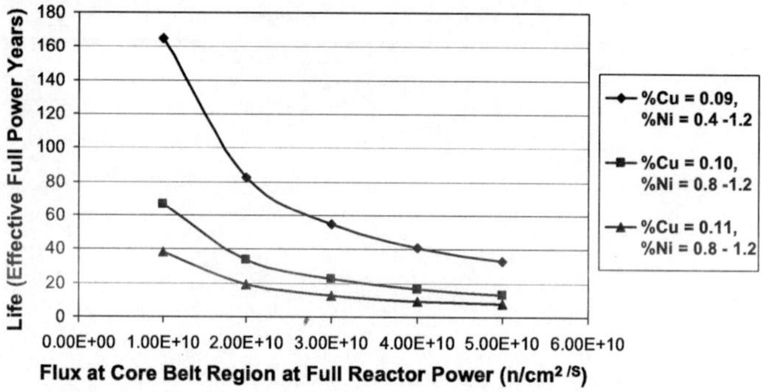

Fig. 20 Predicted life of the model RPV

It can be observed from the figure that the life of RPV strongly depends on the copper content in the RPV steel. Also, it may be noted that life of RPV is a stronger function of material composition and neutron flux at relatively lower levels of neutron flux.

6 Conclusions

Reactor pressure vessel of a nuclear power plant is a key component for deciding the life of the whole plant. Life prediction strategy for a nuclear RPV is discussed and calculations for a typical model RPV are presented. Irradiation embrittlement and cracking is the life-governing aging mechanism for RPV. Due to this reason, fracture mechanics plays a very important role in structural integrity assessment and life prediction of RPV.

For the typical RPV considered, core belt region is shown to be the most critical and life-governing location in terms of loading and material degradation for assuring the integrity of the vessel. Predicted life of the model RPV is presented, both in terms of the end-of-life fluence, and the total years of full power operation, for three different material compositions.

Life of RPV strongly depends on the copper content of the RPV steel. This puts large emphasis on modern steel making technologies to produce steel as clean as possible. The dependence of RPV life on composition and neutron flux is stronger at relatively lower levels of neutron flux.

References

1. English, Colin, Hyde, Jon and Ortner, Sue. Microstructural development in RPV steels. IAEA Technology Nuclear Science, Didcot Oxon OX11 ORA, UK
2. U. S. Nuclear Regulatory Commission, Regulatory Guide 1.99, Revision 2, May 1988
3. ASME Boiler and Pressure Vessel Code, Section III, 2001 Edition, Rules for Construction of Nuclear Facility Components, Nonmandatory Appendix G, "Protection Against Non-Ductile Failure"
4. Sattari-Far I (1996) Constraint effects on behaviour of surface cracks in cladded reactor pressure vessels subjected to PTS transients. Int J Press Vessels Pip 67:185–197
5. Kumar P (1999) Elements of fracture mechanics. Chapter 10, Wheeler Publishing Co., New Delhi, pp. 222–225

Fatigue Crack Growth (FCG) Studies on Landing Gear (LG) Actuating Cylinder of Fighter Aircraft for Life Extension

Gautam Sawaisarje and R. P. Khapli

Abstract Study on life extension was undertaken for actuating cylinder of main and nose landing gear (MLG and NLG) of fighter aircraft. Activities involved for life extension are load estimation on landing gear (using MIL-A-8862) and actuating cylinder, stress analysis of actuating cylinder, fatigue load spectrum generation (using MIL-A-8866C), theoretical fatigue life estimation, and fatigue testing of actuating cylinders. Fatigue testing was carried out on one specimen each of actuating cylinder of MLG and NLG, for load spectrum of 4000 landings, considering scatter factor of 4. Crack was observed on MLG actuating cylinder eye end, one of the major load-carrying member. Therefore, it was decided to carry out fatigue crack growth analysis of eye end to decide the inspection interval in life extended period. Fatigue crack growth analysis is carried out using Paris equation. Suitable stress intensity factor (SIF) for crack geometry and material constants for Paris equation are selected from available literature. It is concluded that, assuming an initial crack of 0.5 mm in eye end at threaded region, it will become critical after minimum 1500 landings. This was practically seen during fatigue test. Crack length reached during fatigue test was around 7 mm; however, unstable crack growth was not observed. This study recommends the thorough inspection of eye end to be carried out at every 400 landings after initial assigned life in number of landings to ensure absence of cracks.

Keywords Actuating cylinder · Landing gear · Fighter aircraft · Crack growth · Life extension · SIF

1 Introduction

Fatigue testing was carried out on MLG and NLG actuating cylinder of fighter aircraft for life extension by 1000 landings. Crack was observed (Fig. 1) during inspection, emanating from grease nipple hole at the end of fatigue testing of 8 blocks of 500 landings each. Therefore, it was recommended to carry out fatigue crack growth

G. Sawaisarje (✉) · R. P. Khapli
AURDC, Hindustan Aeronautics Ltd., Nasik, Maharashtra, India
e-mail: gautam.s@hal-india.com

S. Seetharamu et al. (eds.), *Fatigue, Durability, and Fracture Mechanics*, Lecture Notes in Mechanical Engineering, https://doi.org/10.1007/978-981-15-4779-9_5

Fig. 1 Crack on eye end

(FCG) analysis for eye end to decide the inspection interval during life extended period.

Activites involved in FCG is fatigue load spectrum generation, selection of appropriate material constants and stress intensity factors (SIF) for the crack, stress analysis of eye end, etc. This paper describes the FCG analysis of eye end of MLG actuating cylinder.

2 Load Estimation on Landing Gear

Load estimation on MLG and its actuating cylinder is based on the geometric and landing weight data of the fighter aircraft. The load estimation on MLG is carried out using MIL-A-8862ASG [1] and vertical reaction on MLG is estimated as per standard DEF-STAN [2].

LG has three directional loads, i.e., vertical, drag, and side loads. Loading directions are shown in Fig. 2. Load cases considered for load estimation are given in Table 1.

Fig. 2 Loading directions

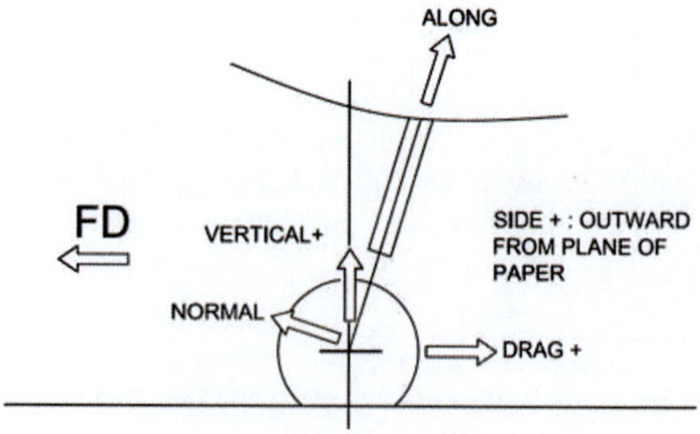

Table 1 MLG load cases

Sl no.	Load case	Sl no.	Load case
1	2P landing MLG (2PLL)	4	3P landing MLG
	Spin-up		Spin-up
	Spring back		Spring back
	Max. vertical		Max. vertical
2	Tail down landing MLG	5	Ground loads MLG
	Spin-up		2P brake roll at landing weight
	Spring back		3P brake roll at landing weight
	Max. vertical		2P brake roll at takeoff weight
3	Drift landing MLG		3P brake roll at takeoff weight
	Drift landing (side inboard)		Un-sym braking
	Drift landing (side outboard)		

2P—2 Point, 3P—3 Point

3 Load Estimation on Actuating Cylinder of MLG

Loads estimated on LG are transferred on actuating cylinder through finite element model using MSC NASTRAN/PATRAN software. FE model for MLG installation is shown in Fig. 3. The cylinder is idealized with rod element in the model. Landing load is applied on LG and forces in actuating cylinder are estimated.

Maximum load on actuating cylinder of MLG is ~25,000 kg (tension) for two-point landing case and −17,500 kg (compression) for two-point brake roll at max takeoff weight.

4 Stress Analysis of Eye End

Stress analysis of eye end is carried out along with actuating cylinder assembly. The result of stress for eye end only is shown in Fig. 4.

Nominal (far filed) stress near grease nipple hole for maximum load of 25,000 kg is 54.3 kg/mm^2. This value is used for crack growth analysis. Stress value will reduce with change in load in different load cycles.

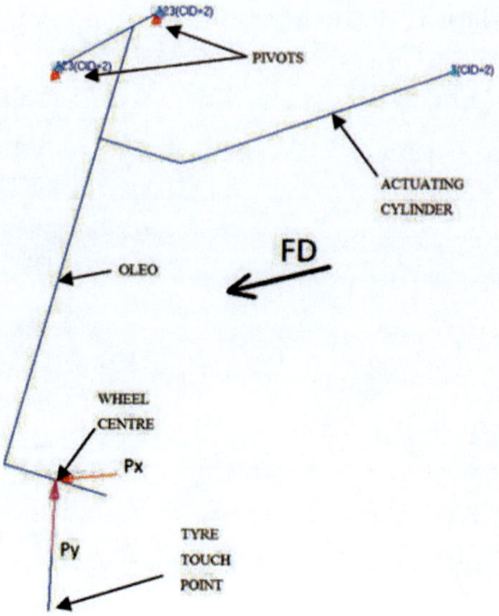

Fig. 3 MLG FE model

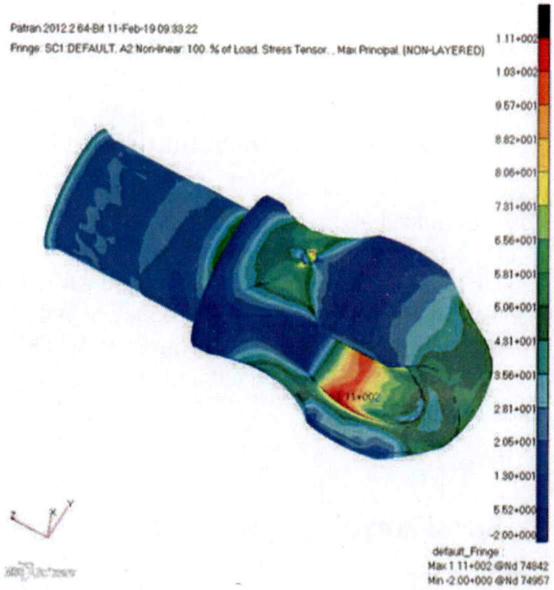

Fig. 4 Stress results of eye end

4.1 Material Properties

Material properties of eye end are given in Table 2.

Table 2 Material properties

Components	Material	E (kg/mm^2)	μ	UTS (kg/mm^2)
Eye end	High strength (HS) steel	21,000	0.3	160

Table 3 SIF solution for crack emanating from hole [3]

a/c < 1	a/c > 1
$\beta(a/c,\ a/t, \theta)\ = \left[M1 +\ M2(a/t)^2 +\ M3(a/t)^4\right] * g1 * g2 * f_{\emptyset}$	
$M1 = 1.08 - 0.03(a/c)$	$M1 = \sqrt{\frac{c}{a}}[1.08 - 0.03(\frac{c}{a})]$
$M2 = -0.44 + 1.06/(0.3 + a/c)$	$M2 = 0.375(c/a)^2$
$M3 = -0.5 - 0.25(a/c) + 14.8(1 - a/c)^{15}$	$M3 = -0.25(c/a)^2$
$g1 = 1 + \left[0.08 + 0.4(a/t)^2\right](1 - \sin\theta)^3$	$g1 = 1 + \left[0.08 + 0.4(c/t)^2\right](1 - \sin\theta)^3$
$g2 = 1 + \left[0.08 + 0.15(a/t)^2\right](1 - \cos\theta)^3$	$g2 = 1 + \left[0.08 + 0.15(\frac{c}{t})^2\right](1 - \cos\theta)^3$
$f_{\emptyset} = \left[(a/c)^2\cos^2\theta + \sin^2\theta\right]^{\frac{1}{4}}$	$f_{\emptyset} = \left[(c/a)^2\sin^2\theta + \cos^2\theta\right]^{\frac{1}{4}}$

5 Fatigue Crack Growth Calculations

Fatigue crack growth is calculated using Paris equation,

$$da/dN = C(\Delta)^m \tag{1}$$

where da/dN = crack growth rate mm/cycle, C and m = material constants,

$$\Delta K = \text{SIF range in MPa}\sqrt{}\text{mm} = \beta\left(\Delta\text{max} - \Delta_{\text{min}}\right)\sqrt{}(\pi a) \tag{2}$$

$\sigma_{\text{max}}, \sigma_{\text{min}}$ = maximum and minimum stresses of fatigue cycle, respectively

5.1 *Solution for β*

Solution for β is given in Table 3, for crack emanating from hole [3, 4]. Crack geometry is shown in Fig. 5.

5.2 *Material Constants for Paris Equation*

Material constants of Paris equation for material HS steel are not available. Therefore, material constants of equivalent material given in Table 4 are used [4].

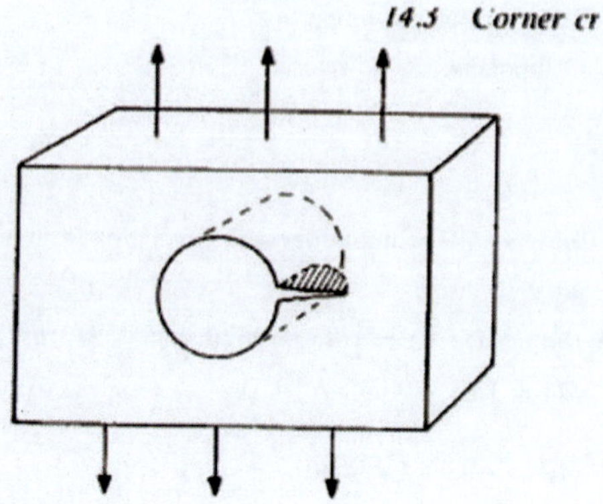

Fig. 5 Geometry of crack emanating from hole

Table 4 Comparison of HS steel with HY-TUF

Material	YS (kg/mm^2)	UTS (kg/mm^2)	Chemical composition				
			C	Mn	SI	Cr	Ni
HS steel	140	160	0.27–0.33	1.0–1.2	0.9–1.2	0.9–1.2	1.4–1.8
HY-TUF [3]	140	160	0.25	1.35	1.5	0.3	1.8

Mechanical properties of HS steel are equivalent to HY-TUF. Therefore, material constants of HY-TUF can be used for crack growth analysis.

Material constants of HY-TUF [4]:

$m = 2.5$ and $C = 1.23 \times 10^{-11}$ in the units MPa$\sqrt{}$mm.
Fracture toughness = 110 Ksi $\sqrt{}$in = 3825 MPa$\sqrt{}$mm.

5.3 *Fatigue Load Spectrum for MLG*

MIL-A-8866C (AS) is used to generate the fatigue load spectrum.

Following fatigue load cycles are considered for fatigue testing and fatigue crack growth analysis.

- Two-point level landing—max. vertical landing impact to parking load (2PLL)—50% cycles,
- Two-point drift landing—inboard side load to outboard side load (DRIFT)—50% cycles,
- Two-point level landing—spin-up load to spring back load (SUSB),
- Taxiing load as per Table 5 (Taxii),
- Turning (inside wheel loads to outside wheel loads) at landing and takeoff weight, and,
- LG extension to retraction (EXTN-RTRN).

Table 5 Number of times per 1000 runway landings that load factor is experienced [5]

Load factor during landing	No. of occurrences
1 ± 0.05	300,000
1 ± 0.15	165,000
1 ± 0.25	27,000
1 ± 0.35	2000
1 ± 0.45	90
1 ± 0.55	4
1 ± 0.65	0.15

Table 6 Distribution of sinking speeds per 1000 landings [5]

Sink speed Ft/s	No. of occurrences
<5	903
6	66
7	22
8	7
9	1.5
10	0.5

Distribution of load factor for taxii and sinking speed (for 2PLL, DRIFT, and SUSB) is given in Tables 5 and 6, respectively [5].

2PLL, DRIFT, and SUSB are divided as per sinking speed. Sinking speed less than 7fps is considered together and named as SP7 and same way SP8 for speed 8 fps.

Fatigue load spectrum (Table 7) is used for crack growth analysis, block by block.

5.3.1 Fatigue Load Spectrum for Crack Growth Analysis

The fatigue loading cycles are derived for MLG actuating cylinder assembly shown in Fig. 6. One block of loading consists of 500 landings. Maximum stress 54.3 kg/mm^2 (533 MPa) corresponding to 25,000 kg is taken for crack growth analysis and minimum is zero in case it is compressive. This far field stress is induced in eye end when maximum stress in actuating cylinder is around 107 kg/mm^2 (SUSB-SP10 case). Stress will reduce proportionally with load/stress values in cylinder assembly.

Table 7 Fatigue load spectrum for MLG actuating cylinder for 1 block (1 block = 500 landings)

MLG cylinder fatigue cycles	No. of cycle in 1 block of 500 landings	MLG cylinder fatigue cycles	No. of cycle in 1 block of 500 landings
2PLL-SP7	247.75	TAXI-M3	13,500.00
2PLL-SP8	1.75	TAXI-M4	1000.00
2PLL-SP9	0.38	TAXI-M5	45.00
2PLL-SP10	0.13	TAXI-M6	2.00
DRIFT- SP7	247.75	TAXI-M7	0.08
DRIFT- SP8	1.75	TURN-WL-M1	1250
DRIFT- SP9	0.38	TURN-WTO-M2	1250
DRIFT- SP10	0.13	BRHD-WL-M1	500
SUSB- SP7	495.50	BRMD-WL-M2	1250
SUSB- SP8	3.5	BRHD-WTO-M3	500
SUSB- SP9	0.75	BRMD-WTO-M4	1250
SUSB- SP10	0.25	PIVOT-M1	166.67
TAXI-M1	150,000	EXTN-RTRN	625
TAXI-M2	82,500		

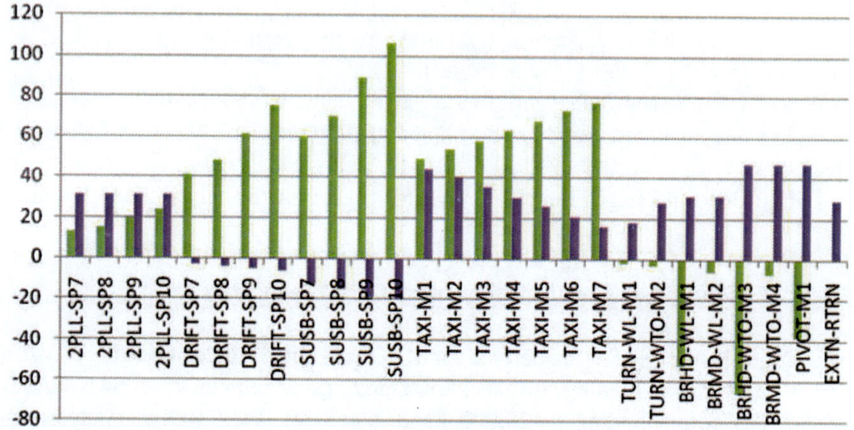

Fig. 6 Fatigue load (Stress) spectrum for actuating cylinder of MLG

5.4 *Maximum Allowable Crack Length*

Critical crack length is through thickness crack, i.e., 8 mm or crack length at which the SIF becomes equal to or more than fracture toughness (3825 MPa$\sqrt{}$mm (crack length whichever is less).

Table 8 FCG results

No. of block loading	Crack growth (mm)	SIF (MPa$\sqrt{mm}$)
Initial surface crack mm	0.5	125
Crack length after Block-1 (500 landings)	1.003	1021
Crack length after Block 2 (1000 landings)	1.896	1444
Crack length after Block 3 (1500 landings)	4.226	2231
Crack length after Block 4 (2000 landings)	13.47 (>8 mm)	4614 (>3825)

5.5 *FCG Results*

Crack growth is calculated with sequence of loading cycles from low stress to high stress. Initial surface crack length of 0.5 mm on surface is propagated to length of 1.003 mm, i.e., growth of 0.5 after one block of loading, i.e., after 500 landings. Same way crack growth is calculated for next blocks of loading.

Brief of block-wise crack growth results is given in Table 8.

6 Discussion

It is observed from calculations that crack will propagate more than given limit of 8 mm and fracture toughness more than 3825 MPa$\sqrt{mm}$ in fourth block. Therefore, crack will not become unstable till 3 blocks or 1500 landings. With scatter factor of 4, safe life between inspections can be ~400 landings.

7 Conclusion

Fatigue crack growth analysis of actuating cylinder eye end is carried out and based on analysis inspection interval of eye end can be selected as 400 landings in life extended period.

References

1. MIL-A-8862ASG (1960) Military specification. Airplane strength and rigidity landplane landing and ground handling loads
2. DEF STAN 00-970, Leaflet 304/1. Design and Airworthiness Requirements for service Aircraft volume 1, Aeroplanes
3. Newman JC Jr., Raju IS (1981) SIF equation for cracks in three dimensional finite bodies. NASA TM 83200 NASA Langley Research centre
4. Bahram F (2012) Fatigue & fracture mechanics of high risk parts. Chapman & Hall, International Thomson Publishing, p 328
5. MIL-A-8866 C (AS) (1987) Standard: airplane strength & rigidity, reliability requirements, repeated loads, fatigue and damage tolerance

Residual Life Assessment of Nuclear Power Plant Components

Lokesh Hegde, Sunil Gadgil, and Ravi Satyanarayana

Abstract Indian pressurised heavy water reactors are pressure tube type reactors using natural uranium as fuel. Primary heat transport system consists of heavy water as coolant at high temperature and pressure flowing on the fuel bundles to remove the fission heat generated. The primary heat transport system consists of 306 Zr–2.5% NB tubes, feeders, headers and steam generators as important pressure boundary components. Breach of any of these components/equipment results in impaired cooling to fuel bundles containing fission products and thus possibility of fuel failure. It becomes mandatory to provide assurance that structural integrity of these components is intact and balance life remains in accordance with the design criteria. A detailed in-service inspection and surveillance programme is in place to monitor the health of systems, Components and structures important to nuclear safety. In service inspection programme on these critical components are devised such that the inspection results do not only demonstrate the healthiness of these components for further safe reactor operation but also predict the balance life of the component is being examined. This in achieved by conducting specialized non-destructive techniques developed of individual critical component. Since these components are located at high radiation field, robotic system is in use to deploy the NDT inspection system probe/sensors to the required area for inspection. The life assessment of critical components like coolant channel, primary piping and stream generator are elaborated in this paper.

Keywords Nuclear reactor · Critical components · Non-destructive testing · Residual life estimation · Coolant channel · Piping and steam generator

---

L. Hegde · S. Gadgil (✉) · R. Satyanarayana
Nuclear Power Corporation, Kaiga Generating Station, Kaiga, India
e-mail: sunilgadgil@npcil.co.in

S. Seetharamu et al. (eds.), *Fatigue, Durability, and Fracture Mechanics*, Lecture Notes in Mechanical Engineering, https://doi.org/10.1007/978-981-15-4779-9_6

1 Introduction

Indian pressurised heavy water reactors are pressure tube type reactor using natural uranium as fuel. Primary heat transport system utilises heavy water as coolant at high temperature and pressure, flowing on the fuel bundles to remove the fission heat generated. The primary heat transport system 220 MW reactor consists of 306 Zr–2.5% Nb pressure tubes, feeder pipes, headers and steam generators as important pressure boundary components [1]. Breach of any of these components/equipment results in impaired cooling of fuel bundles containing fission products and thus possibility of fuel failure. It becomes mandatory to provide assurance that structural integrity of these components is intact and balance life remains in accordance with the design criteria [2]. Hence, residual life assessment is essential on these components due to following reasons

- To avoid unexpected component failure jeopardising plant safety.
- To eliminate unwarranted replacement of components before the end of useful life as it directly affects the cost, time and radiation exposure.
- To establish a sound, defensible basis for plant life extension.
- These assessments are mandatory for renewal of operating licence of nuclear plant by atomic energy regulatory board.
- To reduce costly unscheduled outages there by continuous electricity supply to the grid.

2 Life Limiting Mechanisms for Nuclear Components

Like any other industry, nuclear components also have specific life to perform specific function. Majority of the nuclear power plant components are designed for 30 years of life. There are several identified degrading mechanisms causing ageing of structural materials in nuclear power plant during its service life. Various types of corrosion, fatigue, wear, creep, shock loading are few of them. Microstructural damages on account of these degrading processes keep accumulating in the component. In addition to these factors, degradation of component also takes place due to presence of radiation in nuclear power plant. As a result, its load bearing capacity continues to decrease with In-service time. Failure takes place when it falls below a threshold determined by the component geometry and loading. Premature failure is also probable due to various reasons viz., design deficiencies, improper material selection, process deficiencies, assembly and installation error, operation and maintenance error, environmental impact and loading condition [3, 4].

3 Residual Life Assessment Programme

Followings are the sequence of activities followed at Indian nuclear power stations to assess the residual life of critical components [2].

a. Collection of design data and operation parameters.
b. Identification of degradation mechanism and life limiting factors.
c. Review of base line inspection (PSI) data and manufacturing history.
d. Operation and maintenance history.
e. In-service Inspection (ISI) programme for estimation of degradation trend.
f. Prediction of Residual life.

4 Inspection Techniques

Based on the degradation mechanism of specific component, appropriate Nondestructive Testing (NDT) is conducted on identified component at pre-determined interval throughout its service life to monitor/estimate the rate of degradation. A separate in-service inspection programs exist for all critical nuclear power plant components to measure parameters indicating specific life limiting factors. In-service inspection programme on these critical components are devised such that the inspection results not only demonstrate the healthiness of these components for further safe reactor operation but also predict the balance life of the component being examined. Since these components are located at high radiation field, specialised robotic systems are used to deploy the NDT inspection system probe/sensors to the required area for inspection. The life assessment of critical components like coolant channel, primary piping and stream generator are elaborated in this paper.

Inspection programme also calls for limited Destructive testing (DT) on selected components on sample basis as a deterministic approach.

The inspection methods followed for life assessment are [2].

a. Visual inspection.
b. Dimensional measurement..
c. Surface and sub surface NDT (Dye penetrant Testing and Magnetic particle testing).
d. Ultrasonic Inspection (volumetric inspection).
e. Eddy current Testing. (volumetric inspection of tubes).
f. Helium leak Testing.
g. Sliver sampling.
h. Post Irradiation examination.

5 Residual Life Assessment of Nuclear Plant Components

Primary heat transport system of 220 MW pressurized heavy water reactor (PHWR) consists of 306 coolant channel assemblies which house 12 fuel bundles in each channel. Heavy water coolant passes through these fuel bundles and take away the heat from bundle generated during the nuclear fission process. The high temperature coolant moves further through feeder pipes and header to steam generator where the coolant exchange its heat into light water to form steam. After exchanging of heat, the coolant is recirculated back to coolant channel by primer heat transport system pumps [1]. Schematic circuit of PHT system is shown below. Residual life assessment of these components (coolant channel, piping component and Steam generator) are discussed in this paper (Fig. 1).

6 Residual Life Assessment of Coolant Channel

There are 306 coolant channels assemblies in 220 MWe Pressurized Heavy Water Reactor (PHWR). Each assembly consist of Pressure tube (PT) calandria tube (CT), garter spring (GS) and end fittings (2Nos). PT, CT and GS are made from Zirconium alloy and end fitting made from SS-403. PT rolled at both end to the end fitting. PT located concentrically inside CT with the help of GS. Low pressure and low temperature heavy water moderator circulates around CT and high pressure and high temperature coolant flows inside the PT which houses fuel bundles. CO_2 gas circulated between PT and CT for thermal insulation. PT is considered to be the most critical component of coolant channel assembly as it is exposed to very high radiation, high temperature and high pressure. The tube thickness is to be kept optimum in order to reduce neutron absorption by channel material. These conflicting requirements result in design factor of safety lesser than a conventional equitant component. Hence stringent inspection and accurate residual life estimation is required for Pressure Tube of coolant channel assembly (Figs. 2, 3 and 4).

6.1 Design Predicted Life Limiting Parameters

Design predicted life limiting parameters of coolant channel and its consequences are detailed below. These parameters were monitored at specified interval during its service life. Based on the outcome of the inspection findings, degradation trend and residual life of individual channel is estimated.

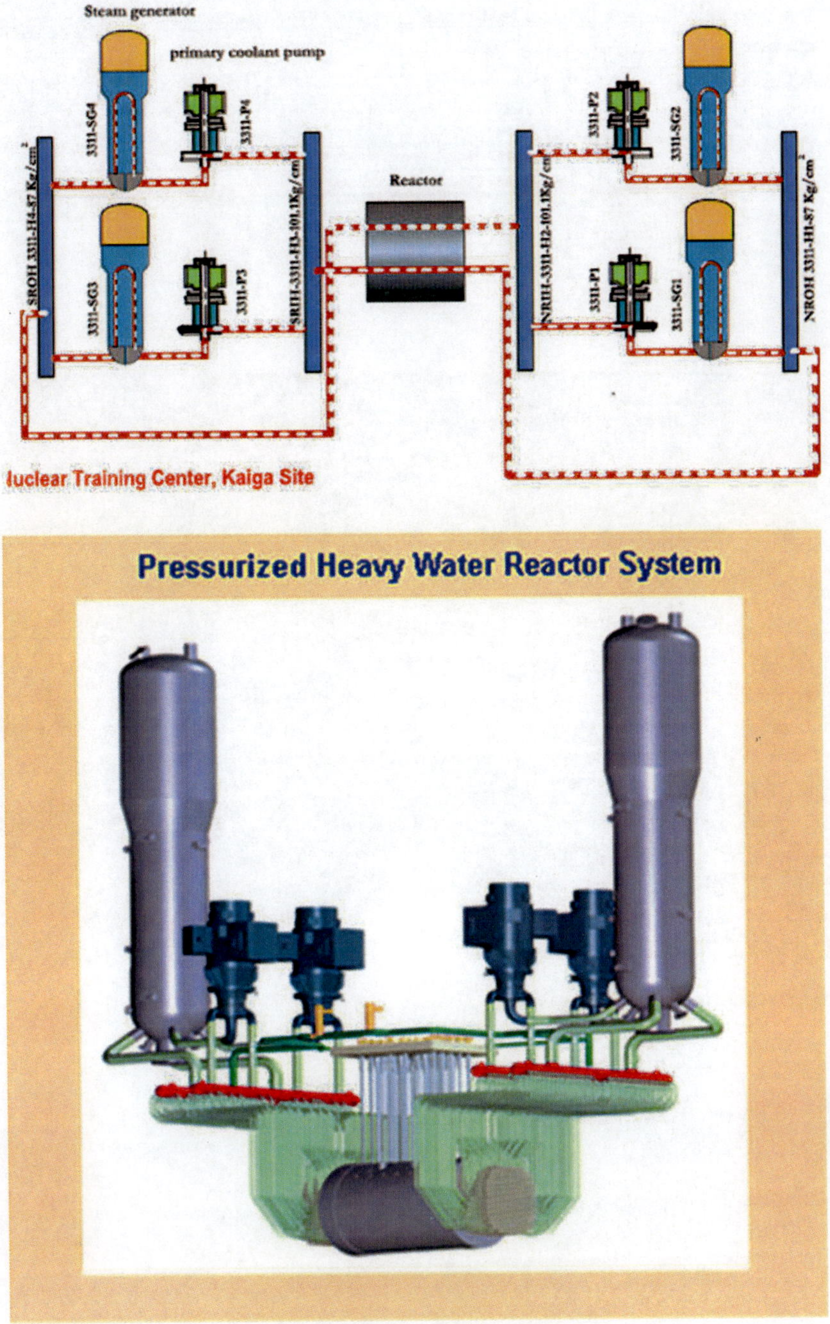

Fig. 1 Schematic view of PHWR type primary heat transport system

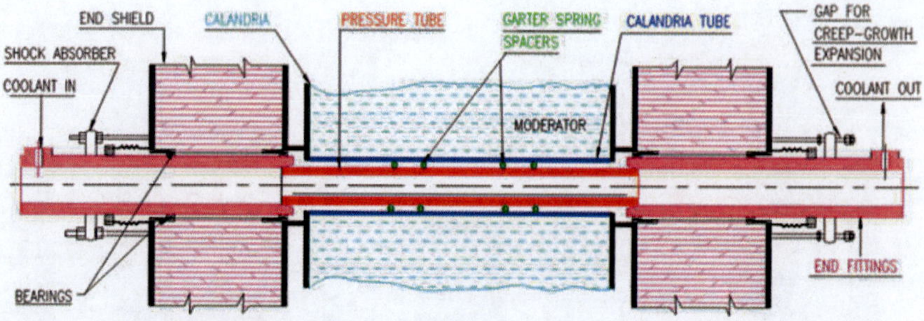

Fig. 2 Isometric view of coolant channel

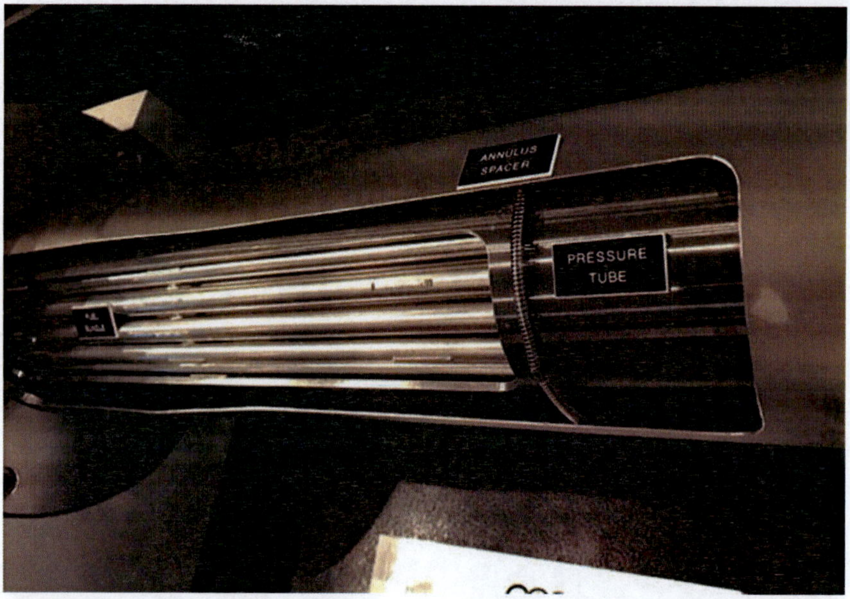

Fig. 3 A cut view of coolant channel

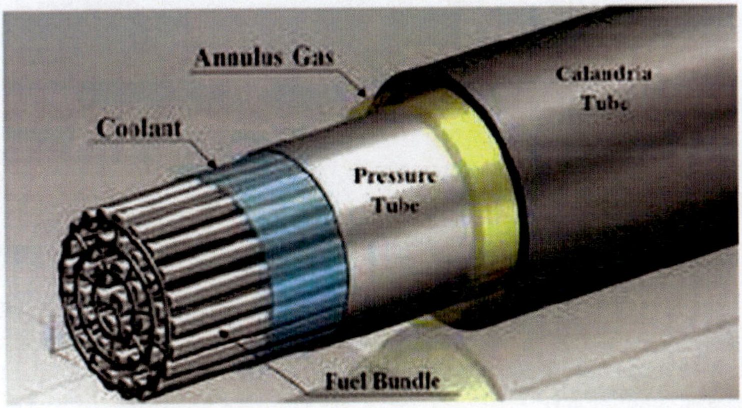

Fig. 4 Fuel bundle inside the pressure tube

Life limiting parameters	Consequences
General corrosion/wear and tear	Reduction in WT
Neutron irradiation degradation	**<u>Dimensional change</u>** ➢ Creep → Axial (length); Circumferential (ID) ➢ Reduction in WT **<u>Change in Mechanical Properties</u>** ➢ Increase in brittleness ➢ Reduction of ductility ➢ Reduction of toughness
Hydrogen induced degradation	• Reduce flaw tolerance • Increase DHC probability • Reduction in fracture toughness
Displacement of garter spring position	• Contact of PT-CT there by hydrogen related issue
Sagging of PT along with CT	On-power fueling related problem
Leak tightness integrity	• Presence of flaw • Nodular corrosion

The major contributor for failure of component in nuclear reactor is neutron irradiation. In case any of above referred parameters of coolant channel is measured to be beyond the specified limit, the said channel is isolated from the service by wet quarantining (WQ). Wet quarantining is a process, in which, fuel bundles are removed from that particular channel to reduce the neutron population there by degradation due to neutron irradiation will significantly reduce.

6.2 *Inspection Systems and Techniques for Coolant Channel Critical Parameter Monitoring*

Following robotic inspection systems are in use to monitor the design predicted life limiting parameters of coolant channel at specified interval. Different NDT/DT methods are applied to monitor specified parameters through these systems [2, 5].

a. **T-MAC (Tool for Measuring Axial Creep) for cannel creep measurement**

This system is designed for measuring axial elongation of coolant channel due to creep. The tool works on LVDT principle and measures the coolant channel length increment with respect to a reference channel not subjected to temperature/pressure/radiation (Fig. 5).

b. **Slivering**

Hydrogen pick-up by pressure tube material during service life is determined in terms of delayed hydride cracking (DHC) and fracture toughness point of view. A remote operated tool is utilized for scooping small amount of zirconium material from inner diameter (ID) surface of pressure tube to analyse the ingress of hydrogen into PT material (Figs. 6 and 7).

c. **BARCIS (BARC Channel Inspection System)**

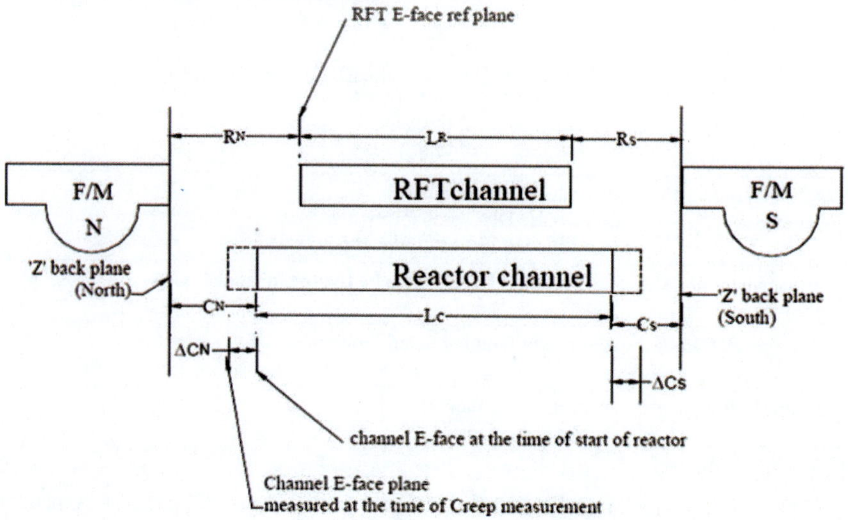

Fig. 5 Methodology for measurement of axial creep of coolant channel

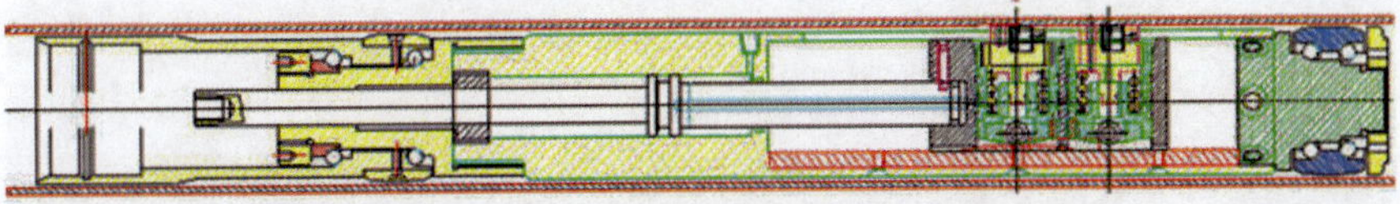

Fig. 6 Cross section view of slivering tool

Fig. 7 Photograph of slivering tool

Due to high temperature coupled with neutron radiation field, pressure tube diameter increases and wall thickness reduces. BARCIS is a robotic system used for measuring pressure tube parameters such as inner diameter, Wall thickness, GS position detection, Volumetric inspection of pressure tube material for detecting service induced flaws, Nodular corrosion monitoring and CT-PT gap measurement. An inspection tool (inspection head) consisting different NDT sensors is deployed into the channel using BARCIS system (Figs. 8 and 9).

d. **PRESAM-220**

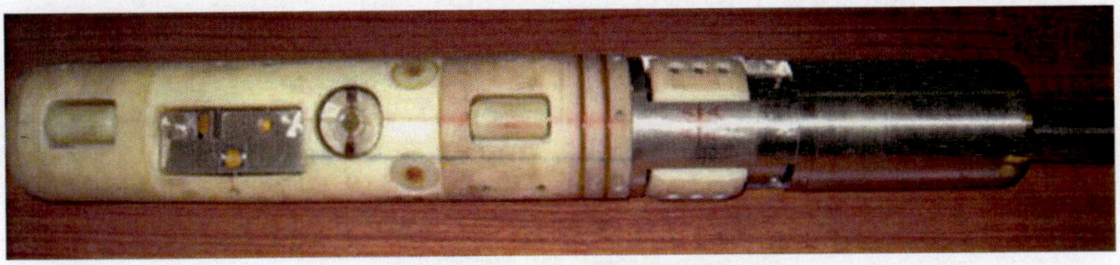

Fig. 8 Photograph of coolant channel inspection heads contains UT & ECT probes

Fig. 9 View of BARCIS drive coupled with coolant channel for ISI

Pressure tube is subjected to Sagging along with calandria tube keeping GS in position due to neutron irradiation and load of fuel and coolant (Fig. 10).

PRESEM-220 tool (inspection Head) works on LVDT principle for sag measurement. The tool is placed inside the channel to measure SAG of PT. BARCIS system is used to deploy the SAG inspection head into the channel for SAG Measurement (Fig. 11).

e. **Post Irradiation Examination (PIE)**

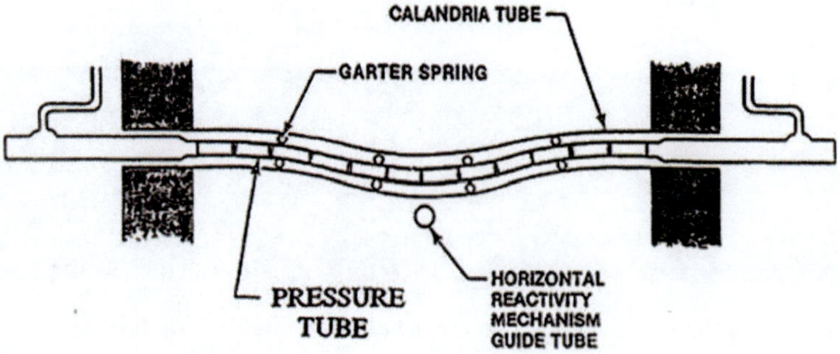

Fig. 10 Sag of pressure tube with calendria tube of coolant channel

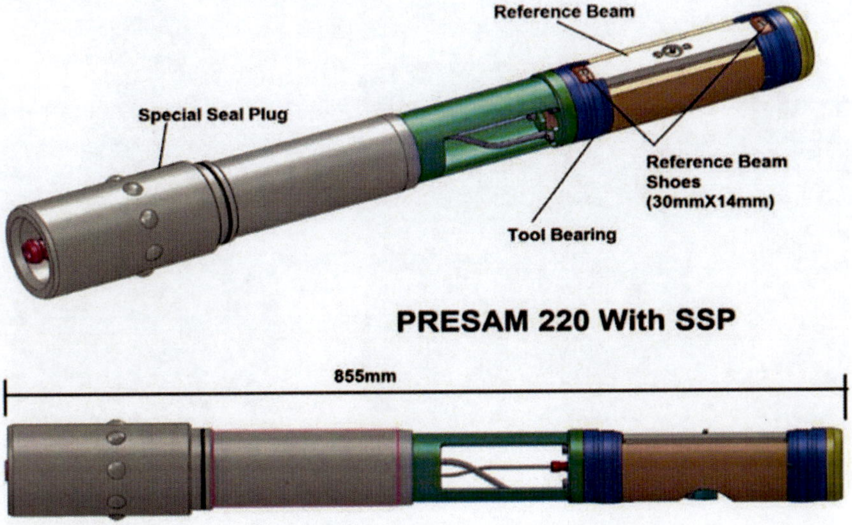

Fig. 11 PRESAM 220 tool for SAG measurement

A representative coolant channel is cut from the reactor core after specified years of operation to assess its mechanical and chemical properties. This is a destructive test to demonstrate requisite mechanical properties such as strength; ductility and fracture toughness are in line with predicted values at particular service life.

7 Residual Life Assessment of Piping Material

Power densities at reactor core are higher than similar components in conventional power plants. This results in higher fluid velocities in certain sections of nuclear piping. In addition, two phase flows are inevitable as steam turbines operate on saturated steam cycle in nuclear plants. Hence FAC (Flow assisted Corrosion) is one of the major degradation phenomena of nuclear power plant piping material. Primary and Secondary heat transport system piping materials of nuclear power plant are subjected to High pressure and temperature during its operating condition. FAC is expected on these piping components and wall thickness reductions are noticed during its service life [1] (Fig. 12).

7.1 Mechanism of FAC

FAC may be defined as corrosion enhanced by mass transfer between a dissolving oxide film and a flowing fluid that is unsaturated in the dissolving species (Fig. 13).

Fig. 12 Failure of piping material due to flow assisted corrosion

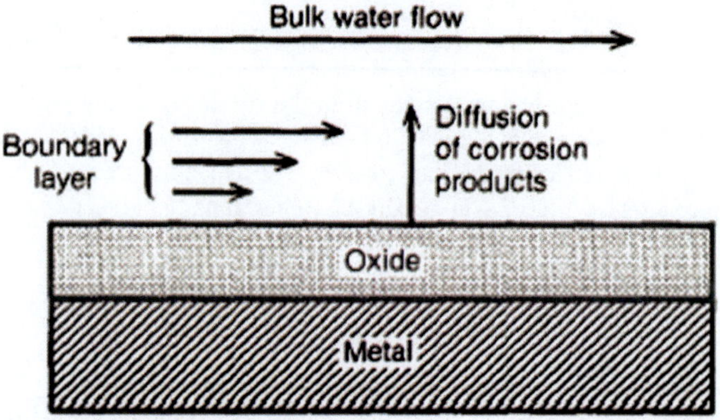

Fig. 13 Mechanism of flow assisted corrosion

7.2 Influential Factors for FAC

- Type of fluid flowing through pipeline → FAC normally seen in duel phase fluid.
- Fluid temperature → FAC rate peak at 130°–150° C.
- Fluid flow/dynamics → FAC rate increases with flow velocity and turbulence factor.
- p^H of Fluid → FAC rate decreases with increased p^H of fluid.
- Piping routing → Normally in change in flow direction (elbow), down stream of restricted Orifice (RO), Controlled valve (CV) etc.
- Material → Presence of chromium in piping material reduce FAC probability.
- Dissolved oxygen → FAC rate mitigated with optimum dissolved oxygen in the fluid.

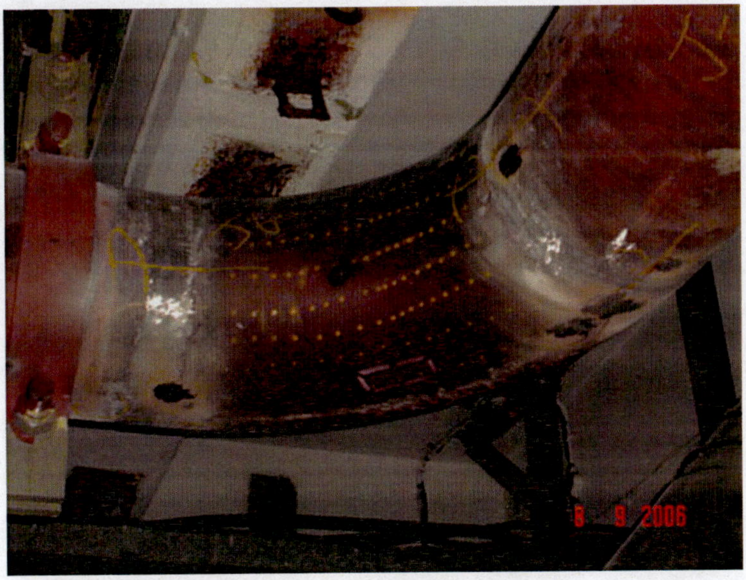

Fig. 14 Permanent marking on pipe material for UT thickness gauging

7.3 Monitoring Method

Wall thickness of the material in the pressure boundary of a system will decide the residual life of piping components [5]. Hence parodic wall thickness monitoring of piping material is required. A comprehensive FAC monitoring program is adopted at NPCIL and material wall thickness of piping material is monitored using UT thickness gauging on predetermined point (permanent grid marking on pipe component) at specified interval. Minimum wall thickness measured during inspection is considered for estimating the balance life of particular component along with all other design parameters (Figs. 14 and 15).

7.4 Material Residual Life Estimation

The residual life is estimated using following formulas.

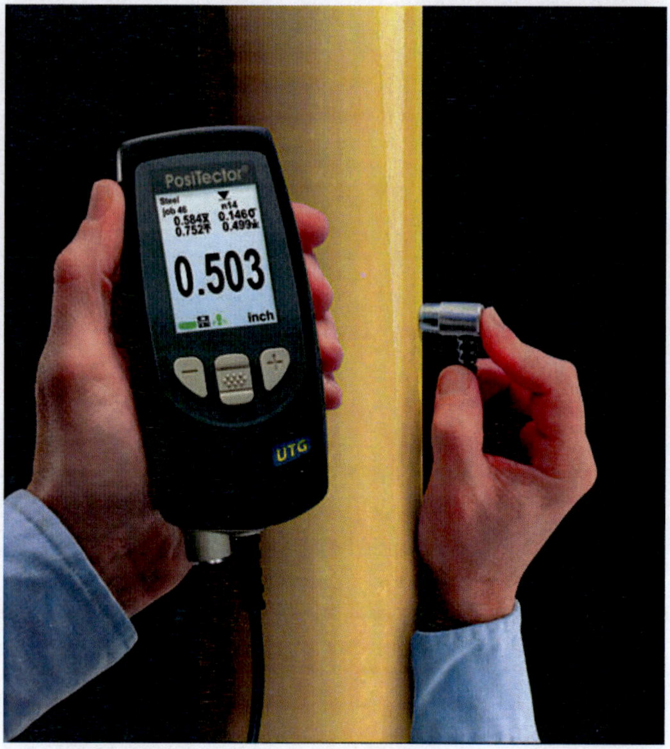

Fig. 15 UT thickness gauging

Erosion rate :

$$= \frac{\text{Reduction in WT between two measurements)} * 1000}{\text{Operating period (years) between two measurement}} \quad \textit{(microns/Year)}$$

$$\textbf{Residual Life} = \frac{(\text{Present WT} - T_{(min)}\text{ required}) * 1000}{\text{Erosion rate}}$$

$T_{(min)}$ = Minimum Thickness required for the system to function.

$$T_{(min)} = \frac{P\,Do}{2(SE+Py)}$$

P : *Design Pressure*
Do : *Outside diameter of Pipe*
S : *Allowable stress at design temperature*
E : *Weld joint efficiency = 1.0*
y : *0.4*

8 Residual Life Assessment of Steam Generator

Steam generator (SG) in PHWR is large shell and tube type heat exchanger that extracts the heat from primary reactor coolant to convert high pressure steam in the secondary side fed to the turbine generator (Fig. 16).

The primary reactor coolant passes through 1 mm thick tubes (1830 Nos) and boils the water on the outside of the tube (secondary side) to make steam. These tubes also act as barrier to confine the radioactivity from neutron activation or fission products with in the primary side during normal operation. Hence safety and integrity of these tubes are vital in view of radioactivity release into the environment.

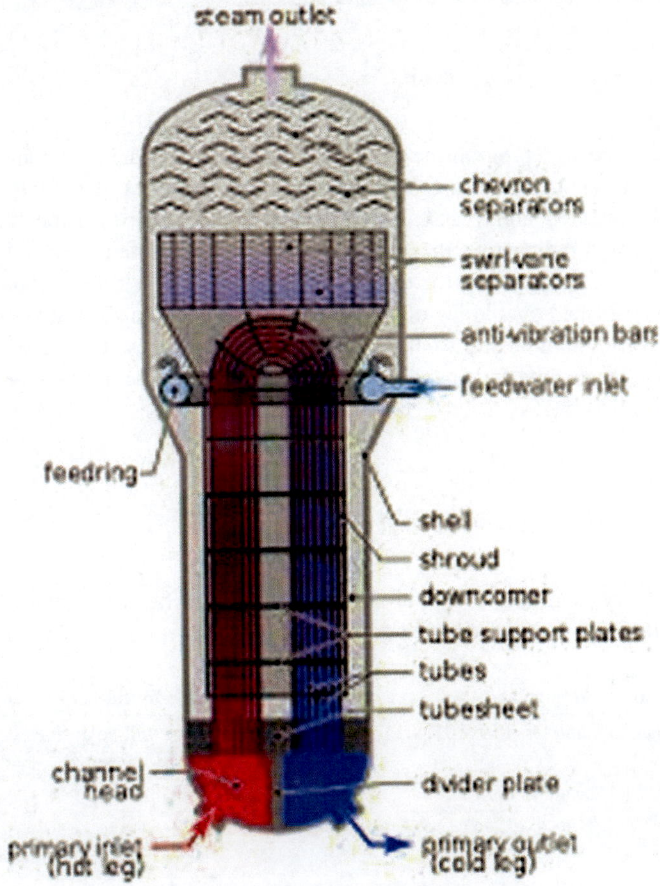

Fig. 16 Steam generator

Fig. 17 Eddy current testing probe for SG tube inspection

8.1 SG Inspection System

A separate inspection programme exists to ensure the healthiness of Steam Generator tubes [5]. Eddy current testing is conducted on these tubes at regular interval to monitor the integrity and to assess the degradation rate. Since the location of these tubes are at high radioactive area, a robotic manipulator system is used to carry the eddy current testing to cover the entire length of tubes. This robotic manipulator is installed inside the hot/cold leg of SG. A pusher puller and a control system will be used along with manipulator to carryout ECT of SG tubing's. The manipulator will identify and locate tube and pusher and puller will perform the probing as per control system commands (Figs. 17, 18, 19 and 20).

8.2 Inspection Analyses and Interpretation

After the inspection and review of inspection findings, fitness of tubes for further continued service is decided based on following criteria

Less than 40% wall thinning	:	Acceptable for further service
Greater than 40% and less than 80% wall thinning	:	Acceptance will be decided based on amplitude of the signal.
Greater than 80% wall thinning	:	Not acceptable

Fig. 18 Manipulator installed inside hot/cold leg of SG for tube identification and alignment

If any indication observed in the tube, the same will to be monitored in the subsequent inspection campaign. Nonacceptable/defective tubes noticed during the inspection shall be isolated from the system by plugging both ends of the tube. Heat removal capacity decreases as the need of tube plugging increases. Prevailing design standards keep 10% margin in number of tubes on account of plugging without affecting its performance.

9 Conclusion

The aging mechanisms for nuclear components and their significance with respect to safety are well understood in Indian nuclear industry. Elaborate inspection programmes are existing in Indian Nuclear power plant to monitor the healthiness and predict then residual life of its components [2]. The robotic systems are extensively used for monitoring the life limiting parameters of different system components to reduce the radiation exposure of inspection personnel.

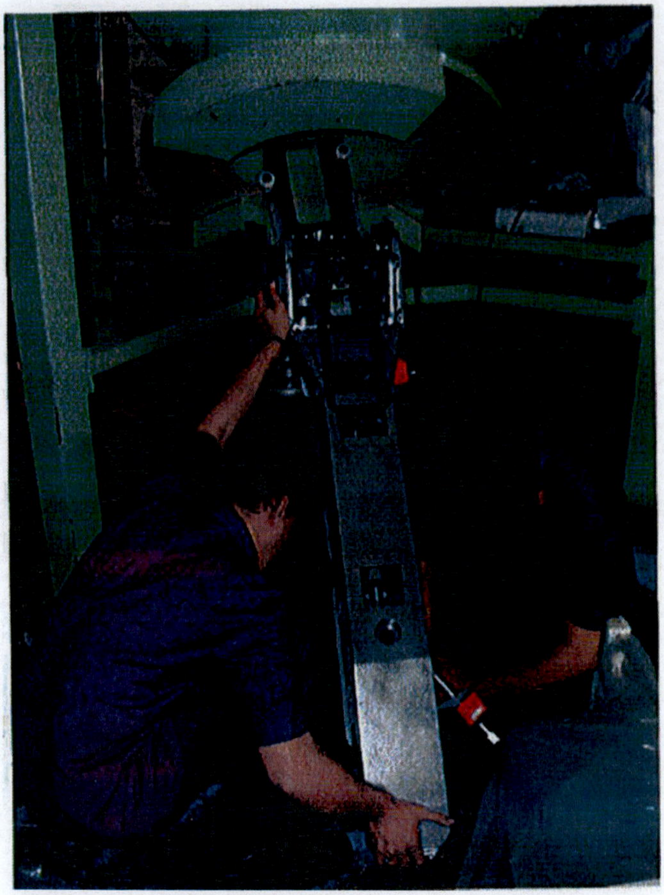

Fig. 19 Assembly of manipulator through SG manhole

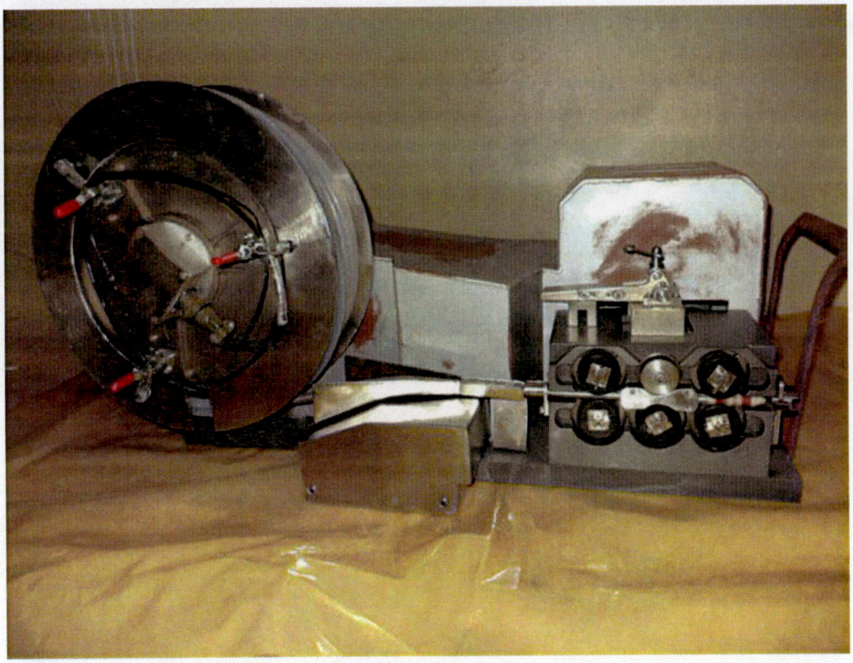

Fig. 20 Pusher and puller which moves the probe to desire position inside the tube

References

1. Design basis reports of Nuclear reactor components
2. Kaiga Generating Station ISI programme documents for Indian PHWR
3. Proceeding of National Conference on Aging Management (NCAM-2004) of Structure, System and Components
4. Proceeding of National Conference on Operating Experience of Nuclear Reactors and Power Plants (OPENUPP—2006)
5. Kaiga Generating Station procedures ISI of coolant channels, Steam Generator and FAC monitoring of piping's

Influence of Ta + Nb Addition on Microstructure and Mechanical Properties of Ferritic Stainless Steel TIG Weldments

P. K. Mandal, K. S. Asutosh, M. S. Harikrishnan, and Joyal James

Abstract Ferritic stainless steel (FSS) is general considered to have poor weldability in comparison to austenitic stainless steel but it has many industrial applications due to their low cost and better corrosion resistance and magnetic nature. In this work a single sided butt-joint (60° V groove) of FSS (430 grade) was selected for welding by using a specific filler wire (ER409Nb Grade) contents Ta + Nb during TIG welding. The optical micrograph studies revealed grain coarsening and martensite and carbide precipitates especially on the grain boundary regions in the WZ. But this feature of micrographs changed gradually if PWHT increases from 500 to 700 °C for 2 h after welding. The PWHT at 700 °C for 2 h decomposed the martensite into ferrite and carbides, eliminated segregation, sensitization and residual stresses, and improved corrosion resistance and tensile properties. The tensile properties was evaluated after TIG welding then PWHT at 500–700 °C for 2 h and subsequently weldments analysed by SEM with EDAX. It has been clearly indicated stabilizer Ta + Nb contents about 1.22 wt% Ta + trace Nb after PWHT. The tensile properties was performed through UTM as per ASTM E8 standard and notable properties observed after PWHT at 700 °C likely to YS of 296 MPa, UTS of 348 MPa, ductility of 8.46%, and RA of 33.47% and YS to UTS ratio of 0.80, respectively. It can be concluded that tensile properties and toughness of weldments increased by using Ta + Nb stabilizers during TIG welding of FSS plate.

Keywords PWHT · SEM with EDAX · Ta + Nb stabilizers · Tensile properties

1 Introduction

Stainless steel is commonly used in sheet metal fabrication, automotive, chemical and rail coach manufacturing for many years due to its excellent corrosion resistance, better strength to weight ratio, and low life cycle cost. The high Cr content (16–18%) FSS is generally considered to have poor weldability than austenitic stainless steel

P. K. Mandal (✉) · K. S. Asutosh · M. S. Harikrishnan · J. James
Department of Metallurgical and Materials Engineering, Amal Jyothi College of Engineering, Kanjirappally 686518, Kerala, India
e-mail: pkmmet@yahoo.in

S. Seetharamu et al. (eds.), *Fatigue, Durability, and Fracture Mechanics*, Lecture Notes in Mechanical Engineering, https://doi.org/10.1007/978-981-15-4779-9_7

and their mechanical properties and corrosion resistance are adversely affected after welding. FSS have good ductility and formability, but poorer mechanical properties than austenitic stainless steel [1, 2]. Because during fusion welding is associated with many problems such as grain growth in both the WZ (weld zone) and HAZ (heat affected zone) and formation of martensite at the grain boundary in the WZ that may adversely affect the ductility and toughness of the weldment [3–5]. TIG welding is the most reliable method for stainless steels welding due to their significant advantages like cleanliness, no spatter, and shielding gas protect the weld-pool, and ability to weld complicated shapes with wide ranges. FSS contain no or limited nickel since this element is costly and using in austenitic stainless steel. FSS in Fe–Cr–C alloys with sufficient Cr, Si, Nb, and Ta are used to prevent the formation of austenite as well as carbide forming elements on heating [6, 7]. Moreover, sensitization is the one of the major problems for stainless steel that the loss of alloy integrity. It may occur within the temperature range 425–815 °C. Intergranular stress corrosion cracking and sensitization may lead to intregranular pitting within the HAZ on exposure to a corrosive environment, as well as low amount of interstitial constituents (C + N) may decrease the susceptibility [8–10]. Specially, HAZ of FSS contains intergranular martensite is brittle and exposed to corrosion attack. During FSS welding Ta + Nb filler wire had used and effectively confirmed by EDAX analysis after PWHT and subsequently microstructural changes during welding. The purpose of the PWHT can improve the toughness without sacrifice the tensile strength, hardness and significant effect of minimizing residual stresses of the weldment. SEM analysis was done after PWHT and found lack of fusion, particularly on the fusion boundary, oxide formation and several defects confirmed over weldments. The main objective of this work has to carefully investigated of FSS (430 grade) weldments by TIG welding with concise parameters through using a special filler wire (ER409Nb Grade) containing stabilizing elements Ta + Nb and subsequently post weld heat treatments (PWHT) effect from 500 to 700 °C for 2 h.

2 Experimental Procedure

The material used in this work was 6 mm thick rolled plate (size: $250 \times 50 \times 6$ mm^3) of AISI 430 grade FSS steel which had a single sided 60° V with 2 mm root gap for TIG welding as shown in Fig. 1a. The chemical composition of base plate and filler wire are provided in the Table 1. In TIG welding, electrode made of 2% thoriated tungsten with diameter of 2.4 mm was used the joints by using ER409Nb filler wire. During welding a number of trail was conducted to optimise the process parameters. Welding joint was completed in two passes by using AC polarity with 140–240 V and 150–230 A current in argon gas atmosphere at flow rate of 10 L per min. Welded plate was sectioned and subjected to solution treatment at 830 °C for 1.5 h then immediately quenched into water, later subjected to ageing at 400 °C for 5 h then air cooling. All samples were mechanically polished, cleaned and etched with Kalling's no. 2 reagent of $CuCl_2$ (0.25 gm) + HCl (5 ml) + CH_3OH (5 ml) for metallography studied. The

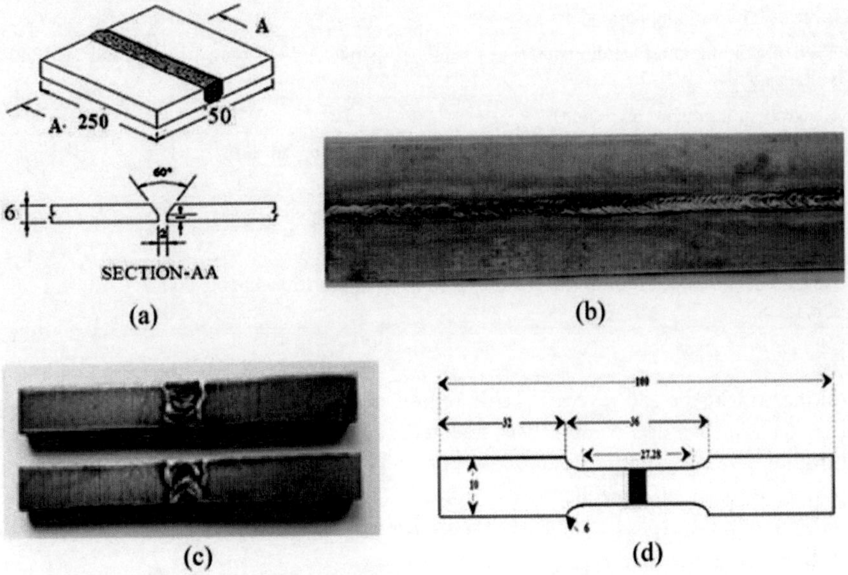

Fig. 1 **a** Schematic sketch of the TIG welded plate (in mm), **b** FSS plate (250 × 50 × 6 mm^3) was TIG welded using argon gas and ER409Nb filler wire, **c** TIG welded tensile pieces, **d** tensile test sample as per ASTM E8-sub size standard

Table 1 Chemical composition of the base material and filler wire (wt%)

Material	C	Cr	Ni	Mo	Si	P	S	Cu	Mn	Nb + Ta	Fe
Base material (AISI 430 Grade)	0.12	16–18	0.50	–	1.0	0.04	0.03	–	1.0	–	Bal.
Filler wire (ER409Nb Grade)	0.019	11.25	0.27	0.04	0.37	0.03	–	0.13	–	0.48	Bal.

GTAW (gas tungsten arc welding) or TIG (tungsten inert gas) welding was carried out using a TIG welding machine (KEMPPI MinarcTig, 180MLP). The heat input was calculated by the following equation, heat input = (V × A × 60)/S; where V is the arc voltage, A is the welding current, S is the welding speed (mm/min), coefficient of thermal efficiency of this welding ~0.6. As well as heat input was measured of 1515.78 kJ/mm. The all welding parameters are given in Table 2. In order to investigate the influence of PWHT on microstructure and mechanical properties of weldment studied in the range of 500–700 °C for 2 h for each condition. The PWHT are mostly preferred for eliminating high temperature embrittlement, tempering of martensite and hydrogen degassing [11]. The PWHT samples were carried out for optical microscopy, SEM with EDAX analysis, Vicker's hardness measurement (5 kg. load), and tensile testing. Vicker's hardness measurement was carried out thoroughly

Table 2 TIG welding parameters

Type of welding joint; welding current (A) and voltage (V)	A single V butt-joint; 150–230 and 140–240
Electrode polarity; arc gap (mm)	AC; 2
Used filler wire and diameter (mm) and Groove angle (°)	ER409Nb; 2.4; 60
Type of electrode and No. of passes	2% thoriated tungsten; 2
Shielding gas; flow rate of gas (Litres/min)	Ar; 10
Heat input (kJ/mm); Pre-heat and PWHT for 2 h, (°C)	1515.78; 150 and 500, 600, 700

on the weldment and revealed HAZ region more prone to brittleness as shown in Fig. 6a. The welded sample was selected for tensile testing which had prepared through EDM wire cutting machine as per ASTM E8 standard. The tensile testing was performed an electro-mechanically controlled Universal Testing Machine (UNITEK-94100) with the cross head speed at 5 mm/min and results are shown in stress-strain curves of Fig. 6b–d.

3 Results and Discussion

3.1 Overview of FSS Welding

Weldability of FSS is challenging and many embrittlement phenomena decreases the controlling heating cycle, heat input, and cooling rate, etc. Typically FSS plates are welded with specified ER409Nb filler wire content Ta + Nb stabilizer that could be controlled intergranular corrosion, hot cracking, and refine grains as the result of enhancement of mechanical properties [12, 13]. The investigation of FSS (430 grade) has been recommended many literatures such as low interstitial elements C + N content $<0.04\%$, Ta $>$ 27.5 (C + N), and Ti $>$ 12.5 (C + N) these parameters are controlled hot cracking and sensitization problems of welding. The sensitization in FSS during welding can be controlled by minimizing interstitial constituents (C + N), ferritic factors, addition of stabilizing elements (Ta, Nb, Ti), and heat input and cooling rate. Typically sensitization problem will be minimized by stabilizers or applying PWHT that improving the corrosion resistance of welded materials. In TIG welding high quality weldments may possible for high degree of control in heat input and filler wire (Ta + Nb) additions separately [14–16]. This welding process can used for both manual and automatic operations as well as in high technological industries like nuclear industry, aircraft, precision manufacturing industry, and automobile industries. Figure 2a shows in Fe–Ta binary system a first eutectic reaction, L $\rightarrow$ δ-Fe + ε at 1442 °C with 7.9 at.% Ta. Figure 2b shows in Cr–C binary system a first eutectic reaction, L $\rightarrow$

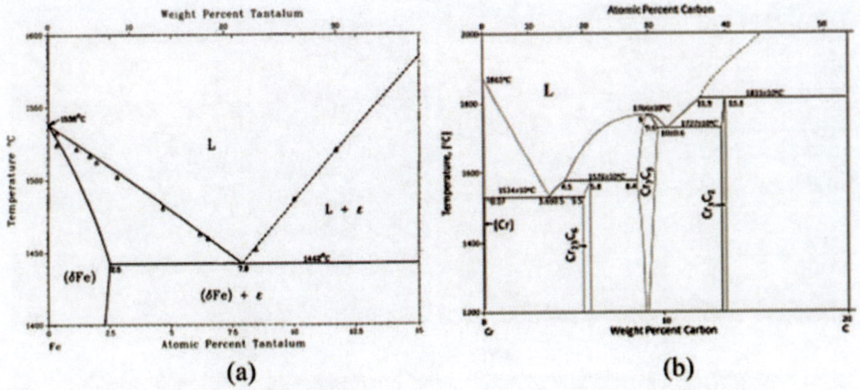

Fig. 2 **a** The Fe rich portion of the Fe–Ta system, **b** the Cr–C binary phase diagram

α-Cr + $Cr_{23}C_6$ at 1534 °C with 3.6 wt% C. Moreover, Cr–Ta alloy system an eutectic structure has a good combination of oxidation resistance and mechanical properties at high temperature (1200 °C) applications. The Cr_2X (X = Nb, Ta, Ti, etc.) Laves phase has an ordered crystal structure so that it shows good oxidation resistance and excellent mechanical properties for high temperature (>1200 °C) applications [17–19]. Laves phase forms vicinity of dislocations and grain boundaries and stabilized ferrites. In combination of Ta + Nb elements can be improved the toughness and sensitization problem due to controlling inclusion with microstructures.

3.2 Microstructure Evaluation of FSS Welding

Figure 3a–c shows the TIG welded micrographs after PWHT of FSS is exhibited remarkable changes in BM (base metal) with irregular distribution of fine grains and eliminated grain boundary segregation and may be several prior austenitic grains are dominated in the matrix. In WZ shows entirely different from BM mostly grain coarsening but precipitates are imbedded on grain boundary regions. Existing precipitates are tempered martensites and carbides (TaC, NbC) and Laves phases Cr_2Ta on the vicinity of grain boundary as shown in the Fig. 3a–b. In HAZ shows finer grains than WZ but all precipitates are restricted on the vicinity of grain boundary. Grains are remarkably changed to fine equiaxed possibly due to Cr_7C_3 and $Cr_{23}C_6$ precipitates rather Laves phases and carbides are separated on the grain boundary regions. Minor tempered martensites are coincided among the precipitates. Figure 3d–f shows the TIG welded micrographs after PWHT of FSS is exhibited irregular coarse ferritic grains with fine carbides in the BM matrix. In WZ shows coarse grains, tempered martensites, several type of carbides and accumulation of Laves phases on the grain boundary. Partially growth of precipitates are dominated owing to faster diffusion in

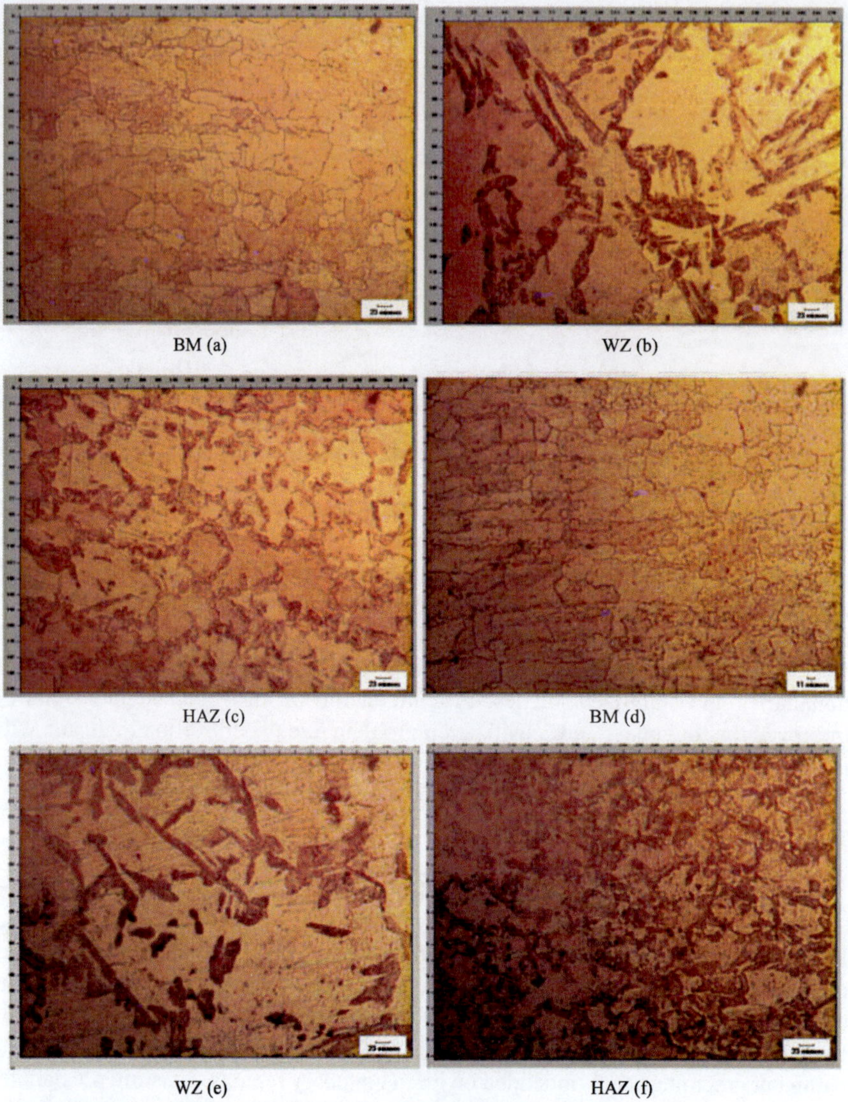

Fig. 3 Illustration of optical microstructures of TIG welded FSS after different PWHT conditions (100X): **a–c** PWHT at 500 °C for 2 h; **d–f** PWHT at 600 °C for 2 h; **g–i** PWHT at 700 °C for 2 h

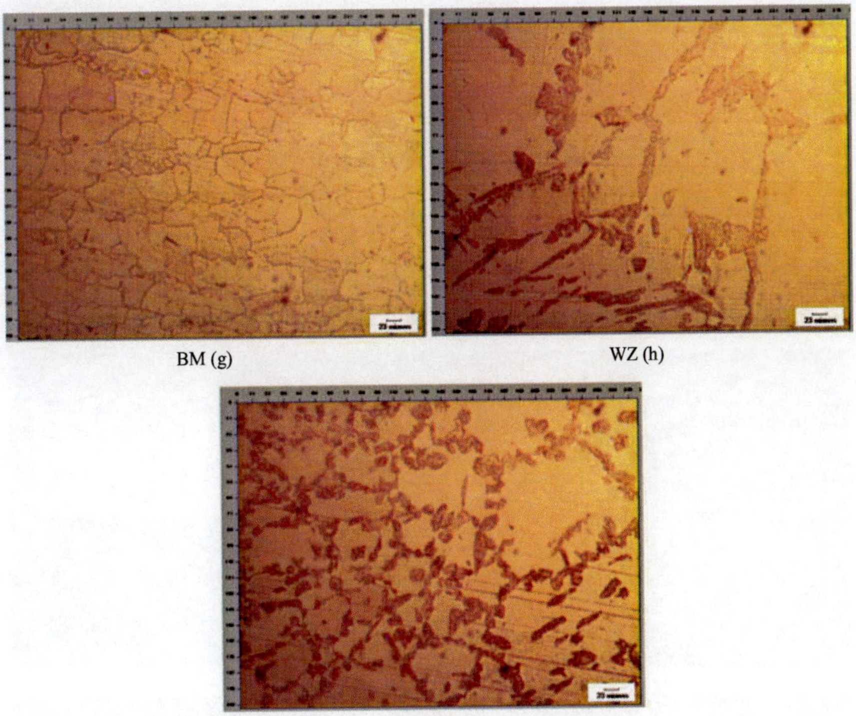

BM (g)

WZ (h)

HAZ (i)

Fig. 3 (continued)

specific direction which reduces number of precipitates in the matrix. It is possibly due to moderate PWHT and carbide stabilizing elemental effects in the matrix. In HAZ shows optimum precipitation forming effect after moderate PWHT exhibited mostly ferritic grains, maximum volume of precipitates, tempered martensites are relatively fragmented on the grain boundary regions. Figure 3g–i shows the TIG welded micrographs after PWHT of FSS is exhibited relatively coarser ferritic grains with mixed few fine grains in the vicinity of grain boundaries due to probably addition of Ta + Nb stabilizers which may appear less concentration at high PWHT condition in the BM matrix as shown in EDAX analysis (Fig. 4c). In WZ shows grain coarsening tendency increases if PWHT temperature increase with tempered martensites, carbides or Laves phases are tends to dissolved, exist needle shape structures and colony of precipitates especially in the vicinity of grain boundaries. In HAZ shows typically finer grains and precipitates are fully imbedded in vicinity of grain boundary. Precipitates are surprisingly arranged segment wise on the grain boundary probably reaction of Ta stabilizer with Laves phase effects in the matrix.

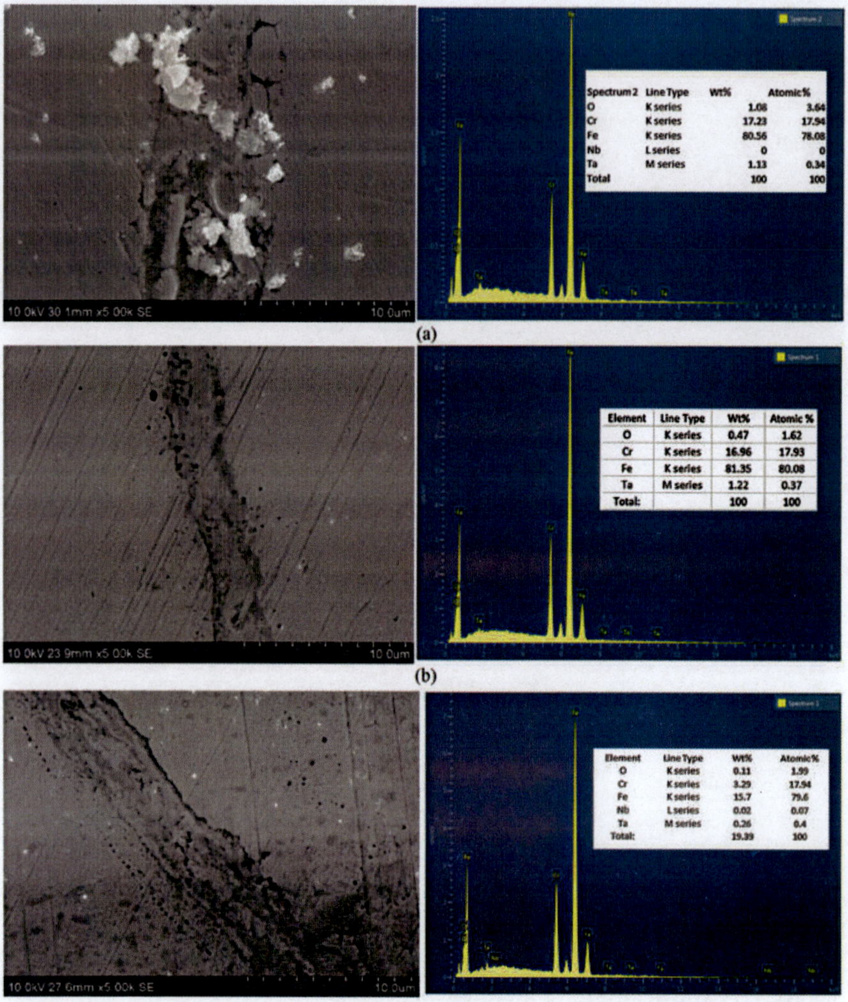

Fig. 4 The SEM with EDAX analysis of TIG welded FSS after different heat treatment conditions: **a** PWHT at 500 °C for 2 h, **b** PWHT at 600 °C for 2 h, **c** PWHT at 700 °C for 2 h

3.3 SEM Analysis of FSS Welding

Figure 4a shows the SEM with EDAX micrograph is exhibited oxide formation, WZ discontinuity with cracks, spattering of oxides, and roughess along the fusion boundary on the weldment after PWHT at 500 °C for 2 h. It has to observed that lack of sound weldment at this condition. From the EDAX analysis it is observed that the high amount of Ta (1.13 wt%) content, and no trace of Nb content in the weldment. Figure 4b shows the SEM with EDAX micrograph is exhibited relatively sound weldment, some of porosity along the fousion boundary and narrow WZ formation at this condition. From the EDAX analysis it is observed that the high amount of Ta (1.22 wt%) content, and no trace of Nb content in the weldment. Figure 4c shows the SEM with EDAX micrograph is exhibited continuous crack formation along the fusion boundary, porosity and inside cracks also observe in the weldment. From the EDAX analysis it is observed that the relatively low amount of Ta (0.26 wt%) content, and trace of Nb (0.02 wt%) content in the weldment. It has to mentioned that at higher PWHT reveal less concentration Ta and trace Nb content in the FSS weldment.

3.4 Hardness and Tensile Testing

Figure 5a shows tensile specimens were cut along the transverse direction of welded plate and machined into geometries of 6 mm thickness and 27.28 mm of gauge length as shown in Fig. 1b, c and subsequently measured to percentage of reduction area (%RA) and percentage of ductility (%El) through Digital Vernier Caliper manually after tensile tested specimens by UTM (Model no.: UNITEK-94100) as shown in Fig. 5b. Vicker's hardness measurement of FSS after PWHT at different temperatures were carried out using 5 kgf. with 15 s dwell time as shown in Fig. 6a. The hardness was taken thoroughly in the specific zones such as base metal (BM), HAZ (heat affected zone) and weld zone (WZ), respectively. Figure 6a shows four segments bar diagram have exhibited without PWHT and with PWHT at different temperatures from 500 to 700 °C. First segment bar diagrams or without PWHT condition show maximum hardness due to non-homogeneous, segregated, and carbo-nitrides type of precipitates and Laves phases and locked the residual stresses in the as-welded structure. Owing to high residual stresses WZ shows about 14% more hardness than HAZ in this condition. But after PWHT conditions in all segments exhibited better hardness in HAZ than WZ in the matrix. Second segment bar diagrams show moderate hardness in HAZ that the about 20% more than WZ due to formation of fine precipitates such as NbC or Nb (C, N) and martensites tempered with fragmentation and Laves phases Cr_2Ta, TaC which are effectively promoted fine grains in the HAZ as shown in Fig. 3b, c. Third segment bar diagrams show uniform hardness in

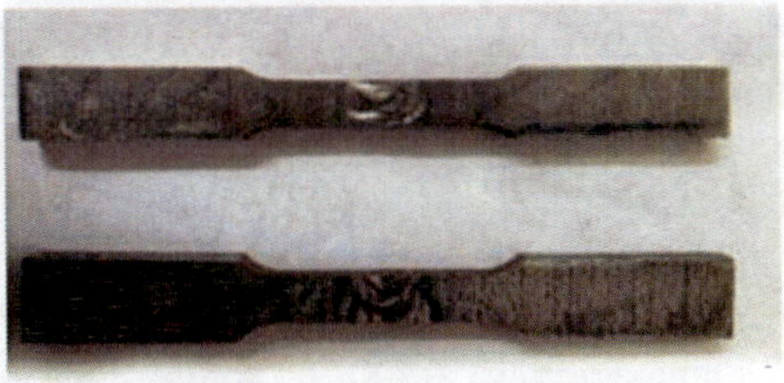

(a)

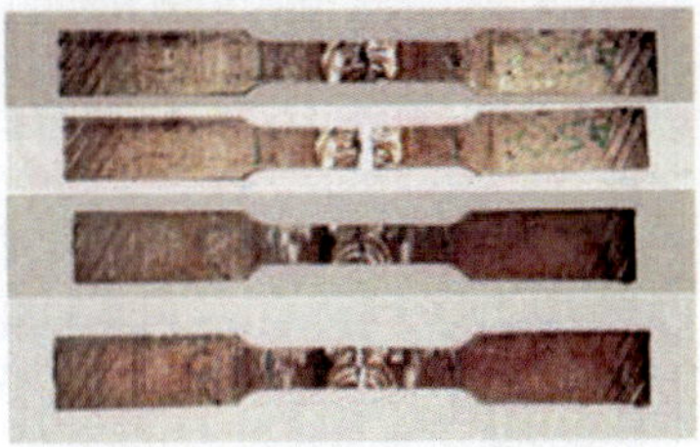

(b)

Fig. 5 **a** Tensile sample prepared by EDM wire cutting machine (ASTM E8), **b** tested tensile samples

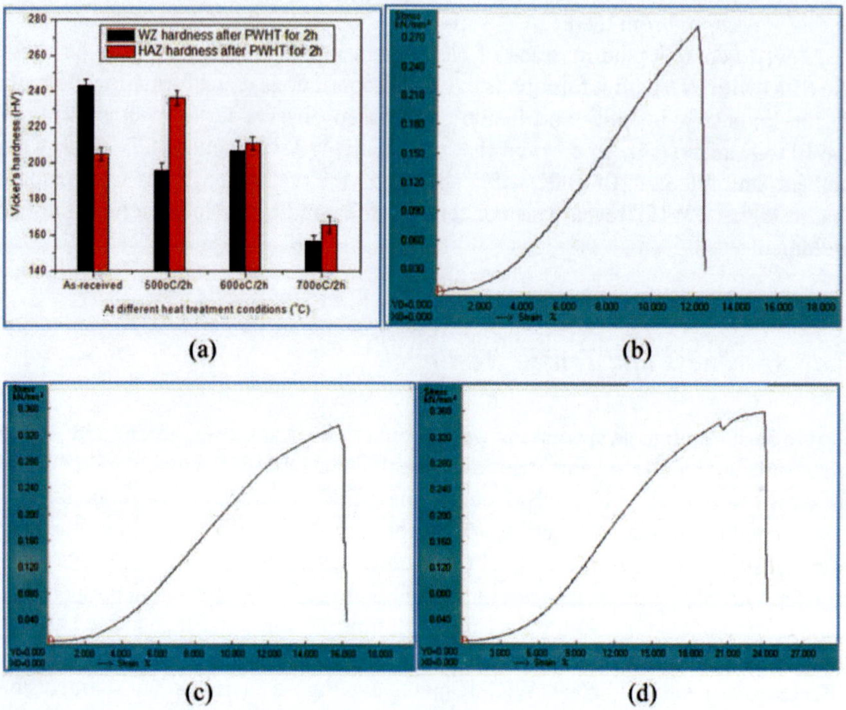

Fig. 6 The illustration of bar diagrams and stress-strain graphs of TIG welded FSS at different PWHT conditions: **a** Vicker's hardness bar diagrams at different conditions, **b** after PWHT at 500 °C for 2 h, **c** after PWHT at 600 °C for 2 h, and **d** after PWHT at 700 °C for 2 h

both of the cases owing to grain coarsening and hard precipitates softening effects prominent in the WZ and HAZ exhibits maximum volume of precipitates than any other PWHT conditions as shown in Fig. 3e, f. Four segment bar diagrams show lowest hardness due to grain coarsening effect prominent and faster dissolution of hardening precipitates (TaC, NbC and Cr_2Ta) for high PWHT as shown in Fig. 3h, i. Figure 6b–d shows the engineering stress-strain curves of tensile specimen after different PWHT conditions, where strength represents the ability of a material to resist external deformation from the constant uniaxial force, and elongation represents the ability of a material to deform without fracture under the same force [20–22]. Hence, the optimum tensile properties have obtained after PWHT at 700 °C for 2 h likely to YS of 296 MPa, UTS of 348 MPa, ductility of 8.46%, and RA of 33.47% and YS to UTS ratio of 0.80, respectively. The enhancement of tensile properties due to refined microstructures, functioning of Ta + Nb stabilizers and Cr_2Ta Laves phases have an ordered crystal structure exhibits good oxidation resistance as well as excellent mechanical properties at high temperatures (>1200 °C) applications. In weldment shows typically homogenised and few precipitates are colonized on the grain boundary regions in WZ and precipitates are orderly arrangements on the grain boundary regions due to Ta + Nb stabilizers synchronized by Laves phases in HAZ of matrix.

4 Conclusions

1. FSS in Fe–Cr-C alloys with sufficient Cr, Si, Nb, and Ta are used to prevent the formation of austenite as well as carbide forming elements on heating.
2. During TIG welding Ta + Nb filler wire had used and effectively confirmed by SEM with EDAX analysis after PWHT and subsequently microstructural changes during welding.
3. SEM analysis was done after different PWHT and found lack of fusion, particularly on the fusion boundary, oxide formation and several defects confirmed over weldments.
4. In WZ micrograph after PWHT (500 °C for 2 h) shows mainly grain coarsening but precipitates are imbedded on grain boundary regions. Existing precipitates are tempered martensites and carbides (TaC, NbC) and Laves phases Cr_2Ta on the vicinity of grain boundary.
5. In WZ shows grain coarsening tendency increases if PWHT (700 °C for 2 h) temperature increase with tempered martensites, carbides or Laves phases are tends to dissolved, exist needle shape structures and colony of precipitates especially in the vicinity of grain boundaries.
6. Vicker's hardness was taken thoroughly in the specific zones such as BM, HAZ and WZ, respectively. Hardness after PWHT at 500 °C for 2 h shows moderate value in HAZ that about 20% more than WZ due to formation of fine precipitates such as NbC or Nb (C, N) and martensites tempered with fragmentation and Laves phases Cr_2Ta, TaC which are effectively promoted fine grains in the HAZ.
7. But hardness after PWHT at 700 °C for 2 h shows lowest hardness due to grain coarsening effect prominent and faster dissolution of hardening precipitates (TaC, NbC and Cr_2Ta) at the high temperature.
8. The stress-strain curves are revealed the maximum tensile properties after PWHT at 700 °C for 2 h likely to YS of 296 MPa, UTS of 348 MPa, ductility of 8.46%, and RA of 33.47% and YS to UTS ratio of 0.80, respectively. The improvement of tensile properties due to Ta + Nb contents, refined microstructures, and formation of Cr_2Ta Laves phases have outstanding mechanical properties and good oxidation resistance for high temperatures (>1200 °C) applications.

Acknowledgements The authors would like to sincere thanks to Mr. Savyasachi N and Mr. Vishnu KC for conducted tensile testing in the Mechanical Testing Laboratory at the Department of Metallurgical and Materials Engineering, Amal Jyothi College of Engineering, Kanjirappally, Kerala, India for providing all facilities.

References

1. Porchilamban S, Amaladas JR (2019) Structural relationships of metallurgical properties influenced by Ni-based fillers on gas tungsten arc welded ferritic/austenitic SS dissimilar joints. J Adv Mech Des Syst Manuf 13(1):1–21
2. Muhammad AK, Shahid Z, Mohd NT, Saeed B, Syed M, Abdoulhdi ABO (2017) Effect of welding phenomenon on the microstructure properties of ferritic stainless steel-a review. J Adv Res Mater Sci 32(1):13–31
3. Arun Mani A, Senthil Kumar T, Chandrasekar M (2015) Mechanical and metallurgical properties of dissimilar welded components (AISI 430 Ferritic-AISI 304 Austenitic stainless steels) by CO_2 laser beam welding (LBW). J Chem Pharm Sci 6:235–338
4. Bhavsar AN, Patel VA (2016) Influence of process parameters of TIG welding process on mechanical properties of SS304L welded joint. Int Res J Eng Technol (IRJET) 03(05):977–984
5. Mallaiah G, Reddy PR, Kumar A (2014) Influence of titanium addition on mechanical properties, residual stresses and corrosion behaviour of AISI 430 grade ferritic stainless steel GTA welds. Procedia Mater Sci 6:1740–1751
6. Khan MR, Pathak AK, Navin N (2017) Practical investigation of weldment of FSS (AISI 430) welded by TIG welding process. Int J Eng Manage Res 7(1):39–44
7. Balsaraf DD, Ambade SP, Patil AP, Puri YM (2013) Literature review on analysis of sensitization and corrosion of ferritic stainless steel (FSS) by different welding processes. Adv Mater Manuf Charact 3(1):263–268
8. Alhegagi FA (2015) 475 °C embrittlement in stainless steels. Int J Sci Eng Res 6(9):145–154
9. Amuda MOH, Mrida S (2011) Effect of energy input on microstructure and hardness of TIG welded AISI 430-ferritic stainless steel. Adv Mater Res 264–265:390–396
10. Amuda MOH, Mrida S (2010) Grain refinement in ferritic stainless steel welds: the journey so far. Adv Mater Res 83–86:1165–1172
11. Srivastava BK, Tiwari Dr SP, Prakash J (2010) A review on effect of preheating and/or post weld heat treatment (PWHT) on mechanical behaviour of ferrous metals. Int J Eng Sci Technol 2(4):625–631
12. Kacar R, Gunduz S (2009) Increasing the strength of AISI 430 ferritic stainless steel by static strain ageing. Kovove Mater-Metal Mater 47:185–192
13. Devakumar D, Jabaraj DB (2014) Research on gas tungsten arc welding of stainless steel–an overview. Int J Sci Eng Res 5(1):1612–1618
14. Jurica M, Kozuh Z, Garasic I, Busic M (2018) Optimization of the A-TIG welding for stainless steels. Mater Sci Eng 329:1–9
15. Subbaiyan K, Kalaiyarasan V, Khan MAG (2017) Experimental Investigations and weld characterization analysis of single pass semi automatic TIG welding with dissimilar stainless steel. Int J Mech Eng Technol 8(5):542–555
16. Kumar A, Sharma V, Baruaole NS (2017) Experimental investigation of TIG welding of stainless steel 202 and stainless steel 410 using Taguchi technique. Int J Comput Eng Res 1(2):98–101
17. Rajkumar VB, Hari Kumar KC (2015) Gibbs energy modeling of Fe-Ta system by Calphad method assisted by experiments and ab initio calculations, CALPHAD. Comput Coupling Phase Diagr Thermodyn 48:157–165
18. Wieczerzak K, Bala P, Stepien M, Cios G, Koziel T (2015) The characterization of cast Fe–Cr–C alloy. Arch Metall Mater 60(2):779–782
19. Dudek A, Lisiecka B, Ulewicz R (2017) The effect of alloying method on the structure and properties of sintered stainless steel. Arch Metall Mater 62(1):281–287
20. Delgodo JA, Ambriz RR, Cuenca-Alvarez R, Alatorre N, Curiel FF (2016) Heat input effect on the microstructural transformation and mechanical properties in GTAW welds of a 409L ferritic stainless steel. Revista De Metalurgi 52(2):1–10

21. Zakaria B, Soumia H, Vincent J (2016) Effect of heat treatment on the microstructural evolution in weld region of 304L pipeline steel. J Thermal Eng 2(6)(5):1017–1022
22. Rajadurai MS, Naveen S, Afnas M, Arun T, Kumar N, Surendhar S (2015) Methods to avoid material sensitization during welding for developing corrosion resistant exhaust system. Int J Recent Dev Eng Technol 4(7):23–36

Development of Reduced Order Strain Model for Life Assessment of a Gas Turbine Rotor Blade

Sri Shanti Potluri, Shyam Kishore, R. Isai Thamizh, and B. V. A. Patnaik

Abstract Compressor rotor blades of Gas Turbine Engines are critical components in the flow path. Their durability is determined by Life and integrity at different operating conditions. These blades experience stresses due to body loads and complex bending stresses due to the aerodynamic work done. While it is essential to establish adequate safety margins in the rotor blade during the design phase, it is equally important to monitor the life that is being consumed in actual operation at different combinations of mass flow rate, pressure and speeds depending on the aero-thermodynamics at various altitudes and manoeuvre conditions. A Meta model is developed with strain as the response surface in terms of mass flow rate and rotational speed as the controlling parameters. Design of Experiments using FEM is conducted for nine combinations of these parameters from which a generic strain model is constructed. This model is capable of predicting the strain in the rotor blade for any engine operating point which in this case is essentially a function of the mass flow and rotational speed of the rotor blade. At the same time, few of these rotor blades have been strain gauged during developmental engine tests for strain/health monitoring. Strain data obtained from engine testing at various operating points is utilized to correlate the aforementioned response model developed wherein good correlation is observed and the error is within 5%. Strain responses predicted from the meta model at respective engine operating points are then used to arrive at the strain ranges the blade would encounter during operation. Rain flow cycle counting method in conjunction with life models are used to arrive at the damage fraction in the rotor blade during each engine test cycle. Damage is cumulated over repeated usages in the engine to arrive at residual life of the rotor blade.

Keywords Response surface method · DOE · Rotor blade · Strain · FEM · Mass flow · Rotational speed

---

S. S. Potluri · S. Kishore (✉) · R. I. Thamizh · B. V. A. Patnaik
Gas Turbine Research Establishment, Defence Research and Development Organization, Bangalore, India
e-mail: shyamkishore@gtre.drdo.in

S. Seetharamu et al. (eds.), *Fatigue, Durability, and Fracture Mechanics*, Lecture Notes in Mechanical Engineering, https://doi.org/10.1007/978-981-15-4779-9_8

1 Introduction

This paper gives details of correlation studies with respect to static strain measurements of a Low pressure compressor (LPC) rotor blade of an aero gas turbine engine vis-à-vis strain response obtained from meta models constructed based on Finite Element stress analysis. General purpose finite element software ANSYS is used for stress analysis and arrive at strain responses.

Earlier, work was carried out in this direction for a High pressure compressor rotor blade [1, 2] and the response surface meta models were correlated with Finite Element analysis results with an error contained within 1.5%. The current work is aimed at developing a similar model for the LPC rotor blade and correlations in this case are made with respect to actual field test data as well. A generic model is developed with strain as the response function [3]. The response surface model (RSM) is based on speed and mass flow as the controlling parameters. Strain estimates obtained from RSM model are compared with real time engine run strain measurements as well as with predicted FE analysis results and it is found that the RSM model is in close agreement with the actual measured strain in the rotor blade. This generalized RSM model is used to predict the state of strain at the critical location(s) of the rotor blade which is input to the life model to arrive at the low cycle fatigue damage in the blade due to the various operating (speed) excursions that the blade is subjected to while under test. Cumulative damage over repeated usage in the engine is accumulated and tracked with in-house software tools.

Methodology and details. The rotor blades are strain gauged to monitor the static strain/stress levels during developmental engine testing at specific strain gage mount location(s). Strain levels at other critical locations of the blade can be correlated with respect to the strain gaged location(s) from the finite element stress distribution or state of stress in the rotor blade. Detailed Finite element stress calculations are performed for the speed and mass flow combinations at design point and for few other limited points in the operating envelope as part of the structural design and analysis activity.

Response surface Meta model developed for the LPC rotor blade as part of this work shall predict the strain state in the blade for any operating condition within the band of extreme operating conditions that the blade shall be subjected to while in operation. Details of this model are given in the following section.

Response Surface Model (RSM) by DOE Method. In this model, application of DOE (Design of experiments) [4] is used to evaluate the strain response surface with Speed (RPM) and Mass flow rate as normalized parameters (between −1 and 1) w.r.t. the limit operating conditions. A total of 'nine' simulations for all the combinations of operating speed (RPM) and mass flow rate are conducted through finite element analysis. The response equation for strain is then formulated from the strain responses obtained through FE simulations and used to predict the strain for any other operating condition in the operating regime. Errors in prediction of strain obtained from the RSM model, FE analysis and engine test measured strain values are then estimated.

2 Low Pressure Compressor (LPC) Blade: Finite Element Simulation and Stress Analysis

General arrangement of the Low Pressure Compressor of a developmental engine is given in Fig. 1a and the strain gauge mount location on the rotor blade during testing is shown in Fig. 1b. The LPC consists of three stages and this work is restricted to first stage rotor blade. The rotor blades made of Titanium based super alloy are assembled in the disk through an axially inclined dovetail root.

Finite element mesh of the blade is generated in HYERMESH software with adequate care to capture the stress critical areas using eight node brick elements having three translational degrees of freedom per node. Primary loads acting on the blade due to aerodynamic flow and rotational speed are simulated on the blade. For displacement boundary condition simulation of the blade-disc attachment, the blade is fixed with respect to a local co-ordinate system on the dovetail root defined for this purpose. Here, nodes on the angled blade dovetail surface are fixed in the normal and sliding directions to simulate rigidity offered by the disc.

Nonlinear Static stress analysis for the operating loads and boundary conditions is carried out with centrifugal stress stiffening and large deformation effects [5, 6]. Initial stress analysis is carried out for two extreme (limit) conditions in the operating envelope of the LPC. Further details of stress analyses required to develop the strain based RSM for the rotor blade as per the DOE matrix of nine load cases are discussed in the following section.

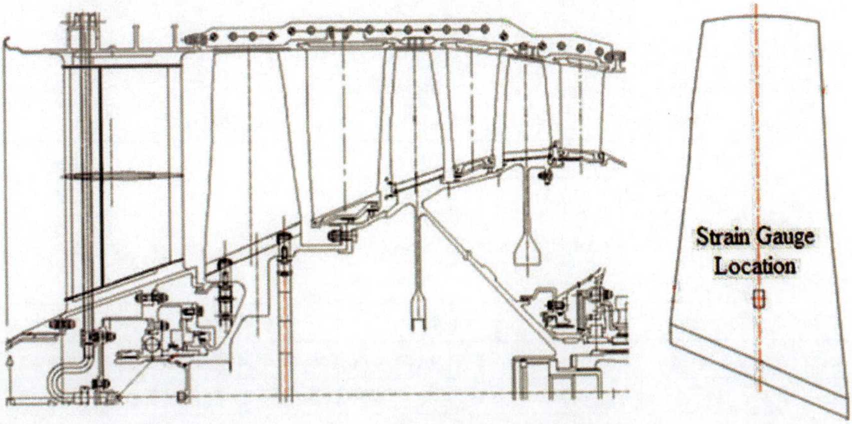

Fig. 1 **a** General arrangement of LPC, **b** Strain gauge mounting arrangement of LPC 1R

3 Response Surface Model by DOE Method

In this work, fractional factorial method is used where strain is treated as the response variable with speed and mass flow rate as the controlling parameters. These settings (speed and mass flow) are normalized between −1 and 1. As aerodynamic data corresponding to the intermediate setting (0, 0) is not available, data corresponding to the operating condition at (0.146, 0.082) is used. The DOE using FE analysis is carried out for a combination of load cases derived from the data settings. The data settings are given in Table 1 from which a DOE matrix consisting of nine load cases are formulated. The DOE matrix for stress analysis is given in Table 2.

Stress analysis is carried out for the nine DOE cases to obtain strain in the Fan blade at the strain gauge node location. The results are given in Table 3 and in Figs. 2, 3, 4, 5, 6, 7, 8, 9, 10 and 11.

The Global Least square fit method enables estimation of the trend of the DOE where curve fitting becomes necessary. The curve fit with minimal deviation from the data points which is strain in this case is desired and the best fit curve is obtained. The general form of the equation is given below:

$$\varepsilon = a_0 + a_1(x_1) + a_2(x_2) + a_3\left(x_1^2\right) + a_4\left(x_2^2\right) + a_5(x_1x_2)$$

where, a_0 to a_5 are the coefficients and x_1, x_2 are the normalized governing parameters. On using this model which is called the Response Surface Model with the strain input from the nine analysis conditions, the coefficients are obtained as given

Table 1 Data sets

Variable parameters		Variable settings (%)		
		+1	(x_1/x_2): 0.146/0.082	−1
x_1	Operating speed (RPM)	100	14.6	−100
x_2	Mass flow rate (kg/s)	100	8.2	−100

Table 2 DOE matrix

Parameters/case no.	1	2	3	4	5	6	7	8	9
x_1	1	1	1	0.146	0.146	0.146	−1	−1	−1
x_2	1	0.082	−1	1	0.082	−1	1	0.082	−1

Table 3 Strain response for the nine DOE cases of FE analysis

Case no.	1	2	3	4	5	6	7	8	9
Micro strain	1587.6	1676.5	1791.7	854.75	961.57	1099.7	89.18	224.48	398.98

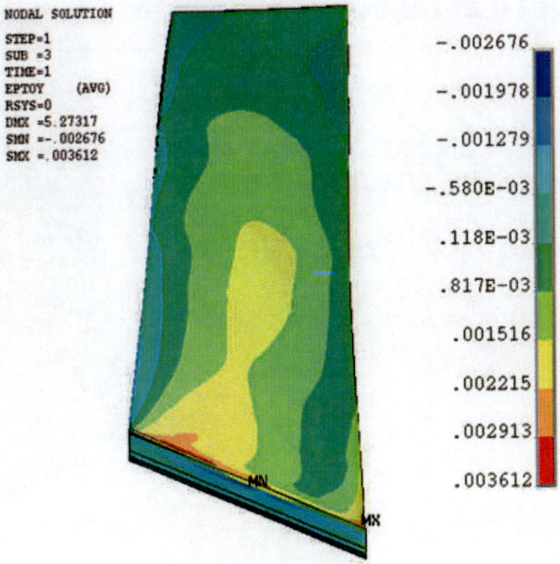

Fig. 2 Strain for case 1

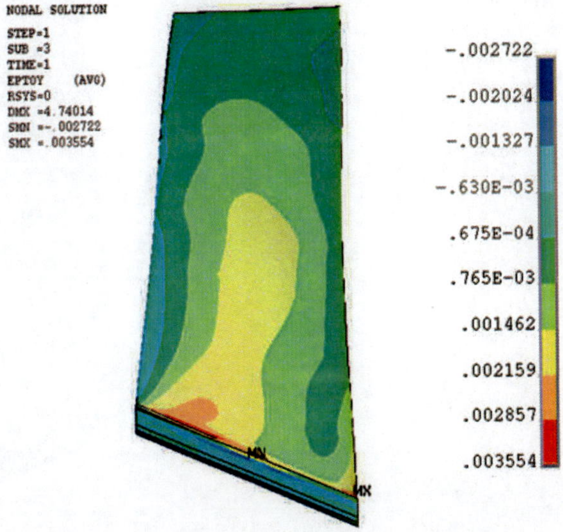

Fig. 3 Strain for case 2

in Table 4. The response model equation for strain in the LPC rotor blade at the strain gage node location is:

$$\varepsilon = 863.863 + 723.134\,(x_1) - 127.767\,(x_2) + 97.1543\left(x_1^2\right) + 5.80319\left(x_2^2\right) + 26.5575\,(x_1 x_2)$$

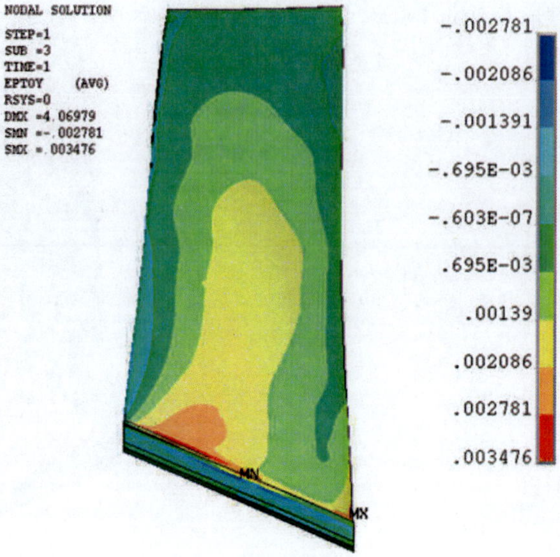

Fig. 4 Strain for case 3

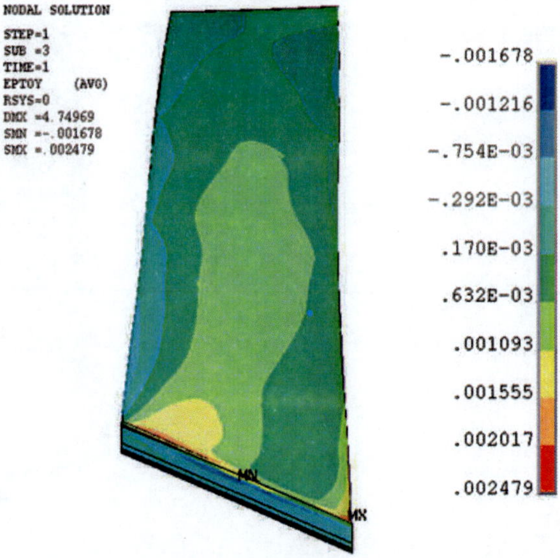

Fig. 5 Strain for case 4

This equation is used to predict the strain at the strain gauge location (as per Fig. 1b) for any other condition in the operating regime and can be in turn utilised to arrive at the state of strain at any other critical location in the blade. The error in strain is estimated from the RSE model vis-à-vis the strain obtained through Finite element method for a verification load condition (case a) and is given in Table 5.

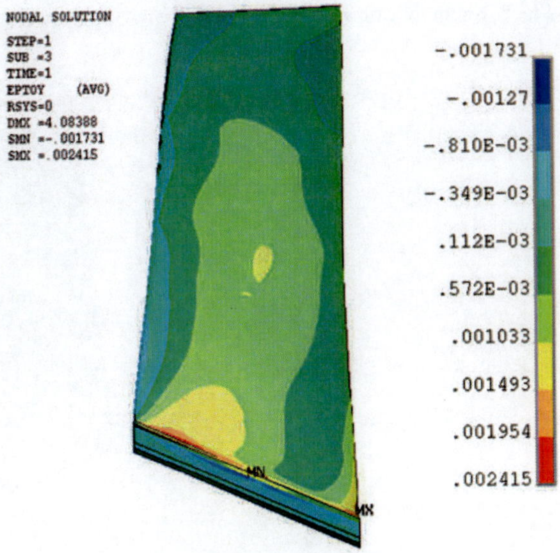

Fig. 6 Strain for case 5

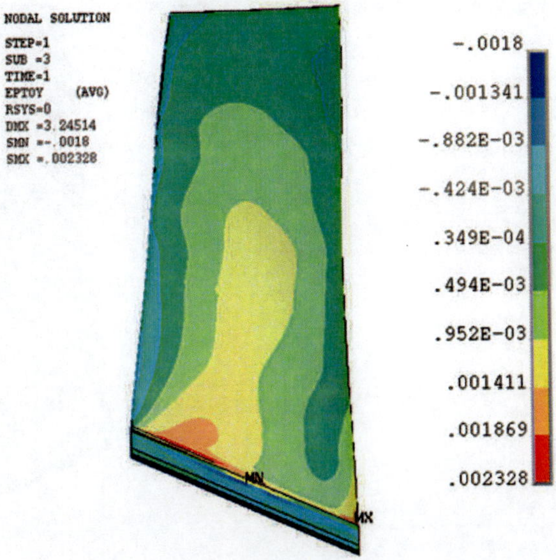

Fig. 7 Strain for case 6

Also, error w.r.t. strain corresponding to engine test condition vis-à-vis the strain obtained through the RSE model for verification cases b, c and d is given in Table 6.

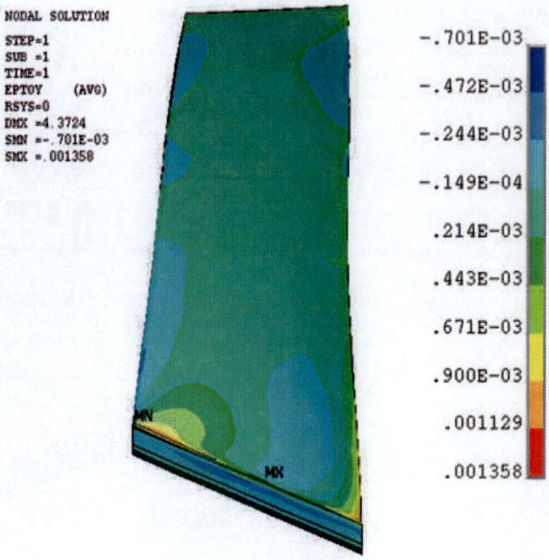

Fig. 8 Strain for case 7

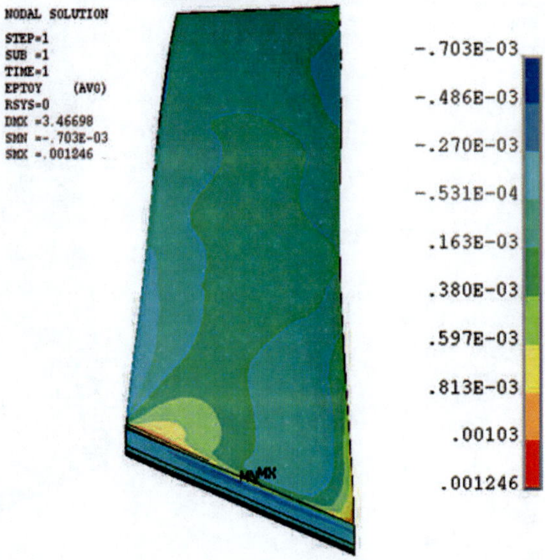

Fig. 9 Strain for case 8

4 Life Monitoring

Real time engine run data is filtered to arrive at the speed/mass flow versus time data and interfaced with the RSM model of the blade. Strain responses predicted from the meta model at respective engine operating points are used to arrive at the strain ranges the blade would encounter during operation. Rain flow cycle counting method

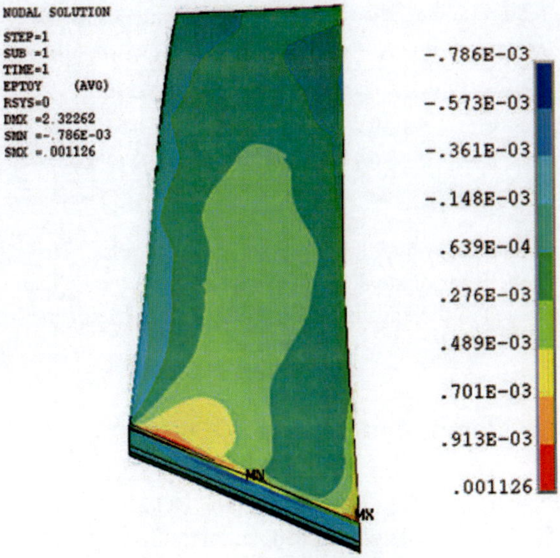

Fig. 10 Strain for case 9

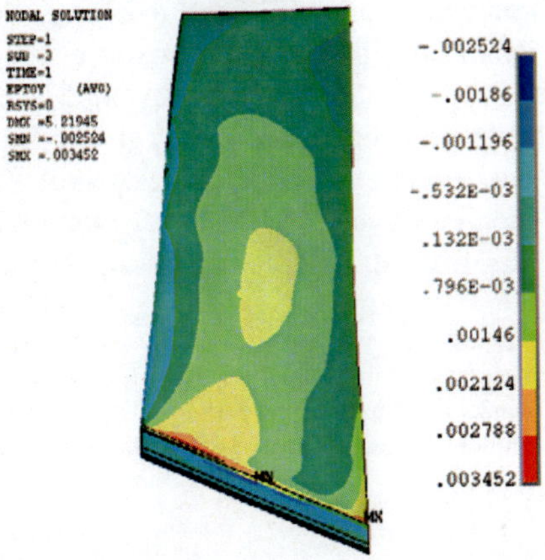

Fig. 11 Strain for case a

Table 4 Coefficients of equation based on best curve fit

Coefficient	a_0	a_1	a_2	a_3	a_4	a_5
Value	863.863	723.134	−127.767	97.1543	5.80319	26.5575

Table 5 Strain response verification case with FE analysis

Verification	x_1	x_2	RSM (Strain)	FEM (Strain)	%error
Case a	0.91366	1.3447	1476.97	1469.7	0.49

Table 6 Strain response verification case with FE analysis

Verification	x_1	x_2	RSM (Strain)	Test data (Strain)	%error
Case b	0.5434	08154	1196.93	1160	3.18
Case c	−0.0881	−0.0475	807.13	810	0.35
Case d	0.9794	1.3044	1542.42	1480	4.22

in conjunction with life and material models are used to arrive at the damage fraction in the rotor blade during each engine test cycle. Damage is cumulated over repeated usages in the engine to arrive at residual life of the rotor blade [7].

5 Conclusion

Response surface method based on DOE is used to evaluate the strain response in the LPC rotor blade with speed and mass flow rate as normalized parameters. Predicted strain from this model is correlated with strain obtained from FE analysis for a validation case and found to be less than 1%. Also, the strain from this model is validated with strain from engine test conditions and the error is found to be less than 4.3%. The RSM equation developed based on DOE can be used to predict the strain state in the rotor blade for any given operating condition i.e. for a set of speed and mass flow parameters. The method serves as an effective tool towards life consumption monitoring of the gas turbine rotor blade undergoing different operating conditions at various altitudes and manoeuvre conditions.

Acknowledgements The authors express their sincere gratitude to the Director, Gas Turbine Research Establishment (GTRE) DRDO for his kind permission to publish this work. Our sincere thanks are due to our Additional Director, colleagues of SMG, Strain Gage group, Compressor Module, Engine testing and all others who have supported this activity.

References

1. Sri Shanti P, Isai Thamizh R, S Shanmugam, Patnaik BVA (2011) Evaluation of stress response in the compressor blade of a gas turbine engine. In: International conference PLMSS, Coimbatore, December 2011
2. Saravanamuttoo HIH, Rogers GFC, Cohen H (2008) Gas turbine theory, 5th edn. Pearson Education
3. Box GEP, Draper NR (1987) Empirical model building and response surfaces. Wiley, New York
4. Montgomery DC (1997) Design and analysis of experiments
5. ANSYS documentation, ANSYS Inc
6. Srinath LS (2009) Advanced mechanics of solids, 3rd edn. McGraw-Hill Education
7. Collins JA (1993) Failure of materials in mechanical design analysis prediction prevention. Wiley, New York

Comparative Study of Controlled and Fractured Concrete Model for FRP Confined Circular Short Columns

Sachin M. Kulkarni and K. G. Vishwanath

Abstract The experimental and analytical study on calculation of ultimate strength and deformation of plain cement concrete strengthened by wrapping with fiber reinforced polymer (FRP) for circular short column is presented. The study emphasizes on comparison of analytical strength obtained by Abaqus® (finite element tool) with theoretical mathematical equations formulated by researchers Mander et al., Girgin and Richart et al. This paper focuses on the validation of deformation and strength by experimental, analytical and mathematical results. The validation assessment is made with virgin and fractured specimen strengthened with carbon FRP (CFRP) and incorporating the similar boundary conditions in analytical tool (Abaqus®). The study highlights on good agreement between analytical model with experimental deformation results and strength that are validated with proposed models proposed by researchers Mander et al. [1], Girgin [2] and Richart et al. [3] The study has helped in evaluating the confinement value (k) for PCC for the development of confined compressive strength using the empirical formula.

Keywords FRP confined concrete strength · Design oriented models · Analytical · Concrete damage plasticity (CDP) · Abaqus®

1 Introduction

Externally bonded FRP are used to enhance the strength and reduce the deformation of axially loaded structural element. This technique is needed, to improve structural efficiency after deterioration and aging of structures.

In such situations, demolition of entire structure or part of the element will be uneconomical and un-safe in-terms of financial and environment constraint. For understanding the behavior of element in such situation, analytical solutions using finite element tools (Abaqus®) plays a vital role. Finite element models needs to be

S. M. Kulkarni (✉) · K. G. Vishwanath
Jain College of Engineering, Belagavi 590014, Karnataka, India
e-mail: sachinkulkarni1@gmail.com

S. Seetharamu et al. (eds.), *Fatigue, Durability, and Fracture Mechanics*, Lecture Notes in Mechanical Engineering, https://doi.org/10.1007/978-981-15-4779-9_9

validated with experimental and mathematical solutions for deformation and strength when subjected to axial compressive load.

A detailed research is carried out by considering the compression element (Column) subjected to axial load, considering suitable boundary conditions in terms of assigning loads, support specification and material assignment for concrete and FRP in Abaqus software. Strength values are validated with proposed mathematical equations by Mander et al. [1], Girgin [2] and Richar et al. [3].

2 Methodology

In the present research, a study is performed to validate the results of experimental, analytical and mathematical approaches. Experimental study is conducted by casting cylindrical specimens of size 150 mm diameter and 300 mm height. The test specimens are subjected to compressive loading under UTM and deformation are measured using displacement dial gauge of least count 0.01 mm. The test setup is as shown in Fig. 1.

The specimens casted, are divided to three categories as controlled/virgin PCC (without wrapping), Controlled/virgin PCC strengthened with CFRP and Fractured PCC with CFRP. All the specimens are studied under compressive loading. The controlled unwrapped specimen will be an example for in situ column sections highlighting deformation and strength values. The strength of controlled PCC unwrapped specimen is enhanced by wrapping the virgin specimen with CFRP of 0.129 mm thickness having modulus of elasticity 230,000 MPa as material detail furnished by supplier Sika India Pvt. Ltd. The third category will be in situ condition where

Fig. 1 Experimental setup of circular PCC specimen

Table 1 The following element properties and parameters are assigned for Abaqus®Analytical model

Plain cement concrete element	Modulus of elasticity = 22360.68 MPa	3D Beam element	Element:: 8 node brick element reduced integration (Designated as C3D8R)	Damage plasticity Parameters: Dilation angle = 31 Eccentricity = 0.1 $\frac{f_{bo}}{f_{co}} = 1.16$ K = 0.667 Viscosity parameter = 0
Carbon fiber reinforced polymer	Modulus of elasticity = 230,000 MPa	Thickness = 0.129 mm	Shell element of S4R family	

element needs to be strengthened after deterioration of structure due to environmental or aging factor. To correlate the same a virgin specimen is subjected to loading to initiate the crack and then unloaded, the same specimen is strengthened by wrapping with CFRP and then reloaded under compressive loading till failure. The strengthened specimens for controlled and fractured will have more strength and less deformation in comparative with uncontrolled specimen because of confinement provided by CFRP and resistance offered by confined element as hoop strength.

2.1 Analytical Model

For analytical (Abaqus®) model, 8 noded brick element (C3D8R) with reduced integration is considered. For modeling of FRP shell element is adopted with thickness of 0.129 mm. One surface end of the cylinder is encastered as per prevailing boundary condition on field and laboratory; the other surface end of the cylinder is subjected with compressive force. The *x*, *y* and *z* coordinate systems are noted as 1, 2 and 3 respectively in analytical tool Abaqus®. As load is applied in negative *z* direction, analytical output results U3 and S3 are deformation and strength in *z*–*z* direction. The individual element properties are as per Table 1 (Fig. 2).

3 Results and Discussion

The comparative studies of categorized specimens are studied after the validation of specimens and models. The results are validated for unwrapped PCC, wrapped PCC and fractured PCC specimens. The validated values are tabulated for comparative study.

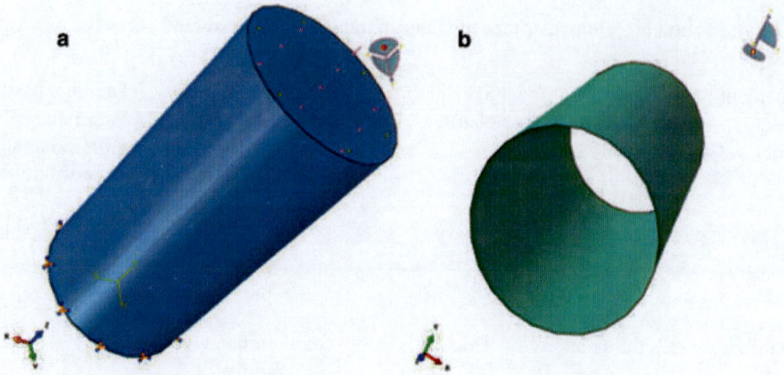

Fig. 2 a,b Concrete and FRP model respectively

The analysis results from Table 2, highlights about the appropriate selection of elements for models adopted in Abaqus® which is evident with validation results of deformation and strength as noted in Tables 2 and 3 respectively. The comparative analysis is represented in Graphs 1 and 2. The validation results of PCC element helps in adopting the same element to study enhancement of capacity of un-strengthened or fractured element.

The study on strengthening of controlled or virgin PCC element is studied by wrapping the specimen externally with CFRP and the same is simulated using the analytical tool (Abaqus®). The contact surface bonding is ensured and the results are validated for deformation. The enhancement of strength is verified with various theories proposed by different researcher depending on their respective experimental findings. The Abaqus® result obtained has good acceptance with the theories proposed. From the model study it is observed that the deformation is restrained in comparison to unwrapped PCC element and strength is enhanced against PCC element. The change in property and enhancement in strength values is exhibited for the reason of confinement provided by FRP which help in resisting dilation effect of concrete, FRP provided around the periphery help in enhancing the strength of the element. From Tables 4 and 5 the values of deformation and strength are listed and verified with mathematical solution as mentioned. The study on validation of deformation values is highlighted in Graph 3. From Graph 4 the values obtained from analytical tool are verified with model proposed and arrived with output with adoption of model that can be conveniently used for further analysis (Table 5 and Graph 5).

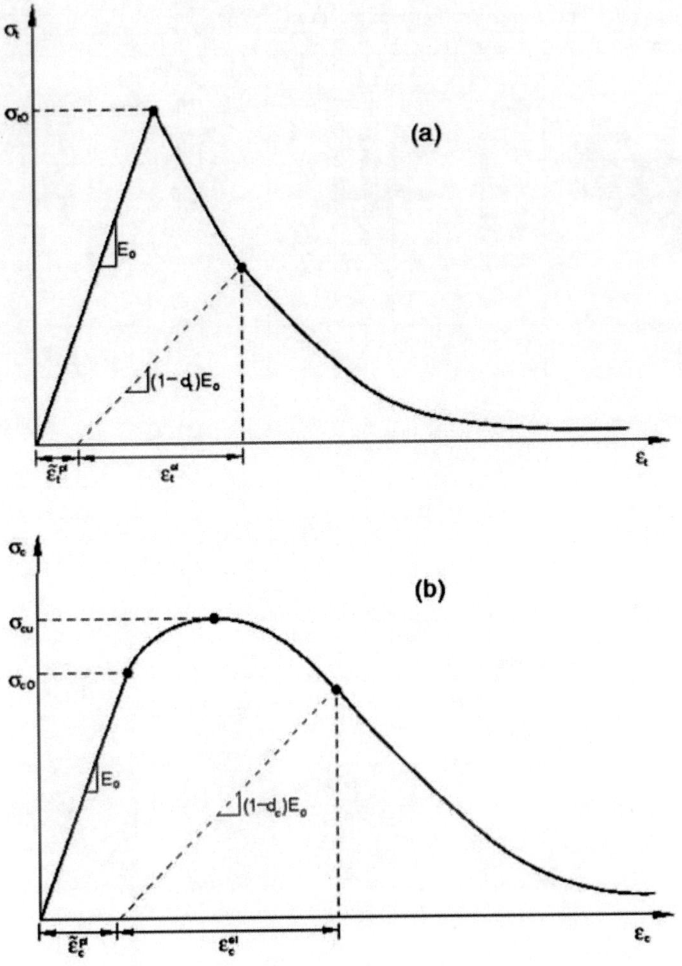

Fig. 3 Concrete under axial tension (**a**) and tension (**b**) strength (Abaqus® user manual) [4]

The study of strengthened failed PCC element is studied by comparing the accuracy of model with mesh seeding, dilation angle and fracture energy of element. The Abaqus® results are validated with the mathematical solutions as per the equation listed in Tables 6 and 7. Concrete damage plasticity (CDP) is one of the possible constititutive models to predict the behavior of concrete by introducing the scalar damage variables as d_c and d_t which can vary from 0 to 1 which is represented in Fig. 3 showing tensile and compressive response of concrete [5]. For undamaged material the degradation coefficient values shall be considered as 0 where damaged material total loss of strength, the values shall be considered as 1. The model has good validation of deformation and strength Experimental and mathematical equations.

Table 2 Validation of deformation (mm) for plain cement concrete

Validation of Deformation (mm) for Plain Cement Concrete (Controlled Specimen without wrapping)				
Load	**100kN**	**200kN**	**300kN**	**360kN**
Abaqus® 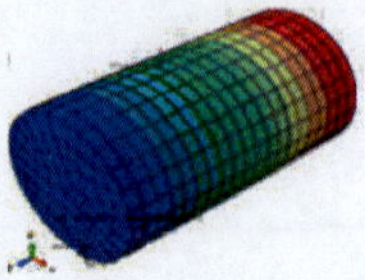	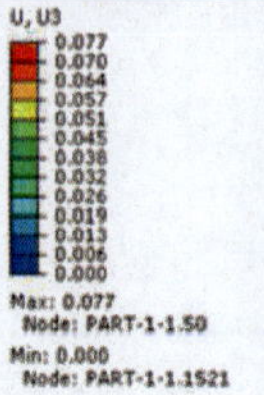	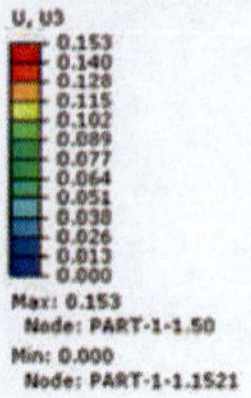	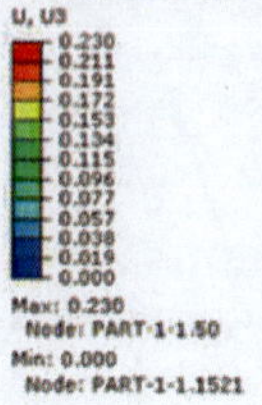	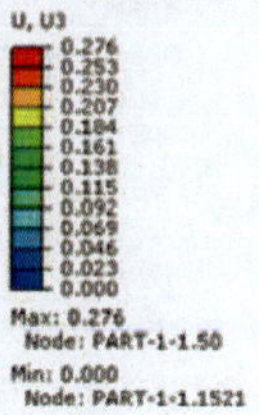
Mathematical	$\Delta=\frac{PL}{AE}$=0.0759 mm P= Load (N) L= 300mm $A=\frac{\pi}{4}\times 150^2$ =17671.46 mm^2 $E=E_c=5000\sqrt{f_{ck}}$ $=5000\sqrt{20}$ =22360.68N/mm^2	0.152	0.23	0.273
Experimental	0.076	0.15	0.22	0.27

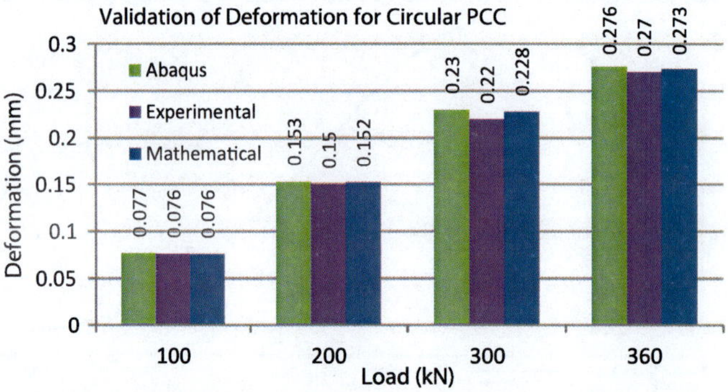

Graph 1 Validation of deformation for circular PCC

In CDP model, 'Ψ' is dilation angle measured in the p–q plane, at high confining pressure, 'ε' is an eccentricity of the plastic potential surface with default value as 0.1, and the ratio of initial biaxial compressive yield stress to initial uni-axial compressive yield stress is represented as 'σ_{bo}/σ'_{co} (Tables 8, 9 and Graphs 6, 7 and 8).

Table 3 Validation of strength (N/mm^2) for plain cement concrete

Validation of Strength (N/mm^2) for Plain Cement Concrete (Controlled Specimen without wrapping)				
Load	100 kN	200 kN	300 kN	360 kN
Abaqus® 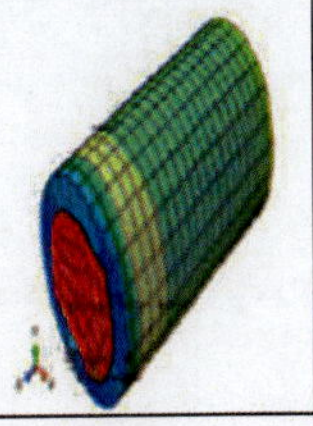	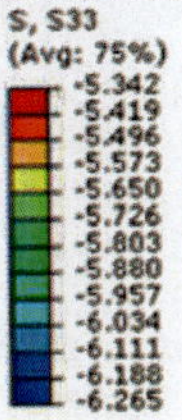	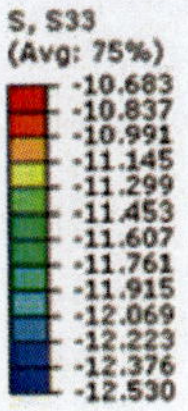	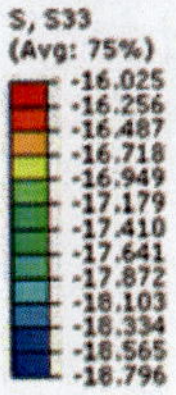	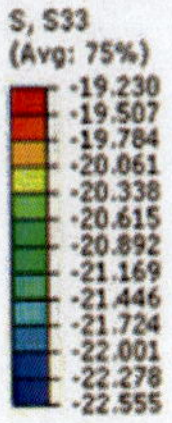
Mathematical	$\sigma=\frac{P}{A}=5.66$ P=Load (N) $A=\frac{\pi}{4}\times 150^2$ =17671.46 mm^2	11.32	16.98	20.37
Experimental	5.65	11.32	16.98	20.37

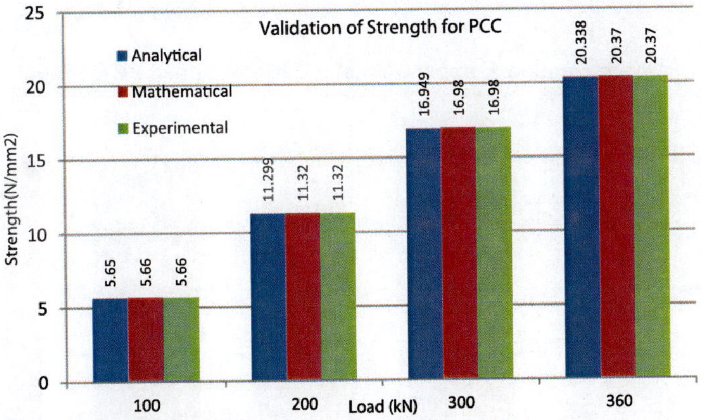

Graph 2 Validation of strength for PCC

4 Summary and Conclusion

The PCC specimens casted and modeled are validated by Experimental, Mathematical and Analytical procedures with minimal variance due to the hypothetical boundary conditions prevailing in case of analytical and mathematical solutions. The validation result indicate that the C3D8R brick element with reduced integration would be the best element that can be adopted for concrete, likewise the selection of shell element of S4R element would be the best choice for FRP. The correct selection of modeling element will help in saving time, material, for studying more iteration cycle under different loading and boundary conditions. From this study , we can

Table 4 Validation of deformation (mm) for wrapped controlled plain cement concrete specimen

Validation of Deformation (mm) for Wrapped Controlled Plain Cement Concrete Specimen					
Load	100 kN	200 kN	300 kN	400 kN	410 kN
Abaqus® 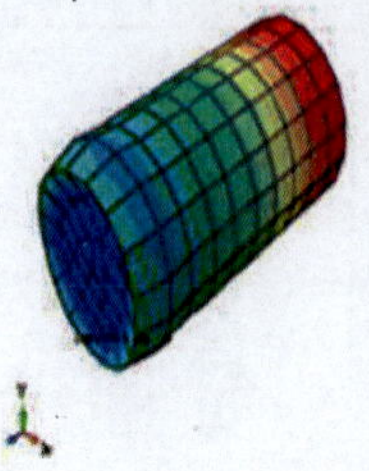	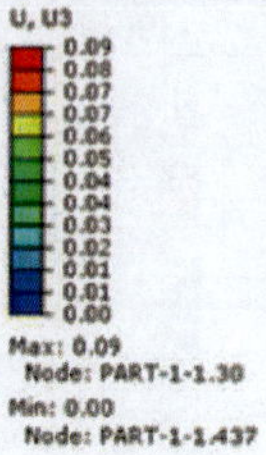	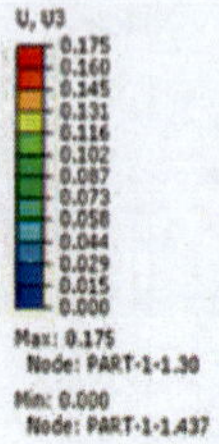	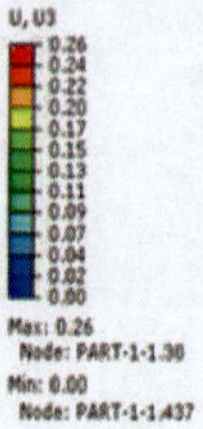	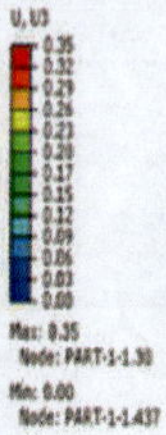	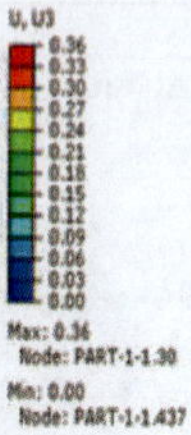
Mathematical	$\Delta=\frac{PL}{AE}= 0.073$ P= Load (N) L= 300mm $A=\frac{\pi}{4}\times 150^2$ $=17671.46\ mm^2$ $E=E_c=5000\sqrt{f_{ck}}$ $=5000\sqrt{20}$ $=22360.68 N/mm^2$	0.146	0.218	0.291	0.298
Experimental	0.075	0.15	0.23	0.3	0.31

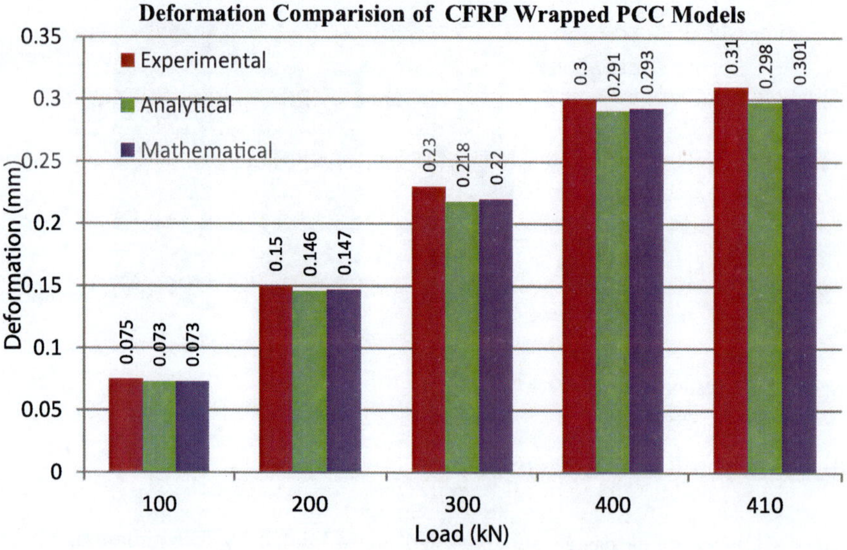

Graph 3 Deformation comparison of CFRP wrapped PCC models

conclude that the model and equations adopted will be the apt procedure to carry out the strengthening of compressive element in absence of experimental results.

Table 5 Validation of strength (N/mm^2) for wrapped controlled plain cement concrete

Validation of Strength (N/mm^2) for Wrapped Controlled Plain Cement Concrete					
Load	**100 kN**	**200 kN**	**300 kN**	**400 kN**	**410 kN**
Abaqus* 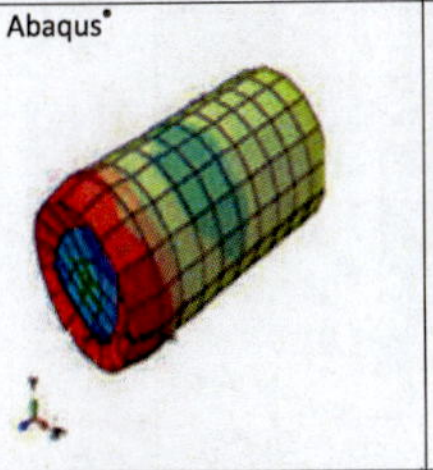	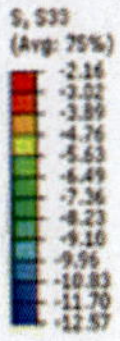				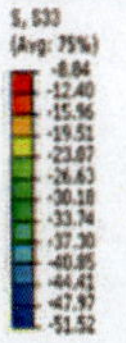
Mathematical Equation's					
ACI (Mandar et al.) $f_{cc'} = f_{c'}[2.25\sqrt{1+7.9\frac{f_l}{f_{c'}}} - 2\frac{f_l}{f_{c'}} - 1.25]$ $f_{c'}=f_{co} = \frac{P}{A}$ $[A=\frac{\pi}{4}d^2]$ $\rho_f=\frac{4nt}{d}$ =0.00344 $f_l = \frac{k_a \epsilon_f \rho_f E_f}{2} = \frac{1 \epsilon_f \rho_f E_f}{2}$ {K_a = 1 for circular section}	8.183	14.517	20.573	26.501	27.09
Girgin $f_{cc}' = f_l + [s f_{co}^2 + m f_{co} f_l]$ $f_{c'} = f_{co} = \frac{P}{A}$ $[A=\frac{\pi}{4d^2}]$ where, s=1 for intact or undamaged concrete m=2.9(f_{co}= 7 to 18MPa) m=6.34-0.076fco(f_{co}=20 to 82 MPa) m=0.1(f_{co}= 82 to 108 MPa)	7.198	13.032	18.68	24.232	27.78
Richar et al. considering k=4.1 as paper fcc=fco+k x fl	7.6	13.73	19.708	25.62	26.21
Experimental Value of k=2.91691	7.04	13.03	18.92	24.76	25.34

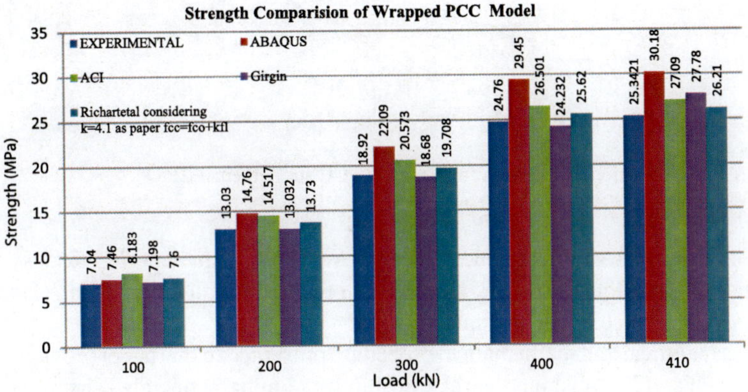

Graph 4 Strength comparison of wrapped PCC model

Table 6 Strength validation for fractured PCC strengthened with CFRP

Deformation (mm) Validation for Fractured PCC Strengthened with CFRP				
Load	100kN	200 kN	300 kN	400kN
Abaqus®	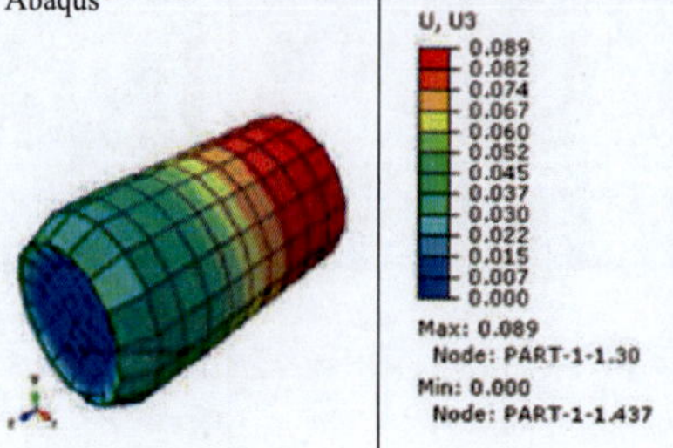	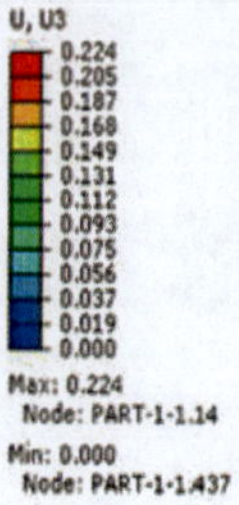	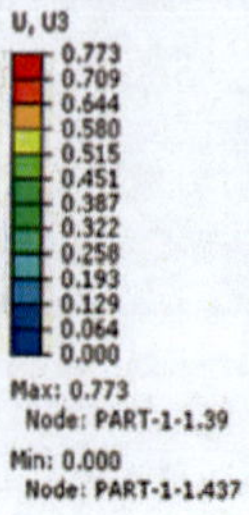	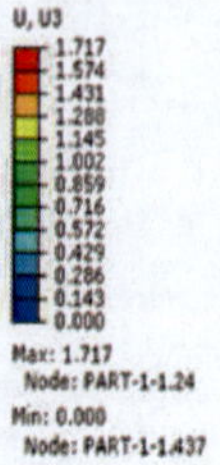
Mathematical	$\Delta=\frac{PL}{AE}=0.073$ P= Load (N) L= 300mm $A=\frac{\pi}{4}\times 150^2$ $=17671.46\ mm^2$ $E=E_c=5000\sqrt{f_{ck}}$ $=5000\sqrt{20}$ $=22360.68 N/mm^2$	0.146	0.220	0.293
Experimental	0.05	0.2	0.24	0.5

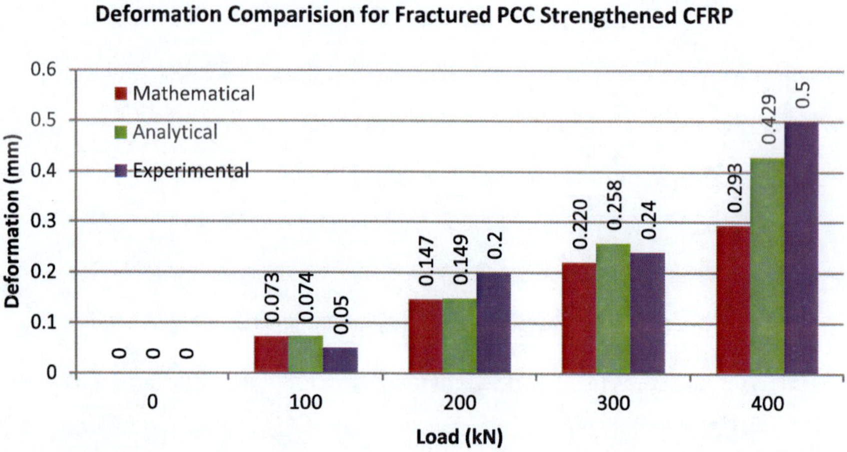

Graph 5 Deformation comparison for fractured PCC strengthened CFRP

The values assigned for concrete damage plasticity element, in Abaqus® shows a good deformation and better strength improvement validation with reference to un-strengthened controlled element.

The analytical and mathematical procedure presented in this paper provides convenient means of establishing the relationship for confined concrete. The procedure is relatively simple to use and general enough to cover the majority of concrete member.

Table 7 Strength validation for failed PCC strengthened with CFRP

<table>
<tr><th colspan="5">Strength (N/mm^2)Validation for Failed PCC Strengthened with CFRP</th></tr>
<tr><th>Load</th><th>100 kN</th><th>200 kN</th><th>300 kN</th><th>400 kN</th></tr>
<tr><td>Abaqus®
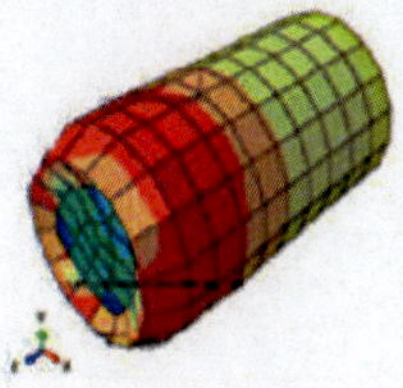</td>
<td>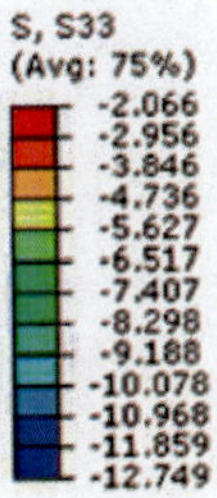
</td>
<td>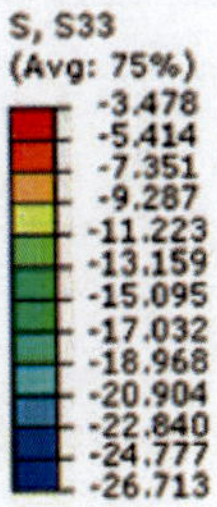
</td>
<td>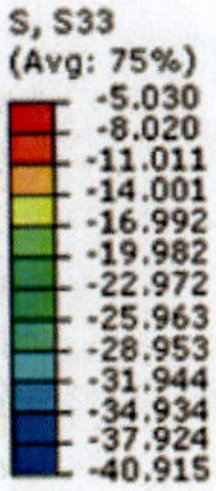
</td>
<td>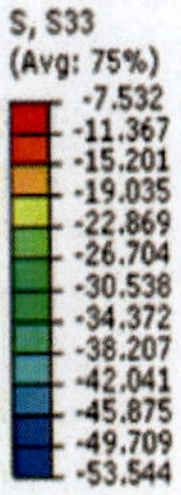
</td></tr>
<tr><td>Mathematical</td><td>$\bar{\sigma}_c = \frac{\sigma_c}{1-d_c}$ =5.5220
$\sigma_c = \frac{P}{A}$
P=Load (N)
A =Area (mm^2)
dc=0.1</td><td>11.0440</td><td>16.5660</td><td>22.0880</td></tr>
</table>

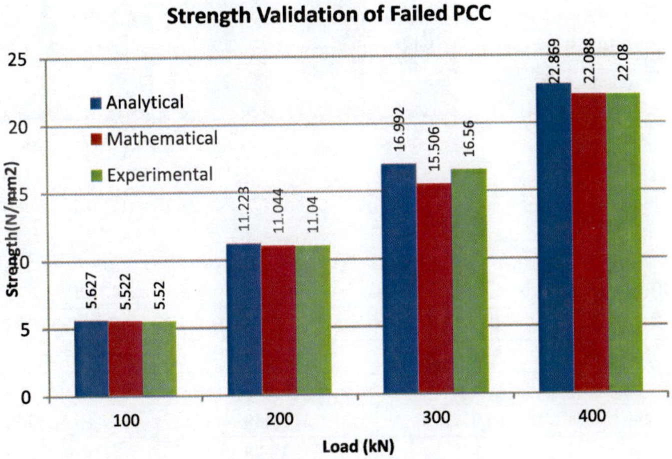

Graph 6 Strength comparision of failed PCC

Table 8 Deformation comparison for controlled and CFRP strengthened specimens

Load	PCC	PCC+CFRP	FPCC+CFRP
0	0	0	0
100	0.076	0.073	0.073
200	0.152	0.146	0.147
300	0.228	0.218	0.220
360	0.273	0.264	0.264
400	-	0.291	0.293
410	-	0.298	-

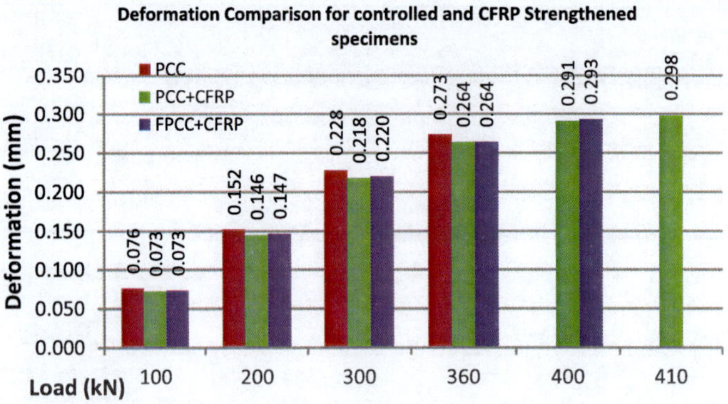

Graph 7 Deformation comparison for controlled and CFRP strengthened specimens

Table 9 Strength comparison for controlled and CFRP strengthened specimens

Load	PCC	PCC+CFRP	FPCC+CFRP
100	5.659	7.36	5.5220
200	11.318	14.72	11.0440
300	16.977	19.48	16.5660
360	20.372	22.73	19.8792
400	-	25.98	22.0880
500	-	30.18	-

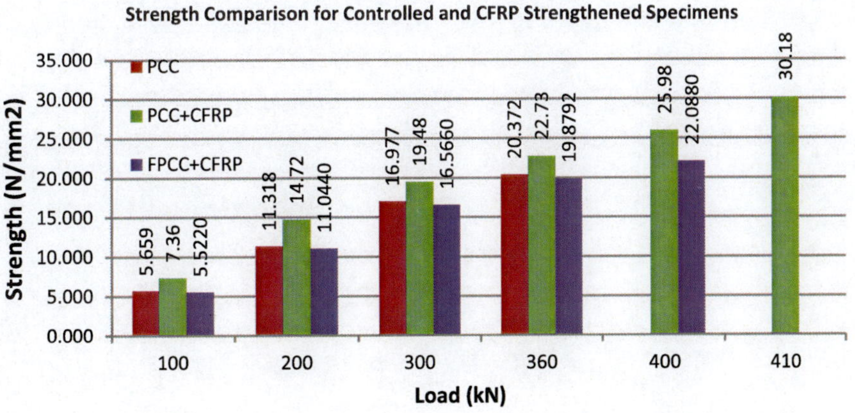

Graph 8 Strength comparison for controlled and CFRP strengthened specimens

Acknowledgements This work has received material assistance from SIKA India Pvt. Ltd. for the supply Carbon Fiber and epoxy for conducting experimental study. The support extended is gratefully acknowledged.

References

1. Mander JB, Priestley MJN, Park R (1988) Theoretical stress-strain model for confined concrete. J Struct Eng 114(8):1804–1826
2. Zehra GC, Girgin K (2015) A design-oriented combined model (7 MPa to 190 MPa) for FRP-confined short columns. MDPI, Polymers. https://doi.org/10.3990/polym7101489
3. Richart FE, Brandtzaeg A, Brown RL (1929) The failure of plain and spirally reinforced concrete in compression. Bulletin No. 190, Engineering Experiment Station, University of Illinois, Urbana, USA
4. Abaqus/CAE (2012) ver 6-12.2, Dassault System Simulia Corp
5. Yusuf S, Muharrem A (2015) Defining parameters for concrete damage plasticity model. Chall J Struct Mech 1(3):149–155
6. Benzaid B, Chikh NE, Mesbah H (2009) Study of the compressive behaviour of short concrete columns confined by fibre reinforced composite. Arab J Sci Eng 34(1B):15–26
7. Chikh N, Mesbah H, Gahmous M, Benzaid R (2012) Performance of externally CFRP confined RC columns with changes in thickness of the wrap, slenderness of the column and shape of the section, vol 2001, pp 0–6
8. Saadatmanesh H, Ehsani MR, Li MW (1994) Strength and ductility of concrete columns externally reinforced with fiber-composite straps. Aci Struct J 91(4):434–447
9. Saadatmanesh H, Ehsani MR, Jin L (1997) Repair of earthquake-damaged RC columns with FRP wraps. ACI Struct J 94(2):206–X
10. ACI 318-02 (2002) Building code requirements for structural concrete. Farmington Hills, American Concrete Institute
11. ACI 440.2R-02 (2002) Guide for the design and construction of externally bonded FRP systems for strengthening concrete structures. Farmington Hills, American Concrete Institute
12. Campione G, Miraglia N, Papia M (2004) Strength and strain enhancements of concrete columns confined with FRP sheets. Struct Eng Mech 18(6):769–790

Review of Various Hypotheses Used to Correct Notch Elastic Stress/Strain for Local Plasticity

Shreebanta Kumar Jena, Punit Arora, Suneel K. Gupta, and J. Chattopadhyay

Abstract In present study, the notch geometry, constraint conditions, loading conditions and peak equivalent strains have been accounted to review classical hypotheses for estimation of localized stress/strain values. Two dimensional and three dimensional FE analyses have been performed on planner and tube geometry for validation of classical schemes based on strain energy conservation on maximum principal stress/strain plane and von-Mises equivalent stress/strain. These geometries (plane 2D and tube) have single central circular hole subjected to three different equivalent peak strain levels. The hypotheses based on conservation of strain energy on equivalent stress and strain values results in better predictions of localized stress for plane stress, plane strain and tube geometries. The hole diameter was also varied to study the effect of strain gradient. It has been brought out that for given equivalent peak strain, the percentage difference between predicted and FE localized stress results is nearly similar for various hole diameters. Further, a three dimensional tube geometry with intermediate constraint level was also analyzed and the suitability of classical hypotheses have been brought out w.r.t. FE outcome.

Keywords multiaxial stress · low cycle fatigue · localized stresses · Neuber · Hoffmann–Seeger

Nomenclature

σ_1^{FEA}	Maximum principal stress evaluated using FEA
$\sigma_1^{\mathrm{Neuber}}$	Maximum principal stress evaluated using Neuber approach
$\varepsilon_1^{\mathrm{Neuber}}$	Maximum principal elastic-plastic strain evaluated using Neuber approach
σ_1^{El}	Pseudo elastic maximum principal stress evaluated using FEA

S. K. Jena (✉)
Homi Bhabha National Institute (HBNI), Anushakti Nagar, Mumbai 400094, India
e-mail: skjena@barc.gov.in; shreebantaodisha@gmail.com

P. Arora · S. K. Gupta · J. Chattopadhyay
Reactor Safety Division, Bhabha Atomic Research Centre, Mumbai 400085, India

S. Seetharamu et al. (eds.), *Fatigue, Durability, and Fracture Mechanics*, Lecture Notes in Mechanical Engineering, https://doi.org/10.1007/978-981-15-4779-9_10

ε_1^{El}	Elastic maximum principal strain evaluated using FEA
σ_{eqv}^{FEA}	von-Mises equivalent stress evaluated using FEA
ε_{eqv}^{peak}	von-Mises equivalent peak strain at notch tip evaluated using FEA
σ_{eqv}^{HS}	von-Mises equivalent stress evaluated using Hoffmann–Seeger approach
ε_{eqv}^{HS}	von-Mises equivalent total strain evaluated using Hoffmann–Seeger approach
σ_{eqv}^{El}	Pseudo elastic von-Mises stress evaluated using FEA
ε_{eqv}^{El}	Elastic von-Mises strain evaluated using FEA
Δ_y^t	Applied remote displacement time history along *y*-direction
θ	Angle between considered path and *x*-axis

1 Introduction

The mechanical components such as pipes, elbows/bends, T-junctions, vessels etc. are subjected to periodic cyclic loading during normal operating conditions as well as during the design basis events and may fail due to the fatigue cycling. Generally, the fatigue damage is evaluated using design codes, which use fatigue design (S–N) curve. These fatigue design curves are derived from uniaxial completely reversible strain–controlled Low Cycle Fatigue (LCF) tests where the test specimen is subjected to uniform stress/strain condition in the gauge region prior to crack initiation. However, the real piping components and vessels experience multi-axial state of stresses/strains due to complex geometry and loadings. The piping/pressure vessels invariably have various geometric discontinuities such as nozzle-shell junction, T-junction, weld fillets etc. to meet process design requirements. These geometric discontinuities result in amplification of stress/strain at notch tip and setup large gradients of stress/strain in the local region. At the notch tip, the peak stress generally exceeds the yield strength of material leading to highly localized yielding. It is well understood and widely reported in literature that strain gradient plays an important role in fatigue failure of a component. Several hypotheses as proposed by Neuber [1], Hoffman and Seeger [2], Molski and Glinka [3] are used to estimate the actual elastic-plastic stress/strain information at the notch tip from the pseudo-elastic stresses/strains. Glinka [4] also proposed the energy density approach to find out the local elastic plastic stress near the notch tip under uniaxial loading condition. Topper et al. [5] applied the Neuber's rule to assess the fatigue lives of notched specimens. Gu and Lee [6] developed a new method to find out the stress and strain in case of non-proportional loading condition. Ali Fatemi and Nicolas Gates [7] found that Neuber's method overestimates the stress-strain ahead of the notch tip which results in undue conservatism in fatigue design.

Subsequent to estimation of localized stress/strain values, various authors have proposed many approaches based on critical distance technique. This critical distance is correlated with material parameter. Neuber [1] suggested point method where a representative strain value results in accurate fatigue life assessment. Peterson [8] proposed a line integral method to consider the effect of strain gradients. Taylor [9] demonstrated that this approach results in reasonable predictions with respect to experimental fatigue life data. On the similar lines, various proposals such as area averaged and volume averaged methods have been introduced over localized zones ahead of disconituity. It is realized that fatigue life assessment under multi-axial state of stress with notches leading to stress/strain gradients is really a challenging task.

From designer's perspective, most of the design codes furnish guidelines for evaluation of linear elastic Finite Element Analyses (FEA) followed by plastic correction using classical hypotheses such as Neuber [1] and Hoffman Seeger [2]. However, very limited literature is available to validate such hypotheses for wide variety of notch sizes, constraint and loading conditions.

In this view of above, extensive analyses have been carried out to validate the applicability of some of the classical and recent hypotheses for estimation of localized stress/strain values. These analyses have been performed to validate the hypotheses w.r.t. elastic-plastic FEA. The constraint conditions cover plane stress, plane strain and tube geometry with central circular hole. The strain gradient was also varied by varying the diameter of hole. The remote loading considers, pure axial and pure torsion state of stress.

2 Finite Element Analyses Details

The accurate assessment of elastic-plastic stress/strain fields can be carried out using finite element analyses. Hence, in order to validate the prediction of localized stress/strain values obtained from Neuber, Hoffmann–Seeger hypotheses, FE results are used for comparison purposes. The FEA have been performed for two dimensional plane stress and plane strain geometries. The plane stress and plane strain structure were modelled with a total length and height as 50 mm each. A central hole in these geometries was modelled. The diameter of the hole was varied from 3 to 8 mm. A quarter symmetric model was used to arrive at elastic-plastic stress/strain field ahead of hole tip. Eight node serendipity element has been used to mesh quarter symmetric geometry/domain. The mesh convergence studies have also been carried out for accurate stress/strain fields. Typical FE mesh for quarter symmetric planar geometry having central hole is shown in Fig. 1.

The tube geometry having outer diameter as 25.4 mm with thickness of 1.7 mm. This tube was modelled with through thickness circular hole of varying diameter to account for different strain gradient effect. The three dimensional model uses 20 node hexahedral element. Along thickness direction eight division has been considered. Typical FE model of tube geometry with circular hole as shown in Fig. 2. The tube was subjected to pure axial and pure torsion loading conditions.

The analyses have been performed for three different levels of loadings such that von-Mises equivalent peak strain (at hole tip) using elastic-plastic FEA are 0.20, 0.25 and 0.30%.

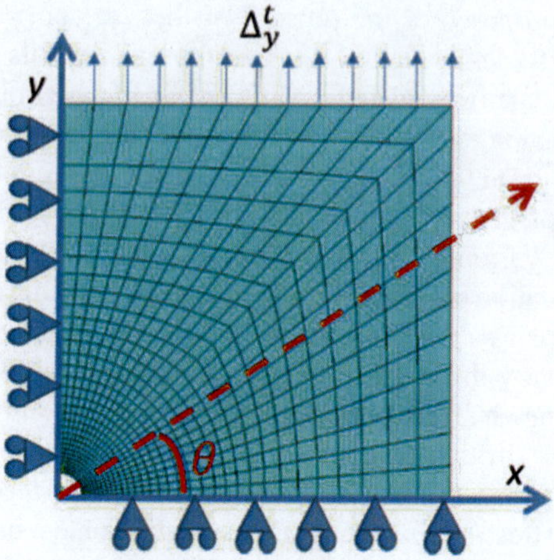

Fig. 1 One quarter FE model with meshing for planner geometry with central hole

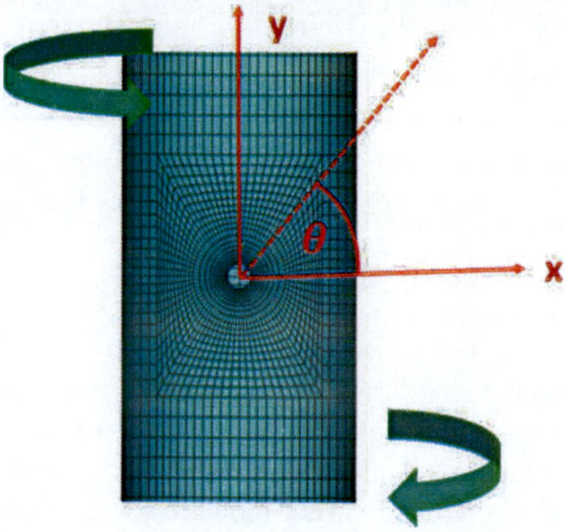

Fig. 2 Detailed 3D modelling of tube geometry with central hole

The cyclic stress-strain curve of SA 333 Gr. 6 material has been used with von-Mises yield criterion and multi-linear kinematic hardening rule. The cyclic stress-strain curve was obtained from pure axial cyclic tests performed on standard solid LCF specimens [10]. The saturated cyclic stress-strain amplitude curve is shown in Fig. 3.

3 Calculation Methodology of Neuber [1] and Hoffmann–Seeger [2] Rules

These hypotheses are employed to correct the localized pseudo-elastic stress and strain values for local plasticity. The localized pseudo-elastic stress and strain values are obtained from finite element analyses of notched geometry considering linear

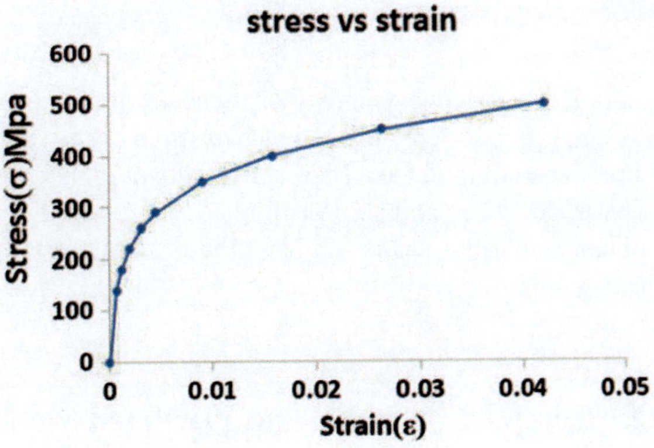

Fig. 3 Stabilized cyclic stress-strain curve of SA 333 Gr. 6 material [10]

elastic material model. These hypotheses assume the strain energy conservation due to maximum principal stress/strain (Neuber's rule) and von-Mises equivalent stress/strain (Hoffmann–Seeger rule) components.

3.1 *Neuber's Rule*

At present most of design codes used for fatigue life assessments for notched components are based on Neuber's rule due to its conservative nature [11, 12, 13, 14]. Neuber method is based on the equivalence of strain energy density owing to maximum principal stress/strain at a material point in linear elastic and elastic-plastic conditions. A hyperbola is plotted for a material point based on elastically calculated stress. The intersection of this hyperbola and material stress-strain curve is the predicted elastic-plastic strain on that material point. The procedure to predict elastic-plastic stress ahead of notch using Neuber's rule is shown in Fig. 4a.

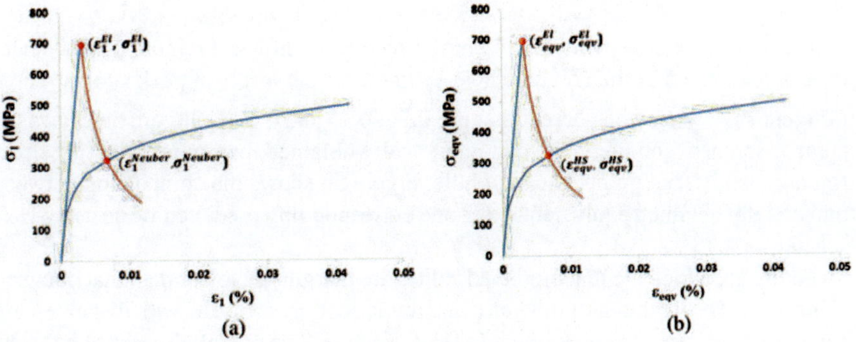

Fig. 4 Estimation of elastic-plastic stress at the tip of the hole from linear elastic FEA results and using **a** Neuber method, **b** Hoffman–Seeger method

3.2 *Hoffmann–Seeger Rule*

In order to improve accuracy of the result obtain by Neuber's rule, Hoffmann and Seeger [2] considered equivalence of strain energy density obtained from equivalent stress and strain values. Like above, a hyperbola considering elastically calculated von-Mises stress and strain is plotted along with the material stress-strain curve to obtain the elastic-plastic von-Mises equivalent stress/strain ahead of notch as depicted in Fig. 4b.

4 Comparison of FE and Predicted Stress Results

The estimated localized stress/strain values (maximum principal or von-Mises equivalent) using Neuber and Hoffmann–Seeger's (H-S) schemes have been compared with the corresponding outcome obtained from detailed elastic-plastic FE analyses.

Following subsections compare FE (maximum principal and von-Mises equivalent) stresses with corresponding predicted stress values using Neuber and H–S schemes respectively.

4.1 *Plane Stress Condition*

The plane stress geometry mainly considers two cases to study Neuber and H-S schemes. The first case accounts the effect of different peak strains at hole tip (or load levels) for a given hole diameter. The second case, considers the effect of notch geometry for a given peak strain.

4.1.1 Effect of Peak Strain

Typical case with hole diameter of 3 mm subjected to different remote displacement control loading (Δ_y^t) such that the hole tip has different levels of peak strains. Three different $\boldsymbol{\varepsilon}_{\mathbf{eqv}}^{\mathbf{peak}}$ were considered, that are 0.20, 0.25 and 0.30%. Figure 5a shows the comparison between predicted and numerically obtained maximum principal stress for various material points ahead of hole. Figure 5b shows the comparison between numerically obtained equivalent stress and corresponding predicted value using H–S scheme.

Figure 5a, b indicate that predicted values are marginally lower than that obtained using FEA. The variation of predicted and numerical stress results with distance from center of hole is in good agreement for H–S scheme. The overall difference between predicted and numerical stress results is smaller for H–S scheme as compared to Neuber's rule.

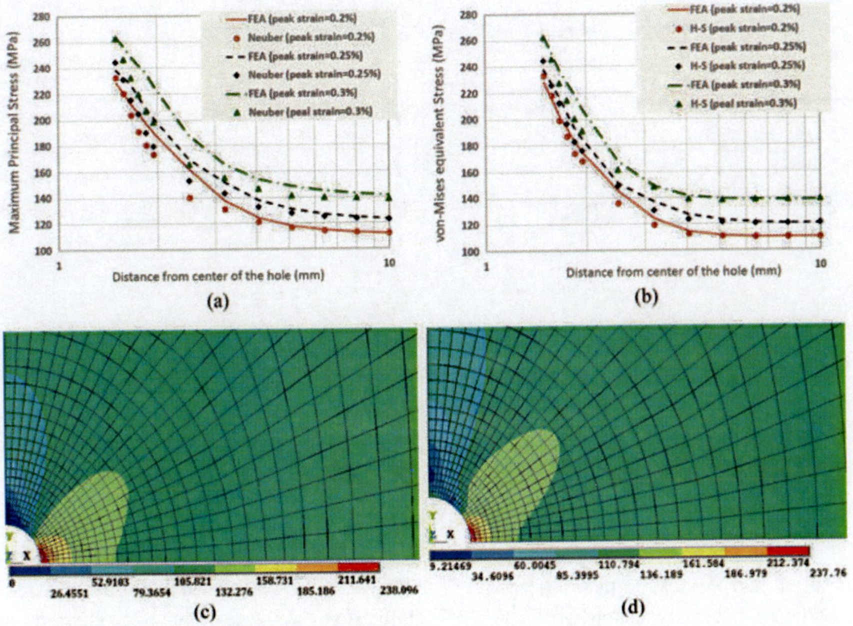

Fig. 5 Plane stress geometry: The variation of stress for a single hole geometry ($D = 3$ mm) at three different equivalent peak strains (0.20, 0.25 and 0.30%) using **a** Neuber method and **b** Hoffmann–Seeger method. **c** Maximum principal stress contours and **d** von-Mises stress contours for $D = 3$ mm and peak strain = 0.25% at hole tip

It is apparent from Fig. 5a, b that for the given diameter of hole, the strain/stress gradient remains nearly same for all peak strains, as indicated from the slopes of stress-versus-distance curves. It is inferred that for a given strain gradient, the difference between numerical and predicted stress increases with increase in peak strain value owing to higher plastification. Figure 5c, d shows the maximum principal stress contour and von-Mises equivalent stress contour for $D = 3$ mm and peak strain = 0.25% at hole tip respectively.

4.1.2 Effect of Notch Geometry

Typical case subjected to peak strain at hole tip as 0.25%, with different hole diameters of 3, 5.5 and 8 mm were considered to study the gradient effect. Figure 6a, b show the variations maximum principal stress and von-Mises equivalent stress with distance from center of hole respectively.

It is inferred from Fig. 6b that H–S scheme results in accurate predictions w.r.t. numerical outcome for all the diameters of hole. However, Neuber's rule results in marginal under-prediction of maximum principal stress value for all diameters of hole (or strain gradients) as shown in Fig. 6a.

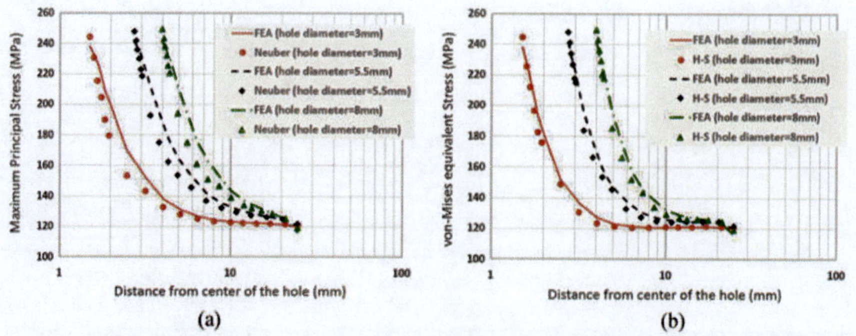

Fig. 6 Plane stress geometry: The variation of stress for three different hole diameter ($D = 3, 5.5$ and 8 mm) for equivalent peak strain of 0.25% using **a** Neuber's method and **b** Hoffmann–Seeger method

4.2 *Plane Strain Condition*

The plane strain geometry also considers two cases to assess the applicability of Neuber and H–S schemes. These cases are similar to that studied for plane stress geometry. FE analyses and prediction of localized stress values have been carried out for three different hole geometries having hole diameter as 3, 5.5 and 8 mm subjected to peak strain values of 0.20, 0.25 and 0.30%.

4.2.1 Effect of Peak Strain

For a given hole diameter of 3 mm, three different $\boldsymbol{\varepsilon}_{\mathbf{eqv}}^{\mathbf{peak}}$ were considered, that are 0.20, 0.25 and 0.30%. Figure 7a shows the comparison between predicted and numerically obtained maximum principal stress for various material points ahead of hole. Figure 7b shows the comparison between numerically obtained equivalent stress and corresponding predicted value using H–S scheme.

Figure 7a shows that maximum principal stress value is under-predicted using Neuber's rule for all peak strains. However, H–S scheme results in comparable predictions w.r.t. numerically obtained von-Mises stress ahead of hole for all peak strain levels (Fig. 7b). Figure 7c, d shows the maximum principal stress contour and von-Mises equivalent stress contour for $D = 3$ mm and peak strain = 0.25% at hole tip respectively.

4.2.2 Effect of Notch Geometry

The case with peak strain of 0.25% on the tip of holes having diameters of 3, 5.5 and 8 mm were considered to study the gradient effect. Figure 8a, b show the variations maximum principal stress and von-Mises equivalent stress with distance from center of hole respectively.

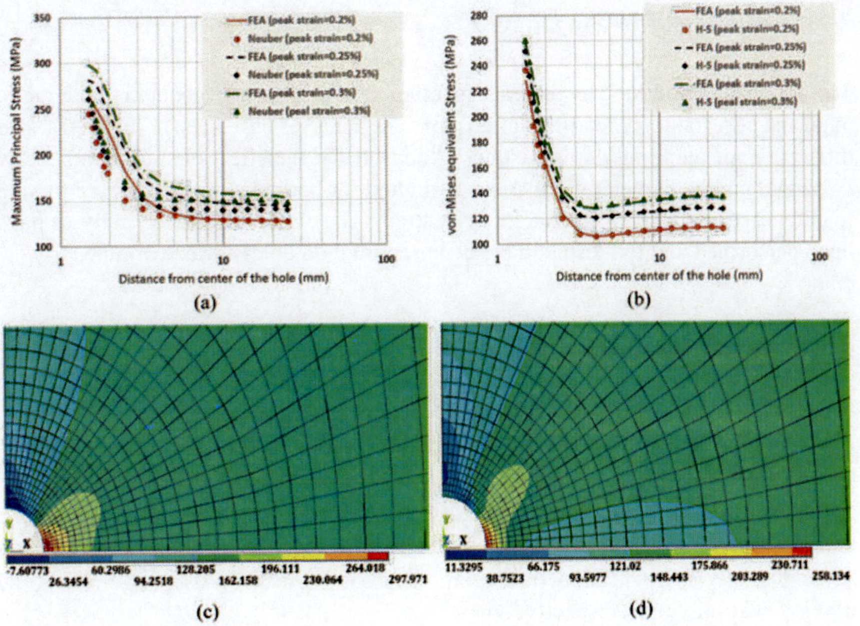

Fig. 7 Plane strain geometry: The variation of stress for a single hole geometry ($D = 3$ mm) at three different equivalent peak strains (0.20, 0.25 and 0.30%) using **a** Neuber method and **b** Hoffmann–Seeger method. **c** Maximum principal stress contour and **d** Von-Mises stress contour for $D = 3$ mm and peak strain = 0.25% at hole tip

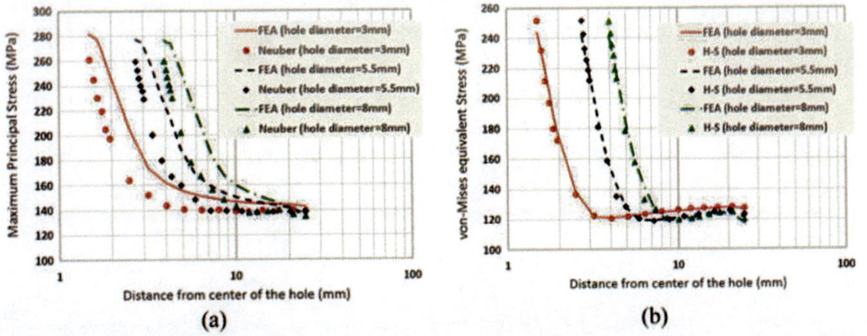

Fig. 8 Plane strain geometry: The variation of stress for three different hole diameter ($D = 3$, 5.5 and 8 mm) at constant equivalent peak strain of 0.25% using **a** Neuber's method and **b** Hoffmann–Seeger method

It is inferred from Fig. 8b that H–S scheme results in accurate predictions w.r.t. numerical outcome for all the diameters of hole. However, Neuber's rule results in significant under-prediction of maximum principal stress value for all diameters of hole (or strain gradients) as shown in Fig. 8a.

4.3 *Tube Geometry*

The applicability of these hypotheses is futher verified on a tube geometry under pure axial and pure torsional loading conditions. The analyses have been performed at two different equivalent peak strain values. Additionally, the effect of strain gradient by varying the hole diameter has also been studied for both pure axial and pure torsion loading cases for a given equivalent peak strain. The elastic-plastic FE analyses have been performed and the results have been reported on outer surface of tube.

4.3.1 Pure Axial Condition

Pure axial loading is performed on the tube geometry having a central circular hole of diameter 5.5 and 8 mm subjected to pure axial displacement (Δ_y^t) control condition. The displacement loading is applied in such a way so that peak strain ($\boldsymbol{\varepsilon}_{\mathbf{eqv}}^{\mathbf{peak}}$) at the notch tip (outer surface) is limited to 0.20 and 0.25% for both hole sizes. Figure 9a, b show the variation of maximum principal stress and von-Mises equivalent stress for a given diameter at two different peak strains stations ($\boldsymbol{\varepsilon}_{\mathbf{eqv}}^{\mathbf{peak}} = 0.20\%$, $\boldsymbol{\varepsilon}_{\mathbf{eqv}}^{\mathbf{peak}} = 0.25\%$) along $\theta = 0°$ direction respectively.

Figure 9a shows that the predicted maximum principal stress is marginally lower than numerically obtained stress value. However, the predicted von-Mises stress values at different distance from center of the hole are in good agreement with corresponding numerical stress values for both the peak strain stations (Fig. 9b). It is inferred from these figures that the state of stress at outer surface nearly resembles with plane stress constraint condition. Hence, the inferences brought out for plane stress condition are consistent with that of material point lying on outer free surface of tube.

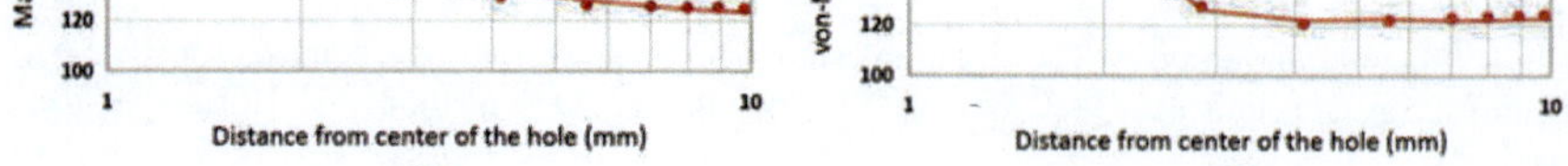

(a) (b)

Fig. 9 Tube geometry: The variation of stress on outer surface of tube having a hole with $D = 3$ mm at two different equivalent peak strains (0.20 and 0.25%) using **a** Neuber method and **b** Hoffmann–Seeger method

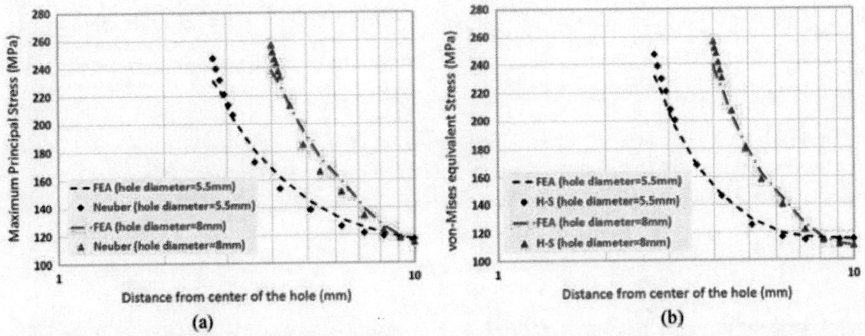

Fig. 10 Tube geometry: The variation of stress for two different hole diameter ($D = 5.5$ and 8 mm) at constant equivalent peak strain of 0.25% using **a** Neuber's method and **b** Hoffmann–Seeger method

Figure 10a, b show the comaprisons between predicted and numerically obtained stress values for different hole diameters subjected to equivalent peak strain at hole tip ($\boldsymbol{\varepsilon}_{\mathbf{eqv}}^{\mathbf{peak}}$) as 0.25%. It is clear from these figures that predicted stress values are comparable with corresponding numerical stress values for both hole dimateres. These inferences are also in line with that brought out for plane stress constraint conditions.

4.3.2 Pure Torsion Condition

The elastic-plastic FE analyses have been carried out at two different $\boldsymbol{\varepsilon}_{\mathbf{eqv}}^{\mathbf{peak}}$ stations of 0.10 and 0.15% accounting two different hole diameters of 5.5 and 8 mm. The numerical stress results obtained from FEA are reported for material point lying on outer surface of tube along $\theta = 45°$ direction.

Figure 11a, b plots the variation of maximum principal stress and von-Mises equivalent stress for a hole diameter 5.5 mm and for two different peak strains stations ($\boldsymbol{\varepsilon}_{\mathbf{eqv}}^{\mathbf{peak}} = 0.10\%$, $\boldsymbol{\varepsilon}_{\mathbf{eqv}}^{\mathbf{peak}} = 0.15\%$) along $\theta = 45°$ direction respectively. Here von-Mises equivalent stress initially is observed to decreases significantly below the remote (far away from notch) corresponding stress value.

Under the pure torsion loading, the state of stress away from notch i.e. remote state of stress would be pure shear (say shear stress τ on outer surface). Hence at remote point, the major and minor principal stresses would be respectively τ and $-\tau$ and would be inclined by $\theta = 45°$ with respect to tube axis. Since, the material point is considered on the outer surface of tube, the out of plane stress would be zero. Hence, the remote equivalent stress would be $\sqrt{3}\tau$ which is higher than remote major principal stress.

However, at the hole tip, the state of stress is uniaxial due to two numbers of free surfaces, that are, outer surface of tube and free surface of hole. This results in same values of maximum principal stress and von-Mises equivalent stress as shown

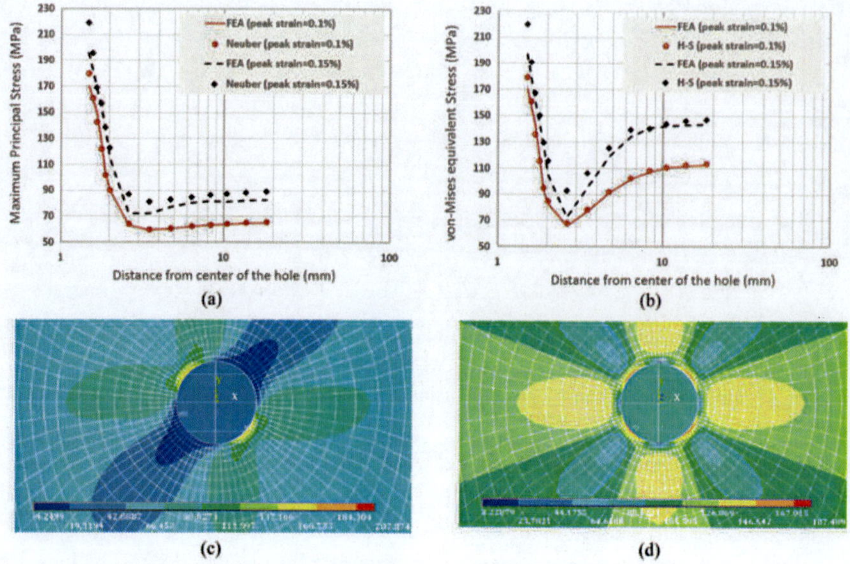

Fig. 11 Tube geometry: The variation of stress for a hole with $D = 3$ mm at two different equivalent peak strains (0.10 and 0.15%) using **a** Neuber method and **b** Hoffmann–Seeger method. **c** Maximum principal stress contour and **d** von-Mises stress contour for $D = 3$ mm and peak strain = 0.10% at hole tip

in Fig. 11a, b. However, with slight increase in distance from hole tip along $\theta = 45°$ direction, the stress state becomes biaxial with tensile nature of minor principal stress. Due to this, maximum principal stress is higher than that of von-Mises stress at that location. With further increase in distance, though the state of stress remains biaxial, however, the minimum principal stress changes its nature from tensile to compressive which results in higher equivalent von-Mises stress than maximum principal stress at remote location. This transition from tensile nature to compressive nature of minimum principal stress causes minima in von-Mises equivalent stress at certain distance.

Figure 11a shows comparable values of σ_1^{Neuber} and σ_1^{FEA} for different $\boldsymbol{\varepsilon}_{\mathbf{eqv}}^{\mathbf{peak}}$ using Neuber's rule. Figure 11b shows that $\sigma_{\text{eqv}}^{\text{HS}}$ is in good agreement with $\sigma_{\text{eqv}}^{\text{FEA}}$ using H–S scheme. Figure 11c, d shows the maximum principal stress contour and von-Mises equivalent stress contour for $D = 3$ mm and peak strain = 0.10% at hole tip respectively.

Figure 12a, b shows the variation of maximum principal stress and von-Mises equivalent stress with distance at a given strain percentage ($\boldsymbol{\varepsilon}_{\mathbf{eqv}}^{\mathbf{peak}} = 0.15\%$) for two different hole diameters. It is evident from these figures that the predicted stresses by both Neuber and H—S method are in good agreement with that of corresponding numerically obtained stress values.

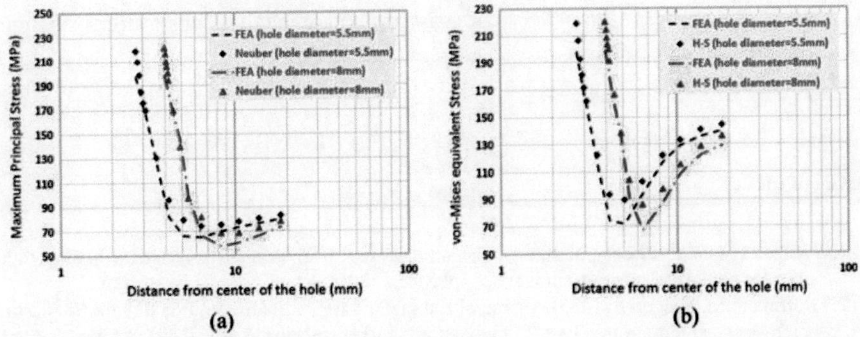

Fig. 12 Tube geometry: The variation of stress for two different hole diameter ($D = 5.5$ and 8 mm) at constant equivalent peak strain of 0.15% using **a** Neuber's method and **b** Hoffmann–Seeger method

5 Conclusions

The finite element analyses have been performed on SA 333 Gr. 6 material to obtain the numerical stress results ahead of notch. The predicted maximum principal and von-Mises stress values using Neuber and Hoffmann–Seeger hypotheses have been compared with corresponding numerical values. The comparative analyses using Neuber and H–S hypotheses for different constraint conditions, peak strain values, strain gradients and loading conditions are summarized as below:

1. For plane stress constrained condition, Neuber's rule results in marginal under-predication of maximum principal stress. However, predicted stress results using H–S scheme are in good agreement with that of FEA outcome for all peaks strain values at hole tip with different strain gradients, considered in present study.
2. Neuber's rule results in significant under-prediction of maximum principal stress ahead of notch for nearly all cases studied for plane strain constraint condition. The predicted von-Mises stress values using H–S scheme is in good agreement with that of corresponding numerical stress values for all peak strain and strain gradient combinations.
3. The predictions have been made using Neuber and H–S schemes on tube geometry with central hole. The predicted results on outer surface of tube (resembling with plane stress constrain condition) are reported and compared with numerical results.

   a. For pure axial displacement control loading condition, the stress values have been predicted/numerically obtained along $\theta = 0°$ direction. Neuber's rule results in marginal under-predication of maximum principal stress, whereas, H–S scheme results in accurate assessment of localized von-Mises stress values.
   b. For pure angle control loading scenario, the stress values have been predicted/numerically obtained along $\theta = 45°$ direction. Both Neuber and

H–S schemes result in reasonable assessment of localized maximum principal stress and von-Mises stress values, respectively.

References

1. Neuber H (1961) Theory of stress concentration for shear-strained prismatical bodies with arbitrary nonlinear stress-strain law. J Appl Mech 28(4):544
2. Hoffmann M, Seeger T (1985) A generalized method for estimating multiaxial elastic-plastic notch stresses and strains, Part 2: application and general discussion. J Eng Mater Technol 107(4):255
3. Molski K, Glinka G (1981) A method of elastic-plastic stress and strain calculation at a notch root. Mater Sci Eng 50(1):93–100
4. Glinka G (1985) Energy density approach to calculation of inelastic strain-stress near notches and cracks. Eng Fract Mech 22(3):485–508
5. Topper TH, Wetzel RM, Morrow J (1967) Neuber's rule applied to fatigue of notched specimens. Def Tech Inf Cent 1–13
6. Gu RJ, Lee Y-L (1997) A new method for estimating nonproportional notch-root stresses and strains. J Eng Mater Technol 119(1):40
7. Gates N, Fatemi A (2016) Notch deformation and stress gradient effects in multiaxial fatigue. Theor Appl Fract Mech 84:3–25
8. Peterson RE (1959) Notch sensitivity. In: Sines G, Waisman JL (eds) Metal fatigue, vol 22, no 3. McGraw-Hill, New York, pp 293–306
9. Taylor D (1999) Geometrical effects in fatigue: a unifying theoretical model. Int J Fatigue 21(5):413–420
10. Goyal S, Gupta SK, Sivaprasad S, Tarafder S, Bhasin V, Vaze KK, Ghosh AK (2013) Low cycle fatigue and cyclic plasticity behavior of Indian PHWR/AHWR primary piping material. Procedia Eng 55:136–143
11. Sakane M, Ohnami M (1986) A study of the notch effect on low cycle fatigue of metals in creep-fatigue interacting conditions at elevated temperatures. ASME J Eng Mater Technol 108(4):279–284
12. Zappalorto M, Kujawski D (2015) Neuber's rules and other solutions: theoretical differences, formal analogies and energy interpretations. Theor Appl Fract Mech 79:2–13
13. Sharpe WN, Yang CH, Tregoning RL (1992) An evaluation of the Neuber and Glinka relations for monotonic loading. J Appl Mech 59(2S):S50
14. A16 guide of RCC-MR for defect assessment and leak before break analysis. Edition 2002

Modeling of Strain Range Memory Effect to Capture Non-linear Behavior in Non-masing Metals to Accurately Predict Life

Prem Andrade, Vinay Carpenter, and Ravindra Masal

Abstract Several metal alloys demonstrate a cyclic plasticity hardening behavior that is higher than the monotonic hardening behavior seen in a uniaxial test. This increased hardening with increase in strain range is known as non-masing behavior. This influences the shape of the stress–strain hysteresis curve and has a direct impact on the failure life for that strain range. This paper numerically models this non-masing behavior and its memory effect by tracking the current strain range hardening region in the plastic strain space. This model dynamically modifies plasticity parameters as the strain range changes as well as captures the memory effect of strain range seen in such materials. The impact of modeling this behavior is then showed by comparing with experimental stress–strain hysteresis curves as well as life numbers available in literature for 316L(N) stainless steel.

Keywords Strain range · Non-masing · Low cycle fatigue · Strain hardening · Strain memory effect · Thermomechanical fatigue

1 Introduction

The push for light weighting, increasing efficiency, decreasing emissions, increasing power density, and enabling variable usage results in challenging duty cycles for components across the aviation, power, transportation, and electronics industries. The engineering challenge in designing components for these industries is to use the appropriate material that can meet all the functional specifications of that component at a minimum cost. Another challenge is to understand how these materials behave at different operating conditions and under different usage scenarios and based on

P. Andrade (✉) · V. Carpenter · R. Masal
ANSYS India Private Limited, Pune, India
e-mail: prem.andrade@ansys.com

V. Carpenter
e-mail: vinay.carpenter@ansys.com

R. Masal
e-mail: ravindra.masal@ansys.com

S. Seetharamu et al. (eds.), *Fatigue, Durability, and Fracture Mechanics*, Lecture Notes in Mechanical Engineering, https://doi.org/10.1007/978-981-15-4779-9_11

that how the life of the component can be calculated. The different duty cycles that components will go through cause cyclic mechanical and thermal loads. These result in cyclic stresses and strain that vary spatially and temporally across the component. To accurately predict the life of the component through simulation, it is critical to model the constitutive material behavior to accurately capture its response to these varying cyclic loads. This paper attempts to study the impact of strain range on the kinematic hardening model for 316L(N) stainless steel.

2 Modeling the Non-linear Kinematic Hardening Behavior of 316L(N) Stainless Steel

Several metal alloys exhibit a non-linear kinematic hardening of rate-independent plasticity when subjected to cyclic loads. This phenomenon is connected to the Bauschinger effect, where the center of the yield surface moves in response to plastic strains and this center represents the back stress in the system [1]. In many metal alloys, this movement has a distinct non-linear behavior and similar to work hardening has a limiting surface to which the yield surface can move to. Implementation of the Chaboche model [2–4] can capture these characteristics accurately. The formulation for the incremental of deviatoric back stress (α) is shown in Eq. 1 as a function of incremental plastic strain ($d\boldsymbol{\varepsilon}^p$) and incremental equivalent plastic strain (dp). As can be seen by the equation, the deviatoric back stress is made up of multiple layers (M) to be able to fit the experimental data of the alloy being studied.

$$d\alpha_i = \frac{2}{3} C_i d\varepsilon^p - \gamma_i \alpha_i dp \tag{1}$$

$$d\boldsymbol{\alpha} = \sum_{i=1}^{M} d\boldsymbol{\alpha}_i \tag{2}$$

Taking a von Mises yield criterion, the yield function (f) can be written as a function of the deviatoric stress ($\boldsymbol{s}$), deviatoric back stress, ($\boldsymbol{\alpha}$) and the current radius of the yield surface in the deviatoric plane (σ_o)

$$f = \sqrt{2(\boldsymbol{s} - \boldsymbol{\alpha}) : (\boldsymbol{s} - \boldsymbol{\alpha})} - \sigma_o = 0 \tag{3}$$

The increment in plastic strain is governed by the associative flow rule flow as follows:

$$d\boldsymbol{\varepsilon}^p = dp \frac{\partial f}{\partial \sigma} \tag{4}$$

$$dp = \sqrt{\frac{2}{3} d\boldsymbol{\varepsilon}^p : d\boldsymbol{\varepsilon}^p} \tag{5}$$

At higher temperatures, rate effects can be seen and this is often modeled by considering a viscoplastic potential above the yield surface [2, 3]. The study in this paper is confined to room temperature response where this behavior is not seen. In addition, the radius of the yield surface (σ_o) evolves and can change based on work hardening and is a function of the accumulated plastic strain. This tends to stabilize over cycles, and it reaches a saturated value which is called the isotropic saturated yield stress [3]. In this study, it is assumed that all hysteresis curves of stress and strain are at its saturated level.

All these material behaviors are observed in 316L(N) stainless steel experiments done at room temperature by Roy et al. [1] and good fits have been obtained against this experimental data for a given strain range-controlled cycling. These models were used in a commercial CAE package ANSYS to model the behavior of 316L(N) stainless steel [5].

3 Strain Range Hardening and Memory Effect

Metal alloys when loaded with a cyclic strain-controlled loads show either of the following two behaviors. One possibility is as the strain range of the cycling increases the hardening curve that is followed follows the same path of the previous strain range and as a result has a very similar path as the monotonic strain-controlled hardening curve. These are known as masing materials. The other possibility is that the material with increased strain range sees a significant increase of its hardening capability and deviates from the previous hardening of smaller strain range cycles. These are called non-masing materials. Experimental work done by Roy et al. show that 316L(N) falls in the non-masing material category [1]. Several non-masing alloys show a characteristic which makes it easier to generate a model to capture this increased hardening behavior. This characteristic is that the different strain ranges have similar kinematic hardening behavior and what deviates is only the isotropic saturated yield stress value (σ_o) [3, 6]. This trend seen is with increasing strain range that the isotropic saturated yield stress value increases and then asymptotes to a maximum value. Another behavior seen commonly is the strain range memory effect. This captures the phenomena observed that these alloys have a memory of the strain range they were previously subjected to. So, even if they are now subjected to a smaller strain range the hardening behavior seen is similar to the largest strain range it has experienced in the past. However, work has shown this memory can fade over time and the hardening behavior can be commensurated with the current strain range after several cycles at that level [3].

To model the strain range effect the following additional equations need to be modeled. A strain range surface needs to be tracked which is similar in form as the kinematic stress yield surface but in the strain space [6–8]. This surface is governed by the strain range function (F) whose radius in the strain space and center are determined by $qand\boldsymbol{\xi}$, respectively, as shown in Eq. (6).

$$F = \sqrt{\frac{2}{3}(\boldsymbol{\varepsilon}^p - \boldsymbol{\xi}) : (\boldsymbol{\varepsilon}^p - \boldsymbol{\xi})} - q = 0 \tag{6}$$

$$\boldsymbol{n} = \frac{(\boldsymbol{s} - \boldsymbol{\alpha})}{(\boldsymbol{s} - \boldsymbol{\alpha}) : (\boldsymbol{s} - \boldsymbol{\alpha})} \tag{7}$$

$$\boldsymbol{n}^* = \frac{(\boldsymbol{\varepsilon}^p - \boldsymbol{\xi})}{(\boldsymbol{\varepsilon}^p - \boldsymbol{\xi}) : (\boldsymbol{\varepsilon}^p - \boldsymbol{\xi})} \tag{8}$$

$$dq = \eta H(F)\langle \boldsymbol{n} : \boldsymbol{n}^* \rangle dp \tag{9}$$

$$d\boldsymbol{\xi} = \sqrt{\frac{3}{2}}(1 - \eta) H(F)\langle \boldsymbol{n} : \boldsymbol{n}^* \rangle \boldsymbol{n}^* dp \tag{10}$$

The increase in radius and the shift of center of this strain surface are governed by Eqs. (7)–(10). In these equations $\boldsymbol{n}$ and $\boldsymbol{n}^*$ represent unit normal to the stress yield surface and the strain hardening surface, respectively.

4 Impact of Modeling the Strain Range Hardening on 316L(N) Stainless Steel

The cyclic hardening behavior of alloy 316L(N) stainless steel at different strain ranges at room temperature has been experimentally found by Roy et al. [1]. In that work, the non-masing behavior of 316L(N) stainless steel has been very well established. The numerical study involved computationally modeling the material behavior as described in Sect . 3 as it underwent different strain range tests for low cycle fatigue. This involved considering a simple coupon numerically and finding the various material coefficients required to fit the experimental data. In this study, the largest strain range experimental curve was used as a reference to fit the material models. Also, since an asymmetricity was noticed in all the experimental curves, the choice was made to closely fit the loading part of the curve. The simulation was done in ANSYS using in-built material models [5] that capture the various aspects that are described in Sect. 2. The closeness of fit of the numerical model with the experimental values [1] for different strain ranges can be seen in Fig. 1. It can be clearly seen that for the large strain range which was used as the basis for fitting the match is good but for the smaller strain ranges the deviation of the peak stress as well as the area contained in the curve is quite significant. This is due to, as described in the previous section, the behavior of non-masing materials where the hardening is different for each strain range.

In the next step, the strain range hardening effect as described in Sect. 3 was modeled by using User Defined Subroutines in ANSYS to track the evolution of the strain range surface as well as the current isotropic saturated hardening yield stress. Again, the base material coefficients were modeled on the basis of the largest

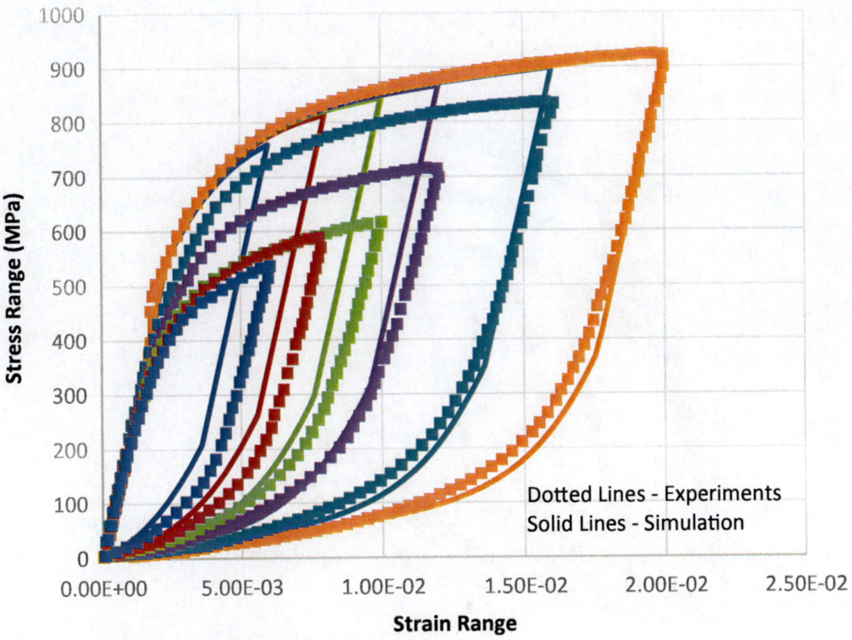

Fig. 1 Comparison of simulation and experiment of isothermal strain-controlled cycling for 316L(N) stainless steel without considering strain range effect

strain range experimental data. However, this time the isotropic saturated hardening yield stress was allowed to be dynamically updated as a function of the strain range surface. The closeness of fit of the updated numerical model with the experimental values for different strain ranges can be seen in Fig. 2. It can be clearly seen that for all the strain ranges the match of the peak stress as well as the area contained in the curve is much closer.

An attempt has been made to quantify the impact by calculating the strain energy in a cycle for the simulations and compare them with experimental values. In Table 1 this comparison can be seen. The average error in capturing the strain energy per cycle without considering strain range effects is 22%. This is brought down considerably to 4% by considering strain range effects. Similarly, the peak stress between simulation and experiment is compared in Table 2. The average error in capturing the peak stress without considering strain range effects is 24.4%. This is brought down very significantly to 0.6% by considering strain range effects.

In this study, data was not available to validate the memory effect. However, the modeling approach does take into account the memory effect.

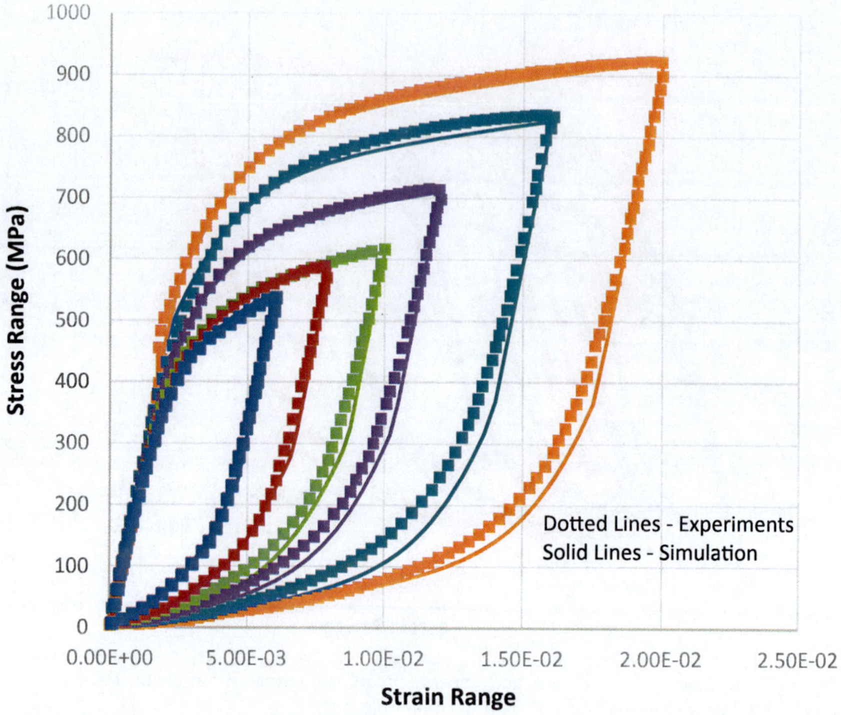

Fig. 2 Comparison of simulation and experiment of isothermal strain-controlled cycling for 316L(N) stainless steel considering strain range effect

5 Conclusions

It was found that the cyclic stress–strain hysteresis cycles of 316L(N) stainless steel across strain ranges could be captured well if strain range hardening seen in non-masing materials was considered. It was also shown that the strain range hardening effect was critical to accurately capture the strain energy per cycle as well as stress range to be able to accurately predict life.

Table 1 Comparison of prediction of strain energy and cycles to failure with and without strain range effect for 316L(N) stainless steel with experimental values [1]

Strain amplitude (%)	Experimental			Simulation (No strain range effect)			Simulation (with strain range effect)		
	Strain energy per cycle (MPa)	Cycles to failure	Total strain energy	Strain energy per cycle (MPa)	% error of strain energy per cycle (%)	Cycles to failure	Strain energy per cycle (MPa)	% error of strain energy per cycle (%)	Cycles to failure
0.3	1.355	14817	20078.66	1.654	22	13132	1.418	5	14158
0.4	2.314	7946	18387.28	2.940	27	6900	2.356	2	7805
0.5	3.047	5838	17787.39	4.382	44	4489	3.380	11	5262
0.6	4.751	2634	12513.95	5.901	24	2355	4.982	5	2512
0.8	8.099	1290	10447.45	9.047	12	1295	8.395	4	1244
1	12.140	765	9286.97	12.250	1	859	12.250	1	758
				Average error %	**22**		**Average error %**	**4**	

Table 2 Comparison peak stresses with and without strain range effect for 316L(N) stainless steel with experimental values [1]

Strain amplitude (%)	Experimental		Simulation (No strain range effect)		Simulation (with strain range effect)	
	Strain energy per cycle (MPa)	Peak stress	Peak stress	% error of peak stress (%)	Peak stress	% error of peak stress (%)
0.3	1.355	537.01	760.04	41.5	544.05	1.3
0.4	2.314	588.35	813.32	38.2	589.91	0.3
0.5	3.047	616.35	844.52	37.0	612.56	0.6
0.6	4.751	713.63	866.4	21.4	710.78	0.4
0.8	8.099	833.28	899.34	7.9	825.89	0.9
1	12.140	922.39	925.96	0.4	924.37	0.2
			Average error %	**24.4**	**Average error %**	**0.6**

References

1. Roy CR, Goyal S, Sandhya R, Ray SK (2012) Low cycle fatigue life prediction of 316 L(N) stainless steel based on cyclic elasto-plastic response. J Nucl Eng Des 253:219–225
2. Chaboche JL (1989) Constitutive equations for cyclic plasticity and cyclic viscoplasticity. Int J Plast 5(3):247–302
3. Chaboche JL (2008) A review of some plasticity and viscoplasticity constitutive theories. J Plast 24:1642–1693
4. Lemaitre J, Chaboche JL (1990) Mechanics of solid materials. Cambridge University Press, Cambridge
5. ANSYS, Inc. (2019) Theory reference for ANSYS 19.2. Canonsburg, PA
6. Kabir SMH, Yeo T (2011) Material parameters of a structural steel with plastic strain memory. J Mech Eng Autom 1:10–17
7. Chaboche JL, Van KD, Cordier G (1979) Modelization of the strain memory effect on the cyclic hardening of 316 stainless steel. L11, IASMiRT
8. Ohno N (1982) A constitutive model of cyclic plasticity with a nonhardening strain region. J Appl Mech 49:721–727

Fatigue Crack Growth Behaviour of Prototype Sized Pipe Bend and Its Equivalent Plate Type Geometry

R. Suresh Kumar, Kaushik, B. N. Rao, and K. Velusamy

Abstract As a part of the stringent safety requirement in the case of the operation of Fast Breeder Reactor, it is necessary to demonstrate that the postulated part-through-crack on the pipe surface shall not become the through-thickness crack during the period between two successive in-service inspection/repair schedules. A fast and accurate estimate of through-thickness crack growth studies plays an important role in demonstrating the structural integrity of the power plant components more effectively. Deployment of an advanced computational technique plays an important role in this direction. 3D numerical estimate on the Fatigue Crack Growth (FCG) behaviour of the prototype piping products is relatively slow and computationally expensive. Towards this end, it is necessary to develop simple computational models without compromising the accuracy. Accordingly, the FCG simulation has been carried out for a full-scale pipe bend, and the results are compared with an equivalent plate type geometry. The salient features adopted in identifying the equivalent plate concept and its comparative performance in the simulation of FCG behaviour are brought out in this paper.

Keywords Fatigue crack growth · Numerical fracture mechanics · Leak-Before-Break

R. Suresh Kumar · K. Velusamy
Indira Gandhi Centre for Atomic Research, Kalpakkam, India
e-mail: suresh@igcar.gov.in

Kaushik (✉)
DHIO Research & Engineering Ltd, Bangalore, India
e-mail: kaushik@dhioresearch.com

B. N. Rao
Indian Institute of Technology Madras, Chennai, India
e-mail: bnrao@iitm.ac.in

S. Seetharamu et al. (eds.), *Fatigue, Durability, and Fracture Mechanics*, Lecture Notes in Mechanical Engineering, https://doi.org/10.1007/978-981-15-4779-9_12

1 Introduction

Sodium-cooled Fast Reactors (SFR) piping systems are subjected to many transients loading conditions during its long service condition which leads to fatigue failure of the piping products. Under this condition, the potential defects can become crack and lead to the growth of crack and leading to leakage of the content due to the occurrence of the through-wall crack [1]. Thus demonstrating Leak-Before-Break (LBB) is one of the important safety requirements towards ensuring the structural integrity of the pressure retaining the piping system in the SFR [2]. The design guideline given in RCC MRx A16 is used for demonstrating the LBB requirements of the SFR piping system [3]. The design guideline rules of RCC MRx A16 is generally conservative [2, 4].

Once the surface crack reaches the other end of the thickness, then the respective crack length is called asymptotic crack (2Cs) length, i.e. the relative crack depth ratio (a/t) becomes unity at this condition. The relative crack depth ratio is the ratio between the crack depth to thickness. Computational fracture mechanics can play an important role in the realistic estimate of Cs value.

Conducting experiments for the estimate of Cs are very expensive and time-consuming. The development of a fast and accurate numerical method can provide an easy estimate of through-thickness crack length (Cs). More realistic prediction on the Cs helps in the economic demonstration of the structural integrity of the power plant components. Deployment of advanced computational technologies such as 3D numerical Fatigue Crack Growth (FCG) studies plays an important role in these considerations. It may be difficult to apply these methods to the prototype full-scale piping system. In this consideration, it is necessary to develop simple computational models without compromising the accuracy of the full-scale components. Accordingly, the FCG simulation has been carried out for a full-scale pipe bend, and these results are compared with its equivalent plate type geometry, and the results are presented in this paper. The salient features adopted in identifying the equivalent plate concept and its comparative performance in the simulation of FCG behaviour are brought out in this paper.

2 Specimen for FCG Studies

One of the prototype sized pipe bends belongs to the secondary sodium main piping circuit of the typical SFR that is identified for the analysis. SS 316 LN is the principal structural material used for the fabrication of the critical piping systems in SFR. The selected pipe bend is also made up of SS 316 LN material. The outer diameter of the pipe bend is 570 mm, the wall thickness is 15 mm and the bend radius is 820 mm. It is subjected to an in-plane closing bending moment with the application of 130 kN force with an offset distance of 1860 mm.

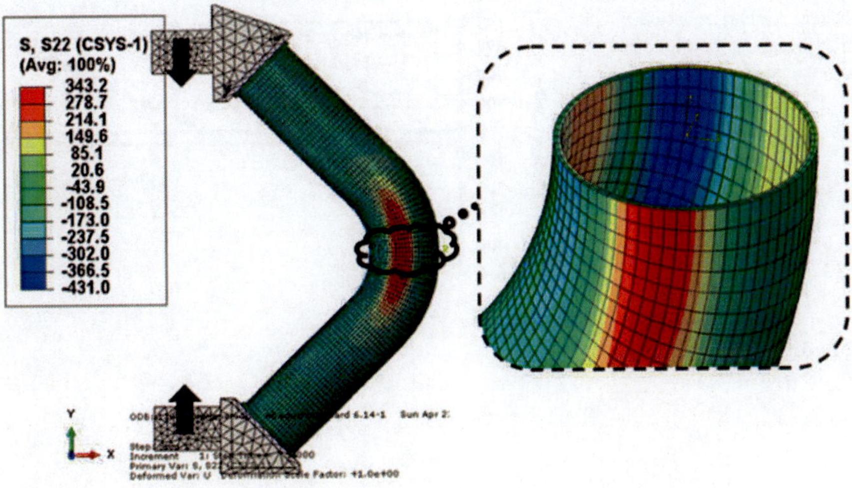

Fig. 1 Hoop stress (MPa) distribution of pipe bend under in-plane bending moment

3 Identification of the Equivalent Plate Concept

3.1 Pipe Bend Stress Analysis

The predominant stress responsible under the application of in-plane bending moment is the hoop stress [2]. The hoop stress distribution of the pipe bend considered for the FCG analysis is shown in Fig. 1. The maximum tensile hoop stress is found at the crown location of the pipe bend. The maximum value of the hoop tensile stress is 343.2 MPa under the application of 130 kN, as shown in Fig. 1. Thus the elliptical surface crack is introduced at this location along the longitudinal direction to cause early crack propagation [5].

Stress Intensification Factor (K_I) is the important parameter which describes the linear elastic fracture mechanics (LEFM) crack propagation behaviour [6]. The value of the K_I depends upon the membrane and bending stress at the location where the crack is going to be inserted [7]. The stress linearization is adopted at the crown location to compute the respective membrane and bending stress responsible for K_I. The pipe bend linearized stress distribution is shown in Fig. 2. Accordingly obtained membrane stress at the crown location is −41.3 MPa and the respective bending stress is 358.7 MPa.

3.2 Equivalent Plate Analysis

The factors considered in the design of the equivalent plate are given below:

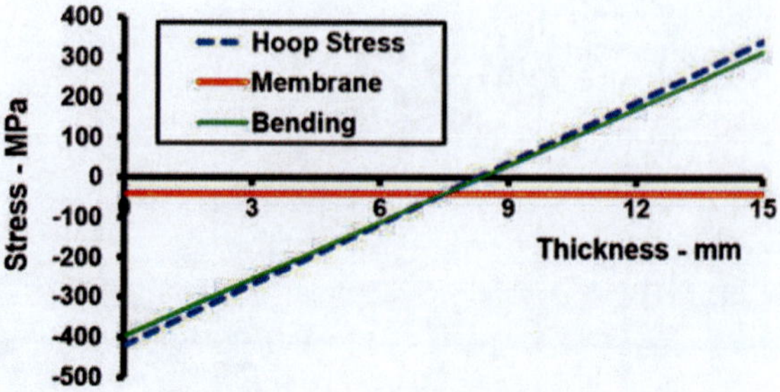

Fig. 2 Linearised stress distribution at the pipe bend crown location

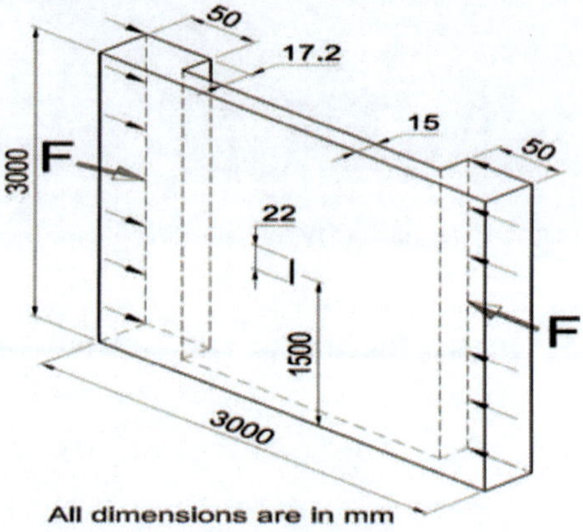

Fig. 3 Geometrical details of the equivalent plate

- The material of construction of the equivalent plate should be the same as the pipe bend.
- The thickness of the equivalent plate is the same as the pipe bend.
- The membrane and bending stress produced at the pipe bend crown location is the same as in case of the equivalent plate at which the crack is likely to introduce. The respective values of the membrane and bending stresses are −41.3 MPa and 358.7 MPa, respectively.

The major dimension of the equivalent plate designed based on the above considerations is shown in Fig. 3. The respective stress distribution, which is responsible for the Mode-I crack propagation on the equivalent plate, is shown in Fig. 4.

4 Fatigue Crack Growth Analysis

Franc3D software is used for the FCG analysis of both the prototype pipe bend as well as the equivalent plate. The capability of Franc3D in predicting crack growth behaviour has been validated using the benchmark analysis [5]. The advancement of the crack front is followed as per Paris law. The governing equation followed by

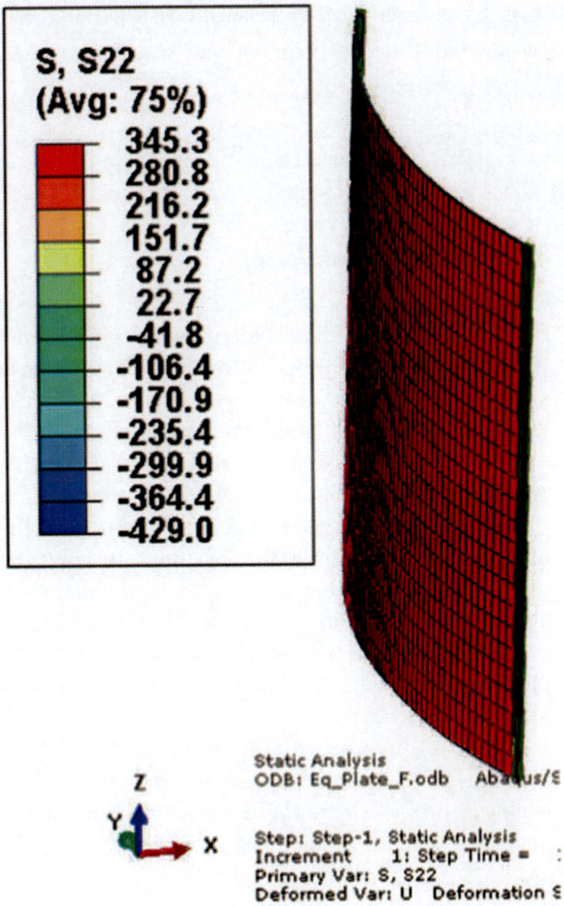

Fig. 4 Stress (MPa) responsible for the Mode-I crack propagation on the equivalent plate

Paris law is given in Eq. 1.

$$\frac{\mathrm{d}a}{\mathrm{d}N} = C\Delta K_{\mathrm{eff}}^{m} \tag{1}$$

where ΔK_{eff} is the effective stress intensity factor. It will take care of the crack closure effect. 'C' and 'm' are the Paris FCG material parameters. The important material parameters used for FCG are given in Table 1. The values of 'C' and 'm' are represented against the unit of 'ΔK_{eff}' in MPa $\sqrt{\mathrm{mm}}$ and '$\mathrm{d}a/\mathrm{d}N$' in mm/cycles. These material properties are taken from RCC MRx A3 (2012) against the material properties of SS 316 LN at room temperature.

Table 1 Important material parameters used in the FCG simulation

Item	E (MPa)	ν	Paris constants		Load ratio ($R = \sigma_{min}/\sigma_{max}$)	ΔK_{th} (MPa$\sqrt{}$mm)	K_{IC} (MPa$\sqrt{}$mm)
			C	m			
Value	195,500	0.3	7.5×10^{-16}	3.75	0.0	190.0	9510.0

4.1 FCG on Pipe Bend

FCG simulation on the prototype sized pipe bend has been carried out for pure bending condition with load ratio (R) = 0. The dimensions of the initial semi-elliptical shaped crack provided on the pipe bend are 22.0 mm length (along the axial direction) and 2.0 mm depth along the thickness direction. The Franc3D crack simulated FE mesh is shown in Fig. 5a. The K_I estimated at the crack surface location (location A and C from Fig. 5a) is 725.4 MPa $\sqrt{}$mm and that of the crack depth location (location B from Fig. 5a) is 431 MPa $\sqrt{}$mm. Both these values are greater than that of the threshold K_I of 190.0 MPa $\sqrt{}$mm. It shows that the surface crack introduced on the pipe bend has the potential to grow under cyclic loading.

The FCG simulation on the pipe bend has been performed, and the results are presented in Fig. 6. The observed crack growth along the surface direction (c) is plotted in Fig. 6a, and the depth directions (a) is plotted in Fig. 6b against the number of fatigue cycles.

The results indicate that the speed at which the Fatigue Crack Growth along the thickness direction is getting reduced and the crack growth along the length direction is increasing under pure cyclic bending condition. This trend is expected behaviour in case of the pure bending load.

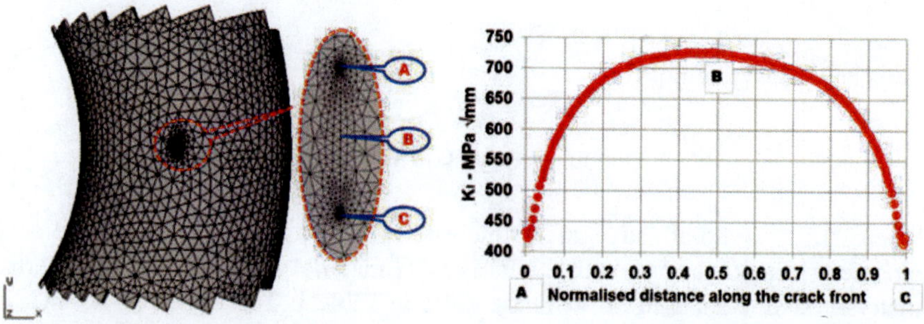

a). Pipe bend FE sub model

b). KI variation along the crack front

Fig. 5 FCG simulation for the pipe bend

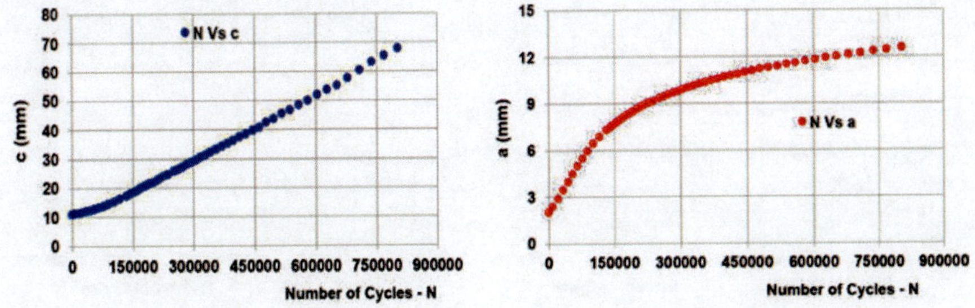

a). Crack length ('c') vs Number of Cycles

b). Crack depth ('a') vs Number of Cycles

Fig. 6 FCG data for the pipe bend

4.2 FCG on Equivalent Plate

FCG simulation on the equivalent plate has been carried out for the same loading condition and the same load ratio ($R = 0$). The same size and shape of the initial surface crack have been introduced on the tensile surface on the equivalent plate. The orientation of the initial crack is provided such that, the under cyclic bending condition, the mode-I crack will be initiated. The Franc3D crack simulated FE mesh is shown in Fig. 7a. The initial SIF variation is shown in Fig. 7b. The K_I estimated for the pipe bend at the same condition also presented in Fig. 7b. The K_I distribution along the initial crack front for both the prototype pipe bend and its equivalent plate is similar. These results indicated that the equivalent plate designed is satisfying one of the fundamental linear elastic fracture mechanics (LEFM) parameters.

The FCG simulation on the equivalent plate has been performed, and the results are presented in Fig. 8. The observed crack growth along the surface direction (c) is plotted in Fig. 8a, and the depth directions (a) is plotted in Fig. 8b against the number of fatigue cycles. The FCG has been performed up to 2 lakhs cycles of fatigue loading.

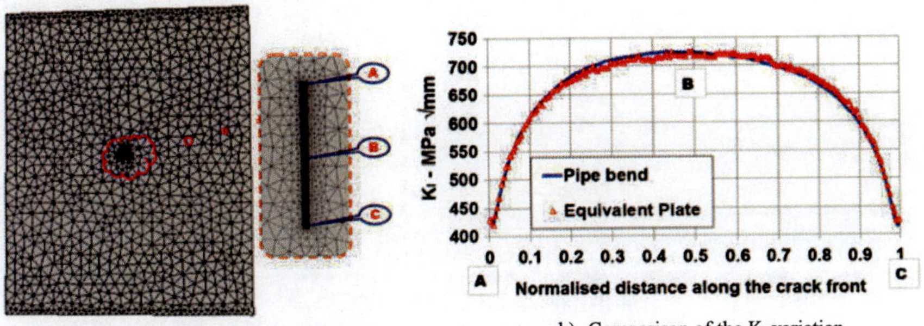

a). FE sub model for the equivalent plate

b). Comparison of the KI variation

Fig. 7 FCG simulation for the equivalent plate

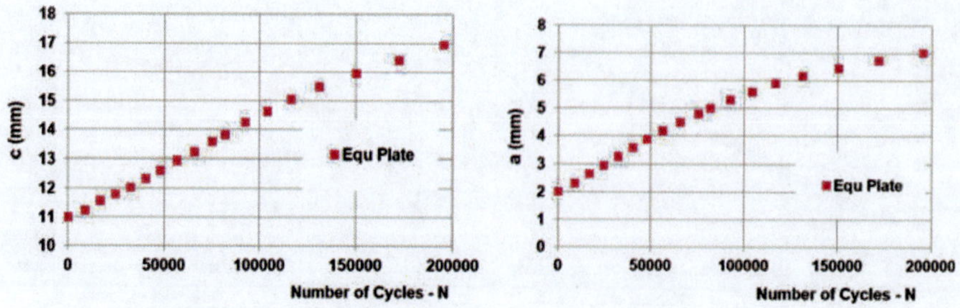

a). Crack length ('c') vs Number of Cycles b). Crack depth ('a') vs Number of Cycles

Fig. 8 FCG data for the equivalent plate geometry

5 Summary and Conclusion

The results of the numerical FCG analysis carried out for the prototype pipe bend as well as its equivalent plate geometry are summarized below:

- Towards simplifying the computational efforts associated with the numerical FCG analysis, the equivalent simplified plate geometry has been designed. The criteria adopted in designing the equivalent plate concept are the same material properties, same thickness, same membrane and bending stress location at which the crack is introduced.
- The analysis indicated that both the pipe bend as well as the equivalent plate had the same initial K_I variation along the initial crack front.
- The crack growth behaviour has been compared and the results are presented in Fig. 9a for the crack growth along the surface direction (c) and Fig. 9b for the crack growth along with the thickness direction (a).

The comparison of the FCG behaviour of the prototype pipe bend and its equivalent plate is given in Fig. 9. The results show that the numerical prediction on the crack length along the surface direction as well as the thickness direction had very good

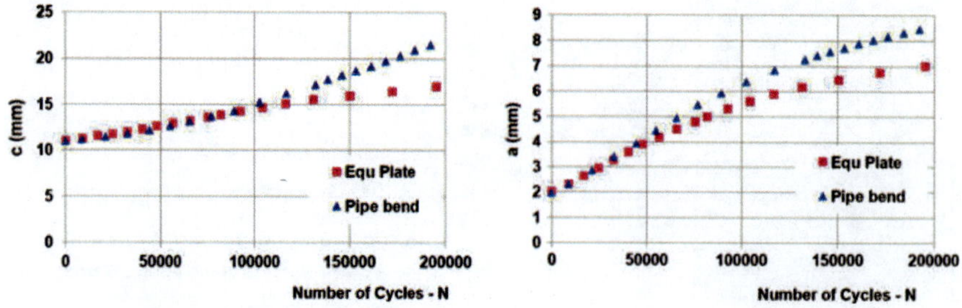

a). Crack length ('c') vs Number of Cycles b). Crack depth ('a') vs Number of Cycles

Fig. 9 Comparison for the FCG behaviour

agreement during the initial condition. Subsequently the trend on both is getting separated. It may be due to the difference in K_I of the equivalent plate and the pipe bend geometry due to the advancement of the crack size. Further investigation is in progress to assess the difference.

References

1. Suresh Kumar R et al (2017) Bridging the gap in predicting the crack initiation of the real life structures by nuclear class-1 piping design philosophies and full scale testing. In: 14th international conference on fracture (ICF 14), Rhodes, Greece, 18–23 June
2. Suresh Kumar R, Rao BN, Velusamy K (2019) Relevance of factor of safety based on number of cycles in the prediction of fatigue crack initiation as per A16 sigma-d approach. Int J Press Vessels Pip 175:103924
3. RCC-MR Section III—Tome 1—Subsection Z—Appendix A16 (2012) Guide for the prevention of fast fracture, Leak Before Break analysis and defect assessment
4. Suresh Kumar R, Rao BN, Velusamy K, Jalaldeen S (2019) Computational assessment of asymptotic crack size of a prototype sized pipe bend: comparison with A16 master curve. In: 25th international conference on structural mechanics in reactor technology—SMiRT25, Aug 04–09, Charlotte, North Carolina, USA
5. Suresh Kumar R, Rao BN, Velusamy K, Jalaldeen S (2019) Specimen level and component level simulation of fatigue crack growth behavior under cyclic bending. Frattura ed Integrità Strutturale. J Ital Group Fract 49:526–535
6. Anderson TL (2017) Fracture mechanics: fundamentals and applications. CRC Press
7. Newman J, Raju I (1984) Nasa technical memorandum 85793. NASA Langley Research Center, Hampton, VA

Fatigue Loads-Its Impact on Optomechanical Systems Considering Thermo-Environmental Effects

B. N. R. Ganesh, M. S. Raviteja, Visweswar Putcha, and N. Sravani

Abstract Optomechanical systems include a set of optical lens mounted in mechanical housing which will be exposed to the external environment. It is cumbersome to hold this mechanism in adverse thermal and vibrational conditions, unlike other metallic materials. However, it is inevitable to make sure this optomechanical combination to withstand fatigue loads. This paper presents Finite Element Analysis (FEA) and analytical calculation for the thermal, vibrational effects, and performance characteristics of optical lens while it is bonded with elastomer in a metallic cell to withstand such stringent conditions without any fracture of optical lens. FEA has been carried out to verify these effects on optical parameters, viz. wavefront errors using Zernike surface deformations polynomials. Both analytical calculations and FEA were carried out to achieve optimal design as well as to ensure for design validation. This paper also determines the fatigue design strength of optical lens for a specified lifetime.

Keywords Optomechanical system · Fatigue loads · Finite element method · Thermovibrational effects

1 Introduction

The optomechanical system comprises of a set of Infrared (IR) lens used to collect the light beam. These highly sensitive lenses are mounted in complex optomechanical housing and mechanical cells.

The challenge lies in the opto mechanical system design is not only to meet specified high-performance accuracy levels also to withstand harsh weather conditions like

B. N. R. Ganesh (✉) · M. S. Raviteja · V. Putcha · N. Sravani
Bharat Electronics Limited (BEL), R. T. Road, Machilipatnam, AP, India
e-mail: ganeshbnr@bel.co.in

M. S. Raviteja
e-mail: ravitejams@bel.co.in

V. Putcha
e-mail: visweswarputcha@bel.co.in

S. Seetharamu et al. (eds.), *Fatigue, Durability, and Fracture Mechanics*, Lecture Notes in Mechanical Engineering, https://doi.org/10.1007/978-981-15-4779-9_13

−20 and +60 °C operating temperature range, high humidity levels, high wind, and dusty conditions. The optomechanical system design should also meet the stringent vibration and shock conditions.

Dealing with fatigue loads in lens materials is critical in accomplishing the required strength when compared to dealing with fatigue loads in metallic materials like cast iron. The nature of the optical materials will further complicate the management of fatigue loads as these optical materials are brittle in nature.

Integrated optomechanical analysis involves the coupling of the structural, thermal, and optical simulation tools in a multidisciplinary process commonly referred to as structural–thermal–optical performance [1]. Designs must provide for positive margin against failure modes that include yielding, buckling, ultimate failure, fatigue, and fracture. Static and dynamic fatigue testings are alternative ways to study environmentally enhanced crack growth in an optomechanical system. Dynamic fatigue testing is an accelerated method where stress is applied to the specimen at a constant rate. Dynamic fatigue testing is performed to find out the fatigue resistance parameter for lifetime estimates. Materials with a larger fatigue resistance parameter have greater resistance to environmentally enhanced crack growth.

2 System Descriptions

The proposed optomechanical system is an *Extender* with 2X magnification, comprises of a set of IR lenses with afocal configuration. This has used double the focal length of objective lens. The extender consists of complex optomechanical housing to hold the lenses. Basically, the two lenses are mounted in a mechanical housing. The first lens (Lens-1) which is exposed to the atmosphere (external environment) is bonded to housing by elastomer (RTV) and is also tightened with a retainer ring in order to firmly hold the lens as well as to meet the vibration, shock requirements. The extender exploded view and optical layout containing the lenses are shown in Fig. 1a, b.

2.1 Optomechanical Design of Extender

Optomechanical design has been carried out and 3D model was built in using Solid works 18. The assembly containing two lenses bonded to mechanical housing using elastomeric adhesive is shown in Fig. 2. Also, these lenses are been tightened with retainer (s) to ensure and enhance the factor of safety of the overall system.

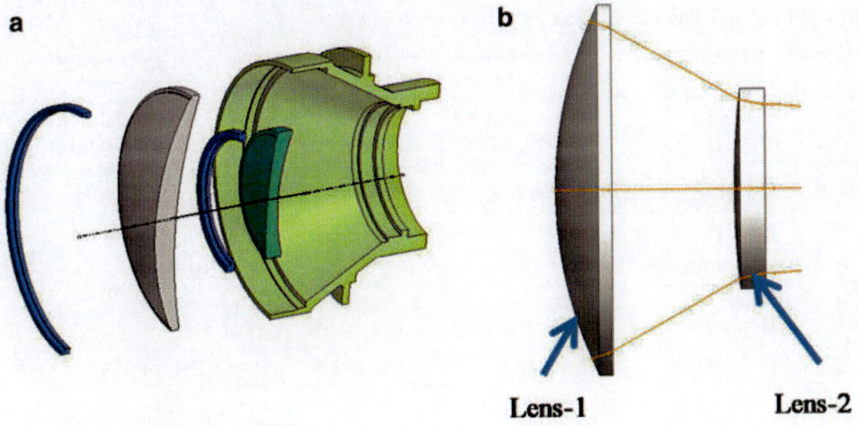

Fig. 1 **a** Extender assembly exploded view. **b** Optical ray diagram

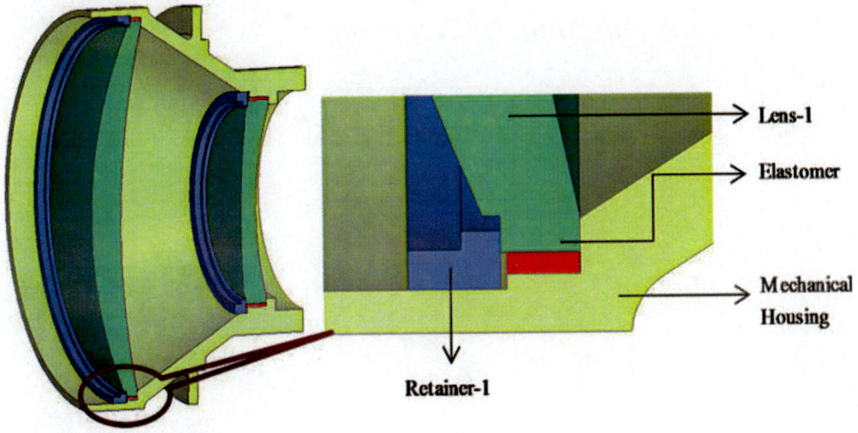

Fig. 2 Section view of the extender assembly

3 Finite Element Analysis of Extender

Nonlinear FE analysis is performed to obtain the stress concentration levels and temperature effects as contains materials like elastomers with nonlinear behavior and to obtain more realistic results when compared with analytical calculations. The solid 3D Model is imported into Ansys-2019-R2 Mechanical Platform to proceed for meshing and preprocessing activities. Step-by-step procedure for carrying out the analysis is explained below.

Table 1 Material properties of all the components

S No	Description	Lens-1 and lens-2	Bonding material (Adhesive)	Housing, retainer-1, and retainer-2
1	Material	Germanium	Silicone rubber compounds (after curing)	Aluminum alloy-6061-T6/T651
2	Young's modulus (E) (Pa)	1.3e+011	1.58e+008	6.916e+010
3	Poisson's ratio (υ)	0.3	0.495	0.33
4	Density ρ (kg m^{-3})	5323	1220	2770
5	CTE (C^{-1})	6.1e−006	3e-003	2.4e−005
6	Yield strength (Pa)	1.5e+008	17.2 e+6	2.75e+008
7	Weight in grams respectively	605 and 154	–	1105, 33, and 23

3.1 *Step-I Defining Material Properties*

Table 1 lists out all the material properties that are been assumed at temperature 25.0 °C. The same properties are added to Ansys Material Library.

3.2 *Step-II Defining Contacts*

A nonlinear static structural analysis has been performed here, and the types of contacts are defined in Table 2.

3.3 *Step-III Meshing*

The 3D CAD model is split into finite elements using meshing as described below.

An automatic meshing method is employed to a maximum extent.

Hex dominant is preferred since accuracy is of the highest concern.

Element sizing and other parameters are refined to get the highest mesh quality.

Meshed model (topological model) is shown in Fig. 3.

3.4 *Step-IV Load Cases*

Analysis s carried out with non-linear effects switched on. Also the large deflection modes are not disabled.

Table 2 Type of contacts between the parts

S No	Contact between parts	Type of contact selected for performing FEA	Inference
1	Lens cell and retainer-1	Frictional (with Coefficient of friction = 0.1)	Non-linear type of contact
2	Lens cell and bonding material (Adhesive)	Bonding	Linear type of contact
3	Lens cell and lens-1	No separation	Non-linear type of contact
4	Lens cell and retainer-2	Frictional (with Coefficient of friction = 0.1)	Non-linear type of contact
5	Lens cell and lens-2	No separation	Non-linear type of contact
6	Retainer-1 and lens-1	No separation	Non-linear type of contact
7	Retainer-2 and lens-2	No separation	Non-linear type of contact
8	Bonding material (Adhesive) and lens-1	Bonding	Linear type of contact
9	Bonding material (Adhesive) and lens-2	Bonding	Linear type of contact

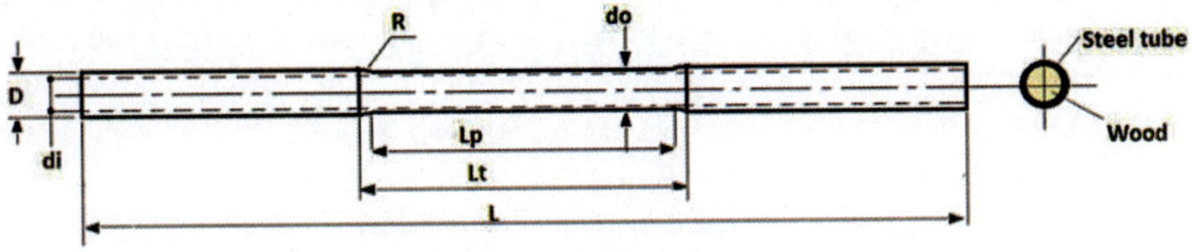

Fig. 3 Topology

The load is applied in both standard earth gravity direction and fixed support cases as shown in Fig. 4a, b. All degrees of freedom of the highlighted face are arrested in a fixed support case.

a. An axial load of 3.0 N is applied through Retainer-1 and 2.0 N is applied through Retainer-2 for tightening of lenses as extra support.

Fig. 4 **a** Standard earth gravity. **b** Fixed support

3.5 Step-V Assumptions

The following are the assumptions taken for carrying out the analysis.

- Lenses are assumed to be linearly elastic even though they are brittle in nature, because any brittle material exposes elastic property when the deformations are in the nanoscale.
- The bonding material defined here is after cured condition.
- All non-linear effects are switched on to analyze the behavior of adhesive (bonding material).
- Coefficient of friction is assumed as 0.1 (empirical and based on our previous experience).
- All contact formulations are set to an automatic process for simplicity.
- All swappable components are meshed using the sweep method for time-saving.

3.6 Step-VI Post-processing

Once solver has finished its solving, the results are assumed to be converged. Since the non-linear types of contacts are selected here, solving takes huge time and hence the maximum number of CPU cores is allotted for solving. The convergence criterion is shown in Fig. 5.

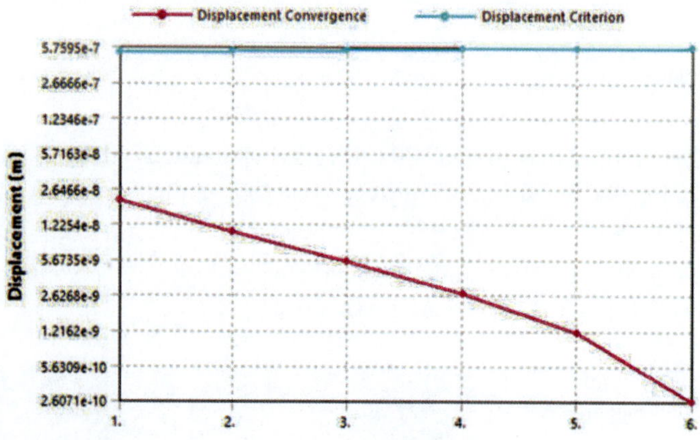

Fig. 5 Displacement convergence

3.7 Acceptance Criteria

The stress levels in each component of the assembly are defined below when they are subjected to different load conditions.

- Stress in Lenses <1.5e+008 Pa
- Stress in Mechanical housing <2.75e+008 Pa
- Stress in Retainers <2.75e+008 Pa
- Stress in Adhesive (Bonding material) <17.2 e+6 Pa
- Natural Frequency of assembly >1000 Hz (since the assembly will be subjected to vibration with frequency range: 5–1000 Hz).

3.8 Thermovibrational Analysis

Thermovibrational analysis was carried out at two different temperature levels and gravitational loads and the results are shown in Table 3.

Free vibration (Modal analysis) is performed for Lens-1, Lens-2, and for the entire Extender Assembly. The results are tabulated below at Table 4.

4 Optical Analysis for Wavefront Deformations Using Zernike Polynomials

Optical surface deformations for the front surface of Lens-1 is calculated from FEA results and the obtained values are taken as an input for generating the deformed surface using Zernike Polynomials in the MATLAB software and are shown in Fig. 6 and Table 5 lists the 36 Zernike polynomials which are useful in defining the magnitude and characteristics of the differences between the image formed by an optical system and original object. The optical design is done to achieve RMS = $\lambda/10$ ($\lambda = 0.6328\ \mu m$) value and obtained result including environmental effects is $\lambda/9.8$ which is quite reasonable and acceptable.

Similarly, Zernike surface deformations are been verified for the remaining three surfaces also.

5 Analytical Calculations

The thickness of the elastomeric bonding (t_e) required around Lens-1 from Yoder [2] is

$$t_e = (D_G/2)(\alpha_M - \alpha_G)/(\alpha_e - \alpha_M)$$

Table 3 Results of stress levels at two conditions

Component	Stress levels at 25 °C	Stress levels at 60 °C and 50 g load
Lens-1	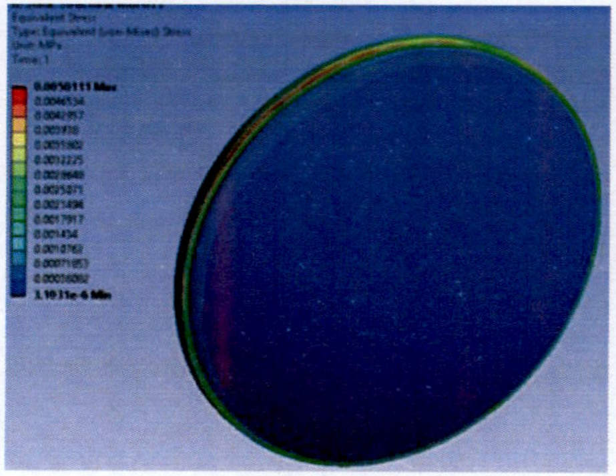	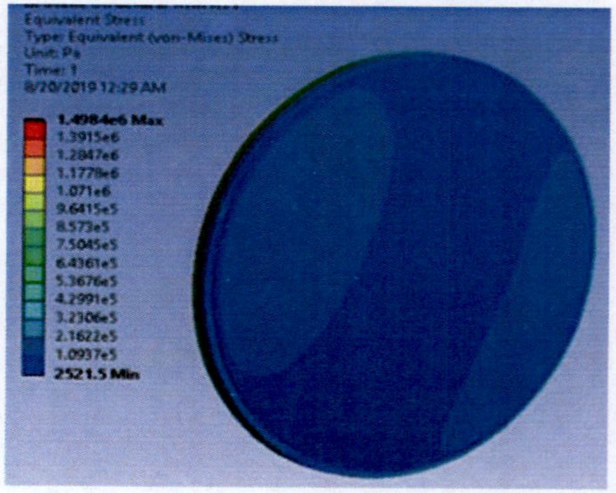

(continued)

Table 3 (continued)

Component	Stress levels at 25 °C	Stress levels at 60 °C and 50 g load
Lens-2	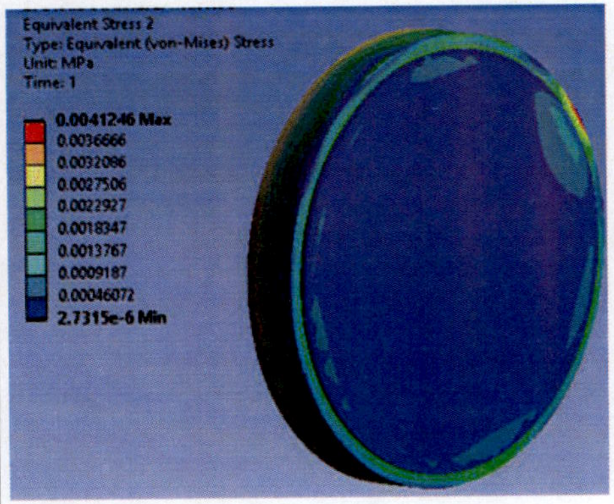	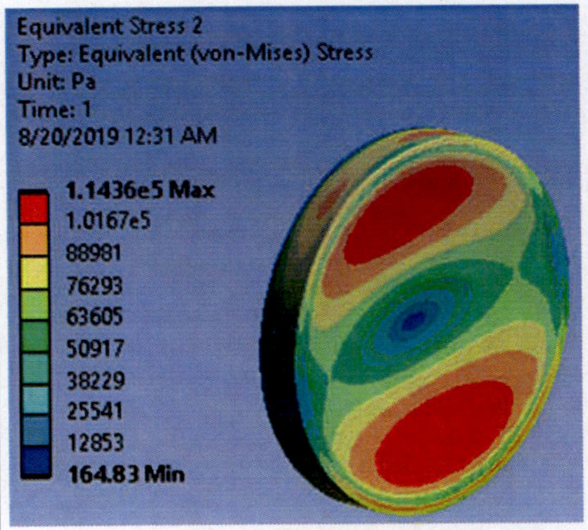

(continued)

Table 3 (continued)

Component	Stress levels at 25 °C	Stress levels at 60 °C and 50 g load
Retainer-1	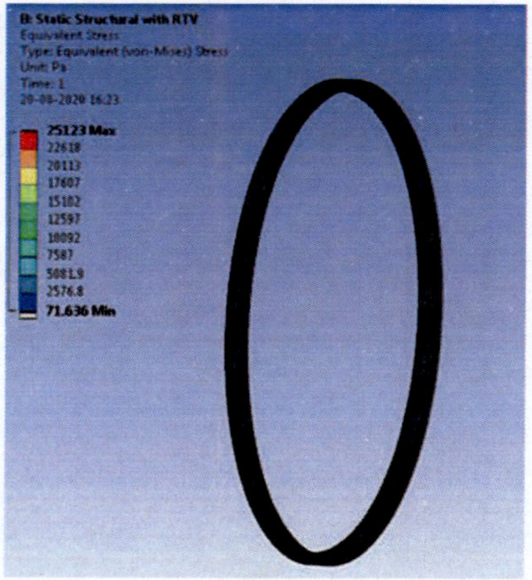	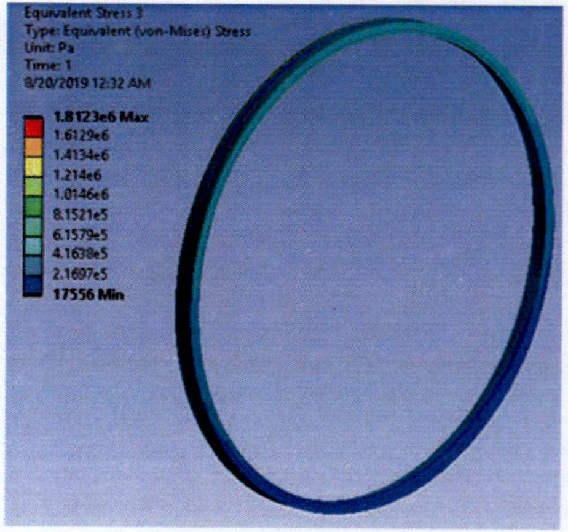

(continued)

Table 3 (continued)

Component	Stress levels at 25 °C	Stress levels at 60 °C and 50 g load
Retainer-2	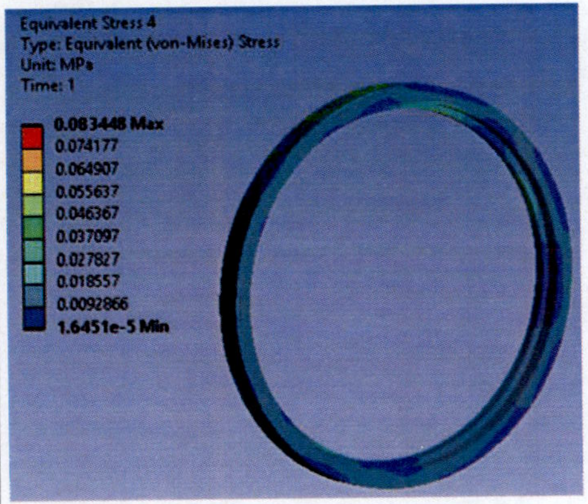	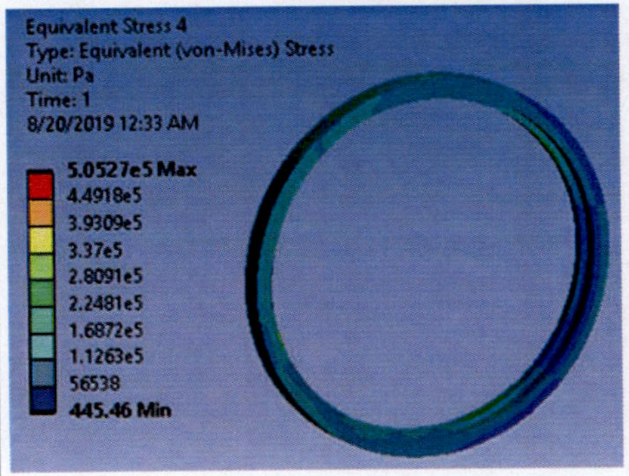

(continued)

Table 3 (continued)

Component	Stress levels at 25 °C	Stress levels at 60 °C and 50 g load
Housing		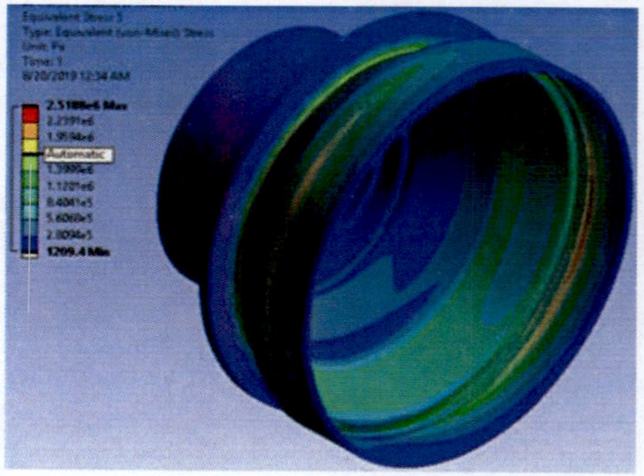
Elastomer-1	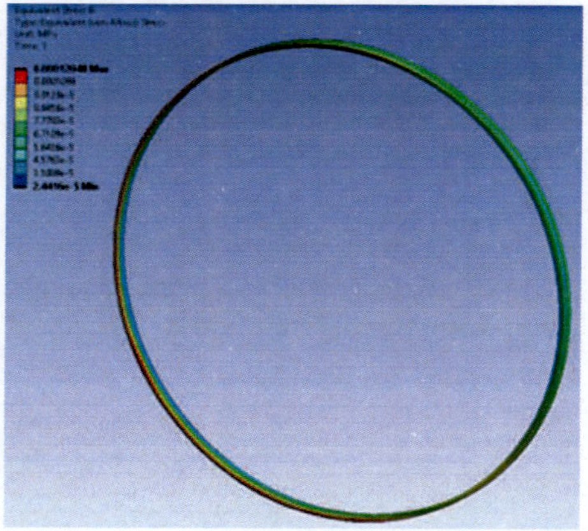	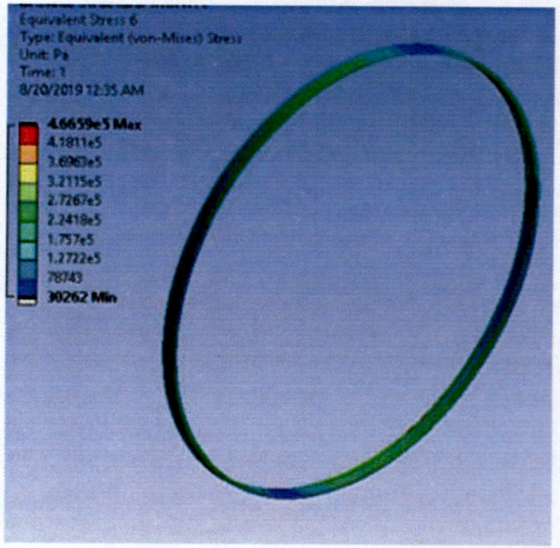

(continued)

Table 3 (continued)

Component	Stress levels at 25 °C	Stress levels at 60 °C and 50 g load
Elastomer-2	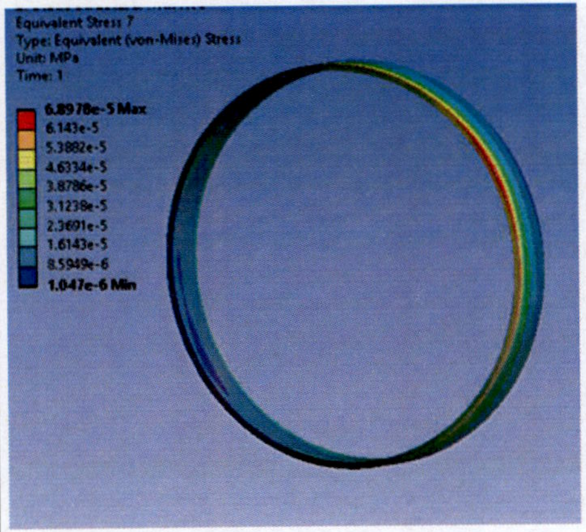	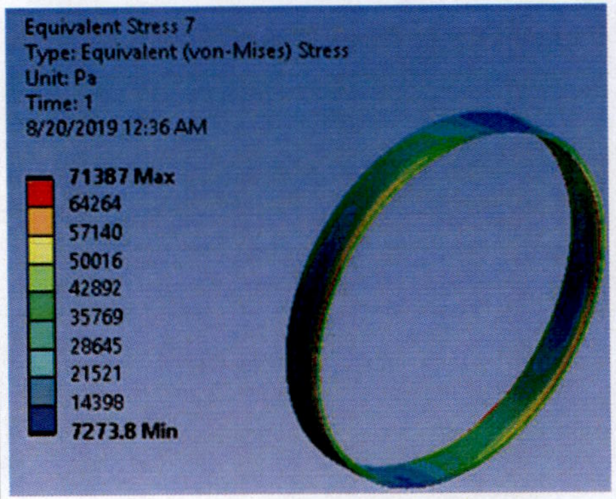

Table 4 First natural frequency for the components

Component	1st natural frequency from FEA results
Lens-1	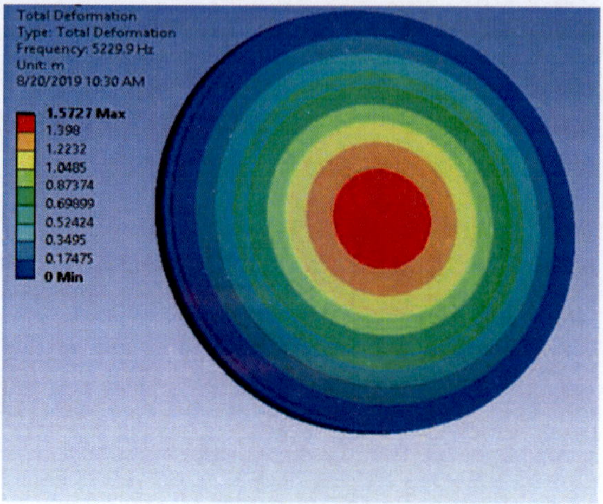 1st natural frequency for Lens-1 is 5230 Hz
Lens-2	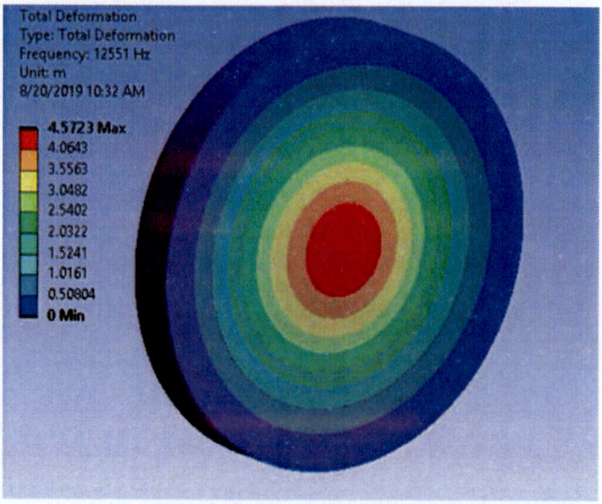 1st natural frequency for Lens-2 is 12550 Hz

(continued)

where α_G, α_e, and α_M are the CTEs of the lens, Elastomer, and mount, respectively.

Here, $D_G = 165$ mm, $\alpha_G = 6.1e{-}6$, $\alpha_e = 3e{-}6$, $\alpha_G = 2.4e{-}5$ pa

From the above formula,

Thickness of Elastomer (t_e) = 0.5 mm

Valente and Richard [3] reported an analytical technique for estimating the decentration, δ, of a lens mounted in a slightly compliant ring of elastomer when subjected

Table 4 (continued)

Component	1st natural frequency from FEA results
Extender assembly	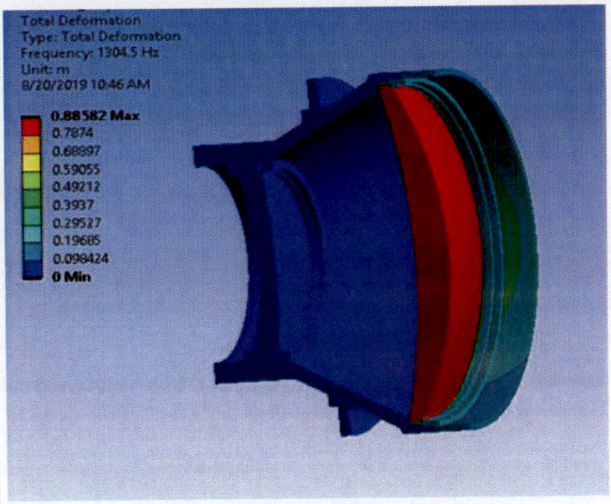 1st natural frequency is 1304 Hz

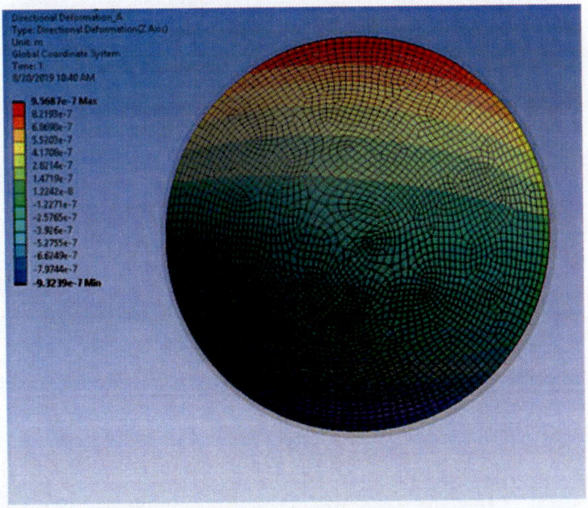

Fig. 6 RMS value for surface of lens-1 is λ/9.8

to radial gravitational loading. The radial acceleration forces by adding acceleration factor a_G as follows:

$$\delta = \frac{2a_G W t_e}{\pi D_G t_E \{[E_e/1 - v_e^2] + S_e\}}$$

$$S_e = E_e/[(2)(1 + v_e)]$$

Table 5 Zernike deformation surfaces and its values

No	Zernike polynomial term	Name	Obtained Zernike	Obtained RMS	Square: obtained RMS
1	1	Piston	2.004E−12	2.004E−12	4.01602E−24
2	$r\cos(\theta)$	Tilt x	−0.002892204	−0.001446102	2.09121E−06
3	$r\sin(\theta)$	Tilt y	−1.447292998	−0.723646499	0.523664256
4	$2r^2 - 1$	Focus	8.94562E−05	5.16476E−05	2.66747E−09
5	$r^2\cos(2\theta)$	Astig x	0.007991596	0.003262556	1.06443E−05
6	$r^2\sin(2\theta)$	Astig y	0.243054096	0.099226419	0.009845882
7	$(3r^3 - 2r)\cos(\theta)$	Coma x	−0.003250973	−0.001149392	1.3211E−06
8	$(3r^3 - 2r)\sin(\theta)$	Coma y	−0.836488338	−0.295743288	0.087464093
9	$1 - 6r^2 + 6r^4$	Pri spherical	0.003143023	0.001405602	1.97572E−06
10	$r^3\cos(\theta)$	Trefoil x	0.00332232	0.001174618	1.37973E−06
11	$r^3\sin(\theta)$	Trefoil y	−0.025053074	−0.008857599	7.84571E−05
12	$(-3r^2 + 4r^4)\cos(2\theta)$	sec.astig x	−0.00089599	−0.000283337	8.02799E−08
13	$(-3r^2 + 4r^4)\sin(2\theta)$	sec.astig y	−0.063412927	−0.020052928	0.00040212
14	$(3r-12r^3 + 10r^5)\cos(\theta)$	sec.com x	0.001045517	0.000301815	9.10921E−08
15	$(3r-12r^3 + 10r^5)\sin(\theta)$	sec.com y	0.192527511	0.055577905	0.003088904
16	$(-1 + 12r^2 - 30r^4 + 20r^6)$	sec.spherical	−0.000154975	−5.8575E−05	3.43103E−09
17	$r^4\cos(4\theta)$	Tetra foil x	0.004957448	0.001567683	2.45763E−06
18	$r^4\sin(4\theta)$	Tetra foil y	0.1257012	0.03975021	0.001580079
19	$(-4r^3 + 5r^5)\cos(3\theta)$	sec.trefoil x	−0.001753016	−0.000506052	2.56089E−07
20	$(-4r^3 + 5r^5)\sin(3\theta)$	sec.trefoil y	−0.169101531	−0.048815407	0.002382944
21	$(6r^2 - 20r^4 + 15r^6)\cos(2\theta)$	Terit.astig x	0.002144977	0.000573269	3.28638E−07
22	$(6r^2 - 20r^4 + 15r^6)\sin(2\theta)$	Terit.astig y	−0.011728979	−0.003134702	9.82635E−06
23	$(-4r + 30r^3 - 60r^5)\cos(2\theta)$	Terit.Com x	9.37862E−05	2.34465E−05	5.4974E−10
24	$(-4r + 30r^3 - 60r^5)\sin(2\theta)$	Terit.Com y	0.007054869	0.001763717	3.1107E−06
25	$1 - 20r^2 + 90r^4 - 140r^6 + 70r^8$	Terit.Spherical	−2.87285E−05	−9.57616E−06	9.17029E−11
26	$r^5\cos(5\theta)$	Penta foil x	−6.68953E−05	−1.9311E−05	3.72915E−10

(continued)

Table 5 (continued)

No	Zernike polynomial term	Name	Obtained Zernike	Obtained RMS	Square: obtained RMS
27	$r^5\sin(5\theta)$	Penta foil y	0.030563569	0.008822942	7.78443E−05
28	$(-5r^4 + 6r^6)\cos(\theta)$	sec.tetfoil x	−9.83929E−06	−2.62966E−06	6.91512E−12
29	$(-5r^4 + 6r^6)\sin(\theta)$	sec.tetfoil y	0.000475197	0.000127002	1.61294E−08
30	$(10r^3 - 30r^5 + 21r^7)\cos(\theta)$	Teri.trefoil x	−1.67967E−06	−4.19917E−07	1.7633E−13
31	$(10r^3 - 30r^5 + 21r^7)\sin(\theta)$	Teri.trefoil y	−2.31903E−05	−5.79758E−06	3.3612E−11
32	$(-10r^2 + 60r^4 - 105r^6 + 56r^8)\cos(2\theta)$	Quat.astig x	4.74625E−10	1.1187E−10	1.25149E−20
33	$(-10r^2 + 60r^4 - 105r^6 + 56r^8)\sin(2\theta)$	Quat.astig y	−7.642E−12	−1.80124E−12	3.24445E−24
34	$(5r-60r^3 + 210r^5 - 280r^7 + 126r^9)\cos(\theta)$	Quat.com x	−6.4E−14	−1.43108E−14	2.048E−28
35	$(5r-60r^3 + 210r^5 - 280r^7 + 126r^9)\sin(\theta)$	Quat.com y	−3.37E−13	−7.53555E−14	5.67845E−27
36	$(252r^{10} - 630r^8 + 560r^6 - 210r^4 + 30r^2)$	Quat.Spherical	0	0	0

where W is the lens weight, t_e the elastomer layer thickness, D_G is the lens diameter, t_E the lens edge thickness, ν_e the elastomer Poisson's ratio, E_e is the elastomer Young's modulus, and S_e is the shear modulus of the elastomer.

For the given data, Decentration, $\delta = 7.59\text{E}{-06}$ mm = 0.007 micron for lens-1

Vukobratovich [4] provided an equation for the natural frequency f_N of vibration for a body that deflects under its own weight when supported on a mount. The lens, whose optical surface sags when its axis is parallel to gravity or a lens mounted in a ring of elastomer with its axis horizontal. This equation is

$$f_N = (0.5/\pi)\sqrt{g/\delta}$$

For the given data, First Natural Frequency, $f_N = 5726$ Hz for the lens-1
Similarly, $f_N = 11325$ Hz for the lens-2
The FEA results and calculated analytical values are compared in Table 6.

Based on the acceptance criteria, it is found that both analytical and FEA values are matching. Frequency of 1304 Hz is far greater than the accepted criteria

Table 6 Comparison of FEA results and analytical values

Component	First natural frequency by analytical method (Hz)	First natural frequency by FEA (Hz)
Lens-1	5726	5230
Lens-2	11325	12550
Extender assembly	Derived value from FEA	1304

values and hence the designed extender is safe in dynamic environment considering thermovibrational effects.

5.1 Apparent Fatigue Resistance and Fatigue Life Calculation

The apparent fatigue resistance is calculated for the Lens-1 which is exposed to the external environment and change in cyclic loadings as it is assumed to be exposed to 100% relative humidity.

An apparent fatigue resistance parameter N' (also known as the residual stress flaw growth exponent) can be computed that accounts for the effects of residual stress [1].

$$N' = (rN + 2)/(r + 1)$$

where r is a measure of the residual stress field. The value of r is often conservatively assumed to represent a residual stress field produced for a point flaw ($r = 3$) that results in the following:

$$N' = \frac{3N + 2}{4}$$

$N = 104$ for the given germanium lens adapted from Table 2 of Strengths of Infrared Materials: Weibull Parameters [5].

Apparent fatigue resistance parameter for Lens-1 is $N' = 78.5$.

The Weibull A-basis Inert strength of the Germanium is 341 MPa.

For deriving time of failure calculations, only the inert strength and the apparent fatigue resistance parameter are require, [1] where RH is the relative humidity (for 100% humidity, $RH = 1$).

$$t = 0.0001RH * \left(\frac{\sigma^* FS_a}{\sigma_i}\right)^{-(N'-2)}$$

$$FS_a = 2 \times 10^{-5}\sigma + 0.98$$

For the desired Mean Time to Failure, $t = 16330$ h,

Applied Stress, $\sigma = 186$ MPa obtained from the above formula

For the time to failure value of 16330 h (Appx. 5.5 Yrs) of operation, the calculated applied stress is 186 MPa.

6 Conclusions

6.1 Optomechanical design of extender was completed. Nonlinear FE simulations using ANSYS workbench for proposed extender were carried out. The extender was designed using a mechanical athermalization concept in which elastomer used between the lens and metallic housing to minimize the relative deflections considering thermovibrational effects.

6.2 It was found that the stresses and the natural frequency in different components were far below the critical values and are within the stress limits. Both analytical calculations and FEA results were compared and validated to attain the optimal design.

6.3 Optical Analysis for deviated wavefront using Zernike Deformation Polynomials was performed and the obtained result is close to the designed value.

6.4 Apparent fatigue resistance and the life calculation of lens which is exposed to the atmosphere were calculated.

6.5 All the components of the extender were fabricated as per the optimized design and analysis parameters. After assembling the parts, a prototype has been made and the realized extender is as shown below.

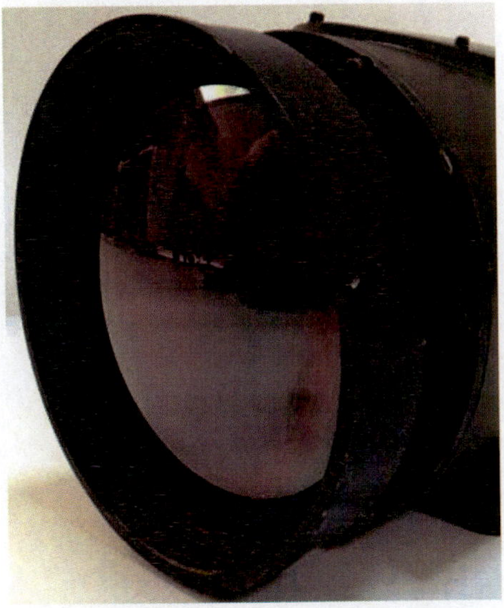

Acknowledgements We would like to express our sincere gratitude to BEL Management for all the help, support, and permission in carrying out the essential work for developing and to present this work.

References

1. Doyle KB, Genberg VL, Michels GJ (2002) Integrated opto mechanical analysis, 2nd edn.
2. Yoder Jr PR (1986) Opto-mechanical systems design, 3rd edn
3. Valente T, Richard R (1994) Interference fit equations for lens cell design using elastomeric lens mountings. Opt Eng 33:1223
4. Vukobratovich D (2003) Introduction to optomechanical design. In: SPIE short course SC014
5. Harris DC (2017) Weibull analysis and area scaling for infrared window materials

Experimental Investigation of Crack Generation in Rubber Compounds Used for Tire Applications

Vikash Kumar, Vidit Bansal, Sharad Goyal, Vipul John, and Sujith Nair

Abstract Tires are the complex composite product of filled-rubber, steel, and fiber cords, made of as many as 20 components. Filled-rubber used for each component has a different formulation, characteristic, and physical property. Each component of tire is designed to have a set of properties to perform a unique functionality. These different components of tire are adhered together by chemical bonding, van der Waals force, or electrostatic attraction. During the motion of the tire, each rubber component is under continuous fatigue loading, which may cause failure at some critical location in the tire. This research study mainly focuses on establishing the experimental framework for failure in rubber components due to crack generation. An experimental study was carried out on machine fatigue to failure testing (FTFT) for evaluating the life of rubber samples. Further surfaces of failed samples have been studied and compared through advanced microscopy.

Keywords Tire · Durability · Crack · Rubber · Experimental · FTFT · Fatigue · SEM

1 Introduction

Tire is the only component used to connect the road and the vehicle. Tire is designed to absorb road shocks, support the weight of the vehicle and goods. It transmits traction, torque, and braking forces to the road surface as well as maintains and changes the direction of travel. To fulfil all these basic functions, tires are made of resilient rubber and inflated with high-pressure air. A tire is an advanced engineering product made with rubber, fibers, and steel cord; it is made of as many as 20 components. Filled-rubber used for each component has a different formulation, characteristic, and physical property. Each component is designed to have a set of properties to perform a unique functionality. These different components of tire are adhered together by van der Waals force, chemical bonding, or electrostatic attraction. The variety of

V. Kumar · V. Bansal (✉) · S. Goyal · V. John · S. Nair
Advanced Engineering CEAT LTD (R&D), Halol, Vadodara 389350, India
e-mail: vidit.bansal@ceat.com

S. Seetharamu et al. (eds.), *Fatigue, Durability, and Fracture Mechanics*, Lecture Notes in Mechanical Engineering, https://doi.org/10.1007/978-981-15-4779-9_14

formulation available in tire for different components acts as a main source of failure in individual components as well as an interface between them during the working condition. Tire purpose fails when a single component of tire fails, which is a loss of resource as well as a safety concern. Therefore, today tire developer's main responsibility is to design having a low probability of crack formation in each component of tire. This job needs a good knowledge of the fatigue behavior of filled-rubber compounds. In the past studies, a variety of approaches were available to distinguish the damage of elastomeric materials due to cyclic loading. Among them, the first approach deals with the prediction of crack growth and it assumes that there is a preexisting initial defect, void, or crack in the materials, but there is no information about the size of crack. In this case, fatigue life directly depends on the growth of preexisting crack. The criteria proposed to understand the growth are maximum principal strain criteria by Cadwell et al. [1] and the strain energy density criteria by Gent et al. [2] and Beatty [3]. Saintier et al. [4, 5] proposed a principal stress criterion in a multi-axial loading condition. Mars [6] introduced an incremental strain energy on a cracking plane. A detailed review of the criteria can be found in Mars [6], Mars and Fatemi [7]. The second approach deals with the use of fracture mechanics for the prediction of crack initiation and growth. The Griffith [8] criteria were well known for crack growth in brittle materials. Rivlin and Thomas [9] extended Griffith criteria for crack growth in rubbery materials. The crack growth rate is governed by tearing energy, which is energy released per unit of incremental area of crack growth. Rivlin and Thomas investigated a number of test pieces and concluded that the tearing energy of cured rubber is a material property. The tearing energy is defined as in Eq. (1).

$$T_{\max} = -\frac{\mathrm{d}U}{\mathrm{d}A} \tag{1}$$

where U is the total stored elastic energy in a sample having crack and A is the area of crack growth surface. According to Lake and Lindley [10], tearing in a non-crystallization rubber occurs in a steady manner, while for crystallizing rubber, it generally occurs in a stick–slip manner, with the increment of force observed during the stick period until rapid failure occurs. The cyclic crack growth equation given by Lake is mentioned in Eq. (2), where c is the crack length, N is the number of cycles, and $T_{\max}$ represents the maximum energy release rate. For the simple extension test, the energy release rate as proposed by Rivlin and Thomas [9] is mentioned in Eq. (3) and crack size variation with cycle proposed by Thomas [11] is mentioned in Eq. (4), where W is the strain energy density of bulk material.

$$\frac{\mathrm{d}c}{\mathrm{d}N} = f(T) \tag{2}$$

$$T_{\max} = 2\pi W_{\max} c \tag{3}$$

$$\frac{\mathrm{d}c}{\mathrm{d}N} = r_c\left(\frac{T_{\max}}{T_c}\right)^F \tag{4}$$

Using this, idea an equation for life cycle calculation as mentioned in Eqs. (5) and (6) was calculated, where r_c and F are constants, N is the total number of cycles for crack growth from length c_0 to a very high value, i.e., ∞ (considering final crack length is very high compare to c_0 for a fixed minimum extension (constant maximum $2\pi W$). Equation 4 can be rewritten as Eq. (7) for the calculation of crack size.

$$N = \int_{c_0}^{\infty} \frac{1}{r_c\left(\frac{T_{\max}}{T_c}\right)^F}\mathrm{d}c = \frac{\int_{c_0}^{\infty}\frac{1}{C^F}\mathrm{d}c}{r_c\left(\frac{2\pi W_{\max}}{T_c}\right)^F} \tag{5}$$

$$N = \frac{T_c^F}{(F-1)r_c(2\pi W_{\max})^F}c_0^{1-F} \tag{6}$$

$$c_0 = \left\{\frac{N(F-1)r_c(2\pi W_{\max})^F}{T_C^F}\right\}^{\frac{1}{1-F}} \tag{7}$$

Models based on fracture mechanics require a good knowledge of initial defects is very restrictive for rubbery compound because these defects are very hard to identify. Other approach is based on continuum damage mechanics (CDM). Kachanov [12] introduced CDM theory, later Lemaitre and Chaboche et al. [13] derived a general theory for metallic materials. Gurtin and Francis [14] extended the idea for elastomeric material in order to simulate the Mullins effect. They studied one-dimension hyperelastic damage. Later, Simo [15] use for general three-dimensional case, which derived a viscoelastic constitutive equation for rubbery materials. Govindjee and Simo [16] improved this model by introducing microscopic concepts. Damage evolution can be found by integrating cycle by cycle. Hence, the damage evolution can be proportional to the number of cycles. This has been used by Ayoub et al. [17] and Ayoub et al. [18].

The primary objective of current work is to establish the test procedure on fatigue to failure test machines for evaluating the life of rubber samples at a fixed strain, then finalizing a method for statistical selection of a single life cycle for each compound in order to calculate the crack size. At last, a microscope has been used to investigate the contribution of void, microcrack, and filler agglomeration on crack growth in the filled-rubber compound during cyclic loading.

Table 1 Ingredients of compound

Compounds	Ingredients (PHR)			
	Natural rubber (RSS4)	Carbon black	ZnO	Sulfur
A	100	50; (N326)	8	5
B	100	50; (N330)	10	5
C	100	55; (N220)	8	3.5
D	100	39; (N326)	4	2

Fig. 1 **a** Standard sample (ASTM D 412) specimens cut from a vulcanized rubber sheet. **b** FTFT machine

2 Experimental Works

2.1 *Materials and Sample Preparation*

The materials and samples preparing processes were done in CEAT R&D laboratories based in Halol, Gujarat, India. To prepare rubber compounds, the ingredients were used to create a compound listed in Table 1. Only a few of the ingredients are mention here, which are required for this study. The compounding formulations which were represented in Table 1 were mixed till reaching a homogenous blend. The curing properties of the prepared compounds were measured with an MDR Rheometer based on ASTM D5289 at a temperature of 160 °C. The resultant rubber compounds were then compression molded to a 90% cure with a hydraulic press at 15 MPa, using a 160 °C cure temperature to produce a vulcanized sample. Dumbbell-shaped samples were cut from vulcanized sheets as shown in Fig. 1a.

2.2 *Result and Discussion*

2.2.1 Fatigue to Failure Test (FTFT)

The cut sample from the vulcanized sheet, as shown in Fig. 1, was tested in the Monsanto FTFT machine (shown in Fig. 1b.) in order to calculate the fatigue life

Table 2 Fatigue life cycle data of nine different sample of each four compound @ 100% strain

Compound details	Nominal strain %	Fatigue life cycle × 100								
		1	2	3	4	5	6	7	8	9
A	100	1354	1270	1274	1063	675	1669	1245	740	1649
B	100	1458	1635	1279	928	1481	1448	1441	1580	1257
C	100	2180	1799	1872	1942	2204	1892	1915	1855	1965
D	100	1689	1039	1453	1188	1628	1086	1686	1195	1117

of the sample. Testing frequency was kept constant throughout the project which was 100 CPM (cycle per Minute). In the starting, we have tested nine samples of compound 'A' from a single sheet of vulcanized sample at a 100% strain, and the result we got is listed in Table 2. While analyzing life cycle data, one can understand that there is a huge variation in the life cycle of data. The minimum life is coming 67,500, whereas maximum is coming 166,900, which is almost two and a half times greater than the minimum value. After this, one may come to a conclusion that FTFT machine is the main culprit behind variation in data. Actually, the variation in life cycle is the reality of filled-rubber compound, which is present due to the inhomogeneity present throughout the sample. Even a very simple rubber compound is made of more than twenty ingredients, which include rubber or blend of rubber, filler (carbon black or white filler which are used for reinforcement purpose), metal oxide, sulfur, fatty acid, processing oil, etc. Therefore, in the current work, it is assumed that life variation is coming due to the ingredients present in sample and not due to the involvement of other factors while preparation and testing. The problem arises here is that we need single life cycle data at a constant strain in order to calculate the crack size on the basis of Eq. (7) which will be used for finite element analysis of tire targeting life prediction. But in reality, we have different life cycle data for a single compound with huge difference.

2.2.2 Criticality of Choosing One Life Cycle at a Fixed Strain

In the previous section, we have got different life cycle data for all the nine tested sample for compound A. In Fig. 2, the impact of different life cycle on crack size calculation has been shown. Equation (7), which shows the relation between life cycle and crack size has been used to draw the graph. The typical values of other parameters considered for calculating the crack size are mentioned in Table 3. Lower crack size corresponds to high fatigue life and vice versa.

We can also see that all the nine sample data have nine different initial crack size varies from 0.008 to 0.019 mm. Maximum crack size is almost more than two times of minimum value. Therefore, in order to use this crack size for simulation, we need to identify the single correct crack size, which will be able to predict the correct life cycle of given compound. Finally, it will help to analyze the weakest section of tire as well as life calculation of whole tire. The above-mentioned reasoning shows the

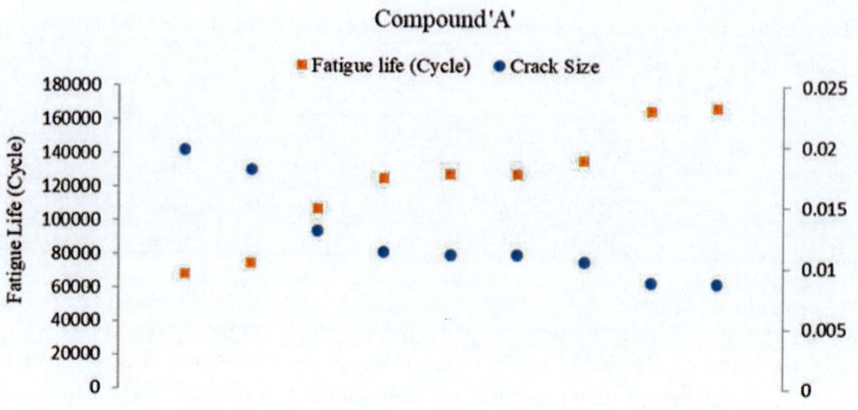

Fig. 2 Variation of crack size with fatigue life of nine different sample of compound '**A**'

Table 3 The typical value used for the calculation of crack size for compound 'A'

Variable	r_c	Tc	F	$W_{\max}$
Value	0.0011 mm/cycle	26 kJ/m^2	2.2	(1.1 mJ/mm^3) (strain)^2

criticality of selecting life of each rubber component corresponding to each strain value.

2.2.3 Statistical Selection of Single Life Cycle Data at a Fixed Strain

To understand the variation pattern in life cycle of different samples, we have tested nine samples of each three compound naming as 'B,' 'C,' and 'D' at same strain. The variation in life cycle is coming in a same manner as it was in the case of compound 'A,' which is random. The only difference observed here is the numerical value of fatigue life cycle, which is maximum for compound 'A' and minimum for Compound 'C.' For detailed understanding of the data variation, the statistical parameters like mean, median, and standard deviation has been calculated for all the four different compounds. The result is mentioned in Table 4.

The standard deviation is coming minimum for compound 'C' and maximum for Compound 'A.' This means most of the data of Compound 'A' is very far from its mean value. Similarly, most of the data of compound 'C' is close to its mean value. The mean and median values of the entire compound are coming very close except for compound 'C,' which is little higher than the other compounds, see Table 4, while comparing the data of life cycle of compound 'A' as shown in Fig. 3. It is clear that mean and median line is very close. Also, it is very near to five data point out of nine data.

The rest four data points are lying away from the mean line. In which two is lying above the mean line, whereas other two is lying below the mean line. The lower life

Table 4 Mean, median, and standard deviation of fatigue life data of four different compounds @ 100% strain

Compound details	Fatigue life cycle × 100			
	Mean	Median	Absolute difference of mean and median	Standard deviation
A	1215.4	1270	54.6	326.9
B	1389.7	1448	58.3	199.6
C	1958.2	1915	43.2	133.2
D	1342.3	1195	147.3	255.2

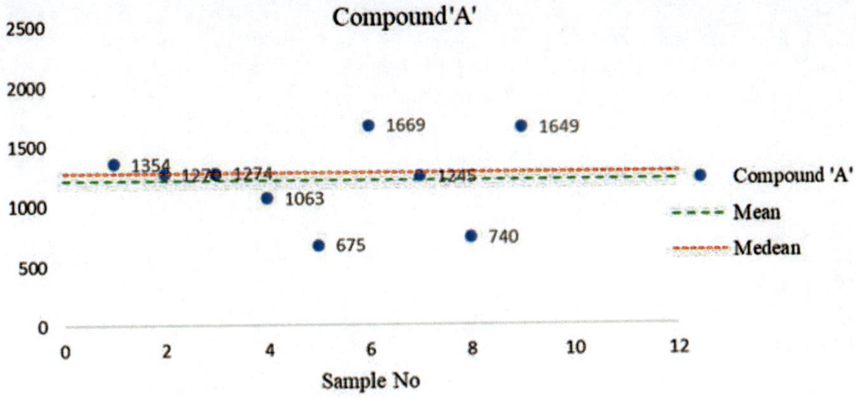

Fig. 3 Life cycle data of nine different samples at a fixed strain of 100% of compound '**A**'

cycle shows that there is high level of discontinuity present in the gauge portion of that particular FTFT sample. That may be due to the presence of filler agglomeration or undispersed zinc oxide and sulfur.

Detailed picture of discontinuity will be discussed in next section with the help of scanning electron microscope (SEM) analysis. Similarly, higher life cycle shows that there is low level of discontinuity present in that particular sample cut from the same vulcanized sheet. Other five data points which are very close to mean line show moderate level of discontinuity. It can be visualized as some agglomerates present in the gauge portion of sample as well as in some portion there is no agglomerate or void. One must note that crack will grow faster if there is large no of agglomerate present. It is because decohesion start at that point and one crack will join another crack and grow accordingly at a higher rate for the case of larger agglomeration present. The low-level discontinuity (safer) can be comprises with moderate level with more safety assurance. Therefore, the probability of moderate-level discontinuity is very high as it included five moderate-level and two low-level discontinuity data points out of nine data points. Hence, selecting mean or median data will be more reasonable for single life selection. Before coming to any conclusion, we have analyses the life cycle data of other three compounds as shown in Figs. 4, 5 and 6. For compound 'B,'

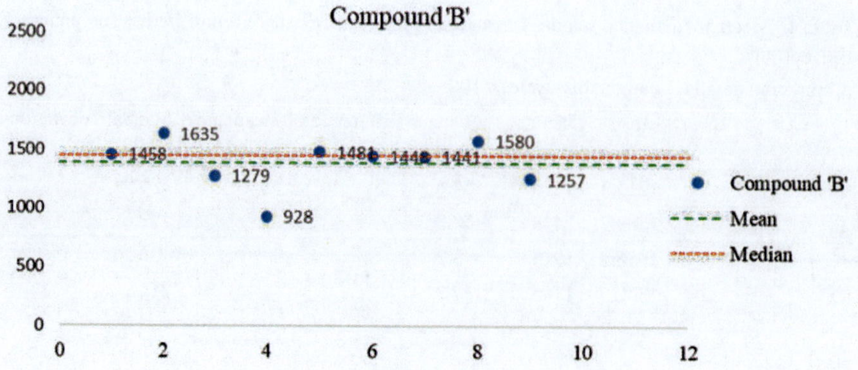

Fig. 4 Life cycle data of nine different sample at a fixed strain of 100% of compound '**B**'

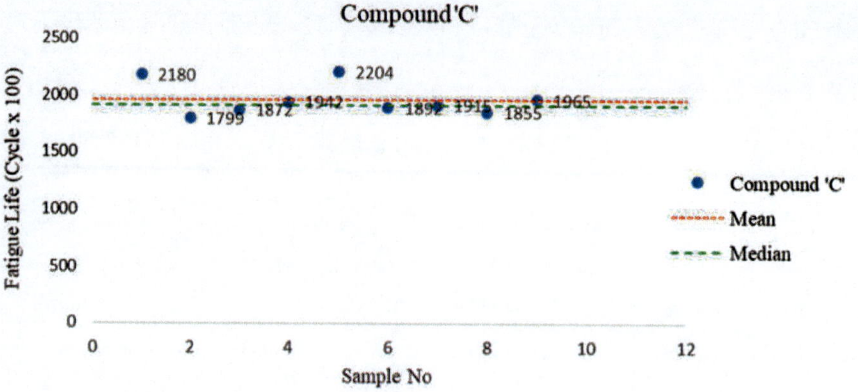

Fig. 5 Life cycle data of nine different sample at a fixed strain of 100% of compound '**C**'

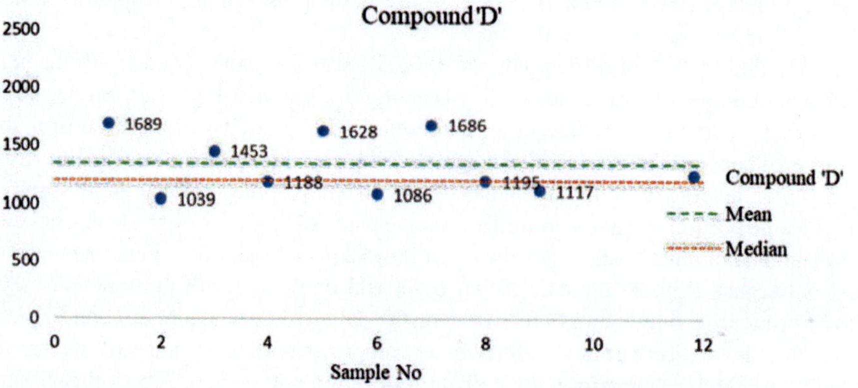

Fig. 6 Life cycle data of nine different sample at a fixed strain of 100% of compound '**D**'

six data points are lying very near to mean and median lines, whereas two are lying above the mean line and one is below the mean line as shown in Fig. 4.

After Appling the methodology adopted for compound 'A,' one can say that eight samples out of nine show the moderate level of discontinuity. Hence, selecting mean and median data will be more reasonable for this compound also. Similarly, for compound 'C' and 'D' all nine data are coming in moderate-level discontinuity as shown in Figs. 5 and 6. Now, question arises on selection between mean and median. For compound 'A,' 'B' and 'C,' selecting either mean or median value will be able to represent the moderate-level discontinuity, whereas in the case of compound 'D' only median value represent the moderate-level discontinuity. Therefore, keeping durability in mind, the best possible life cycle selection will be the median value.

2.2.4 Surface Study of Sample Through Scanning Electron Microscope (SEM)

To understand the level of discontinuity, the surface of FTFT sample was analyzed using SEM. Two types of sample were considered for this analysis, one before fatigue test, i.e., virgin sample and other after fatigue test. Figure 7a shows a large agglomerate at one point, whereas there is no such agglomerate in other point in the matrix. The level of discontinuity assumed in previous section was based on these agglomerates or voids present in virgin sample as shown in Fig. 8. The density of agglomerates or voids decides the level of discontinuity. Higher number of agglomerates or voids present in a given specified area will consider as high level of discontinuity and vice versa. The agglomerate present in Fig. 7a is may be because of filler agglomerates, sulfur or undispersed oxide. For deeper understanding, energy-dispersive X-ray analysis (EDXA) can be done, which is not currently the primary objective of this work. Figure 7b shows an initiation of crack for a sample which has been taken out from an FTFT sample after successfully completion of 30,000 cycles at 100% strain. Probably, it started due to decohesion of filler matrix [19].

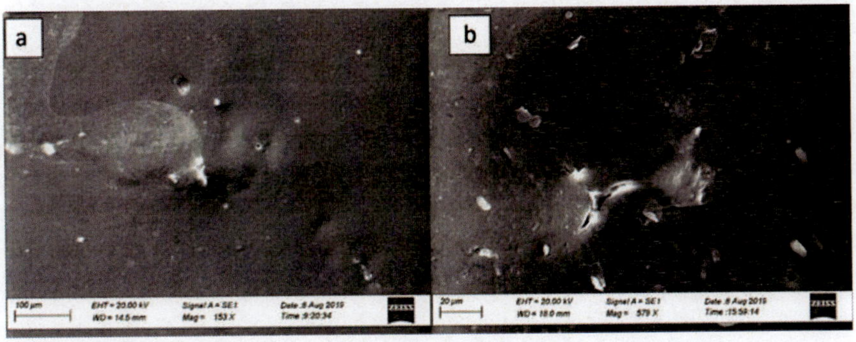

Fig. 7 **a** Before fatigue (Agglomerate). **b** After 30,000 cycles (crack initiation starts due to decohesion of filler and rubber)

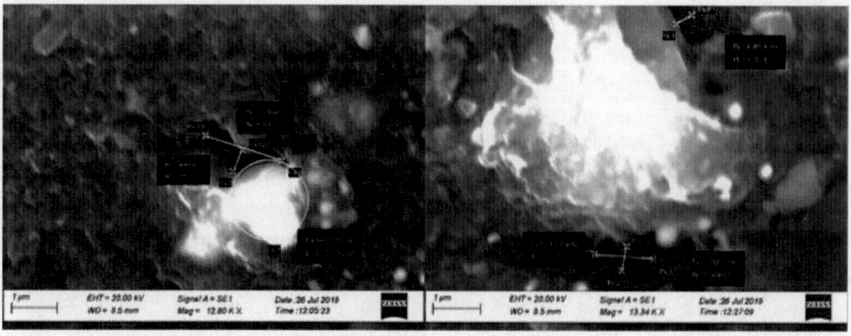

Fig. 8 Before fatigue (Void present near agglomeration)

Figure 8 shows a void near to agglomerate of 0.5–2 μm range. The initial void size present in the matrix before fatigue will increase during fatigue loading and the decohesion of filler and rubber will start from this point. This means it works as a principal source for the crack growth propagation. From above observation, one can say that filler and rubber matrix interaction play a main role on the fatigue behavior.

3 Summary

The life variation within the compound for different samples tested at a common strain was very high, which is due to the level of discontinuity present in the sample. Based on SEM analysis, it can be explained that it is due to the presence of filler agglomeration or undispersed zinc oxide and sulfur. In the present work, four different compounds were made, tested, and variation in FTFT results were justified based on discontinuity levels. Life cycle results obtained from FTFT machine were analyzed statistically and median values will be used for future studies.

4 Future Scope

In future, we will use the developed methodology to all the component of tire at different strains to find out the crack precursor size of each individual. Later, it will be used for finite element analysis, in order to find out the life of tire as well as to identify the weakest section in tire which will help to improve the durability of tire.

References

1. Cadwell SM, Merrill RA, Sloman CM, Yost FL (1940) Dynamic fatigue life of rubber. Ind Eng Chem Anal Ed 12:19–23
2. Gent AN, Lindley PB, Thomas AG (1964) Cut growth and fatigue of rubbers. I. The relationship between cut growth and fatigue. J Appl Polym Sci 8:455–466
3. Beatty JR (1964) Fatigue of rubber. Rubber Chem Technol 37:1341–1364
4. Saintier N, Cailletaud G, Piques R (2006) Crack initiation and propagation under multiaxial fatigue in a natural rubber. Int J Fatigue 28:61–72
5. Saintier N, Cailletaud G, Piques R (2006) Multiaxial fatigue life prediction for a natural rubber. Int J Fatigue 28:530–539
6. Mars WV (2002) Cracking energy density as a predictor of fatigue life under multiaxial conditions. Rubber Chem Technol 75:1–17
7. Mars WV, Fatemi A (2002) A literature survey on fatigue analysis approaches for rubber. Int J Fatigue 24:949–961
8. Griffith AA (1920) The phenomena of rupture and flow in solids. Phil Trans R Soc Lond A221:163–198
9. Rivlin RS, Thomas AG (1953) Rupture of rubber. I. Characteristic energy for tearing. J Polym Sci 10:291–318
10. Lake J, Lindley P (1965) The mechanical fatigue limit for rubber. J Appl Polym Sci 9:1233–1251
11. Thomas AG (1994) Rubber Chem Technol 67:G50
12. Kachanov LM (1958) Time of the rupture process under creep conditions. Izv Akad Nauk SSR Otd Tech Nauk 8:26–31
13. Lemaitre J, Chaboche JL (1985) A continuous damage mechanics model for ductile fracture. J Eng Mater Technol 107:83–89
14. Gurtin ME, Francis EC (1981) Simple rate-independent model for damage. J Spacecr 18:285–286
15. Simo JC (1987) On a fully three-dimensional finite-strain viscoelastic damage model: formulation and computational aspects. Comput Methods Appl Mech Eng 60:153–173
16. Govindjee S, Simo JC (1991) A micro-mechanically based continuum damage model for carbon black-filled rubbers incorporating Mullins effect. J Mech Phys Solids 39:87–112
17. Ayoub G, Nat-Abdelaziz M, Zari F, Gloaguen J, Charrier P (2011) A continuum damage model for the high-cycle fatigue life prediction of styrene-butadiene rubber under multiaxial loading. Int J Solids Struct 48(18):2458–2466. ISSN 0020-7683
18. Ayoub G, Nat-Abdelaziz M, Zari F, Gloaguen J, Charrier P (2012) Fatigue life prediction of rubber-like materials under multiaxial loading using a continuum damage mechanics approach: effects of two-blocks loading and R ratio. Mech Mater 0167–6636(52):87–102
19. Grandcoin J, Boukamel A, Lejeunes S (2014) A micro-mechanically based continuum damage model for fatigue life prediction of filled rubbers. 51(6):1274–1286

Numerical Analysis and Experimental Study for Fatigue Life Behavior of HSLA and UHSS Welded Joints

Rakesh Goyal and Mohamad El-zein

Abstract Welded steel joints are widely used for the structural components in the on/off-highway industry. Due to increasing regulations around CO_2 emissions and competitive demand for improved fuel economy, there is a significant focus on the lightweighting of structures in the machinery components deriving the use of higher strength steel grades. It's given that the lightweighting objectives should be achieved without compromising on functional performance, life-long durability, and safety among many others. HSLA (high-strength low alloy steel) and UHSS (ultra high-strength steel) grades are of specific interest considering their favorable mechanical properties along with excellent weldability. During this study, 345-MPa HSLA and 690-MPa UHSS steel grades were selected. Understanding the fatigue behavior of high-strength steel welded joints is of specific interest because fatigue properties for some of the steel grades can degrade from the thermal effects of welding. Keeping this objective in mind, an extensive and systematic experimental fatigue test study was performed on the double fillet T-joints welded from HSLA and UHSS steel grades. The experimental investigation included detailed metallurgical analysis such as microstructure and hardness variations across three distinct zones of welded joints, residual stress measurements using X-ray diffraction, and weld toe geometry measurements using dental molds. Further, detailed numerical finite element and fatigue analysis were carried out to predict the total fatigue life (crack initiation plus propagation) of the welded joints, based on the total fatigue life approach. The total fatigue life approach is an advanced fatigue analysis method which allows estimating the total life of the structure without the need to divide into arbitrary crack initiation and propagation phases, making this an attractive method for engineering practice. Finally, the outcome from the numerical analysis has been compared with the experimental fatigue test results for the welded joints made from HSLA and UHSS steel grades, showing good correlations.

R. Goyal (✉)
John Deere Asia Technology Innovation Center, Pune 41013, MH, India
e-mail: GoyalRakesh@JohnDeere.com

M. El-zein
John Deere Moline Technology Innovation Center, Moline, IL 61265, USA
e-mail: El-ZeinMohamadS@JohnDeere.com

S. Seetharamu et al. (eds.), *Fatigue, Durability, and Fracture Mechanics*, Lecture Notes in Mechanical Engineering, https://doi.org/10.1007/978-981-15-4779-9_15

Keywords HSLA · UHSS · Welded joints · Crack initiation · Crack propagation · Total fatigue life

1 Introduction

Welded steel joints are widely used for the structural components in the on/off-highway industry. Due to the increasing regulations around CO_2 emissions and competitive demand for improved fuel economy, there is a significant focus on lightweighting of structures in the machinery components deriving the use of higher strength steel grades, e.g., lifting equipment used in the wind turbine industry is moving toward increased usage of higher strength steel grades as the lightweight of the equipment helps with mobility [1]. Marine industry is seriously considering the use of high-strength steel material as more cargo can be carried by lightweighting of the vessel and hence reducing the cost of international freight charges impacting the lives of everyone in the society [2]. The usage of high-strength steel is not only limited to sheet application but includes welded tubular structures as this can significantly reduce construction costs and lower carbon footprints [3]. It's given that the lightweighting objectives should be achieved without compromising on functional performance, life-long durability, and safety among many others. HSLA (high-strength low alloy steel) and UHSS (ultra high-strength steel) grades are of specific interest considering their favorable mechanical properties along with excellent weldability. Understanding the fatigue behavior of high-strength steel welded joints is significantly important because fatigue properties for some of the steel grades can degrade from the thermal effects of welding.

Guo et al. [4] studied the fatigue performance of welded joints made using multiple high-strength steel materials and reported that fatigue strength for the welded joints was lower by around 40% as compared to the base material. Lahtinen et al. [5] reported that hardness gradient and weld geometrical shape are among the main parameters influencing the fatigue behavior of high-strength steel welded joints and responsible for fatigue crack initiation from the weld toe or root. Fatigue damage process starting with crack initiation from the weld toe involves multiple phases such as crack initiation, microcrack propagation, and, finally, macrocrack propagation resulting in final failure [6]. Considering the number of variables which affect the fatigue performance of welded joints, it's important to understand and incorporate influence from these variables in a systematic manner. Measuring actual microgeometrical weld toe features and accounting for the effects of those microscopic weld features is critical to improving the accuracy for fatigue life prediction of welded joints, especially in case of higher strength steel welded joints [7]. Detailed fatigue study was carried out to investigate the influence of various parameters such as stress concentration at weld toe, internal defects, and welding process method on the fatigue behavior of the butt welded joints made from high-strength steel thin sheets [8]. In addition to the geometrical features, measuring and accounting of the weld residual stress is equally important for accurate fatigue life estimation considering that residual stress can significantly influence the mean stress of the cyclic loading history [9]. Effect of constant and variable amplitude loading history on fatigue strength of S355 and S700 high-strength steel welded joints has been studied in detail [10].

Furthermore, numerical estimation methods for fatigue life prediction of welded joints produced using steel grades of varying levels of strength are of even higher scientific interest. Significant literature is available showing continues research efforts being made over many decades to improve the accuracy of stress–life, strain–life, and fracture mechanics-based fatigue life prediction models. IIW has published several documents over many decades regarding fatigue assessment of welded joints most of which are based on advancements of the stress–life-based methods [11–15]. Most of these proposed approaches are applicable for evaluating fatigue life at the weld toe except the effective notch stress (ENS) approach [16] which can be used for analyzing weld toe as well as root failure; however, it was reported that ENS approach can result into unconservative fatigue life predictions for the weld root in case of load-carrying fillet welds made using ultra high-strength steel plates subjected to out of plane loading [17]. For the low-cycle fatigue behavior of welded joints in high-strength steel, ultimate tensile strength has been proposed as a basis to construct a stress–life curve [1]. Relatively, more recent review of various fatigue assessment methods for welded joints can be found in [18]. Lan and Chan [3] provide a review of recent research advances of high-strength steel welded hollow section joints under static and fatigue loadings. Total fatigue damage process (crack initiation and propagation) was modeled within the LCF (Low-Cycle Fatigue) and HCF (High-Cycle Fatigue) range, respectively, using step-by-step discrete crack growth simulation while still using the framework of localized strain-based approach for the fatigue strength assessment [2]. The proposed fatigue model accounts for the effects of variables such as the weld toe radius and material microstructural changes in the form of hardness and grain size for the high-strength steel welded joints. Fatigue crack growth limit curves have been proposed utilizing a combination of two-stage crack growth law and statistical analysis techniques to help with structural integrity of the design and critical fatigue assessment of welded structures made from high-strength steel materials [19]. Research is also focused around extending the fracture mechanics method to model the short cracks propagation which is generally ignored in the standard crack growth law [20]. Chang and Lee [21] showed the importance of accounting for thermal residual stresses during mode-1 crack growth fatigue life analysis of welded joints including the development of a three-dimensional thermo-mechanical finite element-based welding process simulation model to predict the thermal stresses. In their model, the effect of residual stress was accounted for in the form of stress intensity factor through superposition principle of linear elastic fracture mechanics. Coupled stress and energy criterion have been used to evaluate effects of residual stress on the fatigue performance of welded T-joints showing that residual stress plays a significant role while working on an application involving stress ratio, $R < 0.5$ and those effects starts disappearing as the R-value reaches 0.5 and beyond [22]. As a part of the German cluster project IBESS which stands for “integral method of fracture mechanics determination of the fatigue strength of weldments,” the impact of residual stress on the fatigue strength prediction using fracture mechanics approach has been given special focus, further highlighting the importance of research needed to understand the issue of residual stress relaxation and its effect on fatigue performance of welded joints [23]. Material aspects have been specifically evaluated by determining and incorporating properties of the heat-affected zone of the high-strength steel welded joint during the fatigue crack growth analysis showing a good correlation of the estimated lives [24].

It’s clear that several methods have been proposed in the literature for fatigue life assessment of welded joints; however, most of those methods either enable estimation of crack initiation life or crack propagation life with only a few of the recent efforts focused around total fatigue life estimation using a single approach. Therefore, during this work in addition to the traditional strain life-based approach for fatigue life predictions, an effort has been made to apply advanced fracture mechanics-based total life approach [25, 26] for fatigue life predictions of HSLA and UHSS welded joints.

Background of the subject has been provided in this section. Next, Sect. 2 of this paper covers details of the test sample geometry, materials used, and several experimental works such as related to the welded test sample preparation, weld toe geometry measurement, metallurgical evaluations, residual stress measurement, and experimental fatigue testing. Section 3 describes detailed results from the various experimental studies performed along with findings and discussion. Section 4 provides details on the traditional strain–life method-based calculation for crack initiation portion of the fatigue life using experimentally measured strain as input. Section 5 provides details on the numerical fatigue life determination methods, first with traditional strain–life approach using detailed finite element analysis mainly for crack initiation portion of the fatigue life, followed by more advanced fracture mechanics-based total life approach which helps to estimate the total life, i.e., fatigue crack initiation and crack propagation lives. Summary of the research findings during this work along with recommendations for future work is covered in the last section.

2 Experimental Procedure and Details

2.1 High-Strength Steel Base Materials

Two different grades of high-strength steel materials with a large difference in mechanical properties have been used in the investigation presented below. High-strength low alloy steel grades (HSLA-50) and ultra high-strength steel grade (UHSS-100) have been utilized during this work with 345 MPa and 690 MPa minimum yield strength, respectively. HSLA-50 grade has been produced as per ASTM A1011 HSLAS-F 50, whereas UHSS-100 grade is produced to the ASTM A1011 UHSS 100 Type 1 specification. The typical chemical composition of these high-strength

Table 1 Typical chemical composition of high-strength steel grades (wt%) per ASTM A1011

Material	C	Mn	Si	P	S	Cr	Ni	Mo	Cu	V	Nb
HSLA-50	0.15	1.65	–	0.020	0.025	0.15	0.20	0.06	0.20	0.005	0.005
UHSS-100	0.15	2.00	–	0.020	0.025	0.15	0.20	0.40	0.20	0.005	0.005

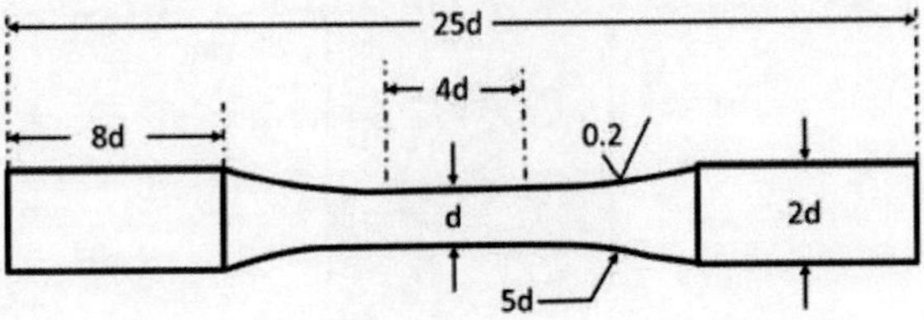

Fig. 1 Drawing of the fatigue test specimen geometry used for strain–life-controlled testing per ASTM E606 of both the high-strength steel materials, where d represents the diameter of the round test specimen

steel materials per ASTM A1011 is shown in Table 1. To characterize the mechanical properties of the base material used for preparing the samples, standard size tensile test samples were prepared from the base material plates and then tested as per ASTM E8 specification. For both steel grades, cyclic stress–strain curves and Manson–Coffin curves were also obtained experimentally as per the standard strain-controlled fatigue test procedure described in ASTM E606. Schematic drawing of the round bar test sample geometry used for cyclic stress–strain testing with a diameter (d) of 6 mm is shown in Fig. 1. The samples were CNC machined to rough shape and ground in the loading direction to a surface finish of 0.2 RMS (microns). All fatigue specimens were carried out at room temperature (23 °C) using a closed-loop servo-hydraulic two post 100 kN load frame. The tests were run in a strain-controlled manner and the stress–strain response was recorded at cyclic intervals. The failure criteria used was a drop in peak stress of 30% from the first cycle. The frequency of the tests was run at 3 Hz.

2.2 *Fatigue Test Sample Geometry*

The fatigue test sample geometry used for this work is in the form of a double fillet-welded T-joint. This geometric shape was selected as this allows the welds to be tested under pure bending loads, which represent more sever load case conditions, in addition to the fact that, bending loads contribute a significant part of the overall loading history for the majority of the structural applications. The thickness of plates used for welding of the T-joint samples was 6 mm. Figure 2 shows the geometrical details of the test sample used in this study.

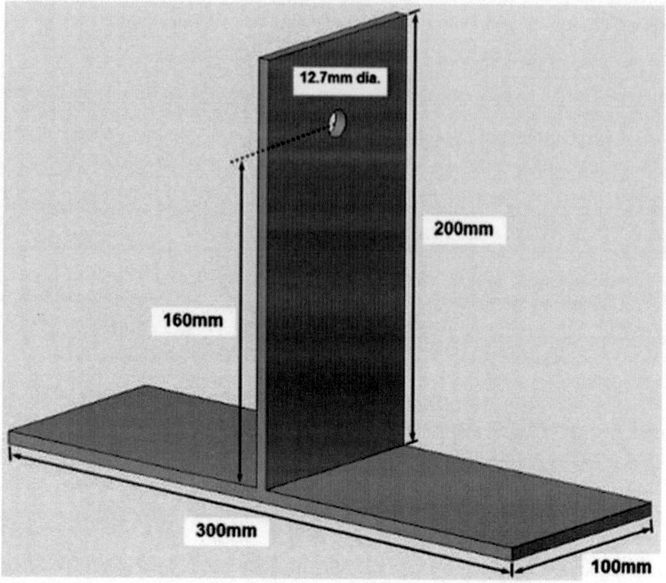

Fig. 2 T-joint welded test sample selected for the fatigue evaluation of high-strength steel welded joints

2.3 Test Sample Manufacturing

The raw material for the test samples was in the form of plates that had a nominal thickness of 6 mm. After receiving the raw stock material from the steel mill, the procedure used to prepare the T-welded joints was to laser cut the plates to the required size and perform double fillet welds with a nominal leg size of 6 mm using gas metal arc welding process (GMAW). All the samples were welded using standard inverter-based GMAW machine, with 1.14-mm-diameter weld wire to the specification AWS A5.18 ER70S-6 using active gas mixture composition of 95% Argon plus 5% Oxygen with a gas flow rate of 33–35 psi using weld power settings of 25.4-V voltage and wire feed speed of 520 ipm. The samples were tack welded in place before making the final welds (Fig. 3). All the samples were marked with 'side-A' and 'side-B' indicating the weld sequence, side-A was welded first followed by welding of side-B. In total, 40 T-joint samples were welded: with 20 samples made from each of the two high-strength steel grades. All the weld joints were visually inspected and found acceptable against AWS D14.3 weld quality standard acceptance criteria.

Fig. 3 Tack welding of the T-joint sample (left) before final welding (right)

2.4 Weld Toe Geometry Measurement

Weld toe microgeometrical features, i.e., weld toe radius and angle were measured using the dental mold putty procedure [27]. Dental molds were prepared to capture the weld toe geometry variations along the weld length. With this objective, three molds were prepared near the weld start, mid-length of the weld, and near the weld end, denoted as 'left,' 'middle,' and 'right,' respectively, for the welds on both sides (side-A and side-B). So, in total, there were six sets of measurements (three on each side) for the weld toe radius as well as angle for each T-welded joint.

2.5 Metallurgical Evaluation

Two welded T-joint samples, one each from the batch of HSLA-50 and UHSS-100, were selected for sectioning and performing metallurgical evaluations. Two sections were made for each of the test samples, roughly at 50 mm (mid-length of the weld) and at 75 mm (3/4th length of the weld) from the weld start, referred to as 'Center' and 'End' sections, respectively, refer Fig. 4. The hardness data were measured on both sides (side-A and side-B) of each weld section and the hardness traverse taken across all three zones of the weld joint (base material, weld HAZ, and weld fusion zone), refer Figs. 5 and 6.

2.6 Residual Stress Measurement

Two welded T-joint samples each of HSLA-50 and UHSS-100 steel were used for residual stress measurement. Residual stress (normal stress component to the weld toe line) was measured close to the weld toe location approximately at the mid-length of the weld, for welds on both sides (side-A and side-B), refer Fig. 7.

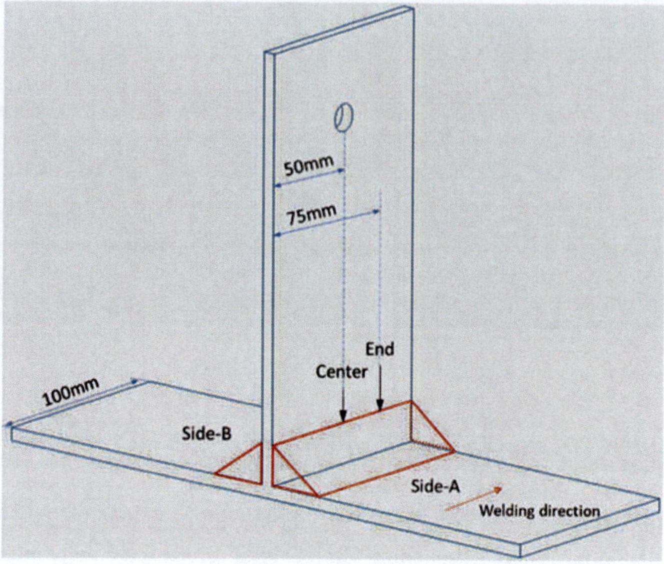

Fig. 4 High-strength steel T-joint welded samples showing the location of 'Center' and 'End' sections taken for metallurgical evaluations on both sides (side-A and side-B)

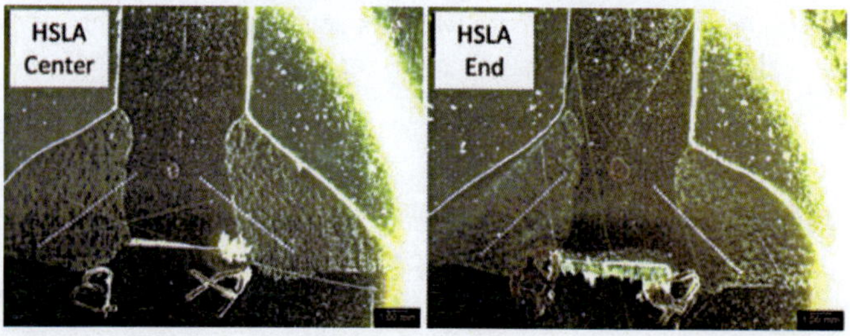

Fig. 5 HSLA-50 T-weld joint hardness traverse for the 'Center' section (left) and 'End' section (right) for welds on both sides (side-A and side-B)

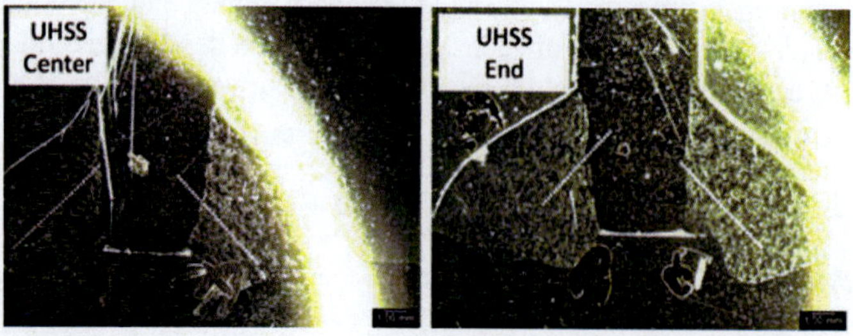

Fig. 6 UHSS-100 T-weld joint hardness traverse for the 'Center' section (left) and 'End' section (right) for welds on both sides (side-A and side-B)

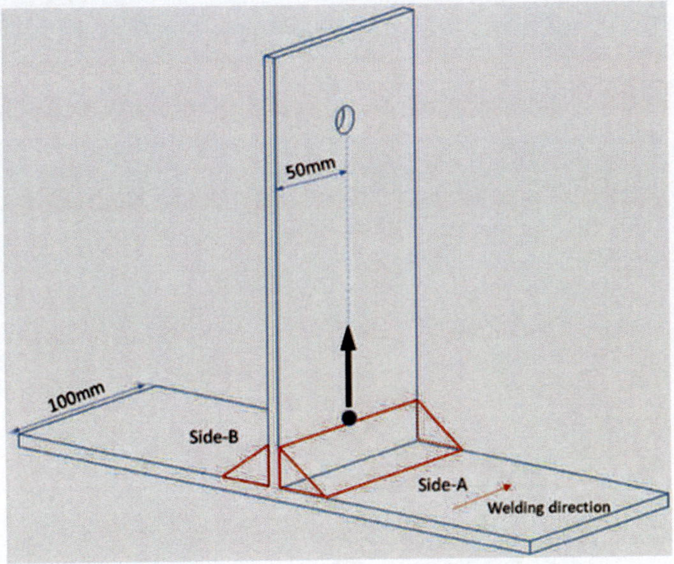

Fig. 7 Location of the residual stress measurement at the weld toe location, represented by dot near the upper weld toe and middle of the weld length. Arrow shows the direction of residual stress measurement (normal to the weld toe line)

2.7 Fatigue Testing Procedure

All the samples were fatigue tested at room temperature (20–23 °C) and relative humidity of 15–25%, using a hydraulic load actuator on a bedplate test system. The tests were run in a load-controlled manner with defined displacement limits (±6 mm) of the actuator. The frequencies of the tests were run at 5 Hz under load control for most of the samples. No excessive heating of the test samples was reported as measured using the infrared temperature gun. All the samples were tested under fully reversed bending load conditions ($R = -1$). The samples were tested until the displacement limit was reached or fracture of the sample into two pieces. In most cases, crack length on the surface reached around 100 mm, i.e., same as the weld length. Efforts were made to measure and record the visible crack length on the surface as a function of the number of cycles at periodic intervals. One test sample each for HSLA-50 and UHSS-100 steel was instrumented with strain gauges to measure the strain levels next to the weld toe (in the normal direction, i.e., normal to the weld toe line) at a nominal distance of 0.4t and 1t away from the upper weld toe line (Fig. 8), where t is the plate thickness as per the recommendation from IIW hot spot stress approach [15].

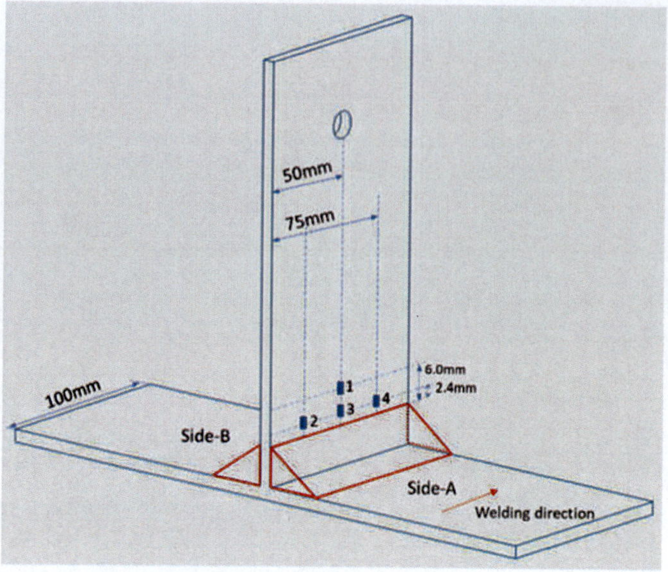

Fig. 8 Schematic showing 4 strain gauges mounted on each side of the T-joint at a nominal distance of 0.4t (2.4 mm) and 1t (6 mm) distance away from the weld toe line; where t is the plate thickness of 6 mm

3 Experimental Results and Discussion

3.1 Weld Toe Geometry Measurements and SCF Calculations

Tables 2 and 3 show the measured data for weld toe radius and angle for the T-welded joints made from HSLA-50 and UHSS-100 steel grades, respectively. The data have been reported as measured for each test sample as well as the calculated value for the minimum, maximum, and averages along with standard deviation for the weld toe radius and angle for welds on both sides corresponding to the start, mid-length, and end of the weld. For an understanding of the variability in the measured weld toe radius and angle for HSLA 50 and UHSS 100 welded joints, the data have been plotted as shown in Figs. 9 and 10, which follows the normal distribution curves. Using this measured data, the stress concentration factors (SCF's) for membrane, $K_{t,hs}^{m}$, and bending, $K_{t,hs}^{b}$, have been derived using Japanese expressions, Eqs. (1) and (2), respectively, [28]—refer Fig. 11 for the nomenclature). Tables 4 and 5 shows the minimum, maximum, and average values for both the SCF's for welds on both sides corresponding to the start, mid-length, and end of the weld for HSLA-50 and UHSS-100 steel welded joints, respectively. Interestingly, the final averaged values of $K_{t,hs}^{m}$ and $K_{t,hs}^{b}$ (averaged over welds on both sides) for HSLA-50 and UHSS-100 steel welded joints are quite similar ($K_{t,hs}^{b}$ of 1.63 vs. 1.58 and $K_{t,hs}^{m}$ of 2.13 vs. 2.06 for HSLA-50 and UHSS-100 welded joints, respectively). This can be attributed to

Table 2 Weld toe radius and angle measurement for HSLA-50 welded joints

Sample ID	HSLA-50 steel T-welded joints—dental mold measurements											
	Radius measurements (mm)						Angle measurements (deg)					
	Side A			Side B			Side A			Side B		
	Left	Middle	Right	Left	Middle	Right	Left	Middle	Right	Left	Middle	Right
HSLA-50 S1	3.32	2.42	2.09	1.39	1.07	0.8	56.81	57.87	59.63	59.10	68.94	76.03
HSLA-50 S2	2.63	2.41	1.95	1.89	1.72	1.24	58.87	54.04	52.04	50.45	62.25	66.52
HSLA-50-1	1.41	1.44	1.73	2.13	2.05	2.47	69.82	62.75	52.59	57.11	53.44	52.91
HSLA-50-2	1.54	2.04	1.17	1.32	2.11	3.24	52.72	50.46	59.75	51.11	55.20	47.10
HSLA-50-3	1.69	1.64	1.54	1.66	2.25	1.80	56.44	56.24	57.26	59.31	57.57	52.62
HSLA-50-4	1.44	1.78	1.17	1.48	2.79	2.05	74.55	63.40	62.05	56.64	58.76	56.14
HSLA-50-5	1.07	1.01	1.35	1.31	2.05	2.39	67.51	61.07	51.89	66.16	58.03	50.45
HSLA-50-6	1.29	1.45	1.34	1.79	1.98	2.66	69.29	60.72	57.46	63.74	57.40	51.15
HSLA-50-7	1.36	2.36	1.28	1.88	3.14	2.32	68.98	60.44	62.30	56.64	54.34	56.34
HSLA-50-8	1.60	1.75	2.74	2.74	2.92	3.82	61.68	61.47	56.57	57.11	52.17	50.79
HSLA-50-9	1.78	1.79	1.80	2.00	2.25	2.25	65.53	61.85	60.89	58.45	59.55	51.57
HSLA-50-10	1.63	1.88	1.88	2.08	2.64	3.26	68.20	61.15	59.54	56.81	48.61	47.29
HSLA-50-11	1.79	2.04	1.91	1.85	3.01	2.67	61.82	59.58	59.59	59.54	51.31	53.42
HSLA-50-12	1.97	1.86	2.04	2.08	2.44	2.51	66.64	62.75	60.46	63.65	56.93	52.45
HSLA-50-13	1.95	1.82	2.03	2.49	2.91	2.98	65.07	61.58	58.38	56.60	56.66	53.95
HSLA-50-14	1.92	2.22	1.70	2.58	2.29	2.79	62.46	61.75	59.97	62.25	54.31	52.51
HSLA-50-15	1.57	1.95	2.00	2.25	2.76	2.64	59.86	55.16	57.39	55.82	56.14	51.06
HSLA-50-16	1.89	2.61	2.49	2.16	2.14	2.63	56.32	57.73	58.28	61.10	55.80	50.61

(continued)

Table 2 (continued)

Sample ID	HSLA-50 steel T-welded joints—dental mold measurements											
	Radius measurements (mm)						Angle measurements (deg)					
	Side A			Side B			Side A			Side B		
	Left	Middle	Right	Left	Middle	Right	Left	Middle	Right	Left	Middle	Right
HSLA-50-17	1.92	2.36	2.41	3.08	3.05	3.95	65.56	51.69	59.79	53.98	49.50	49.81
HSLA-50-18	1.72	2.26	2.25	2.44	2.13	3.26	63.50	58.61	59.19	59.96	54.82	49.89
HSLA-50-19	2.17	2.92	2.49	3.32	2.61	3.30	59.50	55.49	55.35	57.99	56.63	51.62
Minimum	1.07	1.01	1.17	1.31	1.98	1.80	52.72	50.46	51.89	51.11	48.61	47.10
Maximum	2.17	2.92	2.74	3.32	3.14	3.95	74.55	63.40	62.30	66.16	59.55	56.34
Average	1.67	1.96	1.86	2.14	2.50	2.79	63.97	59.15	58.35	58.63	55.11	51.67
Stand Dev.	0.28	0.44	0.48	0.55	0.39	0.57	5.45	3.75	2.79	3.63	3.01	2.41
Average—each side	1.83			2.48			60.49			55.14		
Average—both sides	2.15						57.81					

Table 3 Weld toe radius and angle measurement for UHSS-100 welded joints

Sample ID	UHSS-100 steel T-welded joints—dental mold measurements											
	Radius measurements (mm)						Angle measurements (deg)					
	Side A			Side B			Side A			Side B		
	Left	Middle	Right	Left	Middle	Right	Left	Middle	Right	Left	Middle	Right
UHSS-100 S1	2.57	1.57	2.39	1.03	1.22	1.08	61.01	66.86	53.42	67.12	72.47	72.00
UHSS-100 S2	1.74	2.05	1.84	2.22	1.98	1.99	55.54	60.46	56.37	62.28	64.53	64.86
UHSS-100-1	1.60	2.52	2.63	2.72	2.63	3.43	67.83	66.54	55.74	49.31	50.60	53.81
UHSS-100-2	1.75	2.11	2.29	2.26	2.58	3.45	67.01	56.75	60.19	61.21	52.99	51.36
UHSS-100-3	2.49	2.45	2.42	2.85	3.08	3.51	60.22	61.90	57.50	55.39	53.09	54.09
UHSS-100-4	1.79	1.85	2.41	2.49	2.95	3.17	60.38	53.68	56.14	51.22	52.13	51.14
UHSS-100-5	2.01	2.11	2.76	3.01	2.99	2.98	57.19	55.56	54.38	47.13	52.98	47.74
UHSS-100-6	2.10	2.66	2.41	2.69	3.48	2.99	63.70	54.48	54.32	51.89	52.48	47.49
UHSS-100-7	2.20	3.05	2.92	2.45	2.77	2.35	57.61	53.96	62.70	45.24	46.57	46.59
UHSS-100-8	2.75	2.52	2.99	1.76	2.63	2.32	64.19	56.14	54.83	55.99	50.67	48.58
UHSS-100-9	2.00	1.94	2.99	1.57	2.10	2.91	59.55	53.43	57.99	53.69	49.69	49.57
UHSS-100-10	1.29	1.88	1.63	2.26	2.77	1.95	55.79	54.46	54.80	47.68	47.76	46.56
UHSS-100-11	1.20	1.38	1.57	1.53	1.85	1.72	67.07	55.42	57.53	48.82	45.84	47.43
UHSS-100-12	1.42	2.23	1.78	2.04	2.00	3.43	63.67	57.75	53.79	50.46	50.21	52.59
UHSS-100-13	1.47	2.05	1.95	1.66	2.35	3.07	63.24	56.02	56.45	46.93	46.69	47.52
UHSS-100-14	1.69	2.04	2.04	2.08	3.24	2.16	64.43	55.91	54.44	45.90	52.82	52.27
UHSS-100-15	2.13	2.20	3.24	2.04	1.88	3.02	53.12	59.22	55.43	48.97	48.67	51.44
UHSS-100-16	1.80	1.82	1.57	2.17	2.77	1.91	52.59	48.80	57.85	51.22	51.73	44.76

(continued)

Table 3 (continued)

Sample ID	UHSS-100 steel T-welded joints—dental mold measurements											
	Radius measurements (mm)						Angle measurements (deg)					
	Side A			Side B			Side A			Side B		
	Left	Middle	Right	Left	Middle	Right	Left	Middle	Right	Left	Middle	Right
UHSS-100-17	1.63	2.45	2.35	2.63	2.23	3.26	53.70	60.63	62.13	54.11	45.37	52.93
UHSS-100-18	1.80	2.11	2.52	3.84	3.36	2.49	59.55	51.43	54.49	49.44	55.61	54.16
UHSS-100-19	1.36	2.04	2.85	2.66	4.20	3.01	57.79	54.08	52.40	52.20	49.82	50.05
Minimum	1.20	1.38	1.57	1.53	1.85	1.72	52.59	48.80	52.40	45.24	45.37	44.76
Maximum	2.75	3.05	3.24	3.84	4.20	3.51	67.83	66.54	62.70	61.21	55.61	54.16
Average	1.81	2.18	2.39	2.35	2.73	2.80	60.45	56.11	56.48	50.88	50.30	50.00
Stand Dev.	0.41	0.37	0.51	0.57	0.60	0.57	4.77	3.94	2.78	3.97	2.87	2.90
Average—each side	2.13			2.63			57.68			50.40		
Average—both sides	2.38						54.04					

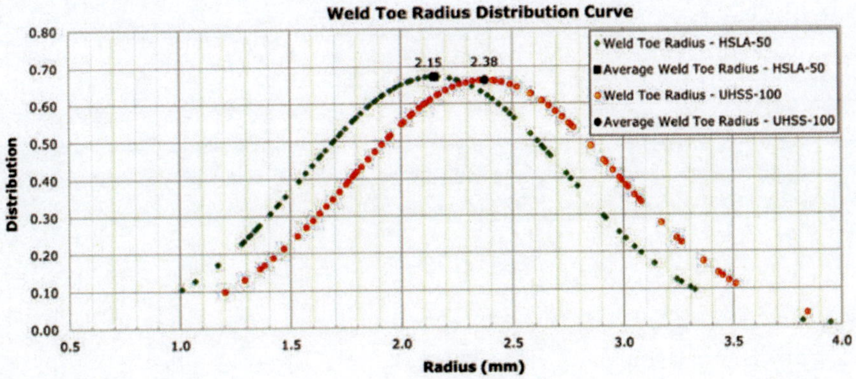

Fig. 9 Comparison of normal distribution curves of weld toe radius for HSLA-50 and UHSS-100 welded joints

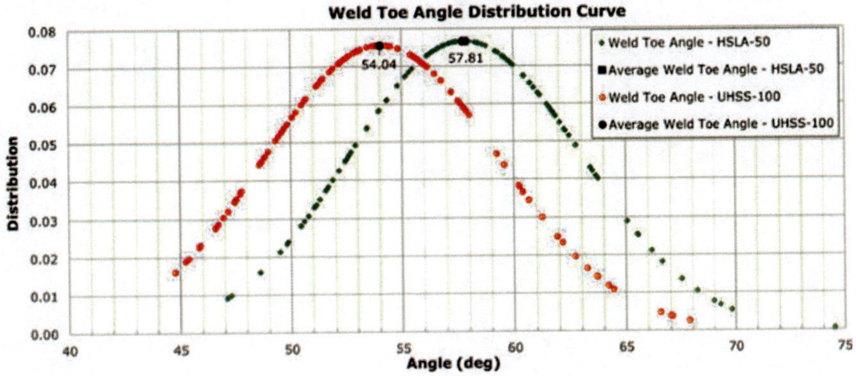

Fig. 10 Comparison of normal distribution curves of weld toe angle for HSLA-50 and UHSS-100 welded joints

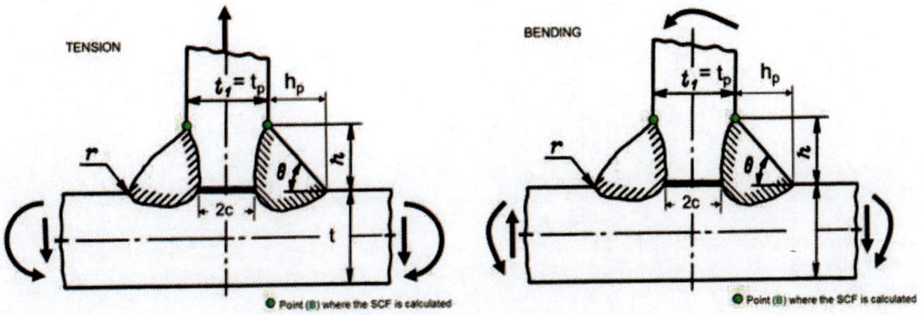

Fig. 11 Nomenclature of the T-welded joints with the vertical member under tension (left) and bending (right). Stress concentration factors (SCF) being calculated at the weld toe marked with dots

Table 4 Weld toe stress concentration factors for HSLA-50 welded joints

Sample ID	HSLA-50—bending stress conc. factor, $K^{b}_{t,hs}$						HSLA-50—membrane stress conc. factor, $K^{m}_{t,hs}$					
	Side A			Side B			Side A			Side B		
	Left	Middle	Right	Left	Middle	Right	Left	Middle	Right	Left	Middle	Right
Maximum	1.93	1.94	1.88	1.82	1.63	1.67	2.56	2.58	2.48	2.39	2.11	2.15
Minimum	1.66	1.53	1.55	1.51	1.51	1.46	2.21	2.01	2.03	1.96	1.96	1.85
Average	1.75	1.67	1.69	1.63	1.57	1.53	2.32	2.19	2.22	2.14	2.04	1.97
Average—each side	1.70			1.58			2.24			2.05		
Average—both sides	1.63						2.13					

Table 5 Weld toe stress concentration factors for UHSS-100 welded joints

Sample ID	UHSS-100—bending stress conc. factor, $K_{t,hs}^{b}$						UHSS-100—membrane stress conc. factor, $K_{t,hs}^{m}$					
	Side A			Side B			Side A			Side B		
	Left	Middle	Right	Left	Middle	Right	Left	Middle	Right	Left	Middle	Right
Maximum	1.87	1.79	1.75	1.72	1.65	1.67	2.47	2.33	2.28	2.23	2.12	2.15
Minimum	1.56	1.53	1.51	1.47	1.45	1.48	2.05	2.00	1.96	1.88	1.83	1.89
Average	1.71	1.62	1.59	1.58	1.53	1.52	2.25	2.11	2.07	2.04	1.97	1.95
Average—each side	1.64			1.54			2.14			1.99		
Average—both sides	1.58						2.06					

the consistency of the weld manufacturing process. Even though the samples were manually welded but single welding operator prepared all the test samples for both the high-strength steel grades, using practically the same welding procedure.

$$\begin{gathered}K_{t,hs}^{m}=\left\{1+\frac{1-\exp\left(-0.9(0.5\pi-\theta)\sqrt{\frac{W}{2h_p}}\right)}{1-\exp\left(-0.45\pi\sqrt{\frac{W}{2h_p}}\right)}\times 2.2\left[\frac{1}{2.8\left(\frac{W}{t_p}\right)-2}\times\frac{h_p}{r}\right]^{0.65}\right\}\\ \times\left\{1+0.64\frac{\left(\frac{2c}{t_p}\right)^2}{\frac{2h}{t_p}}-0.12\frac{\left(\frac{2c}{t_p}\right)^4}{\left(\frac{2h}{t_p}\right)^2}\right\}\\ W=\left(t_p+4h_p\right)+0.3(t+2h)\\ \text{Range of application}: r/t_p=0.025\text{–}0.4,\ h_p/t_p=0.5\text{–}1.0,\ \theta=20^\circ\text{–}50^\circ\end{gathered}\tag{1}$$

$$\begin{gathered}K_{t,hs}^{b}=\left\{\begin{matrix}1+\frac{1-\exp\left(-0.9\theta\sqrt{\frac{W}{2h_p}}\right)}{1-\exp\left(-0.45\pi\sqrt{\frac{W}{2h_p}}\right)}\times\sqrt{\tanh\left(\frac{2t}{t_p+2h_p}+\frac{2r}{t_p}\right)}\\ \times\tanh\left[\frac{\left(\frac{2h_p}{t_p}\right)^{0.25}}{1-\frac{r}{t_p}}\right]\times\left[\frac{0.13+0.65\left(1-\frac{r}{t_p}\right)^4}{\left(\frac{r}{t_p}\right)^{\frac{1}{3}}}\right]\end{matrix}\right\}\\ \times\left\{1+0.64\frac{\left(\frac{2c}{t_p}\right)^2}{\frac{2h}{t_p}}-0.12\frac{\left(\frac{2c}{t_p}\right)^4}{\left(\frac{2h}{t_p}\right)^2}\right\}\\ W=\left(t_p+4h_p\right)+0.3(t+2h)\\ \text{Range of application}: r/t_p=0.025\text{–}0.4,\ h_p/t_p=0.5\text{–}1.0,\ \theta=20^\circ\text{–}50^\circ\end{gathered}\tag{2}$$

3.2 *Monotonic and Cyclic Stress–Strain Results*

Two samples each of HSLA-50 and UHSS-100 steel base material were used for the monotonic tensile test to determine the yield strength, tensile strength, the percent elongation, and the reduction area. Hardness tests were performed on the surface of the sample and the hardness measurement was repeated three to five times on two samples and the average has been reported. The engineering monotonic tensile stress–strain curves obtained from testing of two samples for each of the grades (HSLA-50 and UHSS-100) are shown in Fig. 12.

Around 20 samples were used for the cyclic stress–strain testing. Stabilized stress data obtained from strain–life fatigue tests were used to construct the cyclic stress–strain curve shown in Fig. 13. The cyclic stress–strain curve is described by the Ramberg–Osgood relationship as described by Eq. (3):

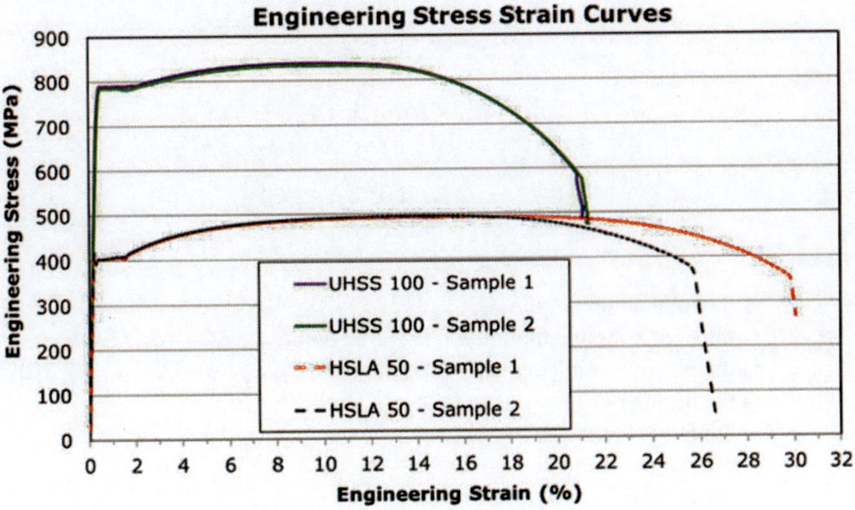

Fig. 12 Monotonic stress–strain curves for HSLA-50 and UHSS-100 steel materials

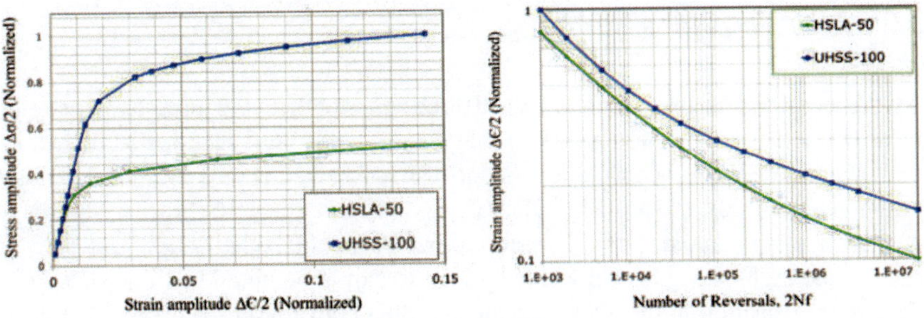

Fig. 13 Cyclic stress–strain (left) and strain life curves of both high-strength steel materials

$$E = \sigma/E + \left(\sigma/K'\right)^{1/n} \tag{3}$$

where ε is the total strain amplitude, σ is the cyclically stable stress amplitude, E is the cyclic modulus of elasticity obtained from the best fit of the above equation to the test data, K′ is the cyclic strength coefficient, and n is the cyclic strain hardening exponent.

Constant amplitude fatigue test data are further used to obtain the strain–life fatigue curve. The stress amplitude corresponding to the peak strain amplitude was calculated from the peak load amplitude at one half of the specimen's life. Constant amplitude fatigue curves for both the grades of high-strength steel are also given in Fig. 13 and can be described by the following equations:

$$\Delta\varepsilon_e/2 = \sigma'_f/E(2N)^b \tag{4}$$

$$\Delta\varepsilon_p/2 = \in_f' \,(2N)^c \tag{5}$$

$$\text{Since } \Delta \in = \Delta\varepsilon_e + \Delta\varepsilon_p \tag{6}$$

$$\Delta \in /2 = \sigma_f'/E\,(2N)^b + \varepsilon_f'(2N)^c \tag{7}$$

where:

$\Delta \in /2$ is the total strain amplitude,
$\Delta\varepsilon_e/2$ is the elastic strain amplitude $\left(\Delta\varepsilon_e/2 = \Delta\varepsilon_{measured}/2 - \Delta\varepsilon_p/2\right)$,
$\Delta\varepsilon_p/2$ is the plastic strain amplitude $\left(\Delta\varepsilon_p/2 = \Delta\varepsilon_{measured}/2 - \Delta\sigma_{measured}/2E\right)$,
2N is the number of reversals to failure,
σ_f' is the fatigue strength coefficient,
b is the fatigue strength exponent,
ε_f' is the fatigue ductility coefficient,
c is the fatigue ductility exponent.

3.3 *Microhardness Results*

All the measured hardness data across all three zones of the weld joint (base material, weld HAZ, and weld fusion zone) for the 'Center' and 'End' sections on both sides (side A and side B) of each weld section for HSLA-50 and UHSS-100 T-welded joints are shown in Fig. 14. To provide a closer look at the data, Fig. 15 shows the hardness data taken just from the 'Center' section for welds on both sides (side-A and side-B) of both the high-strength steel welded joints. It should be noted that microhardness data were measured in the Vickers scale and then converted to Rockwell C for the plots. Following inferences can be drawn from these data:

1. UHSS-100 steel welded joints showed a reduced hardness in the weld HAZ (a maximum drop of 12HRC), contrary to HSLA-50 welded joints wherein increased hardness was observed in the weld HAZ (maximum increase of 13HRC). The observed trend for microhardness drop in the weld HAZ region found during this work agrees with the data as reported in [4].
2. Weld zones have similar hardness levels due to the same weld filler wire used for welding of HSLA-50 and UHSS-100 welded joints. UHSS-100 welded joints appear to have slightly higher hardness in the weld zone compared to the weld zone in HSLA-50 joint, which can be attributed to base metal dilution into the weld fusion zone.
3. Weld on side-A (first weld in the sequence) experienced tempering due to the weld on side-B (second weld in the sequence) for both the high-strength steel materials. Interestingly, though the tempering effect was different for both the steel materials, e.g., weld HAZ hardness on side-A increased for UHSS-100 steel, whereas it reduced for HSLA-50 steel.

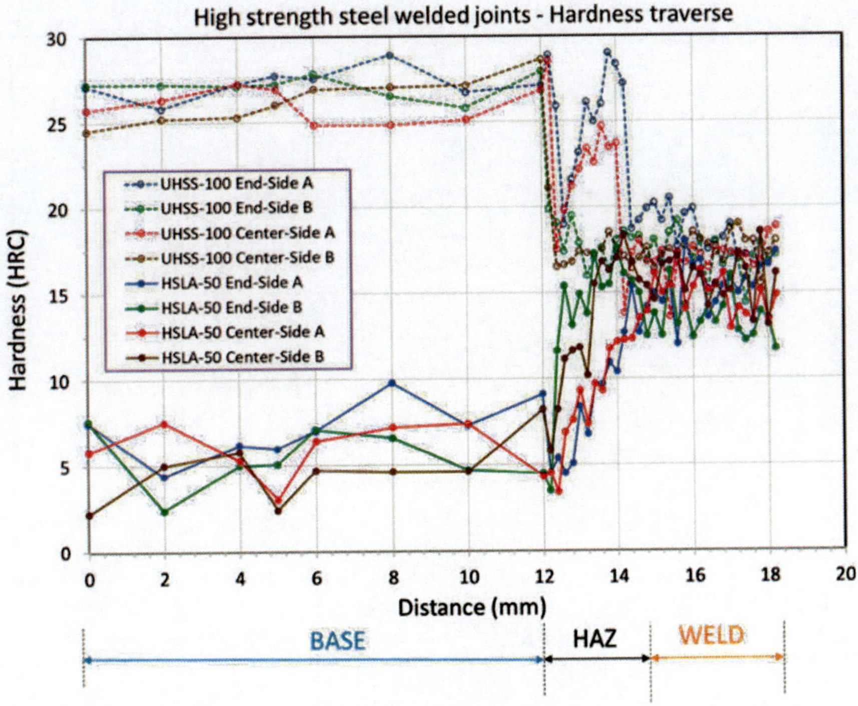

Fig. 14 HSLA-50 and UHSS-100 T-welded joints hardness traverse plots for the 3 zones (base, HAZ, weld)—corresponding to the 'Center' and 'End' sections (refer Figs. 4, 5, and 6) for welds on both sides (side-A and side-B)

3.4 Residual Stress Results

The measured residual stress data consistently showed tensile residual stress values at the weld toe location for all the weld joints. The average values of residual stress on the surface are almost the same for the HSLA-50 and UHSS-100 welded joints (238 MPa and 235 MPa, respectively). Due to the experimental test setup limitations, reported residual stress data could be measured at 0.5 mm away from the weld toe location, rather than exactly at the weld toe location. More detailed experimental residual stress measurements are needed for understanding the residual stress trends as a function of manufacturing process variables and the measurement location on the sample which could not be accomplished within the scope of the current study. Based on the observed behavior of crack growth from the fatigue test experiments (crack grew much faster on the surface than in the depth), it is believed that surface residual stress must be tensile in nature. Further, for the crack growth analysis, the through-thickness residual stress distribution is required as an input, so the self-balancing sinusoidal residual stress profile was 'assumed' as shown in Fig. 16. This

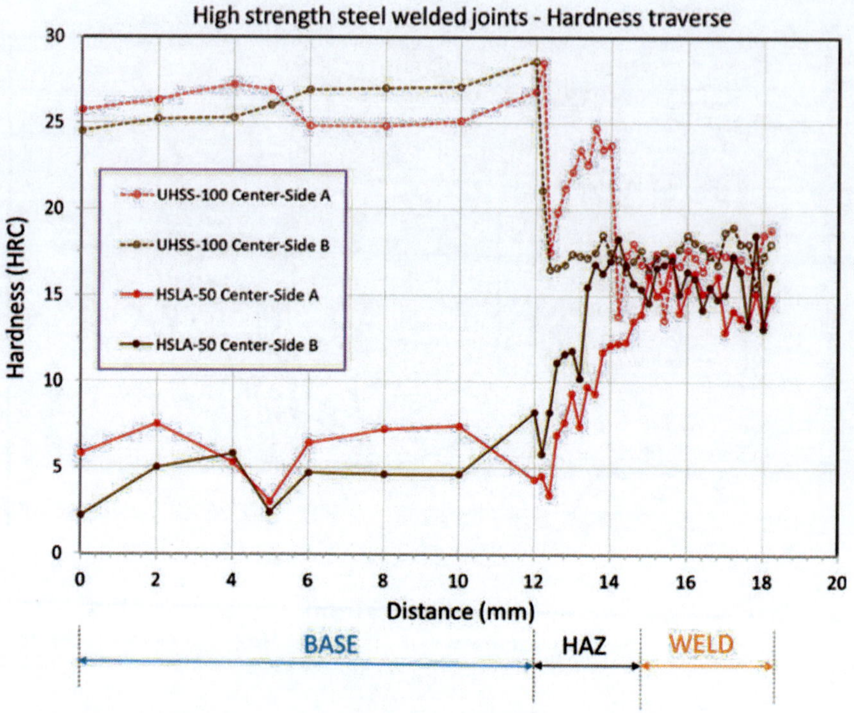

Fig. 15 HSLA-50 and UHSS-100 T-welded joints hardness traverse plots for the 3 zones (base, HAZ, weld)—corresponding to the 'Center' section (refer Figs. 4, 5 and 6) for welds on both sides (side-A and side-B)

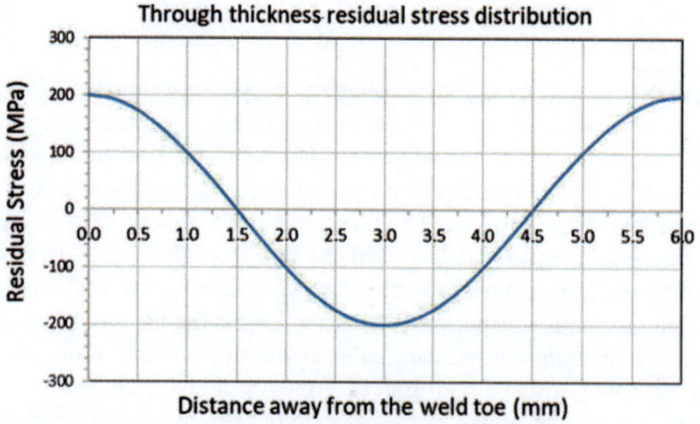

Fig. 16 Assumed through-thickness residual stress distribution for HSLA-50 and UHSS-100 T-welded joints, through the critical hot spot weld toe location

assumption is based on the general observation from the literature [29, 30], and experimental behavior of the failed test samples.

3.5 Fatigue Life Results

Efforts were made to generate fatigue data over a large range of fatigue life regimes (low, medium, and high), specifically in the range of 10 k cycles to 5 million cycles to understand the relative fatigue behavior of the two high-strength steel grades over those life regimes. Summary of all the experimental fatigue test data points along with logarithmic fit curves is shown in Fig. 17, in terms of applied load range versus the number of cycles to failure. The trend line for UHSS-100 welded joints appears to show slightly better fatigue lives than HSLA-50 welded joints, however, considering the experimental scatter as shown in Fig. 17, it would be reasonable to conclude that the welded joints made from HSLA-50 and UHSS-100 welded joints have similar fatigue lives for the same design and fatigue loading conditions. Fatigue life test results are in agreement with the observed hardness behavior of weld HAZ regions for the HSLA-50 and UHSS-100 welded joints (refer Figs. 14 and 15). Other key factors, which could significantly influence the fatigue life of welded joints such as stress concentration factors (SCF's) and residual stress, were found to be quite similar for the welded joints made from HSLA-50 and UHSS-100 steels, refer Tables 2, 3, 4, and 5, again contributing to the observed similar fatigue behavior in both the cases. Further to understand the fatigue crack growth behavior, visible crack length on the surface was recorded after some intervals of the number of applied loading cycles as

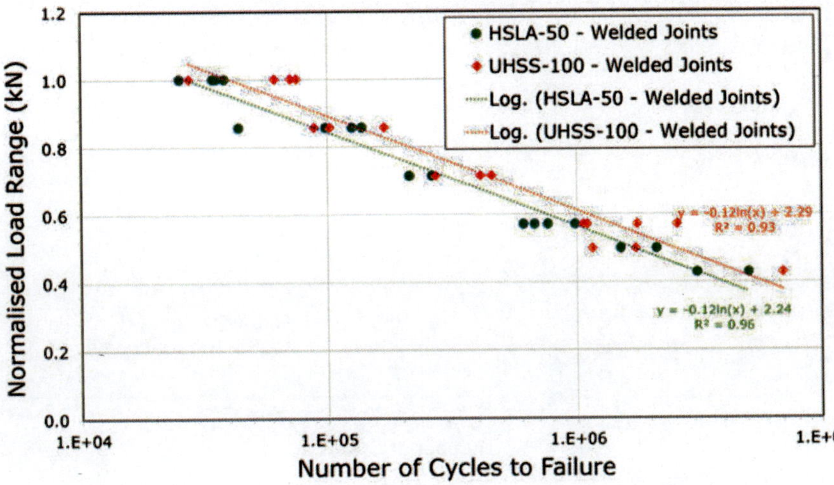

Fig. 17 Number of cycles to failure versus applied load range obtained from fatigue testing of HSLA-50 and UHSS-100 T-welded joints at R = −1

shown in Fig. 18. Fatigue crack length as measured on the surface after certain loading cycles were recorded for HSLA-50 and UHSS-100 steel welded joints is plotted in Fig. 19. This information is quite useful for modeling and validation of the fatigue crack growth behavior of welded joints. It can be concluded that for the test samples with relatively shorter fatigue lives (higher loads), crack initiation happened quite early, meaning that more portion of the life was spent in growing the cracks, while

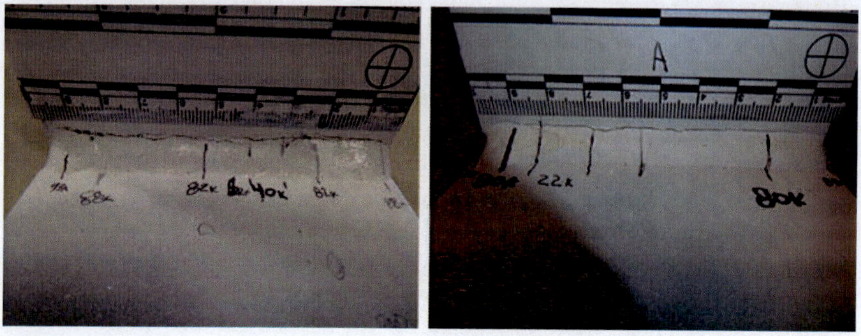

Fig. 18 Fatigue tested samples showing the crack location and recorded crack length on the surface after some intervals of applied loading cycles: HSLA-50 ID-1 sample on the left and UHSS-100 ID-1 sample on the right

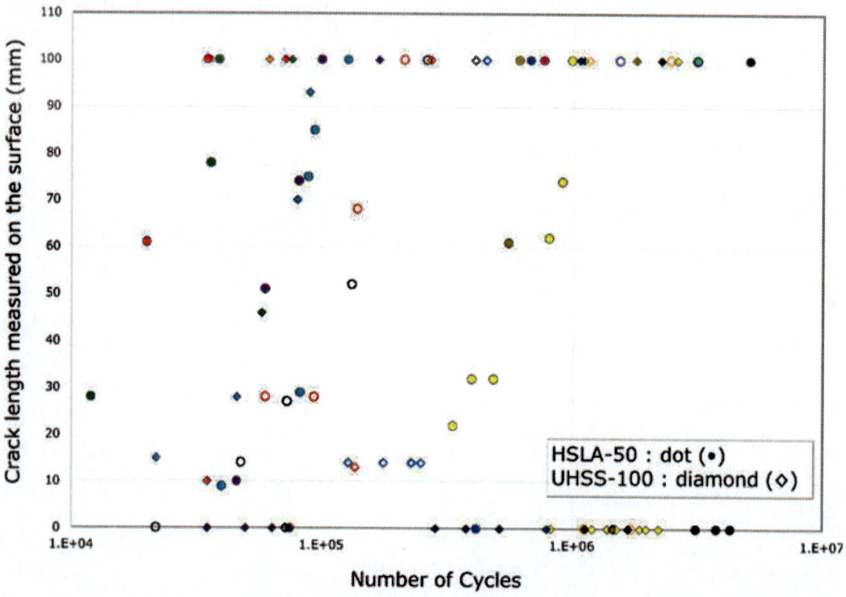

Fig. 19 Crack length measured on the surface after some intervals of applied loading cycles to show the rate of crack growth: HSLA-50 and UHSS-100 welded joints data denoted by circle- and diamond-shaped points, respectively

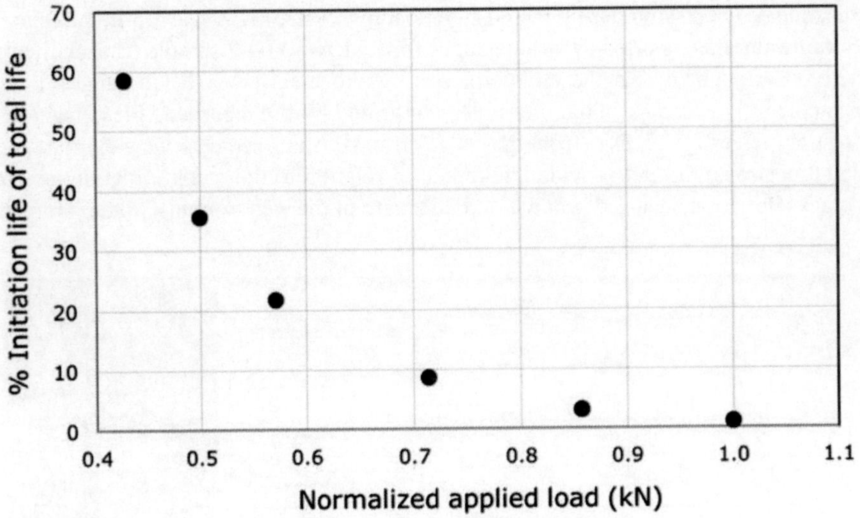

Fig. 20 Crack initiation portion of the total fatigue life reported in percentage as a function of the applied load

the trend was completely opposite for the samples with relatively longer lives (lower loads) as a major portion of their lives was spent in initiating the cracks but once the cracks were initiated, propagation was relatively much faster resulting into fast final fracture, refer Fig. 20. This clearly indicates that designing for crack initiation life might be too conservative for some applications. Furthermore, as the conventional strain–life approach can only help to predict the crack initiation portion of the total life, the benefit of total fatigue life approach becomes even more evident.

4 Fatigue Life Prediction Based on Measured Strain

Strain data near the weld toe were measured for two of the welded joint samples (HSLA-50 ID-2 and UHSS-100 ID-6). Total eight strain gauges were mounted on each test sample (four on each side), with the objective to capture weld toe normal strain values, i.e., strain direction measured normal to the weld toe line. Three strain gauges measured the strain values at a nominal distance of 0.4t (2.4 mm) away from the upper weld toe line, near the weld start, center of the weld, and near the end of weld on each side of the sample. The fourth strain gauge measured the strain at mid-width of the sample mounted at 1t (6 mm) away from the weld toe line.

As expected, measured strains followed the polarities of applied loads, e.g., positive strain values reported by strain gauges mounted on side-A corresponding to the positive cycle of the applied load during which the side-A (on the actuator side) is in tension, while the compressive strains were reported by the gauges on side-B which is under compression during the positive loading cycle, Fig. 21. Drop-in maximum strain values as reported by strain gauges for HSLA-50 ID-2 sample (especially for the ones mounted near the middle of plate width, i.e., gauge 1 and gauge 3) can be correlated to crack initiation as also confirmed by the recorded videos from the testing, refer Fig. 22. Data from the gauges 2 and 4 (mounted near the weld start and end locations) did not provide any direct correlation to the crack initiation, as the cracks initiation occurred near the middle length of the weld for most of the samples,

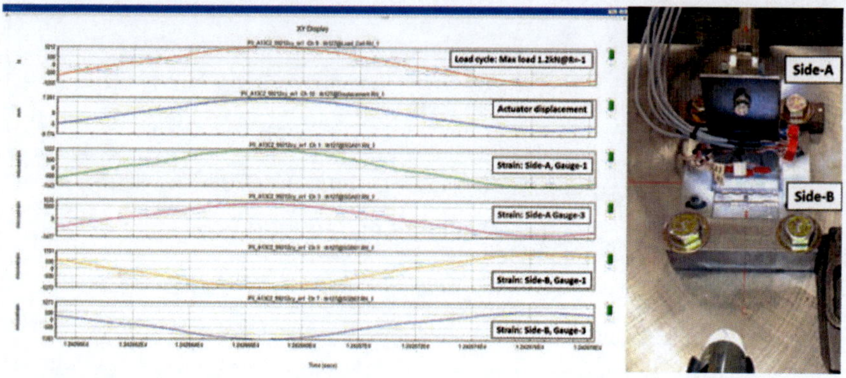

Fig. 21 Strain data for gauge-1 and gauge-3 on both sides (side-A and side-B of T-welded joint) along with applied load and measured actuator displacement for HSLA-50 ID-2 test sample (left picture), side-A is on the hydraulic actuator side (right picture)

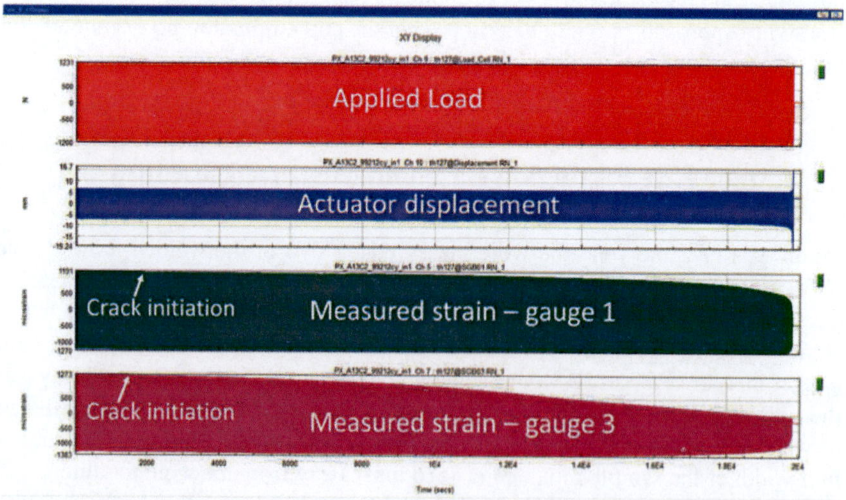

Fig. 22 Strain data for gauge-1 and gauge-3 on side-B of T-welded joint along with applied load and measured actuator displacement for HSLA-50 ID-2 test sample. Reduction in maximum strain values for both the gauges indicates the onset of crack initiation

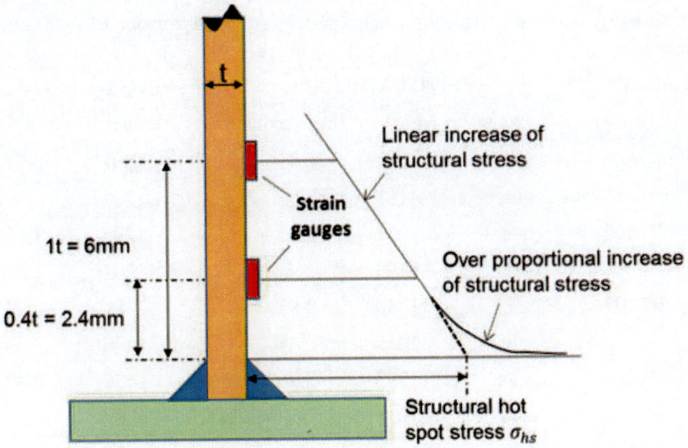

Fig. 23 Location of strain gauges and linear extrapolation of strain data at the weld toe hot spot location based on strain measurements from gauge-1 (located at 1t) and gauge-3 (located at 0.4t), where t = 6 mm. Note that picture is not to the scale and for illustration purpose only

corresponding to the hot spot location, as also identified based on the finite element analysis, Fig. 24.

The strain gauges mounted at distances of 2.4 and 6 mm away from the weld toe helped to capture the stress concentration effect at the macro level, i.e., from the weld itself but are unable to capture the microeffects such as from the weld toe geometry. IIW hot spot stress approach [15] has been utilized to obtain the strain values at the weld toe hot spot location based on the linear extrapolation of measurements at 2.4 and 6 mm away from weld toe, reported by gauge-1 and gauge-3 as shown in Fig. 23. This linear extrapolated strain value at the weld toe hot spot location represents the strain information at the weld toe location but still does not capture the stress concentration effect from the weld toe geometry.

Measured data from strain gauges and load cell is reported in Table 6 for HSLA-50 and UHSS-100 welded joints.

Linearly extrapolated values at the weld toe hot spot location are shown in Table 7. Peak strain values at the weld toe location can be calculated using Eq. (8):

$$\varepsilon_{peak} = \varepsilon_m * K^m_{t,hs} + \varepsilon_b * K^b_{t,hs} \tag{8}$$

As the loading type is known in this case (i.e., full reversed pure bending loads); hence, the membrane, ε_m and bending, ε_b strain values can be calculated from the maximum tensile and compressive strain values at the weld toe hot spot location and are also shown in Table 7.

For HSLA-50 steel welded joints:
$K^m_{t,hs}$ (average of all samples, both sides) = 2.13 [refer Table 4]
$K^b_{t,hs}$ (average of all samples, both sides) = 1.63 [refer Table 4]
$\varepsilon_{\text{peak}} = 24 * 2.13 + 1326 * 1.63 = 2211\ \mu\varepsilon$

Table 6 Strain gauge and load cell measurements during fatigue testing of HSLA-50 and UHSS-100 welded joints

Measured data			HSLA-50 ID-2		UHSS-100 ID-6	
Side	Channel	Range	Value	min/max ratio	Value	min/max ratio
Side-A	SGA01	max	1191	−0.99	1302	−0.97
		min	−1185		−1262	
	SGA02	max	1251	−1.02	1295	−1.01
		min	−1273		−1308	
	SGA03	max	1286	−0.98	1384	−0.94
		min	−1255		−1296	
	SGA04	max	1299	−0.96	1429	−0.89
		min	−1251		−1269	
Side-B	SGB01	max	1163	−1.06	1216	−1.03
		min	−1230		−1256	
	SGB02	max	1206	−1.07	1280	−1.08
		min	−1292		−1381	
	SGB03	max	1239	−1.05	1367	−0.99
		min	−1298		−1349	
	SGB04	max	1318	−0.96	1348	−0.97
		min	−1262		−1304	
Both	Load (kN)	max	1200	−0.99	1194	−0.99
		min	−1191		−1187	

Notes
(a) Strain values are reported as $\mu\mathcal{E}$
(b) HSLA-50 and UHSS-100 data taken from time slice before the crack initiation
(c) SGA01–SGA04: Strain gauges mounted on side-A, SGB01–SGB04: Strain gauges on side-B

For UHSS-100 steel welded joints:
$K^{m}_{t,hs}$ (average of all samples, both sides) = 2.06 [refer Table 5]
$K^{b}_{t,hs}$ (average of all samples, both sides) = 1.58 [refer Table 5]
$\varepsilon_{peak} = 60 * 2.06 + 1379 * 1.58 = 2302\ \mu\varepsilon$

It is worth noting that linear extrapolation of strains to the weld toe location was originally developed for the off-shore welded structures and is valid only if the measured strains at 0.4t and 1t distance away from weld toe are purely elastic in nature, which was the case in this case as well. Furthermore, if the loading condition is not known, then estimating the membrane and bending strain values based on the above approach (i.e., based on measured strain on the surface) may not work well. In that case, it's suggested to use FE analysis strain output results (to understand strain distribution through the plate thickness) along with measured strain gauge data on the surface, to arrive at the appropriate contributions from membrane and bending strain components.

Table 7 Extrapolation of strain values at the weld toe location using IIW hot spot stress approach for the HSLA-50 and UHSS-100 welded joints

Distance from weld toe	Distance from weld toe	Strain gauge	HSLA-50 ID-2		UHSS-100 ID-6	
Fraction of plate thickness (t = 6 mm)	(mm)	ID number	Tensile μɛ	Compressive μɛ	Tensile μɛ	Compressive μɛ
0.4	2.4	SGA03	1286	−1255	1384	−1296
1	6	SGA01	1191	−1185	1302	−1262
0	0	Linearized extrapolated	1349	−1302	1439	−1319
Hot spot location			Membrane ε_m (μɛ)	Bending ε_b (μɛ)	Membrane ε_m (μɛ)	Bending ε_b (μɛ)
Nominal strain at weld toe			24	1326	60	1379
Corresponding stress concentration factors			$K^m_{t,hs}$	$K^b_{t,hs}$	$K^m_{t,hs}$	$K^b_{t,hs}$
Refer Tables 5 and 6			2.13	1.63	2.06	1.58
Peak strain at weld toe			2211		2302	

For performing strain–life fatigue analysis, strain–life material properties are required. Considering the loss of hardness (strength) in the HAZ of the UHSS-100 welded joints, (Figs. 14 and 15), it was decided to use stain life properties of HSLA-50 steel, even while performing fatigue analysis of UHSS-100 steel welded joints.

Residual stress value of 200 MPa (at the weld toe location) was used for the strain–life calculations, refer Fig. 16. Estimated crack initiation life (N_i) based on the strain–life analysis with and without residual stress is shown in Table 8, for the loads used to test HSLA-50 ID-2 and UHSS-100 ID-6 welded joints. The estimated crack initiation lives with RS found to be around 18.5 and 9.2% of the total fatigue lives for HSLA-50 ID-2 and UHSS-100 ID-6 welded joints, refer Table 8. Based on the strain gauge indication from the HSLA-50 ID-2 sample testing (Fig. 22), crack initiation life is expected to be around 7.6%.

5 Fatigue Life Prediction Based on Finite Element Stress Analysis

Peak stress information is required for prediction of fatigue crack initiation life using a strain–life approach, whereas through-thickness stress distribution is required for determining fatigue crack growth life using a fatigue crack growth approach. To determine the peak stress and the through-thickness stress distributions, finite element

Table 8 Fatigue crack initiation life estimated based on the measured strain, GY2 FE analysis and comparison with experimental lives for HSLA-50 and UHSS-100 welded samples (with and without residual stresses)

Fatigue life (cycles)		Estimated based on measured strain N_i (measured)		Estimated based on GY2 FEA analysis N_i (FEA)		Experimental total fatigue life N_T (%)	Crack initiation life based on strain gauge indication (%)
		No RS	With RS = 200 MPa	No RS	With RS = 200 MPa		
HSLA-50 ID-2	Crack initiation life	26740	18310	2683	2245	99212	7500
	Crack initiation life/Total experimental life (%)	27.0	18.5	2.7	2.3	N/A	7.6
UHSS-100 ID-6	Crack initiation Life	22120	15510	2995	2492	168498	No clear indication
	Crack initiation life/Total experimental life (%)	13.1	9.2	1.8	1.5	N/A	N/A

analysis was performed using shell GY2 modeling approach for the welded joints. Refer [31, 32] for details on the GY2 methodology for weld modeling and analysis. Figure 24 shows the details of the overall meshed model along with a closer view of the elements representing weld. Horizontal plate was constrained in all degrees of freedom at both the ends (elements assigned with pink color) to represent the test

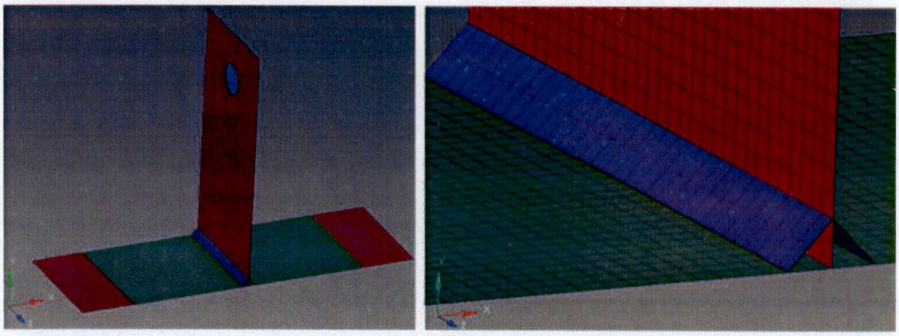

Fig. 24 Shell finite element model of the welded T-joint (left) with a closer view of weld elements assigned with blue color on both sides (right) as per GY2 modeling approach, refer [31, 32] for details on GY2 approach

conditions. Bending load of 1 kN was applied through the 12.7 m diameter hole in the vertical plate along Y-direction (perpendicular to the vertical plate). Displacement contour plot as shown in Fig. 25 matches quite well with the measured displacement of the actuator in Fig. 22, indicating that FE model is a good representation of the experimental test setup. The normal stress component (normal to the crack plane) plays a major role in fatigue life crack initiation and growth, so the stress component S22 was selected for evaluation from the FE analysis. Figure 26 shows the hot spot stress location at the weld toe (matches with the actual crack locations shown in Fig. 18) along with maximum and minimum stress values for S22 component.

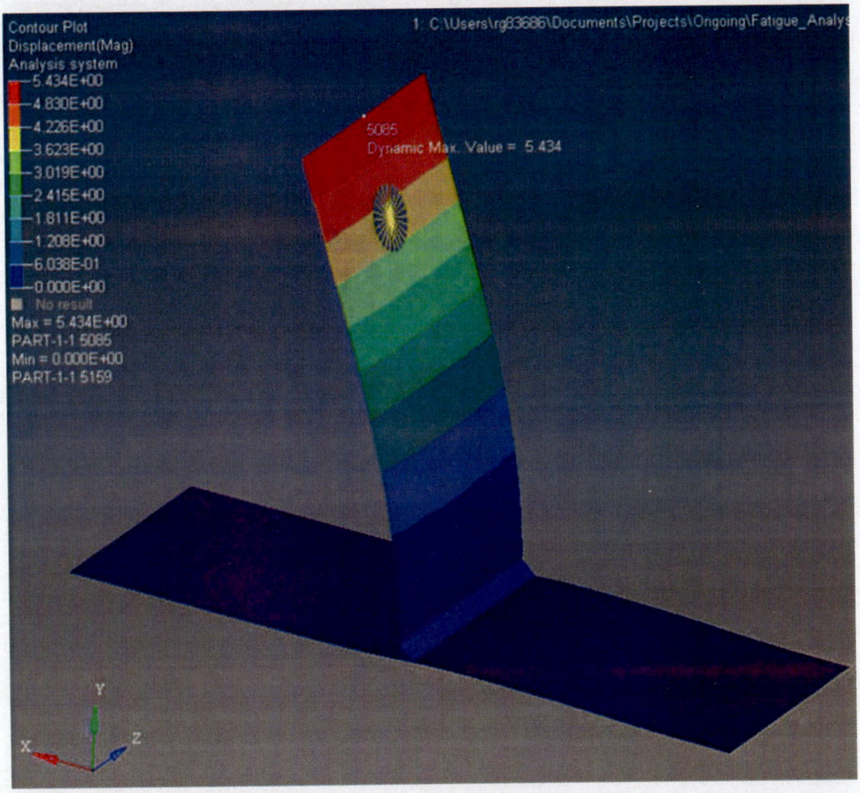

Fig. 25 Displacement contour plot obtained from FE analysis of T-welded joint under 1 kN of applied bending load, showing maximum displacement of 5.43 mm at the free end of vertical plate

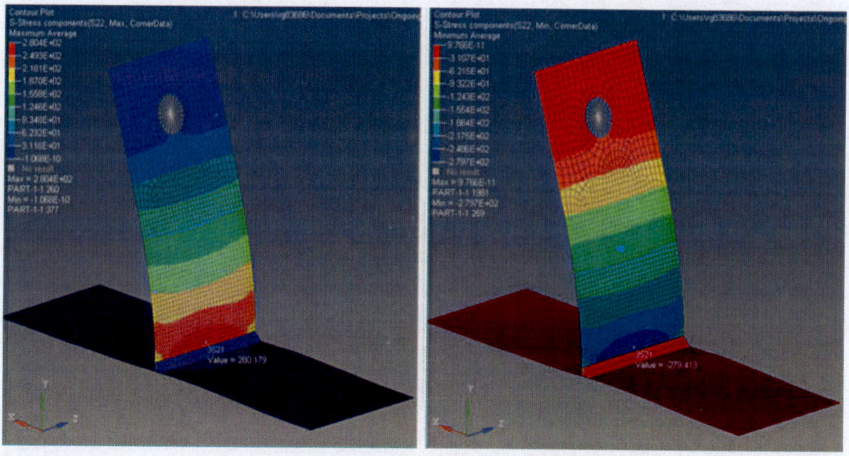

Fig. 26 Normal stress component plots (S22) obtained from FE analysis of T-welded joint under 1kN of applied bending load, showing max and min stress values of 280.2 MPa and −279.4 MPa, respectively at the weld toe hot spot locations

5.1 Fatigue Crack Initiation Life—Strain–Life Analysis Approach

Peak stress information is required for the prediction of fatigue crack initiation life using a strain–life approach. Accordingly, the peak stress at the hot spot weld toe location can be calculated using Eq. (9):

$$\sigma_{peak} = \sigma_m * K_{t,hs}^{m} + \sigma_b * K_{t,hs}^{b} \quad (9)$$

Membrane, σ_m, and bending stress, σ_b, can be obtained quite easily from the shell top and shell bottom stress values at the reference hot spot location from the GY2 shell modeling analysis, refer Fig. 26.

$\sigma_m = \frac{(280.18)+(-279.41)}{2} = 0.38$ MPa
$\sigma_b = \frac{(280.18)-(-279.41)}{2} = 279.80$ MPa

For HSLA 50 steel welded joints:
K_t^m(average of all samples, both sides) = 2.13 [refer Table 4]
$K_{t,}^b$(average of all samples, both sides) = 1.63 [refer Table 4]
$\sigma_{\text{peak}} = 0.38 * 2.13 + 279.8 * 1.63 = 456.27$ MPa

For UHSS 100 steel welded joints:
K_t^m(average of all samples, both sides) = 2.06 [refer Table 5]
K_t^b(average of all samples, both sides) = 1.58 [refer Table 5]
$\sigma_{\text{peak}} = 0.38 * 2.06 + 279.8 * 1.58 = 443.15$ MPa

The strain–life data for HSLA-50 steel as shown in Fig. 13 has been used for fatigue crack initiation life estimation. Residual stress value of 200 MPa at the weld toe location, refer Fig. 16, was used for the strain–life calculations. Estimated crack initiation lives (N_i) based on the strain–life analysis are shown in Tables 10 and 11 without accounting for and with the residual stress, respectively, for all the load cases used to test HSLA-50 and UHSS-100 welded joints. Also, the estimated crack initiation lives (N_i) are shown in Table 8, for the HSLA-50 ID-2 and UHSS-100 ID-6 test samples to provide a direct comparison with prediction based on measured strain, reported in the same table. The estimated crack initiation lives with RS are around 2.3% and 1.5% of the total fatigue lives for HSLA-50 ID-2 and UHSS-100 ID-6 welded joints, refer Table 8.

5.2 *Total Fatigue Life (Initiation + Propagation) Estimation—Total Life Approach*

The total fatigue life (N_T) consists of the fatigue crack initiation life (N_i) and the fatigue crack propagation life (N_p). During this work UniGrow model-based total fatigue life approach [25] has been used to estimate the total fatigue life. One of the main advantages of using this approach is that it does not require arbitrary division of total life into crack initiation and propagation life. Total life approach requires a material characteristic property, (ρ^*) as input rather than the initial crack size. One of the critical pieces of information required for accurate fatigue crack growth analysis is the non-linear through-thickness stress distribution at the critical crack plane. Peak stress as well as through-thickness stress distribution is required for the prediction of total fatigue life, with initial crack size represented by the intrinsic material property (ρ^*) as per the total life approach [26, 33]. Monahan [34] has derived a general expression for the through-thickness stress distribution $\sigma(y)$ at a non-symmetric filet weld as a function of stress concentration factors and the membrane and bending hot spot stress, expressed by Eq. (10).

$$\sigma(y) = \left[\frac{K^m_{t,hs}\sigma^m_{hs}}{2\sqrt{2}}\frac{1}{G_m} + \frac{K^b_{t,hs}\sigma^b_{hs}}{2\sqrt{2}}\frac{1-2\left(\frac{y}{t}\right)^{0.89}}{G_b}\right]\left[\left(\frac{y}{r}+\frac{1}{2}\right)^{-\frac{1}{2}} + \frac{1}{2}\left(\frac{y}{r}+\frac{1}{2}\right)^{-\frac{3}{2}}\right] \tag{10}$$

where:

$$G_m = 1 \quad for \frac{y}{r} \le 0.3$$
$$G_m = 0.06 + \frac{0.94\, exp(-E_m \cdot T_m)}{1+E^3_m T^{0.8}_m \cdot exp\left(-E_m \cdot T^{1.1}_m\right)}\, for \frac{y}{r} > 0.3$$
$$E_m = 1.05\,\theta^{0.18}\left(\frac{r}{t}\right)^q$$
$$q = -0.12\,\theta^{-0.62}$$
$$T_m = \frac{y}{t} - 0.3\frac{r}{t}$$

And

$$G_b = 1 \quad for \frac{y}{r} \le 0.4$$
$$G_m = 0.07 + \frac{0.93\, \exp(-E_b . T_b)}{1+E^3_b T^{0.6}_b \cdot \exp\left(-E_b . T^{1.2}_b\right)}\, for \frac{y}{r} > 0.4$$

$$E_b = 0.9\left(\frac{r}{t}\right)^{-\left(0.0026+\frac{0.0825}{\theta}\right)}$$
$$T_b = \frac{y}{t} - 0.4\frac{r}{t}$$

Equation (10) is valid over the entire thickness in the case of non-symmetric fillet welds and only over half the thickness in the case of symmetric fillet welds. Further, the expression is valid for a range of parameters.

$$\frac{\pi}{6} \le \theta \le \frac{\pi}{3} \quad and \quad \frac{1}{50} \le \frac{r}{t} \le \frac{1}{15} and \quad 0 \le y \le t$$

Figure 11 shows the weld joint geometry nomenclature. For more details, the reader is encouraged to refer [28].

Through-thickness stress distribution obtained for the critical crack plane (at the hot spot weld toe location) using Eq. (10) is shown in Fig. 27.

Total fatigue life approach requires UniGrow model-based material fatigue crack growth property parameters, which have been estimated based on equations proposed by Noroozi and Glinka [3] utilizing experimentally determined strain life material properties. Estimated fatigue crack growth properties are listed in Table 9.

Utilizing total fatigue life approach, estimated total fatigue lives, N_T without and with RS are shown in Tables 10 and 11, respectively, for all the load levels

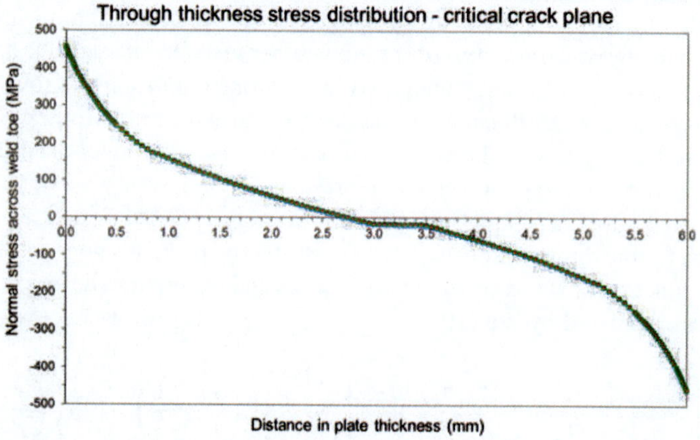

Fig. 27 Through-thickness stress distribution, through the critical crack plane at the weld toe hot spot location, of the normal stress component (S22) obtained using Monahan equation [6] and GY2 model [1, 2]—based FE analysis of T-welded joint under 1 kN of applied bending load

Table 9 Estimated fatigue crack growth properties of HSLA-50 and UHSS-100 steels used during this work

	HSLA-50	UHSS-100
Material block size ρ^*, m	3.9e−05	3.9e−05
Elastic FCG coefficient C_{el}	6.48E−21	4.451e−19
Elastic FCG exponent, γ_{el}	15.38	10.0
Plastic FCG coefficient C_{pl}	7.69E−12	1.408e−11
Plastic FCG exponent, γ_{pl}	2.99	2.833

Table 10 Estimated fatigue crack initiation life (N_i) and total fatigue life (N_T) without residual stress for the HSLA-50 and UHSS-100 steel welded joints

Estimated fatigue lives (without residual stress)					
Material-specimen ID	Normalized load (kN)	R Ratio	Crack initiation life N_i (Estimated)	Total fatigue life N_T (Estimated)	N_i/N_T (%)
HSLA-50-Setup1	1.0	−1	629	75075	0.8
HSLA-50-Setup2	0.9	−1	2683	113674	2.4
HSLA-50-3	1.0	−1	629	75075	0.8
HSLA-50-4	1.0	−1	629	75075	0.8
HSLA-50-11	1.0	−1	629	75075	0.8
HSLA-50-1	0.9	−1	2683	113674	2.4
HSLA-50-2	0.9	−1	2683	113674	2.4
HSLA-50-19	0.9	−1	2683	113674	2.4
HSLA-50-5	0.7	−1	12350	183169	6.7
HSLA-50-6	0.7	−1	12350	183169	6.7
HSLA-50-7	0.6	−1	63713	327635	19.4
HSLA-50-8	0.6	−1	63713	327635	19.4
HSLA-50-9	0.6	−1	63713	327635	19.4
HSLA-50-17	0.6	−1	63713	327635	19.4
HSLA-50-13	0.5	−1	166013	475408	34.9
HSLA-50-14	0.5	−1	166013	475408	34.9
HSLA-50-10	0.4	−1	534079	867546	61.6
HSLA-50-12	0.4	−1	534079	867546	61.6
UHSS-100-Setup1	1.0	−1	693	75075	0.9
UHSS-100-Setup2	0.9	−1	2995	113674	2.6
UHSS-100-4	1.0	−1	693	75075	0.9
UHSS-100-5	1.0	−1	693	75075	0.9
UHSS-100-3	1.0	−1	693	75075	0.9
UHSS-100-6	0.9	−1	2995	113674	2.6
UHSS-100-1	0.9	−1	2995	113674	2.6
UHSS-100-7	0.7	−1	14109	183169	7.7
UHSS-100-8	0.7	−1	14109	183169	7.7
UHSS-100-9	0.7	−1	14109	183169	7.7
UHSS-100-10	0.6	−1	75758	327635	23.1
UHSS-100-11	0.6	−1	75758	327635	23.1
UHSS-100-12	0.6	−1	75758	327635	23.1
UHSS-100-13	0.6	−1	75758	327635	23.1
UHSS-100-2	0.5	−1	203043	475408	42.7
UHSS-100-15	0.5	−1	203043	475408	42.7

(continued)

Table 10 (continued)

Estimated fatigue lives (without residual stress)					
Material-specimen ID	Normalized load (kN)	R Ratio	Crack initiation life N_i (Estimated)	Total fatigue life N_T (Estimated)	N_i/N_T (%)
UHSS-100-14a	0.4	−1	675558	867546	77.9

used to test HSLA-50 and UHSS-100 welded joints. Estimated total fatigue lives using this approach have been compared with corresponding experimental lives in Figs. 28 and 29, without and with residual stress, respectively. It can be observed that estimated lives are within a factor of ±3 times of the experimental lives for the short and medium life regimes. But the estimated lives are conservative by an order of magnitude for the long lives (especially over one million cycles). One of the main reasons could be related to unavailability of accurate fatigue crack growth material properties, especially in the long-life range near the threshold region. Other reason could be related to accurate information of through-thickness residual stress profile, including the residual stress relaxation which could happen during cyclic fatigue loading. It was reported that as much as 45–60% of the maximum residual stresses could be relaxed from the first loading cycle itself due to plastic deformation [35] further as reported and verified experimentally in [9, 36] that the effect of welding residual stress on fatigue life could be overestimated as the actual residual stress could be locally relaxed during cyclic loading. Figure 30 shows the fractured surface of fatigue tested HSLA-50 ID-14 and UHSS-100 ID-14 welded samples. The experimental behavior observed for the crack growth, i.e., much faster crack growth on the surface than in the thickness has been well represented based on the total fatigue life analysis, as shown in Fig. 31.

6 Conclusions and Future Work

The paper describes detailed experimental and numerical fatigue analysis of HSLA-50 and UHSS-100 high-strength steel welded joints. The main conclusions are as follows:

- HSLA-50 and UHSS-100 steel welded joints had similar stress concentration factors both for membrane and bending loads, indicating that material grade type does not contribute to the stress concentration. Rather, this is purely dependent on the weld toe geometry which is based on the consistency of weld manufacturing process.
- UHSS-100 steel welded joints showed reduced hardness in the weld HAZ (a maximum drop of 12HRC), contrary to HSLA-50 welded joints wherein increased hardness was observed in the weld HAZ (maximum increase of 13HRC).

Table 11 Estimated fatigue crack initiation life (N_i) and total fatigue life (N_T) with residual stress for the HSLA-50 and UHSS-100 steel welded joints

Estimated fatigue lives (with residual stress)					
Material-specimen ID	Normalized load (kN)	R ratio	Crack initiation life N_i (Estimated)	Total fatigue life N_T (Estimated)	N_i/N_T (%)
HSLA-50-Setup1	1.0	−1	556	50929	1.1
HSLA-50-Setup2	0.9	−1	2245	72891	3.1
HSLA-50-3	1.0	−1	556	50929	1.1
HSLA-50-4	1.0	−1	556	50929	1.1
HSLA-50-11	1.0	−1	556	50929	1.1
HSLA-50-1	0.9	−1	2245	72891	3.1
HSLA-50-2	0.9	−1	2245	72891	3.1
HSLA-50-19	0.9	−1	2245	72891	3.1
HSLA-50-5	0.7	−1	9239	108025	8.6
HSLA-50-6	0.7	−1	9239	108025	8.6
HSLA-50-7	0.6	−1	37839	173997	21.7
HSLA-50-8	0.6	−1	37839	173997	21.7
HSLA-50-9	0.6	−1	37839	173997	21.7
HSLA-50-17	0.6	−1	37839	173997	21.7
HSLA-50-13	0.5	−1	80274	226030	35.5
HSLA-50-14	0.5	−1	80274	226030	35.5
HSLA-50-10	0.4	−1	186632	319135	58.5
HSLA-50-12	0.4	−1	186632	319135	58.5
UHSS-100-Setup1	1.0	−1	612	50929	1.2
UHSS-100-Setup2	0.9	−1	2492	72891	3.4
UHSS-100-4	1.0	−1	612	50929	1.2
UHSS-100-5	1.0	−1	612	50929	1.2
UHSS-100-3	1.0	−1	612	50929	1.2
UHSS-100-6	0.9	−1	2492	72891	3.4
UHSS-100-1	0.9	−1	2492	72891	3.4
UHSS-100-7	0.7	−1	10410	108025	9.6
UHSS-100-8	0.7	−1	10410	108025	9.6
UHSS-100-9	0.7	−1	10410	108025	9.6
UHSS-100-10	0.6	−1	43529	173997	25.0
UHSS-100-11	0.6	−1	43529	173997	25.0
UHSS-100-12	0.6	−1	43529	173997	25.0
UHSS-100-13	0.6	−1	43529	173997	25.0
UHSS-100-2	0.5	−1	93362	226030	41.3
UHSS-100-15	0.5	−1	93362	226030	41.3

(continued)

Table 11 (continued)

Estimated fatigue lives (with residual stress)					
Material-specimen ID	Normalized load (kN)	R ratio	Crack initiation life N_i (Estimated)	Total fatigue life N_T (Estimated)	N_i/N_T (%)
UHSS-100-14a	0.4	−1	219150	319135	68.7

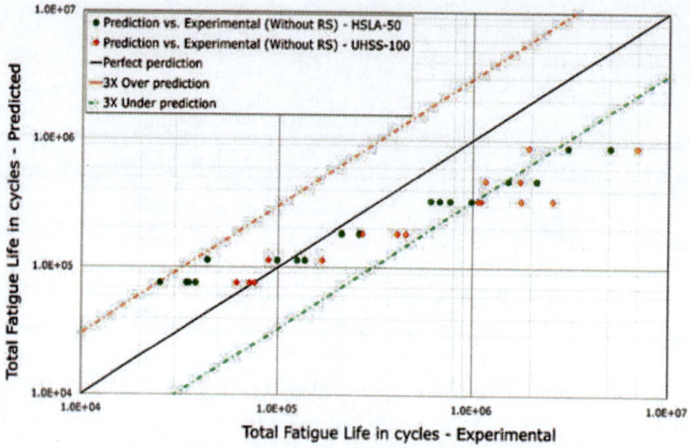

Fig. 28 Comparison of predicted total fatigue lives (Total life method) without residual stress and corresponding experimental fatigue lives for HSLA-50 and UHSS-100 welded samples

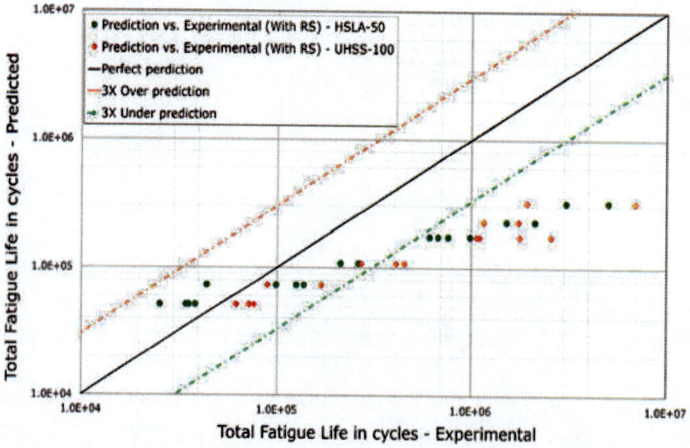

Fig. 29 Comparison of predicted total fatigue lives (Total life method) with residual stress and corresponding experimental fatigue lives for HSLA-50 and UHSS-100 welded samples

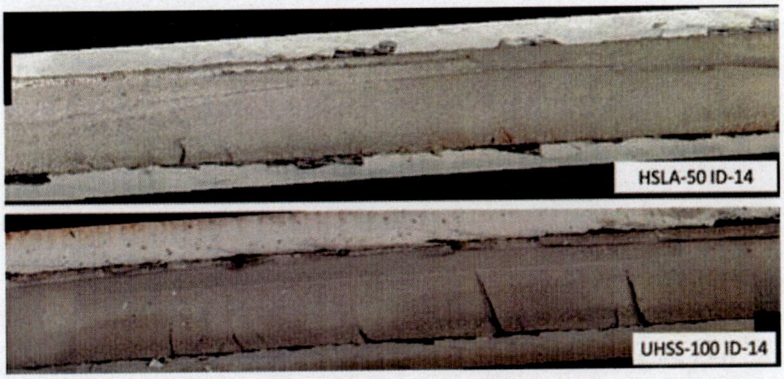

Fig. 30 Fractured surface showing fatigue crack growth through the critical weld toe crack plane for HSLA-50 ID-14 and UHSS-100 ID-14 welded samples

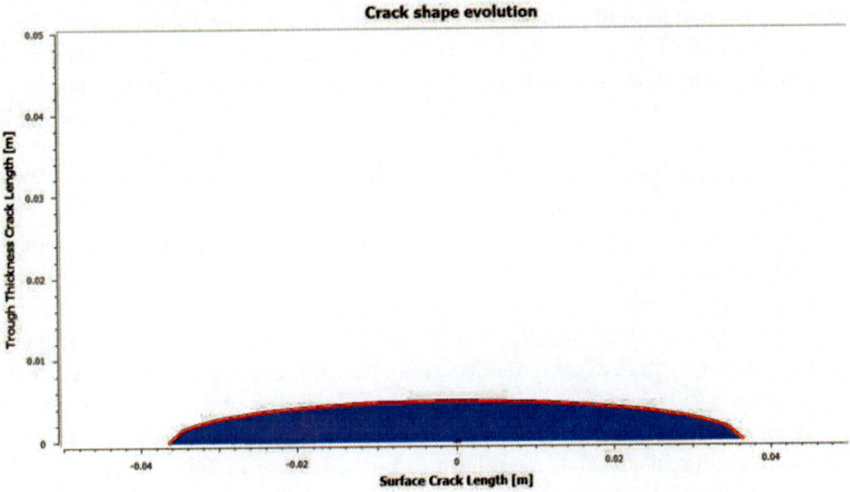

Fig. 31 Evolution of fatigue crack growth and final crack shape through the critical weld toe crack plane obtained from Total life analysis

- Weld zones have similar hardness levels due to the same weld filler wire used for welding of HSLA-50 and UHSS-100 welded joints. UHSS-100 welded joints appear to have slightly higher hardness in the weld zone compared to the weld zone in HSLA-50 joint, which can be attributed to base metal dilution into the weld fusion zone.
- Weld on side-A (first weld in the sequence) experienced tempering due to the weld on side-B (second weld in the sequence) for both the high-strength steel materials. Interestingly though the tempering effect was different for both the steel materials, e.g., weld HAZ hardness on side-A increased for UHSS-100 steel, whereas it reduced for HSLA-50 steel.

- The measured residual stress data consistently showed tensile residual stress values at the weld toe location for all the weld joints. The average values of residual stress on the surface are almost the same for the HSLA-50 and UHSS-100 welded joints (238 MPa and 235 MPa, respectively).
- Welded joints made from HSLA-50 and UHSS-100 welded joints showed similar fatigue behavior for the same design and fatigue loading conditions. These results agree with the observed hardness behavior of weld HAZ regions for both the high-strength steel welded joints.
- For the test samples with relatively shorter fatigue lives (higher loads), crack initiation happened quite early, meaning that more portion of the life was spent in growing the cracks, while the trend was completely opposite for the samples with relatively longer lives (lower loads) as the major portion of their lives was spent in initiating the cracks but once the cracks were initiated, propagation was relatively much faster resulting into fast final fracture. This clearly indicates that designing for crack initiation life might be too conservative for some applications.
- The estimated crack initiation lives predicted based on finite element-based strain life fatigue analysis approach provide reasonably well correlations (though on the conservative side) to the fatigue life predictions based on experimentally measured strains as well as the experimentally obtained fatigue lives.
- The estimated crack initiation lives without and with the included effect from thermal residual stress are around 2.7 and 2.3% of the total fatigue lives for HSLA-50 and 1.8 and 1.5% of the total fatigue lives for UHSS-100 welded joints. This observation again confirms that crack initiation could be representing only a small portion of the total life and result in highly conservative and heavier designs.
- Advanced fracture mechanics based Total life approach showed that estimated total lives are within a factor of ± 3 times of the experimental lives for the short and medium life regimes. Though the estimated lives were found to be conservative by an order of magnitude for the long lives (especially over one million cycles). One of the main reasons could be related to unavailability of accurate fatigue crack growth material properties, especially in the long-life range near the threshold region. Other reason could be related to accurate information of through-thickness residual stress profile, including the residual stress relaxation which could happen during cyclic fatigue loading. These areas are recommended for future research.

References

1. Schjodt T, Andreasen JH (2018) Low cycle fatigue behavior of welded T-joints in high strength steel. Eng Fail Anal 93:38–43
2. Gulyas G (2018) Influence of weld toe radius and steel grade on the fatigue life of fillet welds analyzed by a strain based approach. Master thesis, Aalto University, School of Engineering
3. Lan X, Chan T (2019) Recent research advances of high strength steel welded hollow section joints. Structures 17:58–65
4. Guo H, Wan J, Liu Y, Hao J (2018) Experimental study on fatigue performance of high strength steel welded joints. Thin Walled Struct 131:45–54

5. Lahtinen T, Vilaca P, Infante V (2019) Fatigue behavior of MAG welds of thermomechanical processed 700MC ultra high strength steel. Int J Fatigue 126:62–71
6. Mikulski Z, Lassen T (2019) Fatigue crack initiation and subsequent crack growth in fillet welded steel joints. Int J Fatigue 120:303–318
7. Shiozaki T, Yamaguchi N, Tamai Y, Hiramoto J, Ogawa K (2018) Effect of weld toe geometry on fatigue life of lap fillet welded ultra-high strength steel joints. Int J Fatigue 116:409–420
8. Costa JDM, Ferreira JAM, Abreub LPM (2010) Fatigue behaviour of butt welded joints in a high strength steel. Procedia Eng 2:697–705
9. Zhang W, Jiang W, Zhao X, Tu S (2018) Fatigue life of a dissimilar welded joints considering the weld residual stress: experimental and finite element simulation. Int J Fatigue 109:182–190
10. Leitner M, Ottersbock M, Pubsswald S, Remes H (2018) Fatigue strength of welded and high frequency mechanical impact post-treated steel joints under constant and variable amplitude loading. Eng Struct 163:215–223
11. Niemi E (1995) Recommendations concerning stress determination for fatigue analysis of welded components. IIW document Abington Publishers, Cambridge, UK, pp 1221–1293
12. Fricke W (2002) Recommended hot-spot analysis procedure for structural details of ships and FPSOs based on round-robin FE analysis. Int J Offshore Polar Engg 12(1):40–48
13. Fricke W (2003) Fatigue analysis of welded joints: state of development. Marine Struct 16:185–200
14. Radaj D, Sonsino CM, Fricke W (2009) Recent developments in local concepts of fatigue assessment of welded joints. Int J Fatigue 31:2–11
15. Hobbacher AF (2009) The new IIW recommendations for fatigue assessment of welded joints and components—a comprehensive code recently updated. Int J Fatigue 31:50–58
16. Fricke W (2013) IIW guideline for the assessment of weld root fatigue. Weld World 57:753–791
17. Ahola A, Bjork T, Barsoum Z (2019) Fatigue strength capacity of load carrying fillet welds on ultra-high strength steel plates subjected to out-of-plane bending. Eng Struct 196:109282
18. Fustar B, Lukacevic I, Dujmovic D (2018) Review of fatigue assessment methods for welded steel structures. Adv Civil Eng 2018:1–16. Article ID 3597356
19. Lukacs J (2019) Fatigue crack propagation limit curves for high strength steels based on two-stage relationship. Eng Fail Anal 103:431–442
20. Zerbst U, Madia M, Vormwald M, Beier HTh (2018) Fatigue strength and fracture mechanics—a general perspective. Eng Fail Anal 198:2–23
21. Chang K. Lee C (2011) Finite element computation of fatigue crack growth rates for mode I cracks subjected to welding residual stresses. Eng Fract Mech 78:2505–2520
22. Wang D, Zhang H, Gong B, Deng C (2016) Residual stress effects on fatigue behavior of welded T-joint: a finite fracture mechanics approach. Mater Des 91:211–217
23. Hensel J, Pagel TN, Ngoula DT, Beier HT, Tchuindjang D, Zerbst U (2018) Welding residual stresses as needed for the prediction of fatigue crack propagation and fatigue strength. Eng Fract Mech 198:123–141
24. Kucharczyk P, Madia M, Zerbst U, Schork B, Gerwien P, Munstermann S (2018) Fracture mechanics-based prediction of the fatigue strength of weldments. Material aspects. Eng Fract Mech 198:79–102
25. Noroozi AH, Glinka G, Lambert S (2007) A study of the stress ratio effects on fatigue crack growth using the unified two-parameters fatigue crack growth driving force. Int J Fatigue 29:1616–1634
26. Mikheevskiy S, Glinka G, Cordes T (2015) Total life approach for fatigue life estimation of welded structures. In: 3rd international conference on material and component performance under variable amplitude loading, VAL2015. Procedia Eng 101:177–184
27. Swansom M, Johnson E, Stoytchev A (2010) Automated weld integrity analysis using 3D point data. In: Proceedings of the ASME 2010 world conference on innovative virtual reality, WINVR2010, 12–14 May, Ames, Iowa, USA
28. Iida K, Uemura T (1996) Stress concentration factor formulas widely used in Japan. Fatigue Fract Eng Mater Struct 19(6):779–786

29. Skriko T, Ghafouri M, Björk T (2017) Fatigue strength of TIG-dressed ultra-high-strength steel fillet weld joints at high stress ratio. Int J Fatigue 94:110–120
30. Johnson E (2013) Progress toward a model based approach to the robust design of welded structures. PhD Thesis, Iowa State University, Graduate Theses and Dissertations Paper 13119
31. Chattopadhyay A, Glinka G, El-Zein M, Qian J, Formas R (2011) Stress analysis and fatigue of welded structures. Weld World 55(7–8):2–21
32. Goyal R, El-zein M, Glinka G (2016) A robust stress analysis method for fatigue life prediction of welded structures. Weld World 60:299–314
33. Goyal R, Glinka G (2013) Fracture mechanics based estimation of fatigue lives of welded joints. Weld World 57(5):625–634
34. Monahan CC (1995) Early fatigue cracks growth at welds. Computational Mechanics Publications, Southampton UK
35. Xie X, Jiang W, Luo Y, Xu S, Gong J, Tu S (2017) A model to predict the relaxation of weld residual stress by cyclic load: experimental and finite element modeling. Int J Fatigue 95:293–301
36. Leitner M, Khurshid M, Barsoum Z (2017) Stability of high frequency mechanical impact post-treatment induced residual stress states under cyclic loading of welded steel joints. Eng Struct 143:589–602

Assessment of Cyclic Plasticity Behaviour of Primary Piping Material of Indian PHWRs Under Multiaxial Loading Scenario

Punit Arora, M. K. Samal, S. K. Gupta, and J. Chattopadhyay

Abstract The present study is aimed at assessment of stabilized cyclic elastic–plastic stress/strain response of low C–Mn steel used as the primary piping material of Indian PHWRs. The material model considers von Mises yield criterion, three decomposed non-linear kinematic hardening-based Chaboche model and associative flow rule. Three-dimensional tubular geometry (as tested) has been analysed using in-house developed finite element code. The Chaboche's parameters have been calibrated using the stabilized low-cycle fatigue test loop for a given pure axial strain amplitude. These calibrated parameters have been used to predict cyclic plastic material response for pure axial, pure torsion, in-phase axial–torsion and out-of-phase axial–torsion strain paths. The predicted cyclic stress–strain response under pure axial, pure torsion and in-phase axial–torsion conditions is nearly comparable to test response. However, the material stress response is underestimated in axial and shear directions for non-proportional axial–torsion conditions.

Keywords Non-linear kinematic hardening · Proportional · Non-proportional · Multiaxial · Cyclic plasticity

Nomenclature

E	Young's modulus
σ_y	Tensile yield strength
σ_u	Ultimate tensile strength
%El	Percentage elongations
σ_y^{cyc}	Cyclic yield strength
$\sigma_y^{0.35}, \sigma_y^{0.5}, \sigma_y^{0.75}$	Cyclic yield strengths corresponding to axial strain amplitudes of $\pm 0.35\%$, $\pm 0.5\%$ and $\pm\, 0.75\%$, respectively
$d\alpha_{ij}^k$	Increment in back stress tensor for kth segment of loop

P. Arora (✉) · M. K. Samal · S. K. Gupta · J. Chattopadhyay
Reactor Safety Division, Bhabha Atomic Research Centre, Mumbai 400 085, India
e-mail: punit@barc.gov.in

S. Seetharamu et al. (eds.), *Fatigue, Durability, and Fracture Mechanics*, Lecture Notes in Mechanical Engineering, https://doi.org/10.1007/978-981-15-4779-9_16

$d\varepsilon_{ij}^{p}$	Plastic strain increment tensor
α_{ij}^{k}	Total deviatoric back stress tensor for *k*th segment of loop
dp	von Mises equivalent plastic strain increment
C_k, γ_k	Chaboche's parameters for *k*th segment of loop
ε_x^a	Axial strain amplitude
γ_{xy}^a	Shear strain amplitude
$\varepsilon_{vM}^a, \Delta\varepsilon_{vM}$	von Mises equivalent strain amplitude and range
$\varepsilon_x^t, \gamma_{xy}^t$	Axial and shear strain histories on outer surface of tube
σ_x^t, τ_{xy}^t	Axial and shear stress histories on outer surface of tube
$\gamma_{xy}^a / \sqrt{3}\varepsilon_x^a$	Shear to Axial Ratio (SAR) of strain amplitudes
P^t, T^t	Axial reaction load and torque histories
r, A, J	Outer radius, area of cross section, polar moment of inertia of tube

Acronyms

vM	von Mises
LCF	Low Cycle Fatigue
PHT	Primary Heat Transport
SAR	Shear to Axial Ratio (SAR) of strain amplitudes
PHWR	Pressurized Heavy Water Reactor

1 Introduction

Accurate assessment of fatigue life of realistic piping/vessel components subjected to multiaxial loading conditions is the objective of designer. Extensive literature covers invariant-based [1], plastic strain energy-based ([2, 3]) and critical plane-based ([4–17]) equivalent parameters for complex multiaxial loading scenario. Some of the promising critical plane-based fatigue life models ([15], [17]) have matured to a level for reasonable predictions of fatigue crack initiation life. These models have been benchmarked with respect to various materials and a wide variety of cyclic loading paths. These models, however, need realistic stress/strain histories as input variables to evaluate the equivalent parameter. Hence, it becomes important to first predict the cyclic stress–strain response of material before estimating fatigue life.

The multiaxial cyclic loading is mainly classified into two categories: proportional (in-phase, in general) and non-proportional (out-of-phase, in general) loadings. The proportional loadings are defined on the basis of ratio of any two stress/strain components being independent of time. This leads to fixed principal stress/strain directions and slip activity to occur mainly on primary slip systems. On the other side, the non-proportional loading is constituted due to time-dependent stress ratios. This causes rotation of principal stress/strain planes. This rotation in principal stress axes or maximum shear planes causes time-dependent activation/de-activation of primary slip systems [18]. The slip caused due to dislocation glide on multiple slip systems results in the formation of dislocation sub-structuring leading to additional material hardening under non-proportional loading conditions.

This microstructural phenomenon leading to continuum-level strain hardening needs to be captured using classical constitutive cyclic plasticity material models such as Armstrong and Fredrick [19], Chaboche [20], Ohno and Wang [21], Modified Ohno and Wang [22], Tanaka [23], Benallal and Marquis [24], etc. Hence, a wide variety of cyclic plasticity material models are available in the literature. One model which is most suitable for Primary Heat Transport (PHT) piping material of Pressurized Heavy Water Reactor (PHWR) needs to be investigated. These models mainly use von Mises yield criterion and may use single or multiple yield surfaces [25] to account for memory effect. The hardening rules used in these models can be either pure isotropic or pure kinematic or combined isotropic–kinematic hardening based on Masing/Non-Masing characteristics exhibited under pure axial conditions. Various bilinear, multi-linear and non-linear (single and segmented) kinematic hardening rules are generally used for cyclic applications. Generally, incremental plasticity-based associative flow rule is used to simulate small strain deformations of metallic materials. Amongst, different such combinations of yield surfaces (single or multiple), hardening rule (linear, nonlinear: single or segmented) and associative flow rules, adequate combination for yield criterion, hardening rules and flow rule need to be selected for correct estimation of cyclic stress/strain response of material subjected to proportional/non-proportional multiaxial loading conditions.

Recently, Xing et al. [26] have applied the Ohno–Wang kinematic hardening rule with Marquis isotropic considerations to assess the cyclic response of SS 316L material under complex non-proportional strain paths. Facheris and Janssens [27] proposed modifications in the Chaboche model to capture thermal stress-induced deformation in the primary circuit of light water reactors. Wu [28] carried out an extensive microstructural investigation to correlate damage with Banallal's non-proportionality factor.

2 Scope of Present Work

In present analyses, von Mises yield criterion, three decomposed non-linear kinematic hardening-based Chaboche model and associative flow rule have been considered. An in-house FE-based code has been developed and relevant cyclic plasticity subroutines have been incorporated in this code. The calibration of C_i and γ_i parameters has been performed using pure axial Low Cycle Fatigue (LCF) test data under one strain amplitude. The calibrated Chaboche's parameters from one of the pure axial LCF tests have been used to predict stabilized cyclic material response for other pure axial tests with different strain amplitudes, pure torsion, in-phase axial–torsion and out-of-phase axial–torsion strain paths.

Table 1 Chemical composition in weight percentage for SA333 Gr. 6 material [29]

C	Mn	Si	P	S	Cr	Ni	N	Balance
0.14	0.9	0.25	0.016	0.018	0.08	0.05	0.01	Fe

Table 2 Tensile properties of SA333 Gr. 6 at room temperature [29]

E (GPa)	σ_y (MPa)	σ_u (MPa)	Total elongation (%)
203	307	463	39.1

The predicted cyclic plasticity response of low C-Mn steel has been compared with corresponding multiaxial tests response for different vM equivalent strain amplitudes and Shear to Axial Ratio of strain amplitudes (SAR) as investigated by Arora et al. [29]. The vM equivalent strain amplitude and SAR are given by Eqs. (1) and (2), respectively.

$$\varepsilon_{vM}^{a} = \sqrt{\left(\left(\varepsilon_x^a\right)^2 + \frac{\left(\gamma_{xy}^a\right)^2}{3}\right)} \tag{1}$$

$$SAR = \frac{\gamma_{xy}^a}{\sqrt{3}\varepsilon_x^a} \tag{2}$$

3 Tensile Properties and Chemical Composition

The chemical compositions of primary heat transport piping material of Indian PHWRs with designation as SA333 Gr. 6 (low carbon–manganese steel) in as received condition is given in Table 1.

The uniaxial monotonic tensile properties as determined by conducting a standard test on round tensile specimen following the test procedure of ASTM-E-08 is furnished in Table 2. The specimens were fabricated from the pipe made of SA333 Gr. 6 material.

4 Calibration of Cyclic Parameters

Arora et al. [29] have performed pure axial cyclic tests on round solid LCF specimen and tube specimens using standard test procedures on ASTM-E-606 [30] and ASTM-E-2207 [31], respectively. The pure axial test loops on tube specimens have been used to calibrate cyclic yield strength and Chaboche's parameters.

4.1 Cyclic Yield Strength

To evaluate the stabilized cyclic yield strength under pure axial cyclic conditions, the test loops have been shifted to vM equivalent stress–strain space such that the minimum vM strain–stress coordinate touches the origin (Fig. 1). Two plastic strain offset lines (corresponding to plastic strain of 0.005%) have been drawn that are originating from zero and maximum stress values for each strain amplitude. The loading offset line is emanating from zero stress value and unloading offset line is originating from maximum stress values of stress in each hysteresis loop as shown in Fig. 1.

The intersection between loading offset line and shifted hysteresis loop is twice of cyclic yield strength. The cyclic yield strength is observed to be higher for higher axial strain amplitudes (ε_x^a) as shown in Fig. 1. Further, it is also observed that unloading offset line for a given ε_x^a intersects the shifted hysteresis loop with altitude (along stress axes) of twice of cyclic yield for corresponding ε_x^a. This indicates that, if the cyclic yield strength is considered as a strain range-dependent parameter, then yielding in the reverse direction can be modelled using pure kinematic hardening rule. In other words, the material shows Masing characteristics with a linear drift in hysteresis loops.

It is also found necessary to understand the material behaviour under pure torsion and combined in-phase axial–torsion conditions on shifted vM equivalent stress–strain space. The material also exhibits Masing characteristics with linear drift for pure torsion and in-phase axial–torsion loading conditions (Fig. 2, [29]), viz. the cases under proportional (or in-phase) loading scenario can also be modelled by consideration of vM equivalent strain range-dependent cyclic yield strengths and pure kinematic hardening rule. The loading and unloading lines in Figs. 1 and 2a,

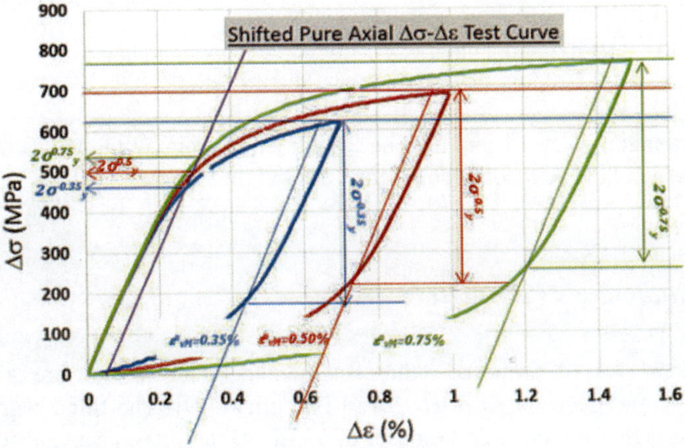

Fig. 1 Typical material test response under pure axial conditions on shifted von-Mises equivalent stress–strain space

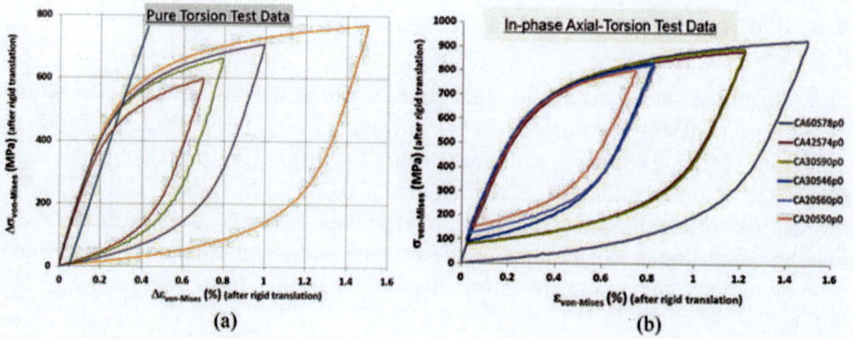

Fig. 2 Stabilized test response under **a** pure torsion and **b** in-phase axial–torsion conditions [29] on shifted vM equivalent stress–strain space

Fig. 3 Variation of cyclic yield strength with vM equivalent strain amplitude

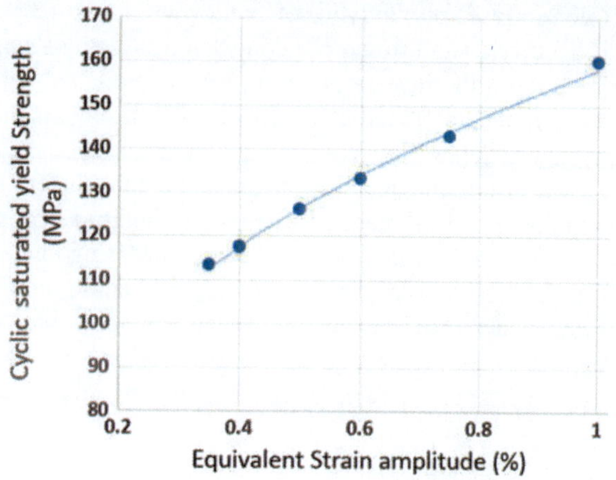

respectively, are for representation purposes and are not to the scale of required offset strain of 0.005%.

The variation of cyclic yield strength with axial strain amplitude as determined from pure axial cyclic loading conditions is given in Fig. 3.

4.2 Chaboche's Parameters

Chaboche et al. [20], proposed a three-segmented version of non-linear kinematic hardening rule given by Armstrong and Frederick [19]. The three segments are; the initial higher modulus at the onset of yielding, the non-linear transition of the hysteresis curve after yielding and until the curve becomes linear and the linear segment of the curve in the range of higher strain values.

Table 3 Calibrated Chaboche's parameters on pure axial saturated cyclic loop with $\varepsilon_x^a = \pm 0.75\%$

σ_y^{cyc} (MPa)	$C_1(MPa)$	γ_1	$C_2(MPa)$	γ_2	$C_3(MPa)$	γ_3
143	7.05×10^5	10,500	1.00×10^5	725	6500	0

$$d\alpha_{ij}^k = \frac{2}{3}C_k d\varepsilon_{ij}^p - \gamma_k \alpha_{ij}^k dp \tag{3}$$

$$\alpha_{ij} = \sum_{k=1}^{3} d\alpha_{ij}^k \tag{4}$$

Chaboche kinematic hardening rule is a superposition of several Armstrong and Frederick hardening rules as indicated by Eq. (4). Here, C_k and γ_k are the cyclic material parameters, k being the variable for the number of segments. For the present case, $1 \le k \le 3$.

The first part of Eq. (3) is identical to Prager's linear kinematic rule [32] and the second part accounts for dynamic recovery in the evaluation of increment of deviatoric back stress tensor ($d\alpha_{ij}^k$). The Chaboche's parameters have been calibrated from pure axial cyclic test loop conducted under axial strain amplitude of ±0.75%. The values of (C_1, γ_1), (C_2, γ_2) and (C_3, γ_3) have been optimized such that predicted material response for ±0.75% as strain amplitude matches well with test hysteresis response. The value of γ_3 is zero for non-ratcheting cases leading to the closure of hysteresis loop. The calibrated parameters are given in Table 3.

5 Prediction of Cyclic Response and Comparison with Test Loops

The calibrated Chaboche's parameters and cyclic yield strengths corresponding to different vM equivalent strain amplitudes have been used for prediction of cyclic response under pure axial (remaining cases other than under $\varepsilon_x^a = \pm 0.75\%$), pure torsion, in-phase axial–torsion and out-of-phase axial–torsion loading cases. The predicted material response in axial and shear directions has been compared with corresponding saturated test response. These analyses account the following considerations,

(a) single-yield surface theory
(b) consideration of cyclic yield strength (under combined axial–torsion cycling) same as that under pure axial conditions for a given vM equivalent strain amplitude
(c) use of three decomposed non-linear kinematic hardening Chaboche model.
(d) Prandtl–Reuss flow rule.

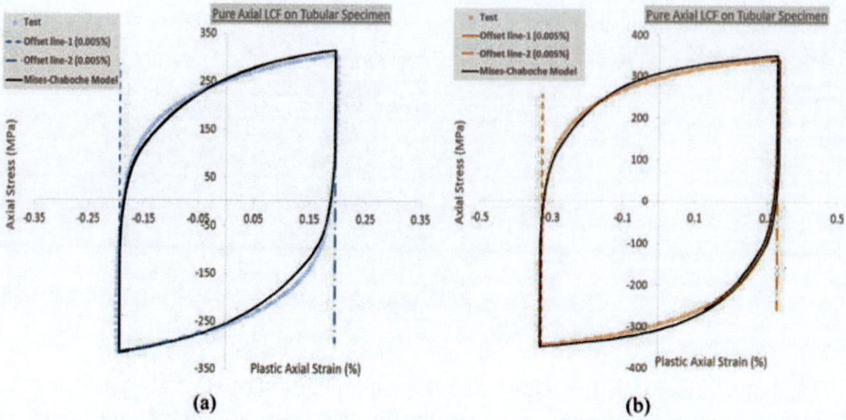

Fig. 4 Comparisons of predicted axial stress-versus-plastic axial strain using FE analyses with test loops for ε_x^a of **a** ±0.35% and **b** ±0.50%

The in-house code uses the B-bar method for stiffness matrix evaluation to avoid volumetric locking [33]. The three-dimensional FE model uses 4800 numbers of 8 node tri-linear elements.

5.1 *Pure Axial Conditions*

The calibrated Chaboche's parameters along with equivalent strain amplitude-based cyclic yield strength (Fig. 3) have been used for predictions of pure axial tests subjected to axial strain amplitudes of ±0.35 and ±0.50%. The predicted axial stress–plastic axial strain response is comparable to test response under the pure axial state of stress/strain conditions (Fig. 4).

5.2 *Pure Torsion Conditions*

Typical comparisons between predicted and actual test responses under pure torsion cyclic conditions for vM equivalent strain amplitude of ±0.50% and ±0.75% have been shown in Figs. 5a and b, respectively. The predicted shear stress–plastic shear strain response compares well with corresponding test response under pure torsion cyclic conditions.

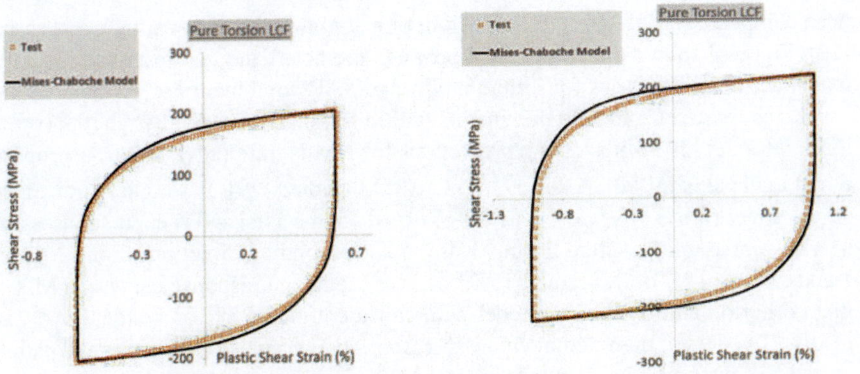

(a) (b)

Fig. 5 Comparisons of predicted shear stress-versus-plastic shear strain using FE analyses with test loops for ε_{vM}^{a} of **a** ±0.50% and **b** ±0.75%

5.3 In-phase Axial–Torsion Conditions

Arora et al. [29] have carried out in-phase axial–torsion LCF tests with several equivalent strain amplitudes ($\boldsymbol{\varepsilon}_{vM}^{a}$) and SARs $\gamma_{xy}^{a}/\sqrt{3}\boldsymbol{\varepsilon}_{x}^{a}$ (for a given $\boldsymbol{\varepsilon}_{vM}^{a}$). All the proportional strain paths have been studied using triangular loading waveforms of axial and shear strain cycling. The slope of proportional strain path, as indicated in Fig. 6b, is the SAR and the half of vector length is equivalent strain amplitude ($\boldsymbol{\varepsilon}_{vM}^{a}$).

The finite element analyses have been performed to predict cyclic behaviour in axial and shear directions. The predicted and test response in both directions have

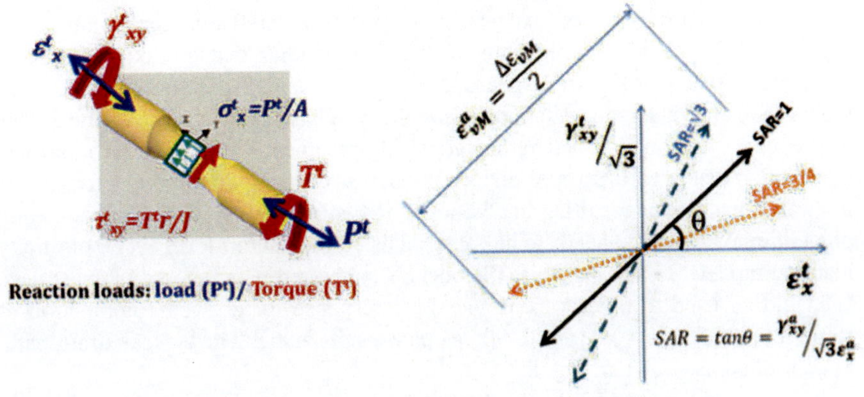

(a) (b)

Fig. 6 **a** schematic of tube specimen subjected to axial and shear strain cycling and **b** proportional strain paths studied by Arora et al. [29]

been compared for the saturated behaviour of material for all proportional load cases [29]. Figure 7 shows the comparison between predicted and test response for both axial and shear directions for different ε^a_{vM} and SAR combinations.

The calibrated Chaboche parameters with different cyclic yield strengths (w.r.t. different ε^a_{vM}) result in accurate predictions for small strain ($\varepsilon^a_{vM} < 0.6\%$) proportional load paths for all SARs ($\sim {}^3/_4$, 1, $\sqrt{3}$) as studied [29]. For higher equivalent strain ranges ($0.6\% \le \varepsilon^a_{vM} \le 1\%$) and SAR ~ 0.75, the axial and shear responses are nearly comparable to actual test response in corresponding directions. For SARs as 1 and $\sqrt{3}$ and ε^a_{vM} of 0.6, and 0.75%, the predicted axial response using von Mises yield criterion and Chaboche model with consideration of ε^a_{vM} dependent σ_y^{cyc}, is nearly 10% lesser than actual $\sigma_x^t - \varepsilon_x^t$ response. For such selective strain paths, accounting distortion in yield criterion might produce reasonable predictions in axial direction. Wu et al. ([34], [35]) have conducted extensive fundamental studies on the determination of shape of subsequent yield surface following various proportional and non-proportional paths on SS 304. They have brought out that yield surface distorts for higher equivalent strain amplitude with proportional/non-proportional load paths. The predicted cyclic torsion response $\tau_{xy}^t - \gamma_{xy}^t$ is in good agreement with corresponding test response for all proportional strain paths with different SARs and ε^a_{vM}. Hence, selective distortion in axial direction would produce better results for SARs as 1 and $\sqrt{3}$ and ε^a_{vM} of 0.6, and 0.75%. However, distortion parameters need to be quantified from few in-phase axial–torsion tests and must be validated for other proportional strain paths where axial response is underpredicted.

5.4 *Out-of-Phase Axial–Torsion Conditions*

Arora et al. [29] conducted 90° out-of-phase axial–torsion tests to understand the effect of non-proportionality for a given combination of axial and shear strain amplitudes (ε^a_x, γ^a_{xy}). It was brought out in their test studies that the axial and shear stress response under 90° out-of-phase axial–torsion cyclic conditions is significantly higher than that observed under corresponding in-phase axial–torsion cycling. This test response at continuum length scale is in agreement with microstructural test observation of the formation of dislocation sub-structuring due to the activation of multiple slip systems resulting in additional hardening [18]. Now, this phenomenon of additional hardening needs to be captured using continuum-level cyclic plasticity material models. In view of this, Chaboche's calibrated parameters (Table 3) have been further used to predict axial and shear response for a given combination of strain amplitudes (ε^a_x, γ^a_{xy}) with 90° as phase shift angle. The typical strain path, as controlled, is shown in Fig. 8.

The predicted axial and shear stress–plastic strain response have been compared with test loops for various ε^a_{vM} and SARs in Fig. 9. It is apparent that for non-proportional strain paths, the use of von Mises yield criterion along with Chaboche's three decomposed non-linear kinematic hardening rule and strain range-dependent

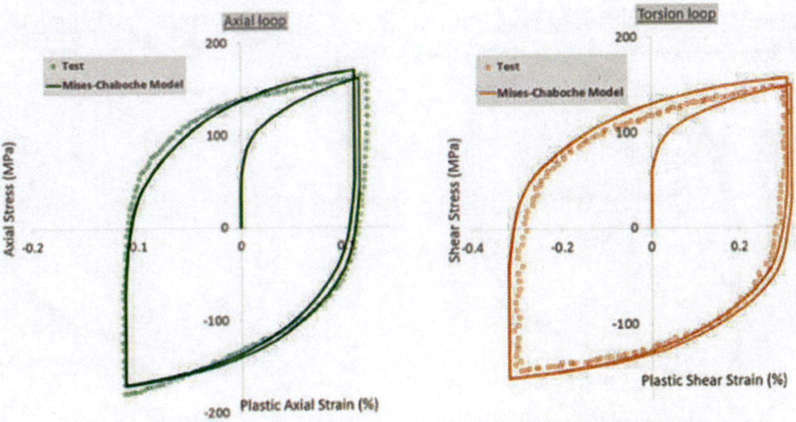

(a) $\varepsilon_{vM}^{a} = 0.35\%$ and SAR=1.43 (Test ID: CA20S50p0 [29])

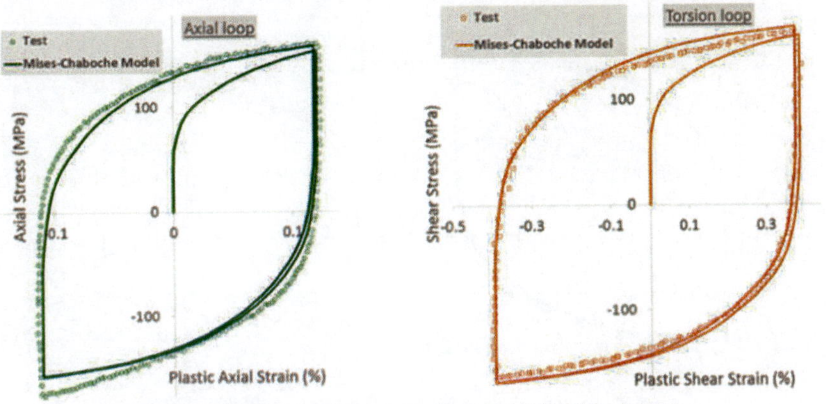

(b) $\varepsilon_{vM}^{a} = 0.40\%$ and SAR=1.73 (Test ID: CA20S60p0 [29])

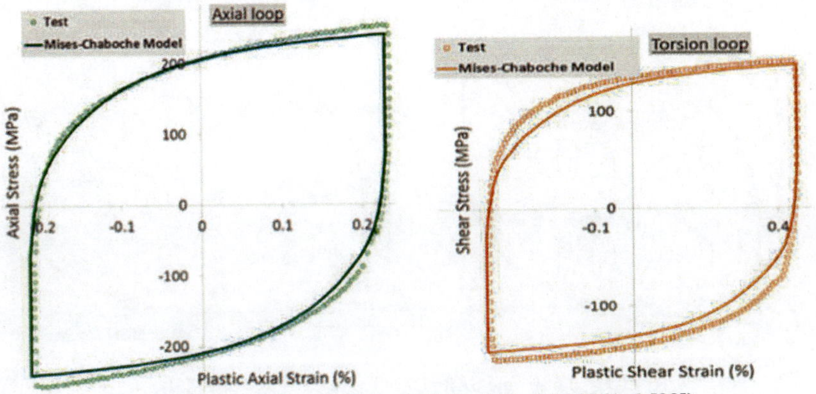

(c) $\varepsilon_{vM}^{a} = 0.50\%$ and SAR=1.00 (Test ID: CA35S61p0 [29])

Fig. 7 Comparison for predicted axial and torsion response of SA333 Gr. 6 material with test results [29] under in-phase axial–torsion loading conditions with different ε_{vM}^{a} and SARs

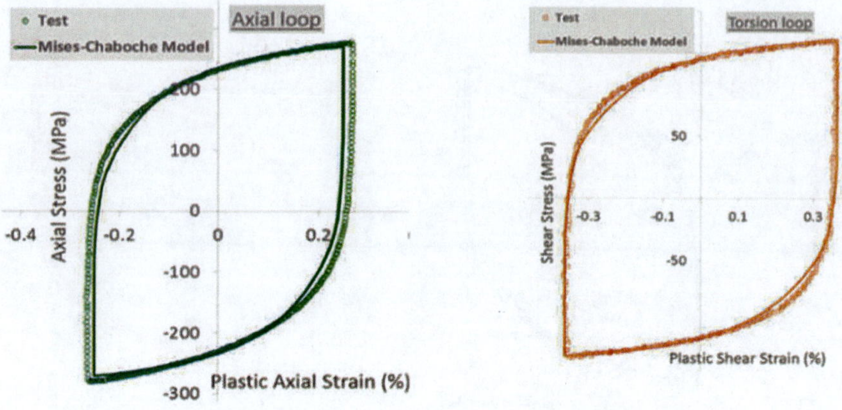

(d) ε_{vM}^{a} = 0.5% and SAR=0.75 (Test ID: CA40S52p0 [29])

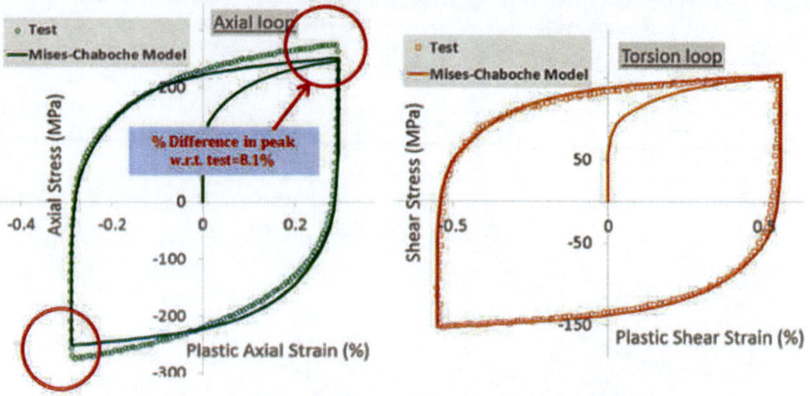

(e) ε_{vM}^{a} = 0.60% and SAR=1.00 (Test ID: CA42S74p0 [29])

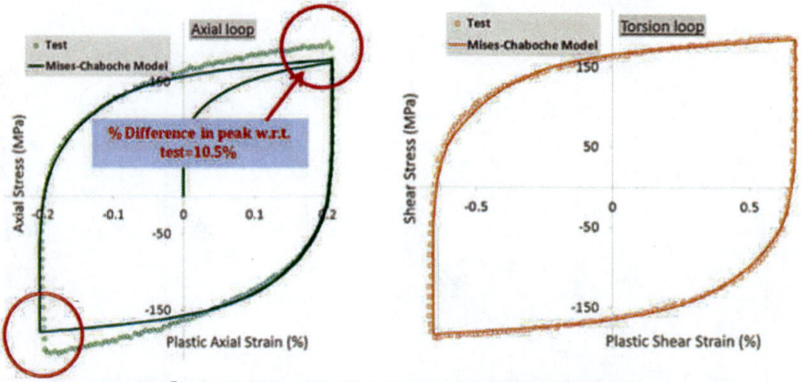

(f) ε_{vM}^{a} = 0.60% and SAR=1.73 (Test ID: CA30S90p0 [29])

Fig. 7 (continued)

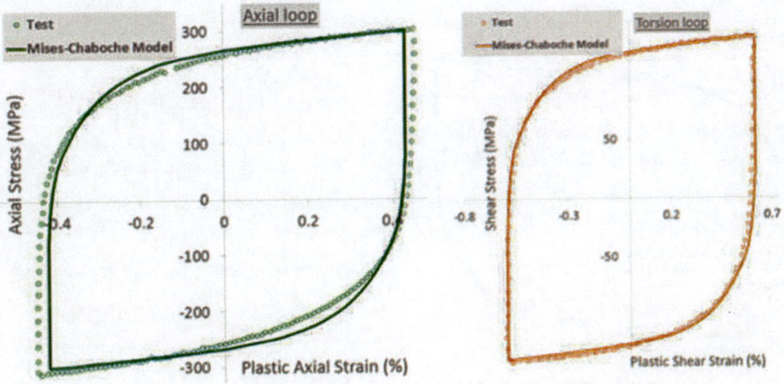

(g) ε_{vM}^{a} = 0.75% and SAR=0.75 (Test ID: CA60S78p0 [29])

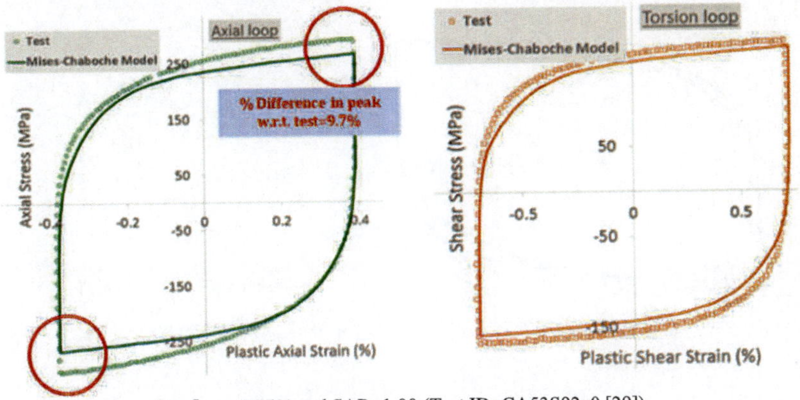

(h) ε_{vM}^{a} = 0.75% and SAR=1.00 (Test ID: CA53S92p0 [29])

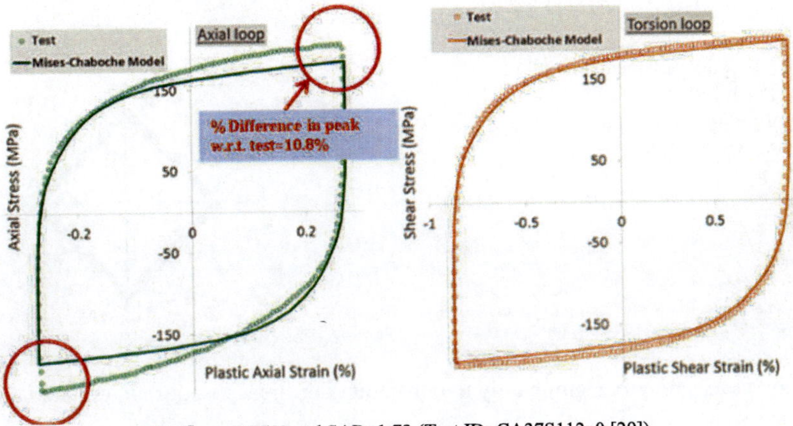

(i) ε_{vM}^{a} = 0.75% and SAR=1.73 (Test ID: CA37S112p0 [29])

Fig. 7 (continued)

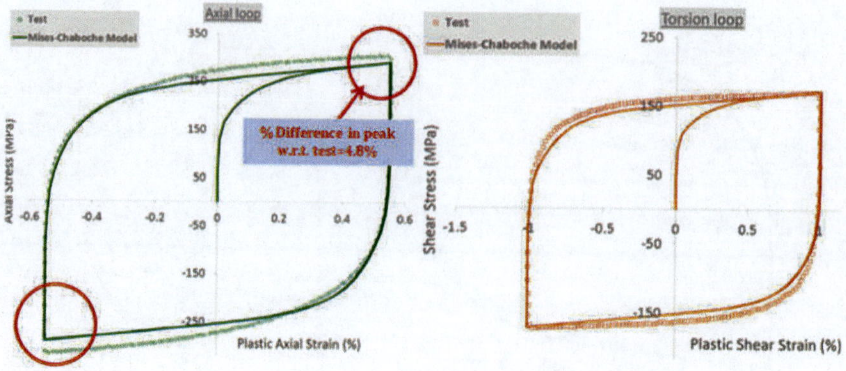

(j) ε_{vM}^{a} = 1.00% and SAR=1.00 (Test ID: CA70S121p0 [29])

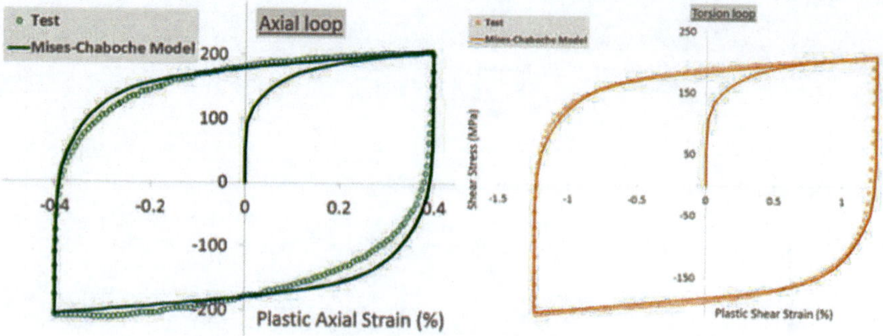

(k) ε_{vM}^{a} = 1.00% and SAR=1.73 (Test ID: CA50S150p0 [29])

Fig. 7 (continued)

Fig. 8 Typical schematic of strain path with 90° out-of-phase axial–torsion cycling with triangular waveform

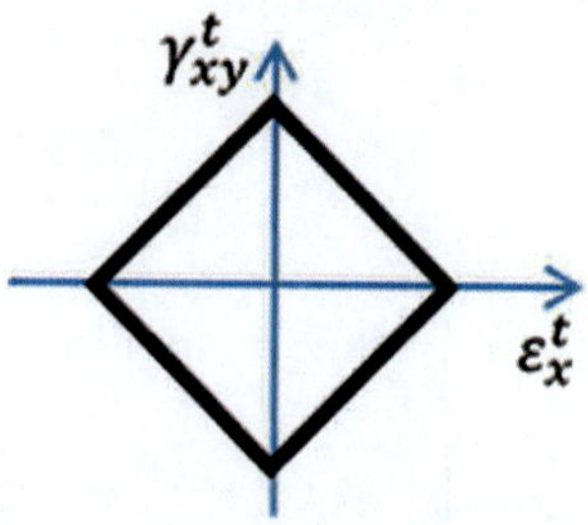

cyclic yield strength, significantly underpredict the stress response in both axial and shear directions.

It is also bought out from the present study that the Chaboche model with considerations of ε_{vM}^{a}-dependent σ_y^{cyc}, predicts little higher hardening in both axial and shear directions as compared to predicted in-phase responses, as indicated in Fig. 10.

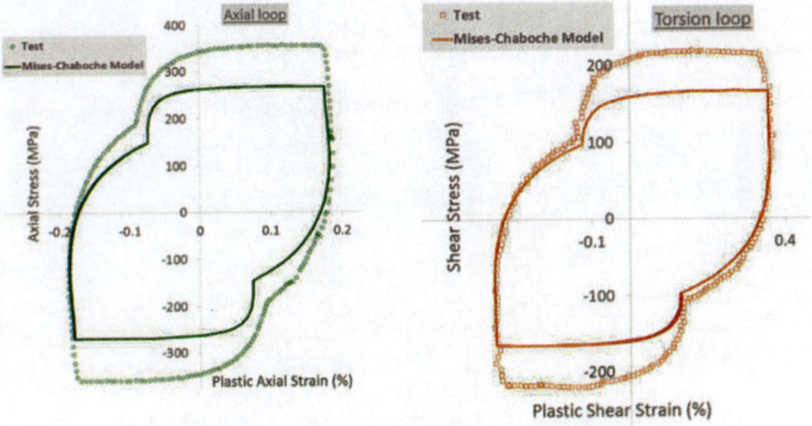

(a) ε_{vM}^{a} = 0.50% and SAR=1.00 (Test ID: CA35S61p90 [29])

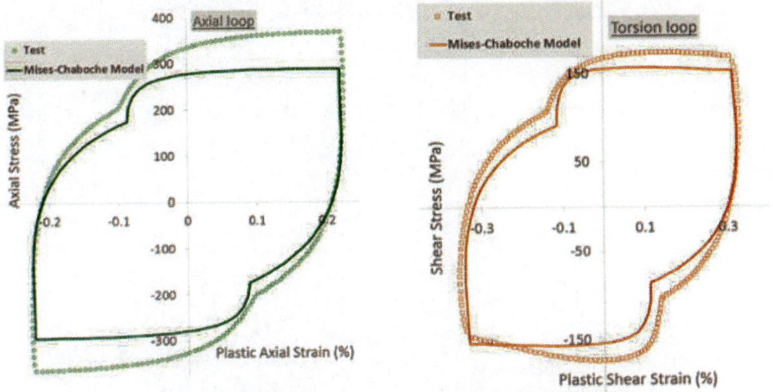

(b) ε_{vM}^{a} = 0.5% and SAR=0.75 (Test ID: CA40S52p90 [29])

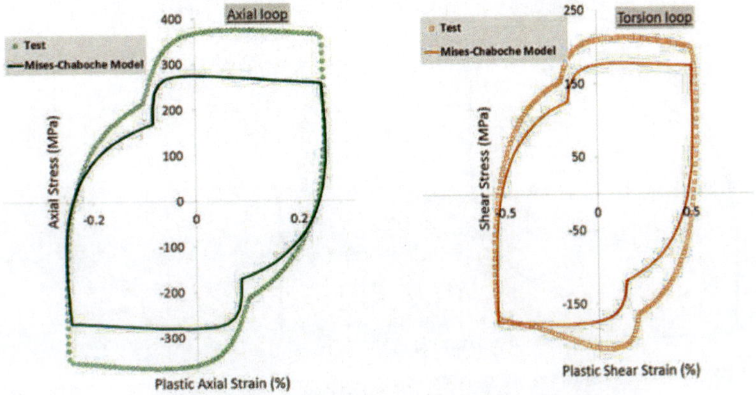

(c) ε_{vM}^{a} = 0.60% and SAR=1.00 (Test ID: CA42S74p90 [29])

Fig. 9 Comparison for predicted axial and torsion response of SA333 Gr. 6 material with test results [29] under 90° out-of-phase axial–torsion loading conditions for different ε_{vM}^{a} and SARs

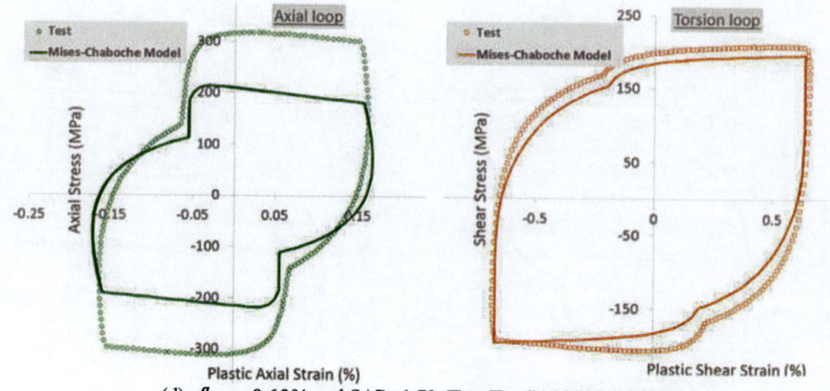

(d) ε^{a}_{vM} = 0.60% and SAR=1.73 (Test ID: CA30S90p90 [29])

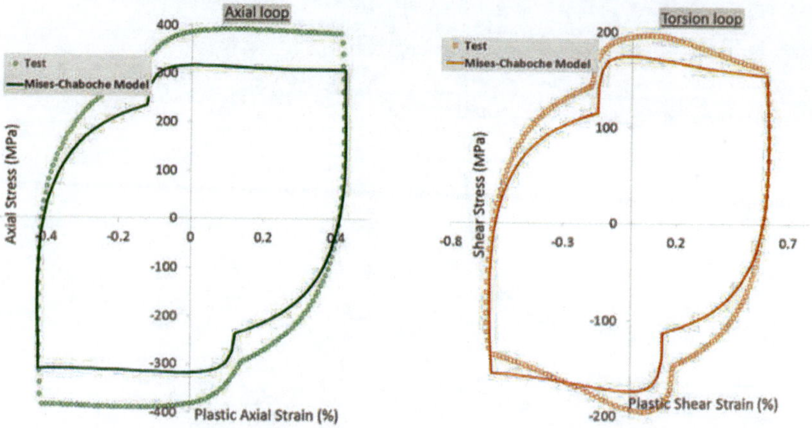

(e) ε^{a}_{vM} = 0.75% and SAR=0.75 (Test ID: CA60S78p90 [29])

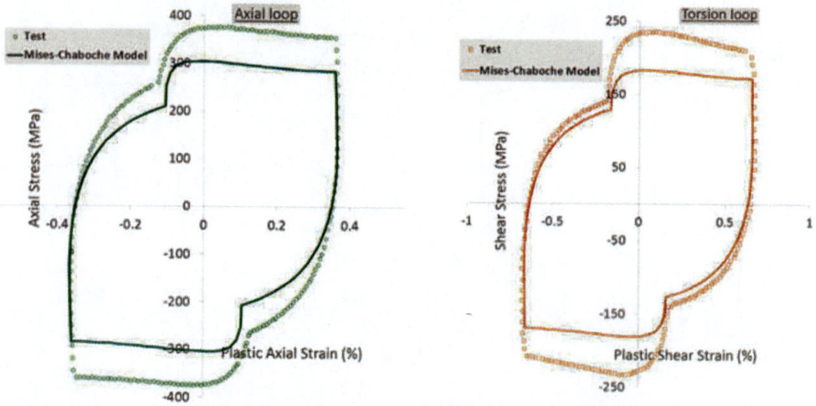

(f) ε^{a}_{vM} = 0.75% and SAR=1.00 (Test ID: CA53S92p90 [29])

Fig. 9 (continued)

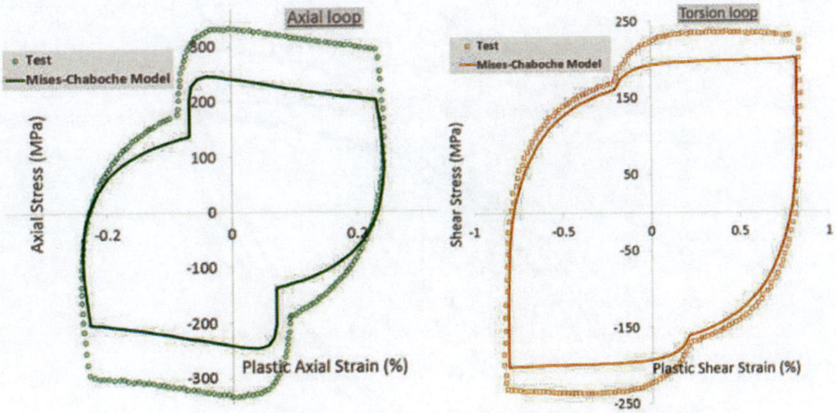

(g) ε^{a}_{vM} = 0.75% and SAR=1.73 (Test ID: CA37S112p90 [29])

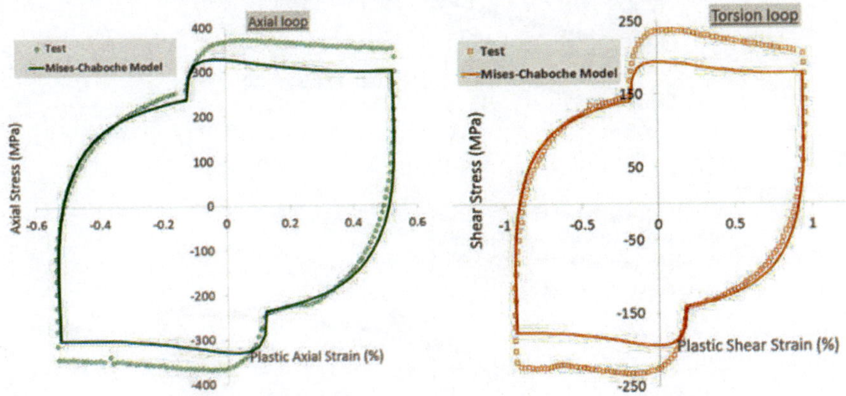

(h) ε^{a}_{vM} = 1.00% and SAR=1.00 (Test ID: CA70S121p90 [29])

Fig. 9 (continued)

However, this higher hardening prediction by the Chaboche model is not adequate and comparable w.r.t. actual test response. Therefore, the use of the Chaboche model might lead to unconservative life assessments for non-proportional load paths. Hence, it becomes important to explore more cyclic plasticity models such as the Ohno–Wang model [21], Modified Ohno–Wang [22], Tanaka [23], Benallal-Marquis [24] for accurate assessment of cyclic response under complex non-proportional loading paths.

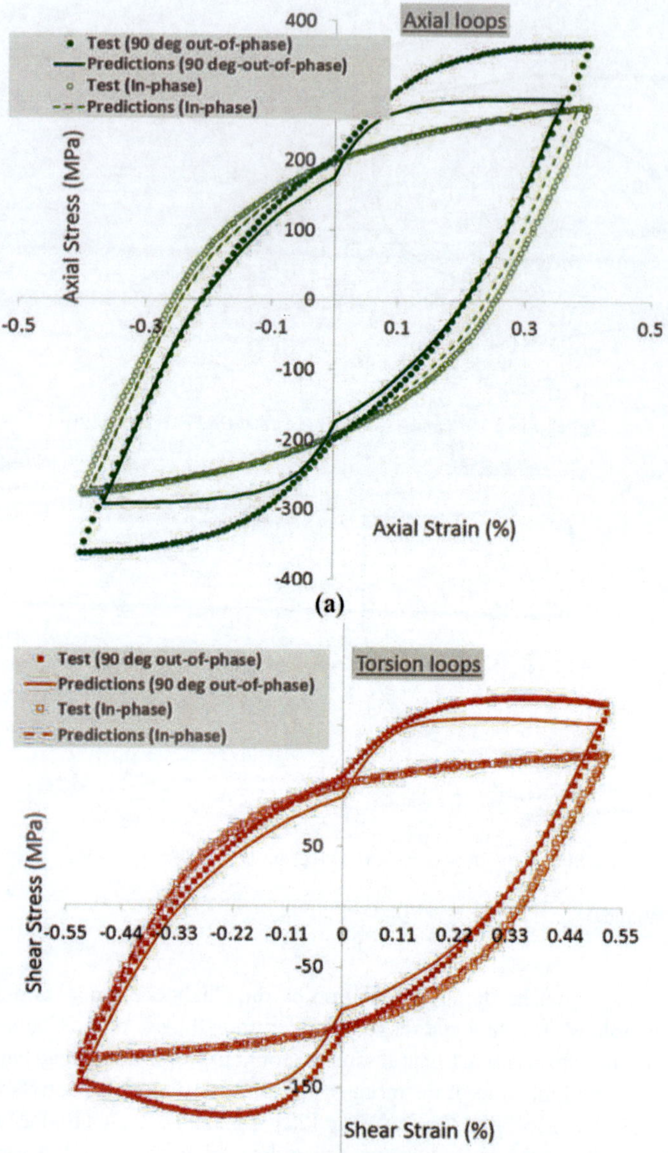

(a)

(b)

Fig. 10 Comparison of predicted and test **a** axial stress and total axial strain response and **b** shear stress and total shear strain response, under in-phase and 90° out-of-phase axial–torsion loading for $\varepsilon_x^a = \pm 0.40\%$ and $\gamma_{xy}^a = \pm 0.52\%$

6 Conclusions

The assessment of cyclic plasticity response for SA333 Gr. 6 material in axial/shear directions under pure axial, pure torsion, in-phase axial–torsion and out-of-phase axial–torsion conditions is summarized as below,

(1) The material shows Masing characteristics with a linear drift under pure axial, pure torsion and in-phase axial–torsion cyclic conditions. This indicates that equivalent strain range-dependent cyclic yield strength with kinematic hardening rule is required for cyclic plasticity material modelling.
(2) The Chaboche's parameters have been optimized using one of the pure axial LCF test loops. The other pure axial and pure torsion loops have been assessed based on these calibrated parameters. The assessed material response for pure axial and pure torsion conditions are comparable to observed test response.
(3) The predicted cyclic response in shear direction is in good agreement with corresponding test response for all proportional load paths. The predicted response in axial direction is comparable to test loops for small equivalent strain amplitudes $\left(\boldsymbol{\varepsilon}^{a}_{vM} < 0.6\%\right)$ and all SARs. However, for larger $\boldsymbol{\varepsilon}^{a}_{vM}$, the axial response is marginally underpredicted.
(4) The Chaboche model with equivalent strain range-dependent cyclic yield strength considerations result in significant underpredictions of axial and shear stress response for all strain-controlled non-proportional paths, as analysed.

References

1. ASME Boiler & pressure vessel code, section III, Division 1, subsection NB, American Society of Mechanical Engineers (2010)
2. Ellyin F, Kujawski D (1984) Plastic strain energy in fatigue failure. J Press Vessel Technol Trans ASME 106:342–7
3. Golos K, Ellyin F (1987) Generalization of cumulative damage criterion to multilevel cyclic loading. Theo Appl Fatigue Mech 7:169–76
4. Kandil FA, Brown MW, Miller KJ (1982) Biaxial low-cycle fatigue fracture of 316 stainless steel at elevated temperatures. Metals Soc, London 280:203–210
5. Ninic D (2006) A stress-based multiaxial high cycle fatigue damage criterion. Int J Fatigue
6. Mc-Diarmid DL (1991) A general criterion for high cycle multiaxial fatigue failure. Fatigue Fract Eng Mater Struct 14(4):429–453
7. Matsubara G, Hayashida A, Kano D (2018) Predicting the multiaxial fatigue limit and the multiaxial high-cycle fatigue life based on the unified equivalent shear stress from axial strength characteristics with various waveforms. Int J Fatigue 112:52–62
8. Maktouf W, Ammar K, Ben Naceur I, Saï K (2016) Multiaxial high-cycle fatigue criteria and life prediction: application to gas turbine blade. Int J Fatigue 92:25–35
9. Fatemi A, Socie DF (1988) A critical plane approach to multiaxial fatigue damage including out-of-phase loading. Fatigue Fract Eng Mater Struct
10. Glinka G, Wang G, Plumtree A (1995) Mean stress effects in multiaxial fatigue. Fatigue Fract Eng Mater Struct 18:755–764
11. Socie DF (1987) Multiaxial fatigue damage models. J Eng Mater Technol 109(4):293–298

12. Smith RN, Watson P, Topper TH (1970) A stress strain function for the fatigue of metals. J Mater 5(4):767–778
13. Chu CC (1995) Fatigue damage calculations using the critical plane approach. J Eng Mater Technol 117:41–49
14. Gupta SK, Fesich TM, Schuler X, Roos E, A critical plane based model for fatigue assessment under fixed and rotating principal direction loading. In: Structural mechanics in reactor technology conference, SMiRT-21, New Delhi, India, Volume: Div-II, Paper ID: 624
15. Ince A, Glinka G (2014) A generalized fatigue damage parameter for multiaxial fatigue life prediction under proportional and non-proportional loadings. Int J Fatigue 62:34–41
16. Zhu S-P, Yu Z-Y, Liu Q, Ince A (2018) Strain energy-based multiaxial fatigue life prediction under normal/shear stress interaction. Int J Damage Mech 1–32
17. Arora P, Gupta SK, Samal MK, Chattopadhyay J (2019) Development of new critical plane model for assessment of fatigue life under multi-axial loading conditions. Int J Fatigue 129, Article 105209
18. Xiaoshan L, Guoqui H, Xiangqun D, Defeng M, Weihua Z (2009) Fatigue behavior and dislocation substructures for 6063 aluminum alloy under nonproportional loadings. Int J Fatigue 31:1190–1195
19. Armstrong PJ, Frederick CO (1966) A mathematical representation of the multiaxial Bauschinger effect, G.E.G.B. report RD/B/N 731, Berceley Nuclear Laboratory
20. Chaboche JL, Dang-Van K, Cordier G (1979) Modelization of the strain memory effect on the cyclic hardening of 316 stainless steel. In: Proceedings of the international conference on SMiRT-5 (1979), pp 1–10
21. Ohno N, Wang JD (1993) Kinematic hardening rules with critical state of dynamic recovery, part I: formulations and basic features for ratcheting behaviour. Part II: application to experiments of ratcheting behaviour. Int J Plast 9:375–403
22. Ohno N, Wang JD (1994) Kinematic hardening rules for simulation of ratcheting behaviour. Eur J Mech, A/Solids 13:519–531
23. Tanaka E (1994) A non-proportionality parameter and a cyclic viscoplastic constitutive model taking into account amplitude dependencies and memory effects of isotropic hardening. Eur J Mech A Solid 13(2):155–173
24. Benallal A, Marquis D (1987) Constitutive equations for non-proportional cyclic elasto-viscoplasticity. J Eng Mater Technol, Trans ASME 109(4):326–336
25. Mroz Z (1967) On the description of anisotropic work hardening. J Mech Phys Solids 15:163–175
26. Xing R, Yu D, Shi S, Chen X (2019) Cyclic deformation of 316L stainless steel and constitutive modeling under non-proportional variable loading path. Int J Plasticity 120:127–146
27. Facheris G, Janssens KGF (2014) An internal variable dependent constitutive cyclic plastic material description including ratcheting calibrated for AISI 316L. Comput Mater Sci 87:160–171
28. Wu H (2018) An empirical non-proportional cyclic plasticity approach under multiaxial low cycle fatigue loading. Int J Mech Sci 142–143:66–73
29. Arora P, Gupta SK, Bhasin V, Singh RK, Sivaprasad S, Tarafder S (2016) Testing and assessment of fatigue life prediction models for Indian PHWRs piping material under multi-axial load cycling. Int J Fatigue 85:98–113
30. Standard Practice for Strain-Controlled Fatigue Testing, (Designation: E 606—04), American Society of Testing Materials (ASTM)
31. Standard Practice for Strain-Controlled Axial-Torsional Fatigue Testing with Thin Walled Tubular Specimens (Designation: E2207—08), American Society of Testing Materials (ASTM)
32. Prager W (1956) A new method of analysing stresses and strains work-hardening plastic solids. J Appl Mech 23:493–496
33. Nguyen TT, Liu GR, Dai KY, Lam KY (2007) Selective smoothed finite element method. Tsinghua Sci Technol 12(5):497–508. ISSN: 1007-0214-01/19

34. Wu HC, Yeh WC, On the experimental determination of yield surfaces and some results of annealed 304 stainless steel. Int J Plasticity 7:803
35. Wu HC, Lu JK, Pan WF (1995) Some observations on yield surfaces for 304 stainless steel at large strain. J Appl Mech 62:626

Characterization of Pretension High Strength Bolted Friction Grip Connections for CFRP-Steel Structure for Tensile Loading Using Finite Element Analysis

Shivaraj Mahajan and Narasimhe Gowda

Abstract This work involves stress analysis of bolted structures using Finite Element Analysis (FEA) with High-Strength Carbon Fiber Reinforced Polymer (CFRP) composite material such as AS4/8552 R134 AW196 which can be replaceable with the steel for the construction of footbridges and railway bridges though it costs more than ancient steel but strength wise most strong approximately four times than steel. Two CFRP-steel models have been created. One with 16 layered CFRP-steel combination with $[0_2/45_2/60_2/90_2]_s$ ply orientation, and other is 8 layered CFRP-steel combination with $[0/45/60/90]_s$ ply orientation. Here, considering the pretension of HSFG bolts and tensile load which was applied to the connected plates, the stress and deformation behaviors and shear failure of bolts were studied. According to macro mechanical analysis of lamina, failure of lamina in terms of strength ratio is studied. Strength ratio versus Layers of graphs are plotted in order to determine the safe and unsafe laminae or order of failure of laminae from safe to unsafe. From the analysis, it can be concluded that weight reduction and hence material and cost saving of CFRP is possible by testing with the half of the thickness of the cover plates and with regular dimensions. From this study, it can be suggested that lesser material of CFRP can be used when compared with steel for the same application of the construction of bridges.

Keywords HSFG bolted joints · FEA · CFRP · Footbridges · Steel structures · Tensile loading · Macro mechanical analysis of lamina

S. Mahajan (✉)
Design Engineering, Dayananda Sagar College of Engineering, Bengaluru, India
e-mail: shivarajsmahajan071@gmail.com

N. Gowda
Department of Mechanical Engineering, Dayananda Sagar College of Engineering, Bengaluru, India
e-mail: ngmechme@gmail.com

S. Seetharamu et al. (eds.), *Fatigue, Durability, and Fracture Mechanics*, Lecture Notes in Mechanical Engineering, https://doi.org/10.1007/978-981-15-4779-9_17

1 Introduction

Fibre-reinforced polymer is one of the sustainable materials with low carbon footprint, lightweight, and resistance to chemicals and corrosion. These are desirable properties for bridge engineering. Although, traditional materials, such as steel and concrete have been used in bridges for 150 years, FRP being relatively new material offers excellent opportunities for bridge engineering [1]. Most of the infrastructures gets reduction in their service life due to corrosion problems. Such problems gravely affecting most of the bridges in coastal areas of Japan. In order to maintain these infrastructures, new materials have been evolved to overcome such problems [2]. In the machine design and construction, the bolted joints are most common connection methods. As a structural component, it is often considered the severe part of an assembly. FEA development gives accurate prediction of the real behavior of such type of joints and since the cost of experimental tests is high [3]. So the aim of this project is to perform the stress analysis in the Steel/CFRP combination of connected plate and cover plate using Finite Element Analysis.

Huang et al. [4] described finite element analysis and experimental study, the mechanical behavior including slip-load relationship, load transfer factors, stress state, and friction stress distribution of HSFG connections in detail. Shishesaz and Hosseini [5] studied from their review that stress distribution of failure of joints can be done either experimentally, analytically, or by FEA. Mulagada and Rao [6] explored the prediction of bolt clamp-up and plate friction for the structural steel using FEA. Sonnenschein et al. [7] presented the mechanical properties and durability of different types of the FRP rebar's and their use in construction of bridges. Nerilli et al. [8] aimed to model the progressive damage of multi-bolted joints connecting structural elements made up of FRP's. Fakih et al. [9] presented the investigations of extended End-Plate connections (EP). Qureshi [1] presented a review of mechanical connection methods for FRP bridges. Slip and fatigue resistance are critical to any bridge design. Miklos et al. [10] presented a case study on assembling using bolted joints, with initial clamping and loaded with transversal forces.

2 Finite Element Method

Finite element method is the numerical method used to solve the complex engineering problems. Complex problems will have complicated boundary conditions and difficult loading conditions. These complex problems can be solved in a short interval of time when the finite element method is aided with the computer. In the finite element method, the complicated problems can be analyzed by using the following steps of FEM:

1. Discretization of domain
2. Approximation of field variables
3. Derivation of element equation

4. Assembly of element equation
5. Application of boundary conditions
6. Solution of the system equation
7. Postprocessing

3 Elements Used in ANSYS APDL [11]

i. SOLID185
ii. CONTA174
iii. TARGE170
iv. PRETS179

4 Case Study on 16 Mm Layered CFRP-Steel Combination

This case study involves the stress analysis and failure of the specimen by making the cover plates with CFRP material, whereas the connected plates and HSFG bolts with the steel material. In this case, it was considered unidirectional laminae of thickness 1 mm each. The ply orientation was chosen like this $[0_2/45_2/60_2/90_2]_s$.

4.1 Finite Element Model Development

All dimensions of the geometry are in mm as shown in Fig. 1a was taken from literature [4]. The CAD Model was prepared in well-known academic CAD software solid edge v19. The saved assembly was in the file extension (.asm). The (.asm) was converted to parasolid format (.x_t) which is one of the neutral format to switch between one CAD/CAE software to other.

In the CAE preprocessing stage, the (.x_t) file was imported in HyperMesh 2017 version. The first step in HyperMesh is to set up the file in millimeters dimensions since in the parasolid format all dimensions will be in meters. Then creating components for all the parts such as two connected plates, two cover plates in which each contains 16 layers of 1 mm thickness, eight bolts and eight nuts. Keeping element size 3 mm for all the parts expect shank part of the bolts which were meshed with 2 mm element size. Meshing was done as per the component list created. The two connected plates were discretized with 213,248 elements, each layer of cover plates was discretized with 7140 elements and eight bolts were discretized with $6808 \times 8 = 54{,}464$ elements. Hence, the total number of elements in this model are $213{,}248 + 7140 \times 32 + 54{,}464 = 496{,}192$ elements, and total number of nodes are 567,998. The mix of quad and tria element were first created in plane and then they were dragged to form hex and tet elements. The model was meshed in HypeMesh as shown in

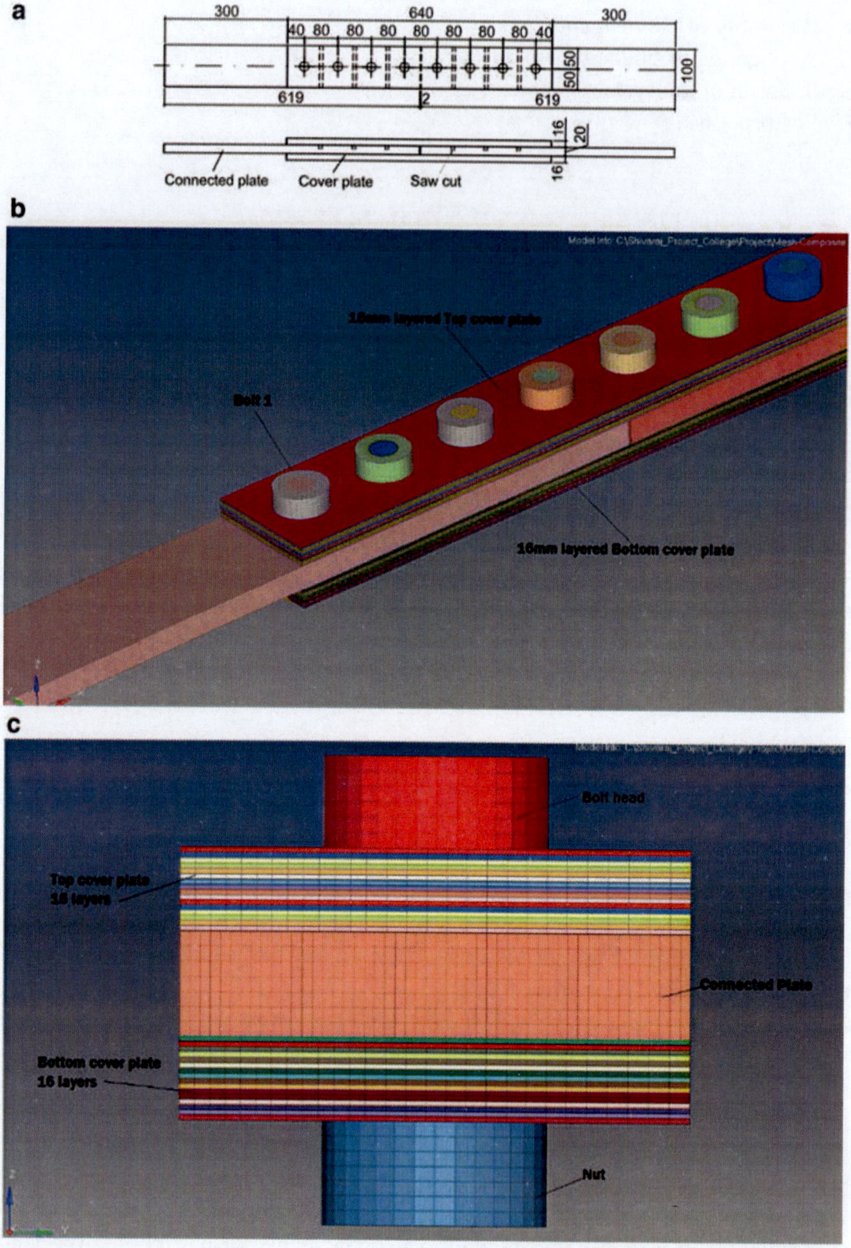

Fig. 1 **a** Geometry of the specimen. **b** Meshed model of 16 mm layered CFRP-Steel cover plate. **c** 16 mm CFRP-Steel cover plate-side view

Fig. 1b and the parts are indicated in the figure and the side view of the specimen with details is shown in Fig. 1c.

The element quality check was done. The aspect ratio of 5, Jacobian of 6.0 was maintained. The check for free edges and T-connections was also done. Hence this ensures a best quality mesh and can be exported to solver.

4.2 Material Properties

Bilinear isotropic hardening properties were used for steel connected plate and steel HSFG bolts [4]. For CFRP layered cover plate the material selected is AS4/8552 R134 AW196 [12]. The material properties of connected plates and HSFG bolts are shown in Table 1 and Table 2, respectively. The material properties of AS4/8552 R134 AW196 used in layers of cover plates are shown in Table 3. The 16 layers of each 1 mm thickness are modeled in HyperMesh whereas its material properties are assigned in the Ansys APDL.

5 Boundary Conditions

The preprocessed model from HyperMesh (.hm) is imported in the Ansys solver deck for solving and postprocessing. The boundary conditions were in literature [4]. One end is fixed and other end carries a tensile load and pretension is given for the bolts while clamping them into connected plates and cover plates. In FEA, the same

Table 1 Material property of connected plates

Definition	Defined values
Steel	16 Mnq
Thickness	16–35 mm
Yield stress	325 MPa
Ultimate stress	490 MPa

Table 2 Material property of bold components

Definition	Defined values
Steel	20 MnTiB
Yield stress	930 MPa
Ultimate stress	1130 MPa
Elastic modulus	210 GPa
Tangent modulus	2.1 GPa

Table 3 Material property of CFRP cover plates: AS4/8552 R134 AW196

Description	Symbol	Value
Density (g/cm^3)	ρ	1.58
Longitudinal modulus (GPa)	E11	141
Transverse modulus (GPa)	E22 = E33	9.75
Poisson's ratio	ν12	0.267
Shear moduli in 1–2 plane (GPa)	G12 = G13	5.2
Shear moduli in 2–3 plane (GPa)	G23	3.19
Longitudinal tensile strength (MPa)	XT	2200
Longitudinal compressive strength (MPa)	XC	1500
Transverse tensile strength (MPa)	YT	81
Transverse compressive strength (MPa)	YC	260
In-plane shear strength (MPa)	S	80

boundary conditions are applied, i.e., one end (Right) is fixed and other end (Left) is applied with a tensile load 200 kN (see Fig. 2). The pretension applied for each bolt is 200 kN. As discussed in the previous section "Elements used in Ansys APDL", The SOLID185 element is used for the modeling, CONTA174 and TARGE170 elements are used for contact behavior, and PRETS179 element is used for pretension of bolts.

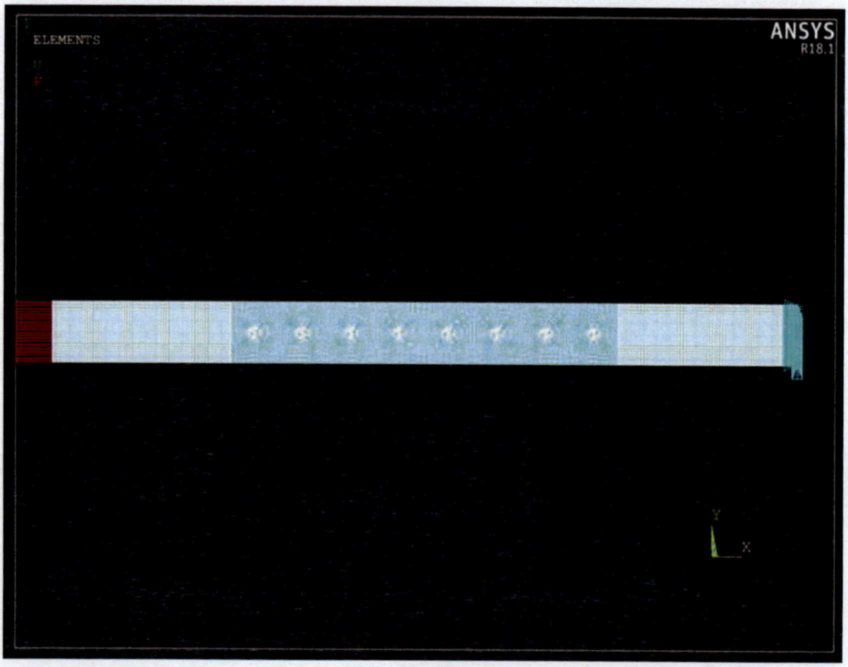

Fig. 2 Loading conditions

A total of 29 contacts pairs are formulated in this model as shown in Table 4. The name of the contact pairs clearly describes the contact behavior. The contacts are standard and surface-to-surface.

Table 4 Contact pairs formulated

S.No.	Contacts	Targets	Short names assigned in HyperMesh	
			Contacts	Targets
1	CONNECTED PLATE_1_TOP	COVER PLATE_1_BOTTOM	C1	T1
2	CONNECTED PLATE_2_TOP	COVER PLATE_1_BOTTOM	C2	T2
3	CONNECTED PLATE_1_BOTTOM	COVER PLATE_2_TOP	C3	T3
4	CONNECTED PLATE_2_BOTTOM	COVER PLATE_2_TOP	C4	T4
5	CONNECTED PLATE_1_SIDE FACE_RIGHT	CONNECTED PLATE_2_SIDE FACE_LEFT	C5	T5
6	BOLT HEAD_1_CONTACT	BOLT HEAD_1_TARGET	C6	T6
7	SHANK_1_CONTACT	SHANK_1_TARGET	C7	T7
8	NUT_1_CONTACT	NUT_1_TARGET	C8	T8
9	BOLT HEAD_2_CONTACT	BOLT HEAD_2_TARGET	C9	T9
10	SHANK_2_CONTACT	SHANK_2_TARGET	C10	T10
11	NUT_2_CONTACT	NUT_2_TARGET	C11	T11
12	BOLT HEAD_3_CONTACT	BOLT HEAD_3_TARGET	C12	T12
13	SHANK_3_CONTACT	SHANK_3_TARGET	C13	T13
14	NUT_3_CONTACT	NUT_3_TARGET	C14	T14
15	BOLT HEAD_4_CONTACT	BOLT HEAD_4_TARGET	C15	T15
16	SHANK_4_CONTACT	SHANK_4_TARGET	C16	T16
17	NUT_4_CONTACT	NUT_4_TARGET	C17	T17
18	BOLT HEAD_5_CONTACT	BOLT HEAD_5_TARGET	C18	T18
19	SHANK_5_CONTACT	SHANK_5_TARGET	C19	T19
20	NUT_5_CONTACT	NUT_5_TARGET	C20	T20
21	BOLT HEAD_6_CONTACT	BOLT HEAD_6_TARGET	C21	T21
22	SHANK_6_CONTACT	SHANK_6_TARGET	C22	T22
23	NUT_6_CONTACT	NUT_6_TARGET	C23	T23
24	BOLT HEAD_7_CONTACT	BOLT HEAD_7_TARGET	C24	T24
25	SHANK_7_CONTACT	SHANK_7_TARGET	C25	T25
26	NUT_7_CONTACT	NUT_7_TARGET	C26	T26
27	BOLT HEAD_8_CONTACT	BOLT HEAD_8_TARGET	C27	T27
28	SHANK_8_CONTACT	SHANK_8_TARGET	C28	T28
29	NUT_8_CONTACT	NUT_8_TARGET	C29	T29

5.1 Results and Discussion

As the objective of this work was to find out the strength ratio of each laminae and hence to find the safe and unsafe laminae, or to find out the order of failure of laminae from safe to unsafe, to find the total deformation, and to find the mass of the specimen. The same is found out as follows.

5.1.1 Strength Ratio of Laminae

In order to find the safe and unsafe laminae according to macro mechanical analysis of lamina [13]: Strength Ratio, the 1st principal stress (σ_{11}), 2nd principal stress (σ_{22}), and shear stress in plane 12 (τ_{12}) for each lamina has to be determined. As per the definition of strength ratio,

$$SR = \frac{\text{Maximum load which can be applied}}{\text{Load applied}}$$

Since in this case, the tensile load is applied at one end as shown in Fig. 2 so the bolts will have shear failure, which means shear stress in plane 12 (τ_{12}) has to be considered for the failure analysis of the laminae. Table 5 and Table 6 show the values of 1st principal stress (σ_{11}), 2nd principal stress (σ_{22}), and shear stress in plane 12 (τ_{12}) for each lamina of top cover plate and bottom cover plate, respectively. The stress units are in MPa. It shows two values of stresses, i.e., the minimum and maximum values obtained from the results. The only magnitude of the stresses has to be considered for value of load applied and maximum load applied will be taken from CFRP material properties as shown in Table 3, i.e., Maximum load applied = In-plane shear strength = S = 80 MPa.

The Strength ratios versus Layers graph for the top cover plate and bottom cover plate have been plotted in Fig. 3 and Fig. 4, respectively. The plot shows the number of laminae which are safe and unsafe. Those laminae having the strength ratio value more than 1.0 are safe and those laminae having the strength ratio value less than 1.0 are unsafe. So it means the lamina that has strength ratio = 1 will have failure shear stress value.

From graphs, it can be predicted that the safe laminae in top cover plate are UL_2, UL_3, UL_4, UL_5, UL_6, and UL_7. Similarly in bottom cover plate are LL_10, LL_11, LL_12, LL_13, LL_14, LL_15.

Those unsafe layers, which have the strength ratio less than 1, the load, needs to be reduced by factor of SR.

Table 5 Stresses in 16 mm layer-top cover plate

16 mm layer-top							
Name of the layer	σ_{11} (MPa)		σ_{22} (MPa)		τ_{12} (MPa)		Orientation (Degrees)
	Min	Max	Min	Max	Min	Max	
UL_1	−63.21	462.46	−139.18	61.40	−133.32	90.41	0
UL_2	−38.06	273.93	−124.80	28.51	−56.74	69.52	0
UL_3	−27.28	225.81	−102.49	29.59	−53.46	54.36	45
UL_4	−25.68	187.74	−62.42	27.04	−50.40	65.56	45
UL_5	−21.22	167.15	−50.62	26.65	−49.12	62.04	60
UL_6	−12.32	173.17	−36.89	27.16	−56.62	73.45	60
UL_7	−3.75	177.14	−35.19	31.41	−58.60	74.61	90
UL_8	−4.65	179.60	−31.81	26.86	−64.01	81.31	90
UL_9	−3.79	190.40	−30.91	30.91	−64.62	80.79	90
UL_10	−5.94	193.26	−28.23	33.50	−68.73	86.83	90
UL_11	−5.96	200.17	−27.42	36.09	−69.14	85.63	60
UL_12	−9.05	198.03	−25.34	39.38	−73.91	91.17	60
UL_13	−9.58	204.24	−24.49	46.96	−75.19	89.26	45
UL_14	−13.32	210.33	−24.21	47.14	−79.61	94.97	45
UL_15	−16.12	211.85	−31.46	63.83	−83.39	90.43	0
UL_16	−19.13	310.70	−46.70	77.23	−106.63	111.90	0

5.1.2 Deformation

In this 16 mm layered cover plate problem, the applied tensile load on one end is 200 kN and the pretension applied for each bolt is 200 kN which is more in value. So obviously one can predict that deformation will be very less but slightly more compared to steel cover plate. The deformation obtained from FEA is only 0.73 mm and shown in Fig. 5.

5.1.3 Mass

The mass of this specimen obtained from Ansys results is found to be 27.26 kg (see Fig. 6). This will be helpful later for comparison of mass.

Table 6 Stresses in 16 mm layer-bottom cover plate

16 mm layer-bottom							
Name of the layer	σ_{11} (MPa)		σ_{22} (MPa)		τ_{12} (MPa)		Orientation (Degrees)
	Min	Max	Min	Max	Min	Max	
LL_1	−19.06	311.42	−46.47	76.88	−112.08	106.64	0
LL_2	−15.94	211.50	−31.32	64.01	−90.17	83.21	0
LL_3	−13.25	210.02	−24.15	46.84	−94.69	79.37	45
LL_4	−9.58	203.88	−24.46	47.06	−88.97	74.99	45
LL_5	−9.03	197.69	−25.33	39.05	−90.87	73.63	60
LL_6	−5.98	199.68	−27.43	36.15	−85.30	68.89	60
LL_7	−5.95	192.91	−28.23	33.10	−86.50	68.42	90
LL_8	−3.79	189.81	−30.93	30.95	−80.42	64.32	90
LL_9	−4.65	179.17	−31.83	26.35	−80.93	63.62	90
LL_10	−4.61	176.54	−35.23	31.38	−74.15	58.24	90
LL_11	−13.00	172.56	−36.93	27.07	−72.98	56.30	60
LL_12	−21.81	167.45	−51.26	26.08	−61.44	48.67	60
LL_13	−25.76	189.27	−63.81	26.94	−65.99	50.02	45
LL_14	−27.33	227.76	−104.12	29.66	−55.18	54.00	45
LL_15	−37.97	275.70	−124.83	28.23	−70.71	57.02	0
LL_16	−61.95	466.22	−138.58	61.25	−93.63	137.43	0

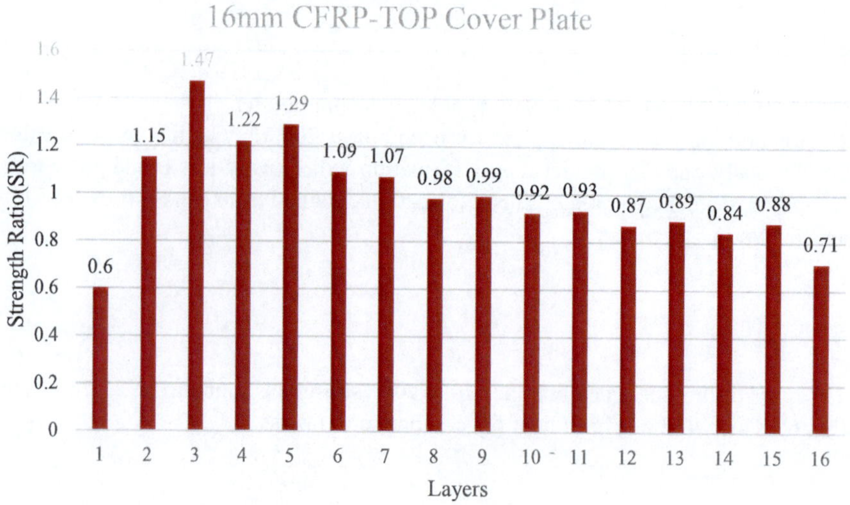

Fig. 3 Plot of SR versus layers in 16 mm top cover plate

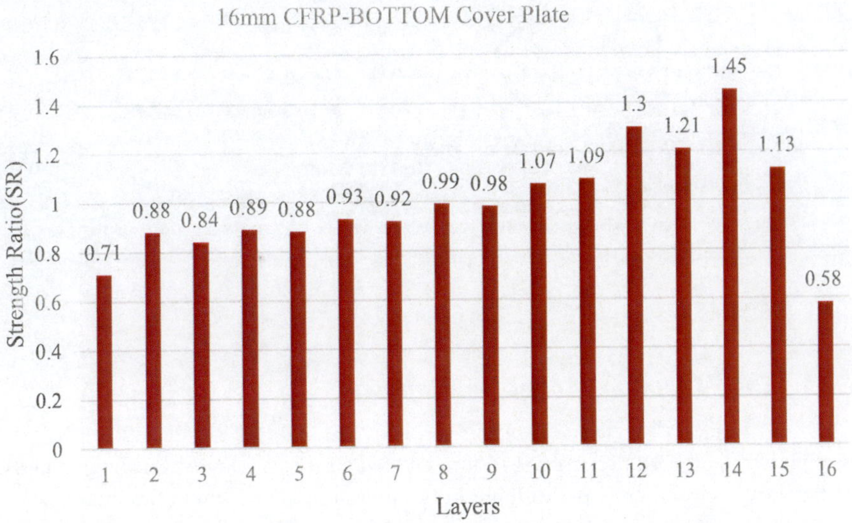

Fig. 4 Plot of SR versus layers in 16 mm bottom cover plate

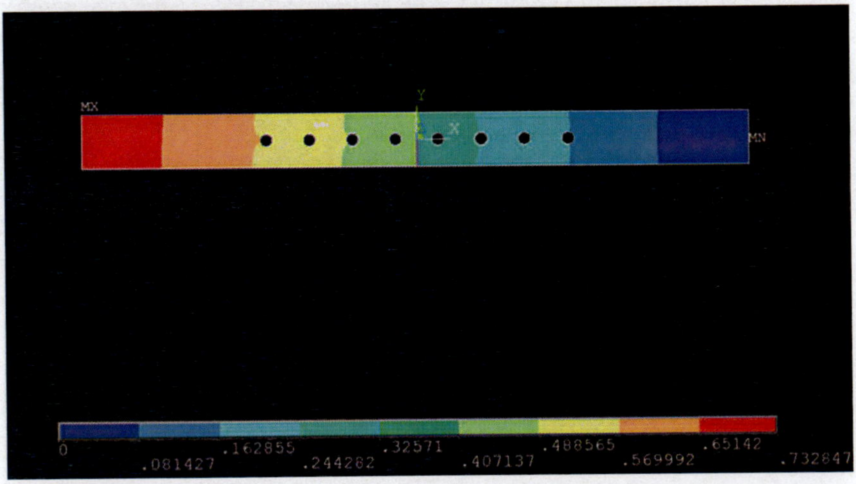

Fig. 5 Deformation in mm

6 Case Study on 8 mm Layered CFRP-Steel Combination

This case study involves the cover plate with 50% reduction in thickness, i.e., here, 8 mm thickness is considered so that 8 layers of each 1 mm thickness. The ply orientation was chosen like this $[0/45/60/90]_s$.

```
TOTAL MASS = 0.27269E-01
  The mass principal axes coincide with the global Cartesian axes

CENTER OF MASS (X,Y,Z)=   0.73832E-06   0.20046E-05  -0.34326

TOTAL INERTIA ABOUT CENTER OF MASS
      26.390        -0.14086E-06  -0.58916E-06
    -0.14086E-06     2769.1        0.14568E-06
    -0.58916E-06    0.14568E-06     2782.1
  The inertia principal axes coincide with the global Cartesian axes
```

Fig. 6 Mass in kg

6.1 Finite Element Model Development

Similar to the procedure done for 16 mm layered CFRP case study, for the geometry of the specimen which was available in the literature [4] in which the thickness of cover plates is considered to be reduced by 50%, i.e., the new thickness of cover plate will be 8 mm each. The two connected plates were discretized with 96,698 elements, each layer of cover plates was discretized with 7426 elements and eight bolts were discretized with $12{,}112 \times 8 = 96{,}896$ elements. Hence, the total number of elements in this model are $96{,}698 + 7426 \times 16 + 96{,}896 = 312{,}410$ elements and total number of nodes are 362,340. The model was meshed in HypeMesh as shown in Fig. 7a, and the parts are indicated in the figure and the side view of the specimen with details is shown in Fig. 7b.

As in the previous case of 16 mm layered cover plate, the same materials properties and boundary conditions are applied to this 8 mm layered cover plate problem also. Therefore, they have not shown it again in this paper.

6.2 Results and Discussion

The same procedure is followed as like previous case study of 16 mm CFRP-steel cover plate. The results are found out as follows.

6.2.1 Strength Ratio of Laminae

Table 7 and Table 8 show the values of 1st principal stress (σ_{11}), 2nd principal stress (σ_{22}), and shear stress in plane 12 (τ_{12}) for each lamina of top cover plate and bottom cover plate, respectively. The stress units are in MPa. It shows two values of stresses, i.e., the minimum and maximum values obtained from the results. The only magnitude of the stresses has to be considered for value of load applied and

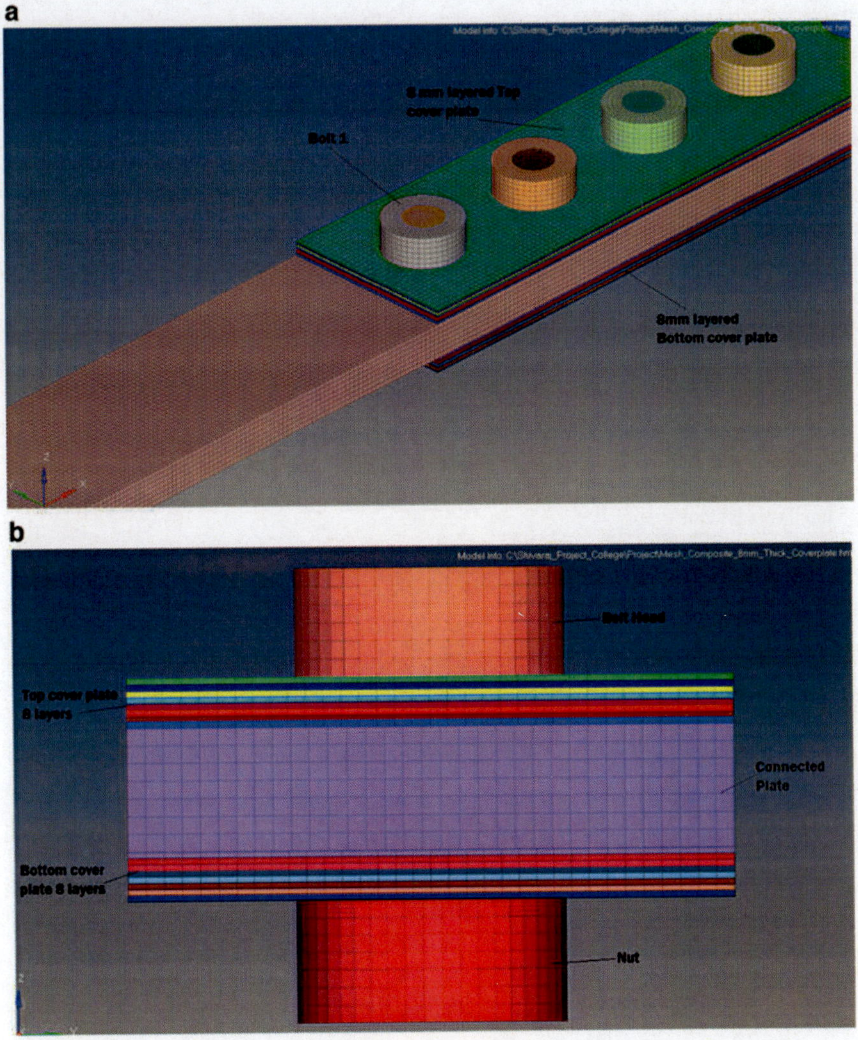

Fig. 7 **a** Meshed model of 8 mm layered CFRP-steel cover plate. **b** 8 mm CFRP-steel cover plate-side view

maximum load applied will be taken from CFRP material properties as shown in Table 3, i.e., Maximum load applied = In-plane shear strength = S = 80 MPa.

The Strength ratios versus Layers graph for the top cover plate and bottom cover plate have been plotted in Fig. 8 and Fig. 9, respectively. The plot shows the number of laminae which are safe and unsafe.

From graphs, it can be predicted that the order of failure of laminae in top cover plate is UL_3, UL_5, UL_2, UL_4, UL_7, UL_6, UL_1, and UL_8 in the safe to

Table 7 Stresses in 8 mm layer-top cover plate

8 mm layer-top							
Name of the layer	σ_{11} (MPa)		σ_{22} (MPa)		τ_{12} (MPa)		Orientation (Degrees)
	Min	Max	Min	Max	Min	Max	
UL_1	−71.35	795.64	−177.36	68.58	−171.28	87.05	0
UL_2	−44.72	495.47	−115.06	43.45	−80.72	136.33	45
UL_3	−28.00	461.77	−77.73	36.50	−69.90	83.92	60
UL_4	−16.22	427.95	−46.72	47.61	−90.61	140.05	90
UL_5	−14.45	502.92	−38.76	45.24	−98.34	112.25	90
UL_6	−6.24	485.12	−32.02	58.91	−116.18	167.98	60
UL_7	−21.87	581.40	−37.59	58.83	−119.69	144.71	45
UL_8	−21.71	540.19	−56.99	121.13	−158.71	201.73	0

Table 8 Stresses in 8 mm layer-bottom cover plate

8 mm layer-bottom							
Name of the layer	σ_{11} (MPa)		σ_{22} (MPa)		τ_{12} (MPa)		Orientation (Degrees)
	Min	Max	Min	Max	Min	Max	
LL_1	−23.30	702.42	−74.97	124.80	−169.49	150.58	0
LL_2	−14.89	389.58	−32.84	58.23	−131.99	135.38	45
LL_3	−7.88	503.29	−35.12	43.11	−132.64	107.08	60
LL_4	−15.69	348.97	−41.20	51.20	−102.22	104.70	90
LL_5	−20.91	475.70	−45.64	27.83	−108.09	86.85	90
LL_6	−26.78	420.62	−89.86	35.53	−74.43	92.93	60
LL_7	−43.22	512.86	−110.63	31.02	−84.17	66.20	45
LL_8	−61.74	667.06	−160.41	79.39	−116.46	223.92	0

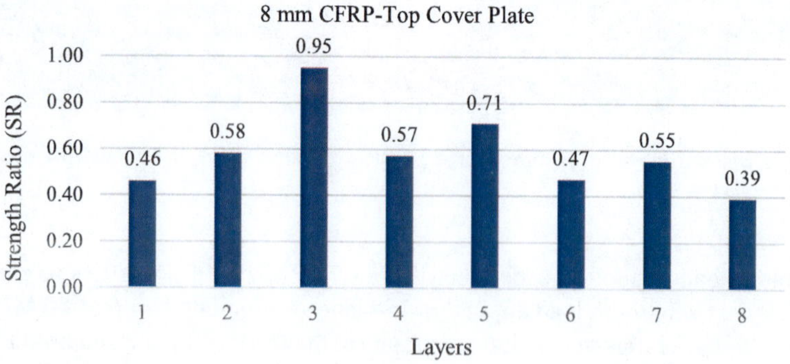

Fig. 8 Plot of SR versus layers in 8 mm top cover plate

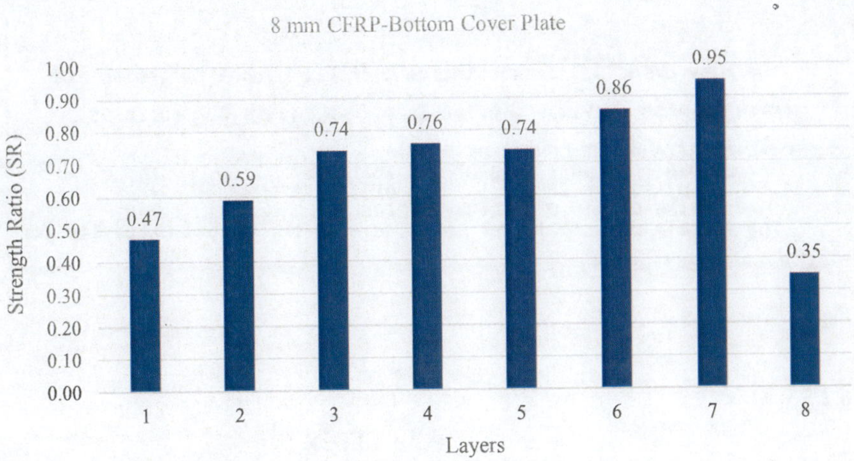

Fig. 9 Plot of SR versus layers in 8 mm bottom cover plate

unsafe order. Similarly in bottom cover plate are LL_7, LL_6, LL_4, LL_3, LL_5, LL_2, LL_1, and LL_8 in the safe to unsafe order.

6.2.2 Deformation

The deformation will be very less but slightly more compared to 16 mm CFRP cover plate and steel cover. The deformation obtained from FEA is only 0.89 mm and shown in Fig. 10.

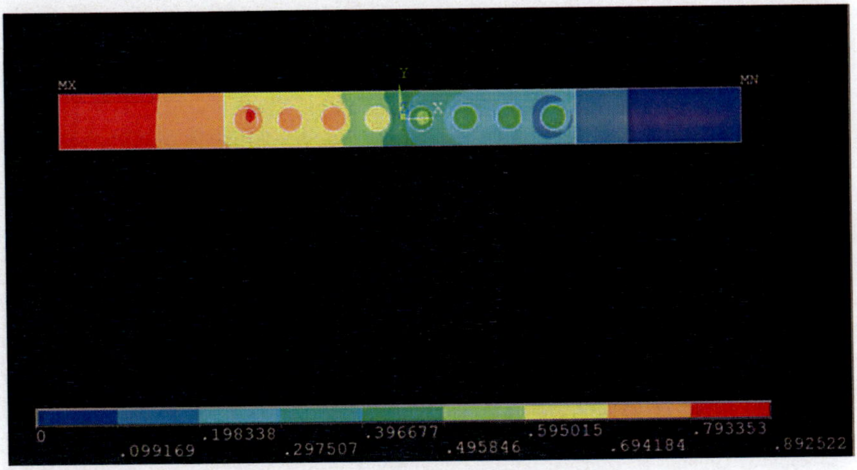

Fig. 10 Deformation in mm

```
TOTAL MASS = 0.25154E-01
  The mass principal axes coincide with the global Cartesian axes

CENTER OF MASS (X,Y,Z)=    0.34853E-05  -0.14087E-06  -0.30595

TOTAL INERTIA ABOUT CENTER OF MASS
       22.038         0.35503E-07   -0.26818E-07
     0.35503E-07      2694.2         0.97375E-08
    -0.26818E-07     0.97375E-08     2708.5
  The inertia principal axes coincide with the global Cartesian axe
```

Fig. 11 Mass in kg

6.2.3 Mass

The mass of this specimen obtained from Ansys results is found to be 25.15 kg (see Fig. 11). This will be helpful later for comparison of mass.

7 Conclusion and Future Scope

From the finite element analysis, results show that steel cover plate specimen had a mass of nearly 39 kg and got deformation about 0.51 mm. The 16 mm CFRP cover plate specimen had a mass of nearly 27 kg and got deformation about 0.73 mm, and 8 mm CFRP cover plate had a mass of nearly 25 kg and got deformation about 0.89 mm. So if it is used CFRP instead of steel cover plate, there will be saving of 12 kg material. The 50% reduction in thickness of cover plate, i.e., 8 mm thick cover plate of CFRP gives slightly more deformation about 0.16 mm but still the total deformation is less than 1 mm only. This analysis shows that even 50% reduction in thickness of cover plate the deformation will not increase very much which means that the CFRP material can be saved so mass is reduced and hence cost can be saved.

One can research by taking different stacking sequences of laminae which becomes one future work. One can research by taking other than CFRP material which is equivalent to steel also becomes one of the future works.

References

1. Qureshi J (2018) Bolted joints in fibre reinforced polymer bridges. Res Develop Mater Sci 6(2):1–5. Crimson Publishers. ISSN: 2576-8840
2. Manalo AC, Mutsuyoshi H, Asamoto S, Aravinthan T, Matsui H (2008) Mechanical behavior of hybrid FRP composites with bolted joints. ResearchGate Publication, 277834390
3. Krolo P, Grandić D, Bulić M (2016) The guidelines for modelling the preloading bolts in the structural connection using finite element methods. J Comput Eng, Hindawi Publishing Corporation, Article ID 4724312, 8 pp

4. Huang Y-h, Wang R-h, Zou J-h, Gan Q (2010) Finite element analysis and experimental study on high strength bolted friction grip connections in steel bridges. J Constr Steel Res 66:803–815
5. Shishesaz M, Hosseini M (2018) A review on stress distribution, strength and failure of bolted composite joints. JCAMECH 49(2):415–429
6. Mulagada HK, Rao KS (2018) Experimental and finite element analysis of stresses due to external forces for symmetrical and unsymmetrical bolted joints. Int J Sci Eng Technol Res (IJSETR) 7(7). ISSN: 2278-7798
7. Sonnenschein R, Gajdosova K, Holly I (2016) FRP composites and their using in the construction of bridges. In: World multidisciplinary civil engineering-architecture-urban planning symposium 2016, Procedia Engineering, 2016, vol 161, pp 477–482
8. Nerilli F, Marino M, Vairo G (2015) A numerical failure analysis of multi-bolted joints in FRP laminates based on basalt fibers. In: XXIII Italian group of fracture meeting, procedia engineering, 2015, vol 109, pp 492–506
9. Fakih KA, Chin SC, Doh SI (2018) Behavior of extended end-plate steel beam to column connections. Open Civil Eng J Bentham Open 12:250–262
10. Miklos IZ, Miklos CC, Lie CIA (2011) Computer aided design possibilities of bolted connection. Univ "Politehnica" Timisoara 1(2):114–119
11. Ansys Inc./Ansys Help Viewer/Ansys Documentation/Mechanical APDL/Element Reference/7. Element Library
12. Calik E, Ersoy N, Oz FE (2018) Experimental and numerical progressive failure analysis of corrugated core type composite sandwich structure. In: ECCM18—18th European conference on composite materials, Athens, Greece, 24–28th June 2018, pp 1–8
13. Kaw AK, Mechanics of composite materials, 2nd edn, © 2006 by Taylor & Francis Group, LLC, ISBN 0-8493-1343-0

Quality-Reliability-Risk-Safety Paradigm—Analyzing Fatigue Failure of Aeronautical Components in Light of System Safety Principles

Johney Thomas, Antonio Davis, and Mathews P. Samuel

Abstract Reliability and durability of aeronautical components and systems are directly influenced by fatigue, which is perhaps the most prominent mode of their failure. Quality aspects like surface integrity affect component reliability, and in turn determine system safety. Quality–reliability–risk–safety paradigm is thus self-reinforcing, either as a vicious or a virtuous cycle. System safety principles, originally proposed by Joseph H. Saleh, are high-level, domain-independent and technologically agnostic principles built on the notion of level of safety hazard and its escalation to accident sequence. They include (i) fail-safe safety principle, (ii) safety margin principle, (iii) ungraduated-response principle, (iv) defence-in-depth principle, (v) observability-in-depth principle; to which (vi) human factors principle and (vii) integration-in-totality principle can be added. Reliability of the component, and safety of the system, can be effectively examined by applying these cardinal principles. The technical article brings out a few case studies related to fatigue failure of aero-engine and aircraft components, analyzing the progression of failure in light of the system safety principles. Suitable remedial measures are recommended to obviate the destructive manifestation of fatigue damage, like improving stiffness and fatigue strength, enhancing attenuation/damping, monitoring fatigue life consumption, isolating contributory factors, etc. so that crack initiation and propagation can be controlled by early anticipation, regular monitoring and timely actions. Fatigue fracture failure of airborne systems can be effectively analyzed using Functional Resonance Analysis Method (FRAM) propounded by Erik Hollnagel. The paper demonstrates the use of 'FRAMED-IN-FRAM® Diagram' developed by the authors, an improved version of the 'FRAM Diagram', for understanding the combinational, resonant and temporal aspects related to the emergence of a safety hazard

J. Thomas (✉)
LCA-Tejas Division, Hindustan Aeronautics Limited, Bengaluru, India
e-mail: johney.thomas@hal-india.co.in

J. Thomas · A. Davis
Intl. Inst. for Aerospace Engg & Mgmt., Jain (Deemed-to-Be University), Bengaluru, India
e-mail: antonio.davis@jainuniversity.ac.in

M. P. Samuel
Regional Centre for Military Airworthiness (Engines), CEMILAC, DRDO, Bengaluru, India
e-mail: drmatmail@gmail.com

S. Seetharamu et al. (eds.), *Fatigue, Durability, and Fracture Mechanics*, Lecture Notes in Mechanical Engineering, https://doi.org/10.1007/978-981-15-4779-9_18

consequent to fatigue deterioration of the component. The authors also introduce a new concept of 'Reflexive Resonance' using FRAMED-IN-FRAM® Diagram for analyzing and depicting the vulnerability/durability of fatigue-damage-prone components and systems.

Keywords Fatigue · Reliability · Failure · System safety principles · FRAM · FRAMED-IN-FRAM · Reflexive resonance

1 Introduction

Modern aircraft and airborne products are complex systems with cutting edge technology. For example, advanced aero gas turbine engines, with critical and highly stressed components, operate at the limits of compressor pressure ratios, rotor speeds, operating temperatures and service conditions. Aeronautical components are exceptionally vulnerable to fatigue fracture failure due to initiation and propagation of cracks under the prevalence of cyclic loading, vibrations, thermal distress, etc. Manifestation of fatigue phenomenon in an aeronautical component directly impacts the reliability and durability of the component, and in turn failure of the related system, and is hence having major flight safety ramifications.

Safety is the most important aspect that needs to be addressed and controlled in an aircraft or airborne system. Failures occur in spite of the most accomplished prevention efforts in design, validation, manufacture, assembly, testing, operation and maintenance. No human endeavour or human-made system can be guaranteed to be absolutely free from hazard. Complete elimination of defects is unachievable in aviation. It is therefore important to enhance reliability and safety by mitigating the risks, with quality as a potent weapon, rather than a problem to be resolved. Hence there is a necessity to view the issue of enhancing aviation system safety from the perspective of Strategic Quality Management, which calls for a profound understanding of what quality means to the customer in any given context, making use of its tools and techniques.

Functional Resonance Analysis Method (FRAM), propounded by Erik Hollnagel, is a powerful technique for understanding and resolving safety-related issues in complex systems [1]. Fatigue fracture failure of airborne systems can be effectively analyzed with the help of FRAM for understanding the combinational, resonant and temporal aspects related to emergence of a safety hazard consequent upon fatigue deterioration of a component.

The authors have developed FRAMED-IN-FRAM® Diagram, a novel version of the conventional 'FRAM Diagram', which can be conveniently deployed for depicting safety incidents including fatigue fracture failures. The authors would like to introduce the new concept of 'Reflexive Resonance Response' using FRAMED-IN-FRAM® for analyzing and depicting the vulnerability/durability of fatigue-damage-prone components and systems, whereby the self-reinforcing nature of manifestation of fatigue vulnerability can be illustrated, understood and demonstrated.

This technical article will provide a few examples of 'Reflexive Resonance Response' as part of the few case studies presented here in connection with the application of system safety principles for analyzing fatigue fracture failures.

2 Quality-Reliability-Risk-Safety Paradigm

2.1 *Quality—The Evolution Towards Strategic Quality Management*

Quality is the harbinger of sustainable growth and prosperity, achieved by way of systemic approach, improvement cycles and risk-based thinking. Feigenbaum defined Quality as follows: 'Quality is the total composite product and service characteristics, through which the product in use will meet the needs and expectations of the customers … Concept of quality must start with identification of customer quality requirements and must end only when the finished product is placed into the hands of the customer who remains satisfied through various stages of relationship with the seller'. The International Organisation for Standardisation (ISO) defined Quality as 'the totality of features and characteristics of a product or service that bears on its ability to satisfy given needs'.

Quality as an organizational function has evolved over the years, from Inspection to Quality-Control to Quality-Assurance to Company-wide-Quality-Control to Total-Quality-Management to Strategic-Quality-Management. In the process, the tenets of Quality also got enlarged with a snowballing effect, to encompass Product, +Process, +System, +People, +(Improvement) Cycle, +Risk, as depicted in Fig. 1 [2].

Quality function has grown accordingly from the 'small q' to the 'Big Q', as theorized by Juran and Godfrey [3], the Big Q subsuming everything in its path, the organization, the customer and the outside-world, as illustrated in Fig. 2 [2]. There have been different approaches to understand Quality, like transcendent, product-based, user-based, production-based and value-based approaches. Garvin defined eight attributes of Quality, viz., performance, features, reliability, conformance, durability, serviceability, aesthetics and the so-called perceived-quality [4].

2.2 *Reliability and Fatigue Durability*

Reliability in engineering is defined as the probability that a component satisfies its specified behavior requirements over time and under given conditions, i.e. it does not fail. It is the probability that a component will perform the intended function without failure under specified conditions for a stated period of time. Quality plays a vital role in determining the reliability of a component. Quality and reliability were being considered as two different entities with their own set of theories, Quality

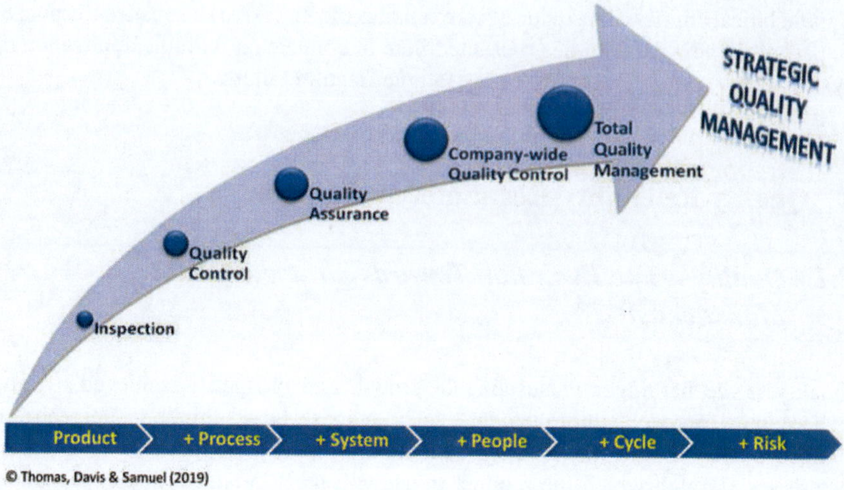

Fig. 1 Evolution of quality over the years

Fig. 2 The expanding role of quality function

generally symbolized by the normal curve and reliability by the bath-tub curve. Reliability is all about having unblemished functional-durability of a component, without introducing a safety hazard to the system, thereby causing the associated risk. A failure occurs when a component does not satisfy the specified requirements. According to Leveson [5], a failure is the non-performance or inability of the system or component to perform its intended function for a specified time under specified environmental conditions. Thus failure is the anti-thesis of reliability.

2.3 A New Risk Management Framework

Reliability of a component is a direct determinant of the risk associated with the functionality of a component within a system. Increased vulnerability of a component for fatigue failure is indicative of reduced reliability and in turn increased risk, the lack of fatigue durability acting as a potential safety hazard.

Nevertheless, the lack of reliability and fatigue durability of a component, its fatigue vulnerability, can be considered as a risk hazard for the system safety, and hence an accident pre-cursor or an accident pathogen.

Accident pre-cursor is defined by NASA as 'an anomaly (off-nominal occurrence or condition) that signals the potential for more severe consequences that may occur in the future, due to causes that are discernible from its occurrence today' [6, 7]. An accident pre-cursor can be understood as an abnormal initiating event which gives rise to more hazardous events and/or states along the path of accident sequence, finally resulting in the accident or serious incident. Likewise, an accident pathogen is defined as 'an adverse latent or pre-existing condition, passive or with no impact on system output until activated or triggered by other adverse occurrences. When compounded with other factors, an (activated) accident pathogen can further advance an accident sequence, precipitate an accident, or aggregate its consequences' [6, 8].

In safety management system under the aegis of International Civil Aviation Organization (ICAO), safety risk is determined based on the risk criticality based on the two factors, namely (i) Severity of potential consequences of a safety hazard, and (ii) Probability of occurrence of the hazard itself [7]. Gnoni and Saleh proposed a qualitative risk assessment in the context of Near-miss Management System (NMS) wherein the safety hazards or accident pre-cursors or accident pathogens are evaluated based on five factors using a Radar plot [6]. The authors would like to propose a six-factor qualitative risk assessment framework connecting the reliability aspect of fatigue durability and the other technical, human and latent factors, which can lead to an accident sequence leading to an accident or a serious incident, the six qualitative risk factors being:

(i) Risk Severity (Significance of potential consequences or impact of a safety hazard)
(ii) Risk Probability (Likelihood or frequency of possible hazard occurrences)
(iii) Risk Detectability (Observability of the safety hazard; added to the original matrix by Gnoni and Saleh)
(iv) Risk Proximity (Closeness to the ultimate safety event)
(v) Risk Preventability (De-escalatability through safety barriers)
(vi) Risk Potentiality (Applicability of the lessons learned) (Fig. 3).

2.4 Safety—Linkages to Reliability and Quality

According to ICAO [9], 'Safety is the state in which Risk (of personal harm or property damage) is reduced to and maintained at or below an acceptable level, through

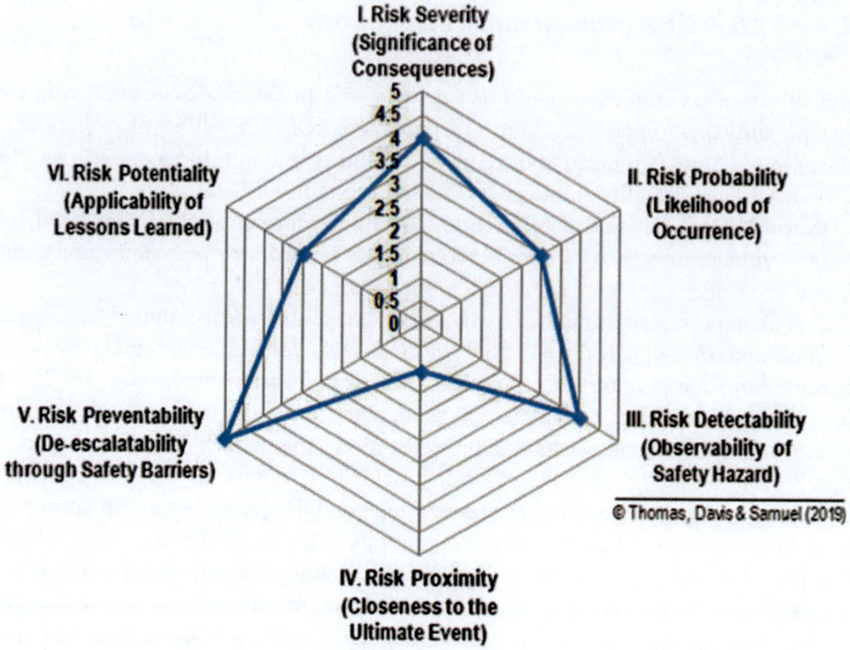

Fig. 3 A radar diagram depicting the six qualitative risk factors (adapted from [6])

a continuing process of Hazard identification and Risk management'. It has been traditionally assumed that safety is enhanced by increasing the reliability of the individual system components. However, recent literature in the field of safety suggests that safety and reliability are different system properties, and one does not imply nor require the other. Leveson suggested that safety is a system property, not a component property, and must be controlled at the system level, not the component level, since the system can be unsafe even if the components are reliable [10]. According to the systems theory, safety is viewed as a control problem (of enforcing the safety constraints) rather than a failure or reliability problem. However, limitations of reliability and durability of a component pose a major hazard against system safety, and hence is required to be identified early on, and managed within a Risk management framework.

Safety of the component is a pre-requisite for Quality of the system/sub-system, which in turn ensures the reliability, the risk-profile and the safety at the higher levels of the system. Thus there is a self-reinforcing circular relationship between Quality, Reliability, Risk and Safety, which can either be a positive virtuous cycle or a negative vicious cycle, which the authors would like to call 'Reflexive Resonance Response'.

2.5 *The Quality-Reliability-Risk-Safety Paradigm*

In view of the foregoing discussions, the authors would like to present the Quality-Reliability-Risk-Safety paradigm as a 'FRAM Diagram' (FRAM: Functional Resonance Analysis Method), in the form of a 'Reflexive Resonance Cycle' illustrated in the FRAMED-IN-FRAM® Diagram developed by the authors [11], which is given in Fig. 4.

3 Fatigue Durability and System Safety Principles

Saleh et al. described the five basic principles of system safety built on the notion of hazard level and its escalation and the Accident sequence [6, 12]. The following are these system safety principles:

(i) The Fail-Safe principle [13] mandates that the system design should be such that it prevents or mitigates the unsafe consequences in case of the failure of the system;
(ii) The Safety Margin principle [14] calls for maintaining the operational conditions, and the consequent hazard level, sufficiently below the accident triggering critical threshold of the hazard, considering the worst-case scenario;
(iii) The Ungraduated Response principle [15] posits that the possibility of completely eliminating the hazard should be the first course of action to be explored for accident prevention, regardless of the extent of severity of the consequence;

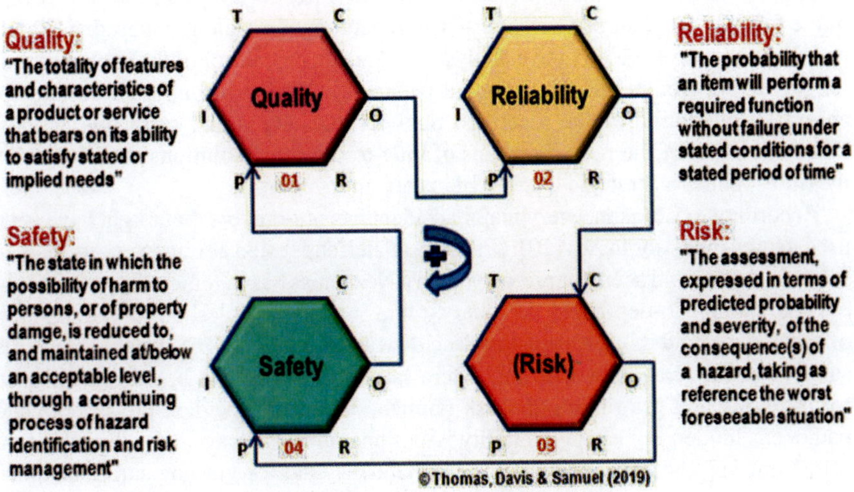

Fig. 4 The quality-reliability-risk-safety paradigm

(iv) The Defence-in-Depth principle [8, 16, 17] requires that multiple layers of defences or safety barriers are to be incorporated along the potential path of accident sequence so as to ensure safety protection;

(v) The Observability-in-Depth principle [17] calls for putting in place various features within the system for observing and monitoring the system state and detecting the safety-degrading events or states early on, to enable necessary remedial actions based on the feedback.

In addition to the above 'basic five' related to the technical factors, the authors propose to include two more system safety principles, to cover the human factors and the systemic aspects, to get a complete set of system safety principles. These 'additional two' system safety principles are

(vi) The Human Factors Principle [9, 18, 19], universally accepted as a cardinal principle of system safety, and

(vii) The Integration-in-Totality principle, which is based on the 'D-C-P Diagram' developed by Stoop [20–22], 'System-Theoretic Accident Model and Processes (STAMP)' proposed by Leveson [5, 10, 23] and the 'Functional Resonance Analysis Method (FRAM)' propounded by Hollnagel [1, 24–27].

4 Analyzing Fatigue Durability in Light of System Safety Principles—Case Studies

The relationship between Quality, Reliability, Risk and Safety was presented as a paradigm in the form of a FRAMED-IN-FRAM® Diagram earlier. It was brought out that fatigue durability, which determines the reliability of a component, can have a direct, though not necessarily proportional, effect on safety of the corresponding system. It was also explained how fatigue vulnerabilities act as system hazards, acting in the form of accident pre-cursors, accident pathogens or latent conditions. Hence it is imperative to have proactive analyses and resulting control of the probability of fatigue failure from a risk management perspective. It is also important to attempt necessary mitigation of the identified risks through necessary corrective actions, and elimination of the risks by means of long-term design solutions and preventive measures, under a Strategic Quality Management framework.

According to Leveson, every major accident has pre-cursors that might have been used to prevent major losses [10]. Gnoni and Saleh suggested a framework to analyze safety incidents and accident pre-cursors in a Near-miss Management System in light of observability-in-depth and the other system safety principles, built on the notion of hazard level and its escalation to an accident sequence [6, 28]. A deep dive into the safety literature can show that the system safety principles can be effectively used for assessing and prioritizing the risk implications of the accident pre-cursors and pathogens hidden in fatigue durability issues and other such system vulnerabilities, for identifying the failure generating mechanisms, and for guiding safety insights, interventions and improvements.

In the next few sections, the authors would like to showcase a few case studies on failures of aeronautical components and systems which can be related to fatigue durability and reliability, and also the associated concepts of Quality, Risk and Safety, in light of the '7S', the seven system safety principles.

5 The Fail-Safe Principle (S1)

The Fail-Safe Safety principle mandates that the system design should prevent or mitigate the unsafe consequences of the failure of a system. If and when a 'fail-safe' system fails, it should still be safe or at least no less safe than when it was operating correctly, as illustrated in Fig. 5 [6, 12, 13].

Fail-Safe safety is a concept which is fundamentally common between the system safety principles and the fatigue durability and reliability literature. It has linkages with other safety principles like ungraduated-response principle and observability-in-depth principle.

There have been four well-acknowledged design criteria developed over the years, viz.,: (i) Infinite Life; (ii) Safe Life; (iii) Fail Safe and (iv) Damage Tolerant [29].

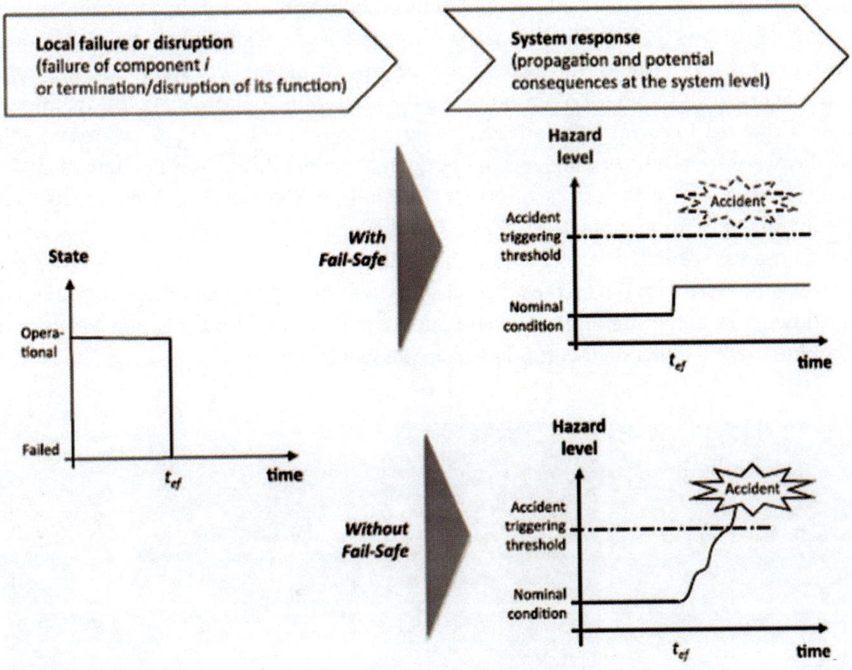

Fig. 5 Fail-safe principle: system behaviour after local failure, with and without its implementation (*Source* [12])

The Infinite Life criterion targets an endless endurance, providing a high safety factor based on stress history and endurance limit of the material. The 'Safe Life' criterion, on the other hand, targets structural integrity for a limited amount of time, translated into a specific number of cycles that needs to be endured by a component, calculated through a strain-based approach (for High Cycle Fatigue) or a stress-based approach (for Low Cycle Fatigue) [30]. The 'Fail Safe' and the 'Damage Tolerant' criteria are based on the idea of allowing a crack to nucleate and propagate under appropriate monitoring and control [31]. A case study is now presented here, wherein the Safe-Life approach was successfully used as a substitute for Fail-safe design criterion to enhance safety to tide over fatigue vulnerability.

5.1 *Case Study: Failure Analysis of LP Compressor Blade of a Low Bypass Turbofan Engine*

Background. While opening the Starboard engine throttle to 'Max dry' at the lineup of a twin-engine aircraft, a loud bang was heard, and turbine gas temperature of the engine was found overshooting. The engine was switched off immediately and the aircraft was subjected to single-engine landing. The engine was taken up for defect investigation. The engine had completed more than 800 h since the last overhaul.

Observations during Receipt Inspection and Strip Examination. During bulk strip examination, one of the first stage compressor blades was found missing and the headed locating pin used for mounting the blade was found sheared off into three pieces, and the locking plate and retaining plate were found missing. The remaining blades were found to be damaged. A portion of the headless pin was retained at the hole location of the Low Pressure (LP) Compressor Stage-1 Disc, which came out freely during strip examination. Stage-1 and Stage-2 LP Stator vanes and Stage-2 LP Compressor blades were found to have damaged. Further, Stage-3 and Stage-4 blades also were found damaged. Metal spray was noticed all along the compressed air flow path, including High Pressure and Low Pressure Nozzle Guide Vanes and Turbine blades. The failed components are shown in Fig. 6.

Fig. 6 Front portion of the aero-engine during receipt inspection, and the debris collected

Laboratory Analysis. Metallurgical failure mode analysis carried out using Scanning Electron Microscope (SEM) revealed the evidence of fatigue crack at the fillet radii of the blade lug at the blade root area, as shown in Fig. 7.

The locking pins fastening the blade to the compressor disc was found to have fractured under bending overload which could be of secondary failure, as shown in Fig. 8.

The fillet radii of the failed blades were found to have sharp corners instead of the smooth radius required as per design specifications. The laboratory analysis suggested that the reduced fillet radius at the blade lug corners as the underlying reason for initiation and propagation of cracks along the blade lug, presumably due to the prevalence of stress concentration due to the sharp corners, as shown in Fig. 9. However, further investigation had to be carried out to pin-point the exact root cause.

Stress Analysis. Stress analysis by Finite Element Method was carried out on blades (i) with root radius as per the drawing, and (ii) without sufficient root radius.

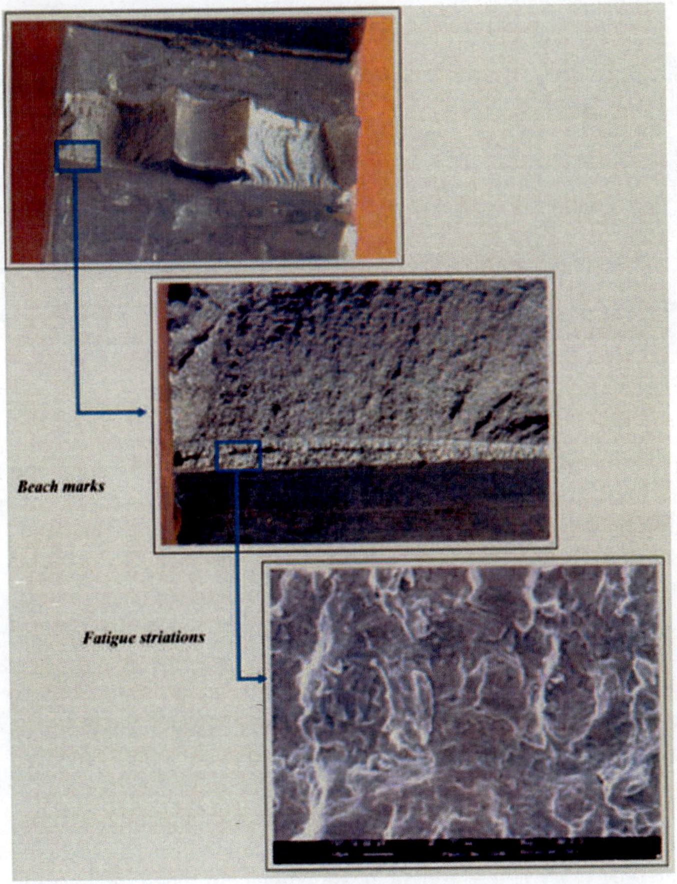

Fig. 7 Presence of fatigue crack at fillet radius of blade lug

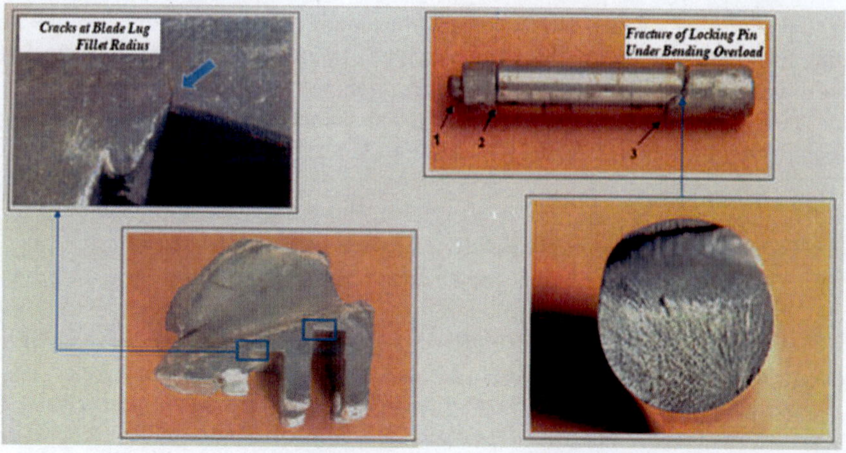

Fig. 8 Fatigue cracks at blade lug fillet radius, and fracture of locking pin under bending overload

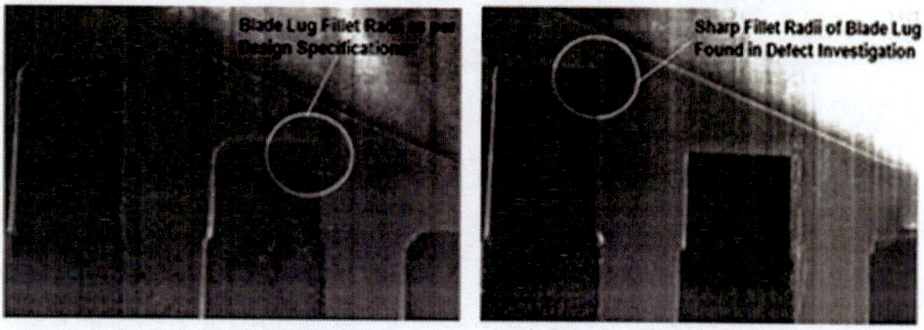

Fig. 9 Reduced blade lug fillet radius

It was found that there was no change in maximum stress location and the stress magnitude remains the same at lug fillet location for both the cases (with or without root radius), and there was no change in the stress contours, as shown in Fig. 10.

Pre-existing Crack. However, subsequent failure analysis could establish the presence of pre-existing crack at the pinhole location of the blade, which got propagated further in a definite sequence, and eventually led to the blade failure, as shown in Fig. 11.

Conclusion. The preliminary laboratory investigation report on the fractured components suggested that the reduced blade lug fillet radii could be the cause for the blade failure. All the other damages were construed to be secondary in nature. However, the stress analysis report inferred that the stress developed at the blade lug location is the same for both the blades, with or without sufficient blade lug fillet radii. Further detailed investigation confirmed that the blade had failed due to pre-existing Frettage crack at the pinhole location, which is the most stressed location as found out in the stress analysis report.

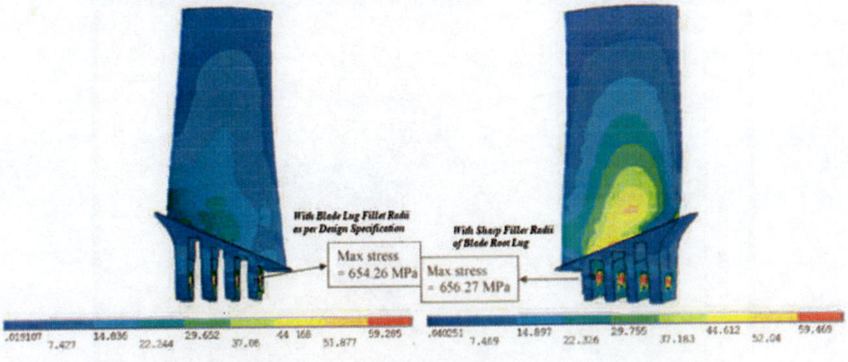

Fig. 10 Von-Mises stress plots of blades with and without sufficient lug fillet radii

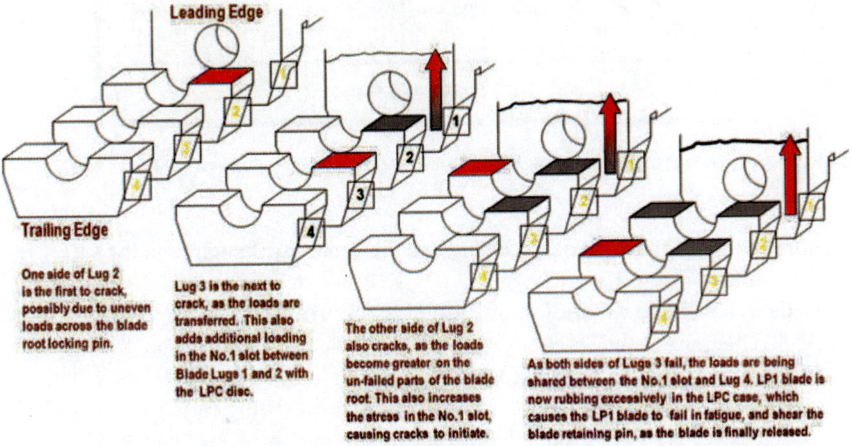

Fig. 11 Sequence of propagation of crack and progression of failure along the four blade lugs

Preventive Action. Eddy current inspection was introduced at every 600 h in the pin locating holes of both first and second stage compressor rotor blades and compressor discs for early detection of initiation and propagation of cracks, as a remedial measure. The stator and rotor modules were withdrawn at half the time between overhaul of 1200 h of the aero-engine, and both the compressor discs and rotor blades are subjected to laboratory examination of the pin locating holes by eddy current inspection. In case micro-cracks due to frottage are observed in the holes, the disc and the blade are rejected and removed from service. Thus, the safe-life approach was introduced as a substitution for the fail-safe approach, taking into consideration the time taken from crack initiation for propagation of the crack till manifestation of the failure. This approach of inspection at reduced intervals to extend the disc and blade life, based on the assessment of the time taken for manifestation of fatigue

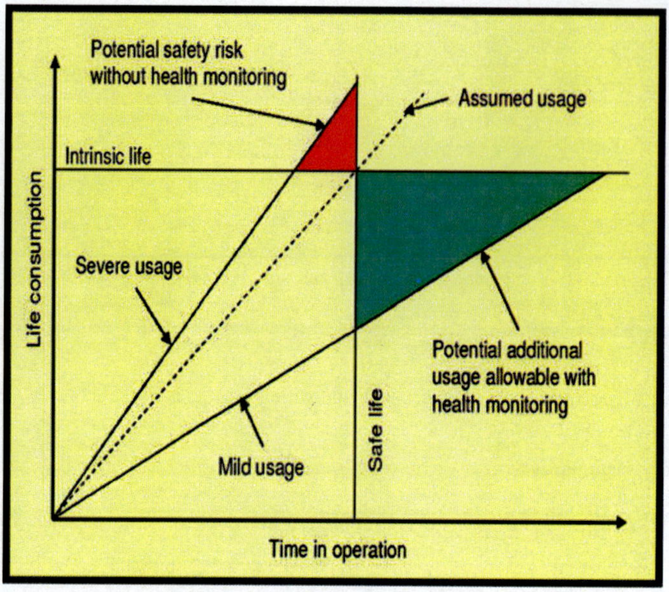

Fig. 12 Assessment of safe life based on severity of service usage (*Source* [31])

fracture failure, is the right balance between the economic benefit and the safety risk [31], as shown in Fig. 12.

Reflexivity in Manifestation of Fatigue Fracture Failure. The FRAMED-IN-FRAM® Diagram of the safety event is illustrated in Fig. 13. On the right side, the self-reinforcing destructive manifestation is shown in the form of a vicious cycle of negative spiral of FRAM Cells, starting from the pre-existing frettage cracks on the pin surfaces (Cell #01) to the safety event of single-engine landing of the aircraft (Cell-16). On the left side, prevention of fatigue fracture failure by eddy current inspection at reduced intervals is shown with 'Reflexive Response' of a positive virtuous cycle of safety.

6 The Safety Margin Principle (S2)

The safety margin principle calls for maintaining the operational conditions and the consequent hazard level sufficiently below the accident triggering critical threshold of the hazard, considering the worst-case scenario. This is achieved by providing adequate 'Safety margin' between the two, which will take care of the uncertainties in both understanding the critical triggering threshold and managing the operational conditions within the system, so that the hazard level will never exceed the real but unknown critical value at any given point of time [6, 12, 14] (Fig. 14).

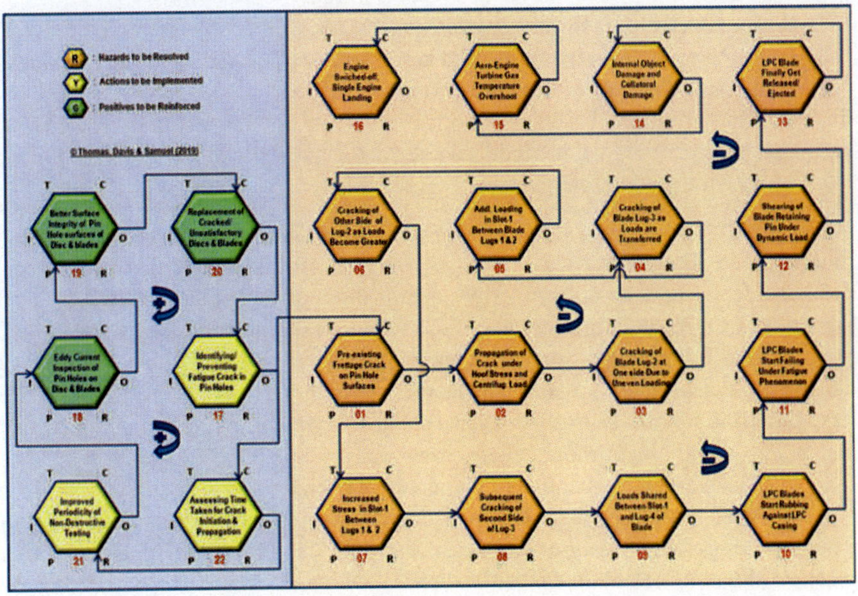

Fig. 13 ‘FRAMED-FRAMED-IN-FRAM® diagram’ showing sequence of propagation of crack and progression of failure

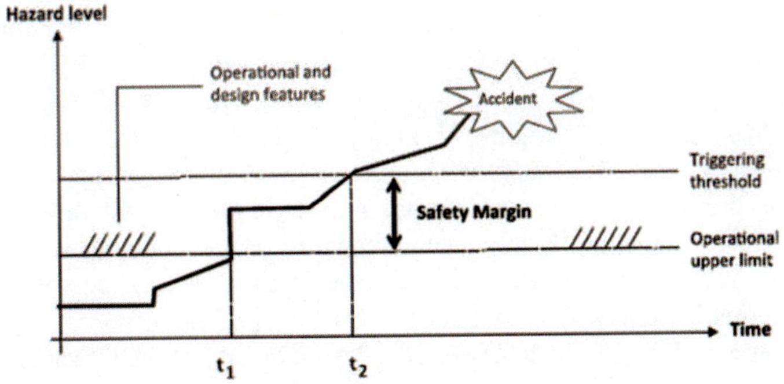

Fig. 14 Safety margin principle: accident trajectory from normal operating condition to an accident (*Source* [12])

The concept of Safety Margins, which is essentially a measure of capability in excess of the requirement, has three distinct views, depending upon the way it is perceived and applied:

(i) ‘Design Factor’ based on ‘Requirement Sense’: It is the ratio of the material strength capability to the intended load for which the component is designed. The design factor is provided in advance by the regulatory requirements, by specification/standard and/or policy/law.

(ii) 'Safety Factor' (or 'Reserve Factor', or 'Factor of Safety') based on the 'Realization Sense': It is the ratio of material strength capability to the maximum possible applied load under actual operational conditions.
(iii) 'Margin of Safety' based on 'Verification Sense': It shows if the part is loaded to the maximum load it would ever see in service, how much more such loads it can withstand before failure.

The Safety Margin should take care of the 'Worst Case' rather than the 'Credible Likely Case'. The 'Worst Case' is a sequence of events/actions that cause accidents/incidents. It does not normally contain the element of probability. It is an extreme case, rare situation, very pessimistic no-data model like Murphy law model. According to Taleb [32], a simplistic view of the statistics of Gaussian normal curve does not reflect the actual risks; and there are risks with 'fat tails' that would normally be expected to be rare or extreme event, but actually having a far higher chance of happening, mostly with more severe, even catastrophic, consequences, which he called a 'Black Swan event'. Fatigue failure is one such risk with 'fat tails'.

The 'Worst-Case' scenario is considered during the Design phase for Design assurance, during the Safety-oriented analysis. The 'Credible likely case' assumes a practical situation within the operational phase, with an optimistic input meeting the specification requirements, which is meant for performance-oriented and reliability-based analysis.

A Probabilistic Design methodology is employed in the present days for incorporating the concept of Safety Margins to take care of safety and reliability of a product [33]. Probabilistic Design considers the effects of random variability upon performance of the engineering system during the design phase. Each variable is considered as a distribution rather than an absolute value or a value with a narrow tolerance band. By considering the variability in the load spectrum and the strength capability, considering the scatter factor inherent in various input characteristics like material properties, and the unforeseen possibilities in the output manifestations, the designer can envisage the 'Extreme case Scenario' of the load/stress exceeding the strength at the extremities of the load/stress and strength normal curves where they cross each other, as shown in Fig. 15.

Many a times, the mismatch between the design intent, the manufacturing realization and the operational manifestation can result in unforeseen worst-case scenarios due to resonance, temporal and combinational factors at the system level. A case study is presented here to illustrate how fatigue durability and reliability are directly impacted by certain beyond the obvious phenomenon like resonance vibration and Backward Whirl, and also the various issues related to design, manufacturing and operation.

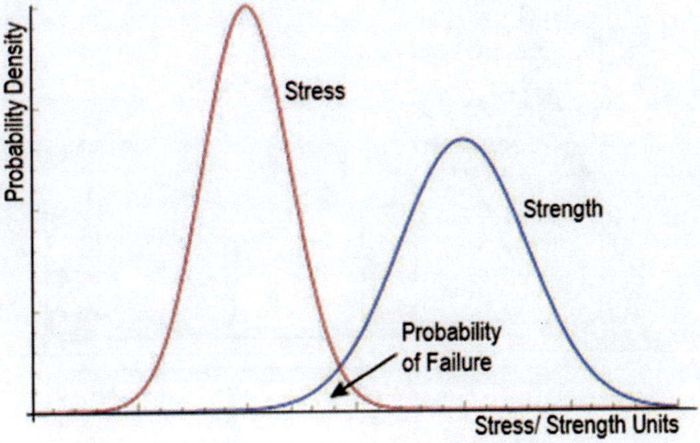

Fig. 15 Probabilistic design philosophy

6.1 Case Study: The Curious Case of Drive Shaft Failure

The authors had presented, in an earlier technical article, a case study on failure of the Drive shaft of the blower fan of the Oil Cooling System (OCS) of a Turboshaft engine [11]. The Shear neck of the Drive shaft was of Φ5.25 mm, operating at a rated speed of 415 revolutions per second, and is designed for maximum breaking torque strength of 21.3–40.5 Nm. In case of any restriction on free rotation of the blower fan assembly, the Shear neck of the drive shaft is intended to break and thus protect the gear trains of the aero-engine gearbox [34]. A view of the Drive shaft arrangement is given in Fig. 16. The Drive shaft, which is the stem portion of the OCS Drive Gear, was found ruptured at the Shear neck in a few of the rotorcrafts. Fractographic examination of the rupture surfaces of the broken OCS shafts revealed two distinct surfaces in the Shear neck rupture surfaces, as shown in Fig. 17.

A few improvement actions like removal of nitriding on Drive shaft stem, improved non-destructive testing methods, use of a bespoke assembly tool, improvement of machining process for the shear neck groove to improve surface integrity, etc. did not yield the desired results.

An Instrumented Engine Test was carried out in the aero-engine test-bed, as part of a detailed Vibration analysis, with an aim to experiment and understand the vibratory behavior of the Drive shaft assembly, and to characterize the vibrations. In the Instrumented test, in the Shaft mode, signature of Split resonance with associated Backward Whirl phenomenon induced by unbalance in the adjacent breather gear driving OCS gear was detected. The Campbell Diagram, the theoretical Frequency Versus Speed Graph, is shown in Fig. 18, which revealed that high amplitude resonance response was happening at the intersection point of the Backward Whirl mode line and the Breather gear 1st order speed line, at about 73% of the rotational speed of gas generator (called NG Speed) [35–37].

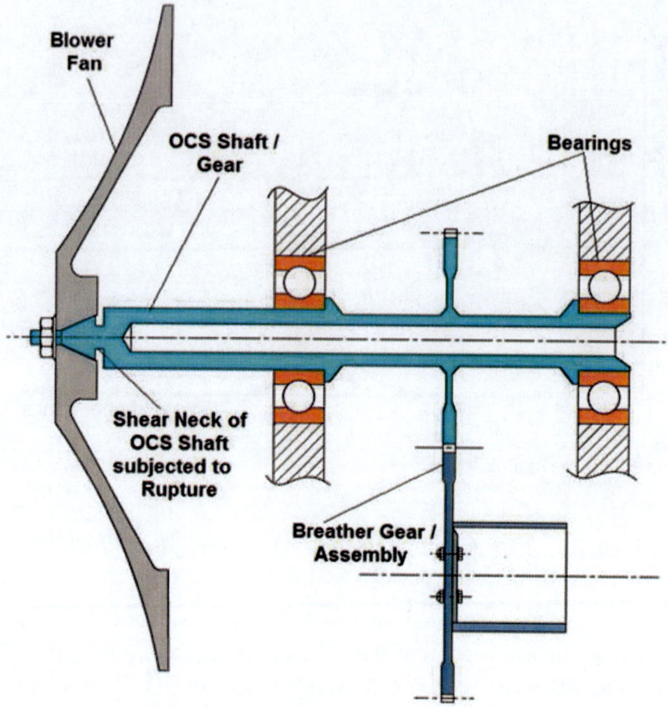

Fig. 16 Schematic view of drive shaft arrangement of the oil cooling system

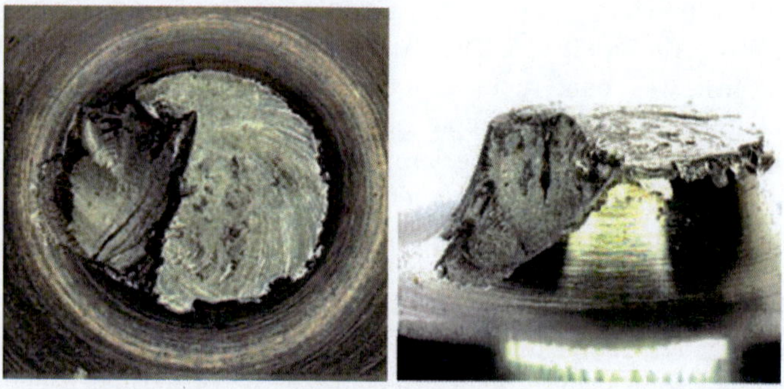

Fig. 17 Surface features of the rupture shear neck

The Radial displacement of the blower fan due to excitation of the Backward Whirl created a wobbling/wave effect on the OCS shaft, which caused alternating compressive and tensile stresses at the shear neck of the Drive shaft, the highest stress level being obtained when crossing the breather gear speed, at about 73% of Gas generator speed. The stress evolution in the shear neck section with reference to time is given in Fig. 19.

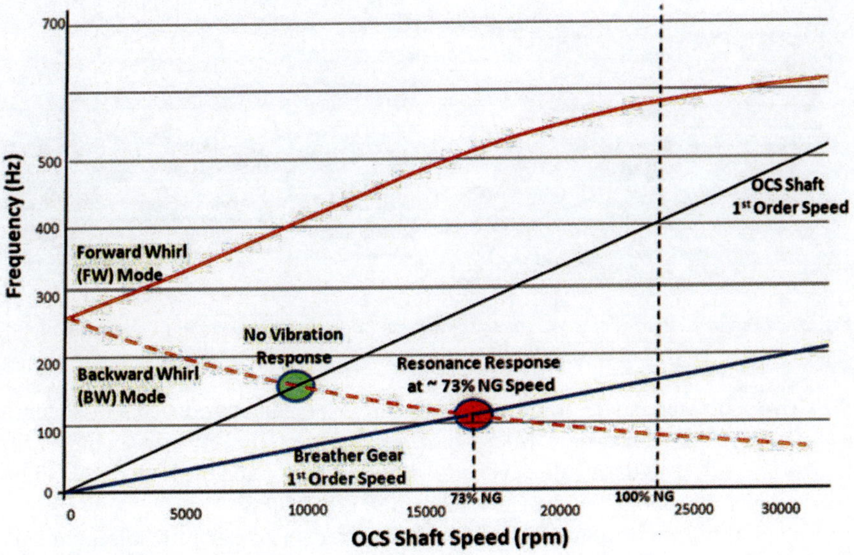

Fig. 18 Campbell diagram: high amplitude resonance response due to backward whirl mode

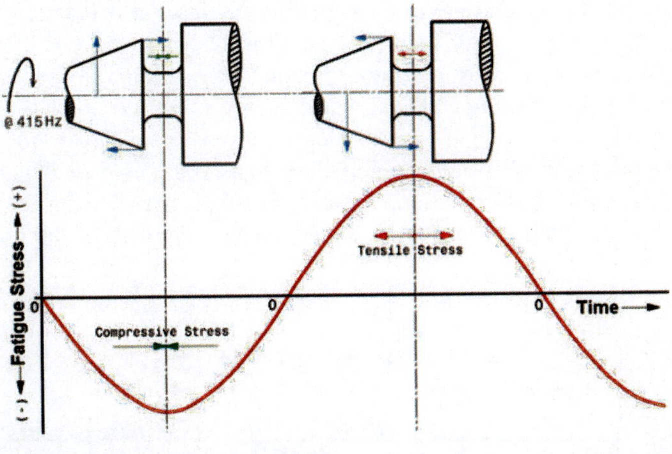

Fig. 19 Stress evolution at shear neck section

The progression of the events leading to the rupture of the Drive shaft, and the influencing sources of the crack and rupture at the component, aero-engine and aircraft levels (micro, meso and macro levels, respectively) are diagrammatically shown in Fig. 20.

Further improvements like defining and avoiding the Gas Generator speed range at which resonance vibration gets excited, removal of unbalance in breather gear assembly by Dynamic balancing, Micro-shot peening at the Shear Neck Section of

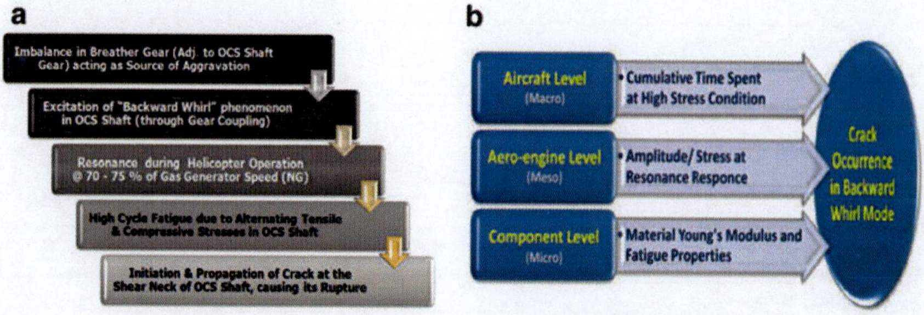

Fig. 20 **a** Progression of events leading to rupture, and **b** Influencing sources of crack and rupture

the Drive shaft were some of the additional improvements introduced subsequently as defences and safety barriers. Micro-shot peening was found effective in improving fatigue strength to withstand the cyclic loading conditions and the high cycle fatigue prevailing at the time of resonance, as shown in Fig. 21 [38].

However, a final breakthrough was achieved for a permanent resolution of the problem by shifting the cross-over point of the Backward Whirl mode Resonance vibration frequency outside the maximum possible Gas generator speed, or beyond the complete operating range of the aero-engine and the aircraft, by enhancing the bending stiffness of the Drive shaft by increasing the Shear neck diameter. The principle of shifting of Backward Whirl/Breather gear Resonance beyond the operating range of the aero-engine Gas generator through stiffness enhancement of the OCS shaft is depicted in the Campbell Diagram as shown in Fig. 22.

Additionally, steel ferrules were introduced in the bearing-seating surfaces of the Gearbox casing made of magnesium alloy, so as to increase the damping effect against Backward Whirl excitation. The two measures of enhancing the structural stiffness, and damping the excitation source, effectively resolved the problem for ever.

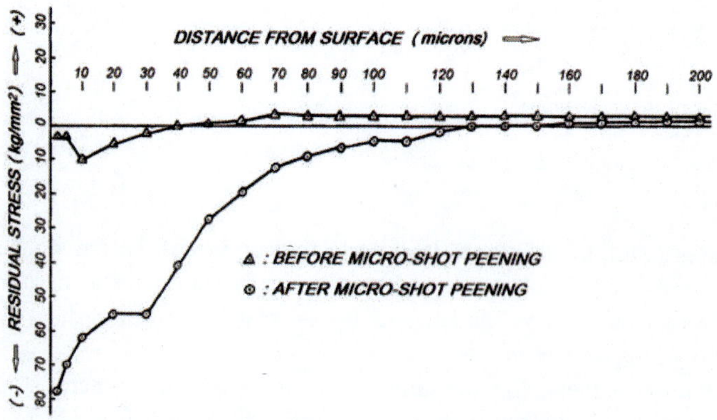

Fig. 21 Residual compressive stresses induced by micro-shot peening

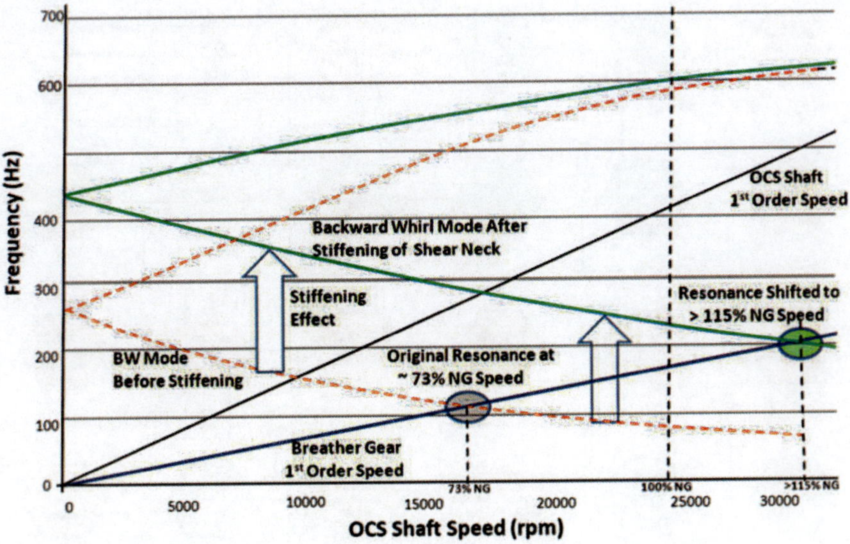

Fig. 22 Campbell diagram showing shift in resonance range with stiffened OCS shaft

The 'FRAMED-IN-FRAM® Diagram', integrating all the combinational, resonant, temporal and control aspects of the problem, is presented in Fig. 23. The case study highlights the importance of the safety margin principle, and shows that establishing the real root cause of a safety event may require looking beyond the obvious, and searching for factors remaining hidden and inconceivable [39].

7 The Ungraduated Response Principle (S3)

The ungraduated response principle posits that the first course of action to explore for accident prevention and mitigation is the possibility of eliminating a hazard altogether, regardless of the extent of its belligerence, using creativity and technical ingenuity. This 'kill first' attitude propounding lethal use of force against hazards is contrary to the normal rule of engagements of a graduated response proportional to the extent of demonstrated belligerence [6, 12, 15] (Fig. 24).

7.1 Surface Integrity—The Armour of Aeronautical Components

In recent years, on a world-wide basis, the designers and manufacturers or critical and highly stressed components have been engaged in a serious re-assessment

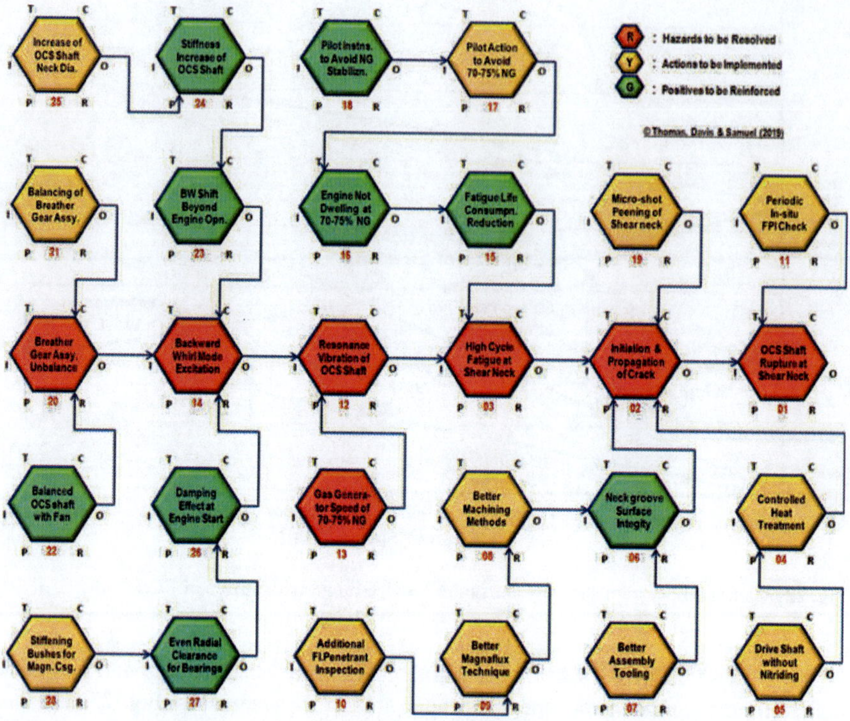

Fig. 23 Combined FRAMED-IN-FRAM® diagram for the case study on drive shaft

of the pronounced influence of manufacturing methods on mechanical properties and service performance, as a result of the type of the surface produced. This has come about because of increased demands for reliability and safety, unusual service requirements and the development of new manufacturing processes, as in the field of manufacture of aeronautical components.

The service performance of a manufactured component can be seriously affected by the surface conditions, in addition to surface finish, generated during material removal. New designs of aeronautical components are now calling for higher stress and lighter weight, and hence reduced reserve factors and safety margins. It also calls for resistance to more severe environmental conditions including fatigue and stress corrosion at elevated temperatures. The heat generation, plastic deformation, chemical reactions, etc. during material removal processes such as grinding, drilling, milling, electro-discharge-machining, etc. cause detrimental surface alterations which extend below the surface, which are usually not detectable even by the non-destructive-testing techniques.

'Surface Integrity' is the subject covering the description and control of the many possible alterations produced in a surface layer during manufacturing, including their effect on component performance. The following are some of the typical surface layer alterations associated with machining processes, causing lack of surface integrity:

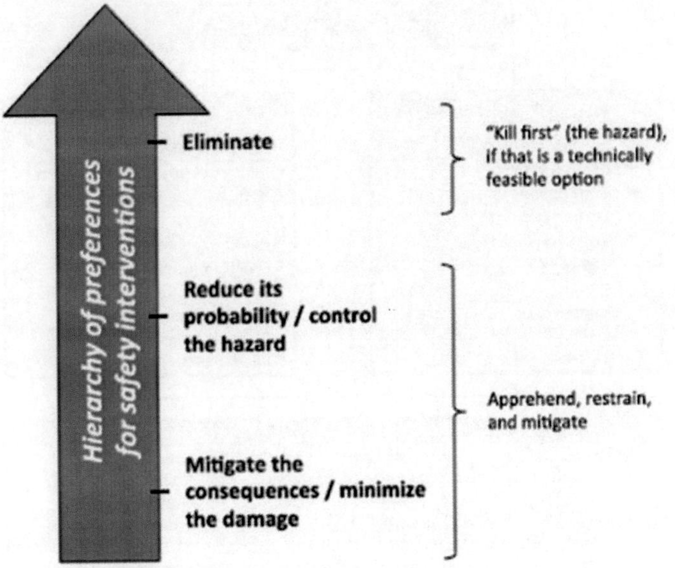

Fig. 24 Ungraduated response principle: hierarchy of preferences for safety interventions (*Source* [12])

(i) Residual stresses induced in the surface layer by machining, causing distortion, fatigue and stress corrosion;
(ii) Tears and laps, and crevice-like defects due to built-up-edge;
(iii) Plastic deformation due to hot or cold work;
(iv) Micro-cracking and Macro-cracking, like Grinding cracks;
(v) Recrystallization, phase transformations and change in surface layer hardness, for example, 'grinding burns' on high-strength components like transmission gears;
(vi) Embrittlement due to chemical absorption of elements like H_2, Cl_2;
(vii) Inter-granular attack and preferential solution of micro-constituents as in electro-chemical-machining;
(viii) Alterations like spattered coating or re-melted metal deposited on the surface, as in electro-discharge-machining, causing lowering of fatigue strength.

'Surface Integrity' is defined as 'the unimpaired or enhanced surface condition developed in the component by controlled manufacturing processes' [38]. Surface integrity has two broad aspects requiring consideration in machining, the first being 'Surface Topography' which describes the 'lay' or 'texture' of the outermost layer of the workpiece, and the second being 'Sub-surface Metallurgy' which describes the nature of the altered layers below the surface, known as 'Altered Material Zones'. The classification and main elements of surface integrity, and the way it affects fatigue durability and component reliability is highlighted in Fig. 25.

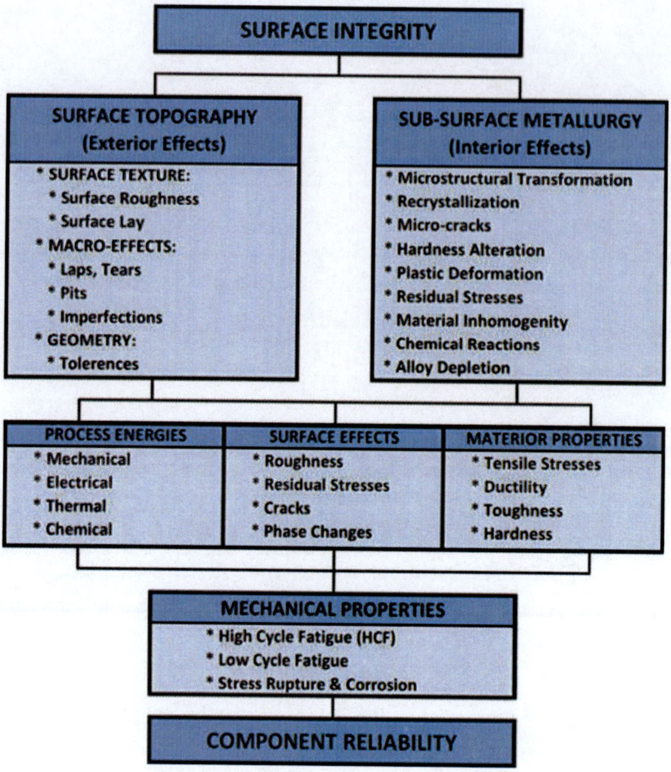

Fig. 25 Surface integrity and component reliability

Surface integrity is an essential requirement in the manufacture of aero-engine components, in addition to the geometrical tolerances and other technical specifications. However, due to a lack of awareness of the importance of surface integrity considering the disproportionate severity of the potential consequences, and/or since surface integrity control generally results in increased manufacturing costs and decreasing production rates, the concept is not always applied to even critical parts or to critical areas of given parts, relaxing the requirements of surface topography, including surface finish in non-critical areas. It is imperative to discuss and select suitable technological processes or methods necessary to improve and ensure surface integrity during machining, post-processing and inspection of aero-engine components. The 'Critical Class-A' components of an aero-engines, for example, are required to be 'pristine', without any blemish like nicks, dents, scratches, handling damages, etc. to be ensured by way of extreme care and control during machining, bench-work, handling and inspection. A typical source of compromise on component integrity is the use of form tools to machine critical and stressed areas like diaphragm transition radii, recesses, etc. of critical rotating components like compressor and turbine discs. Even though the risk has reduced to a large extent with the advent of

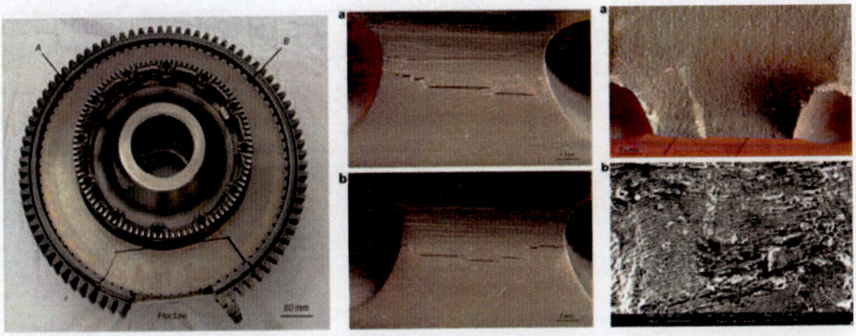

Fig. 26 Fatigue failure related to surface topography (*Source* [40])

Computer-Numerical-Controlled machining, if the various aspects of machining are not properly controlled, the surface integrity can get seriously compromised. These aspects include the machining parameters like speed, feed and depth of cut, the tool geometry, tool trajectory, tool replacement periodicity and even the machining programme, coolant used and rigidity of the machine-fixture-tool-component system, nowadays ensured by the implementation of the so-called 'Fixed Process'.

7.2 Fatigue Failure Related to Surface Topography—Case Study of a Turbine Disc

Sujata et al. [40] presented the case of an aircraft accident wherein High Pressure Turbine Rotor (HPTR) disc of an aero-engine, made of nickel-based superalloy, fractured and flew-off. Metallographic analysis using scanning electron microscope revealed that the disc failure was by fatigue mechanism. A number of discrete fatigue cracks were found to be originated at the fillet radius between the disc diaphragm and the disc rim, aligned along the shallow machining marks in the fillet radius. Further analysis showed that the early fatigue crack initiation happened under over-speeding and consequent over-temping of the aero-engine. However the surface imperfections at the fillet radius had acted as the source of fatigue vulnerability causing the turbine disc to fail, as shown in Fig. 26.

7.3 Fatigue Failure Due to Sub-surface Metallurgy—Case Study of a Gear Failure

Bhaumik [41] gave a case study on the failure of a gear under spalling, causing continuous chip warning in flight, basically due to decarburization at the carburized and hardened surface layers of the gear. Spalling is a self-reinforcing reflexive

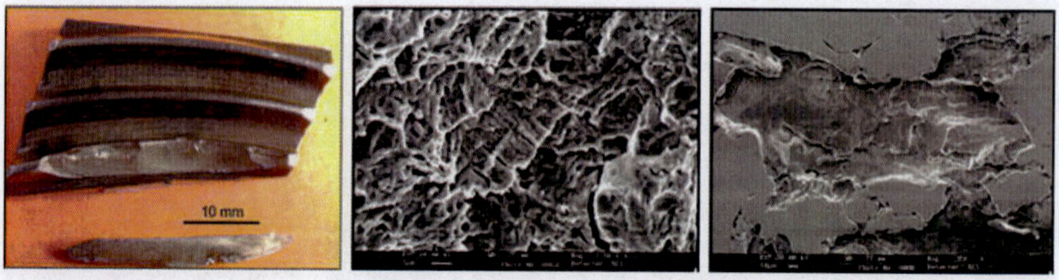

Fig. 27 Fatigue failure of a gear related to sub-surface metallurgy (*Source* [41])

phenomenon (as shown in Fig. 32, in a subsequent case study). Decarburization and associated microstructure transformation, hardness alteration and alloy depletion can happen because of process deficiencies like abusive gear grinding, and repeated hardening as in the present case.

The gear tooth fragment captured by the Electric Chip Detector of a rotorcraft caused chip warning in the cockpit. Metallurgical failure mode analysis revealed that the gear tooth had failed under fatigue. Multiple fatigue cracks had initiated at the dedendum of the gear flank, resulting in the so-called 'spalling' of the gear tooth surfaces, as shown in Fig. 27. The case depth across the tooth flank length varied widely, and the central region had the hardness profile typical of a decarburised layer, with the associated vicissitudes of microstructural transformation, recrystallisation and alloy depletion, which was established to be the root cause of the gear tooth failure.

7.4 Fatigue Failure of Due to Improper Engraving Technique—Case Study of Failure of a Large Gear

In a unique case of failure and seizure in air of a Turboprop engine, leading to single-engine landing of the twin-engine aircraft, the Bull-gear inside the Transmission Gearbox of the engine was found broken into pieces. Fractographic studies confirmed that the gear fracture occurred by fatigue mechanism. The metallurgical failure mode analysis established that the fatigue crack had initiated at the place of inscription of Gear Serial number by 'Electro-spark Etching' on the gear diaphragm surface, as shown in Fig. 28. The spark erosion at the place of etching caused localized melting, as evidenced by discolouration on the fracture surface at the crack initiation point. This not only created crevices caused by removal of material by localized meting which acted as mechanical stress-concentrators, but also introduced metallurgical abnormalities in the form of molten and re-solidified crystal structure associated with a large number of pre-existing micro-cracks.

Examination by scanning electron microscope showed distinct zones demonstrating crack propagation under fatigue loading, viz., crack-origin region, beach

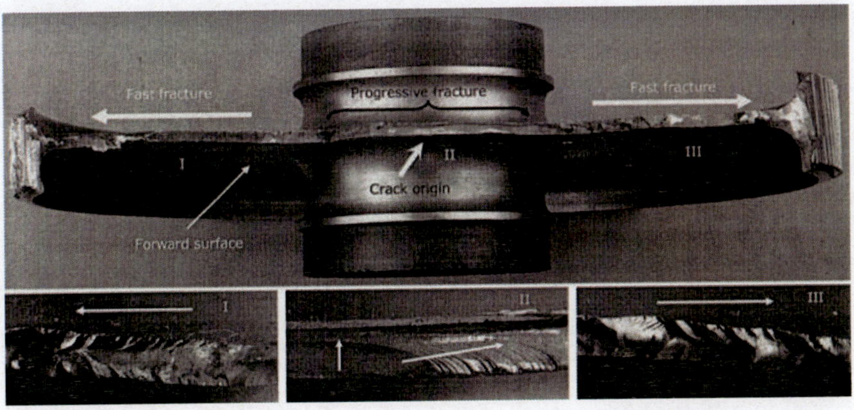

Fig. 28 Fatigue fracture failure of a large gear due to improper engraving technique

marks on either side of the crack origin, fatigue striations in the progressive crack propagation zone and dimple rupture in the fast fracture zone, as shown in Fig. 28.

Evidently, the use of an improper method of inscription on the component surface, instead of a light engraving technique like electrochemical etching, compromised on the surface integrity requirements of the rotating and load-carrying gear component, leading to the fatigue fracture failure.

7.5 Fatigue Failure Due to Pressure Pulsation—Case Study

Body leakage from a crack developed in the Port-cap was noticed in hydraulic pumps of an aircraft. The Original Equipment Manufacturer (OEM) carried out Defect Investigation, and attributed the crack to 'Fatigue failure' due to 'Pressure Pulsation' ('High pressure Impulse levels inside Control unit').

The OEM introduced an Un-Anodized Port-cap, instead of the Anodized part, to have better fatigue resistance of the surface layers of the component. This was implemented in all the new units. However, since this improvement alone could not obviate the problem, the OEM subsequently incorporated an 'Attenuator' as part of the pump to provide damping of the pressure pulsation. The modified unit was subjected to necessary Qualification Tests at the OEM facilities. As a prudent measure, the Airworthiness Certification agency restricted the life by way of reduced Time Between Overhaul (TBO). A FRAMED-IN-FRAM® Diagram of the problem is presented in Fig. 29.

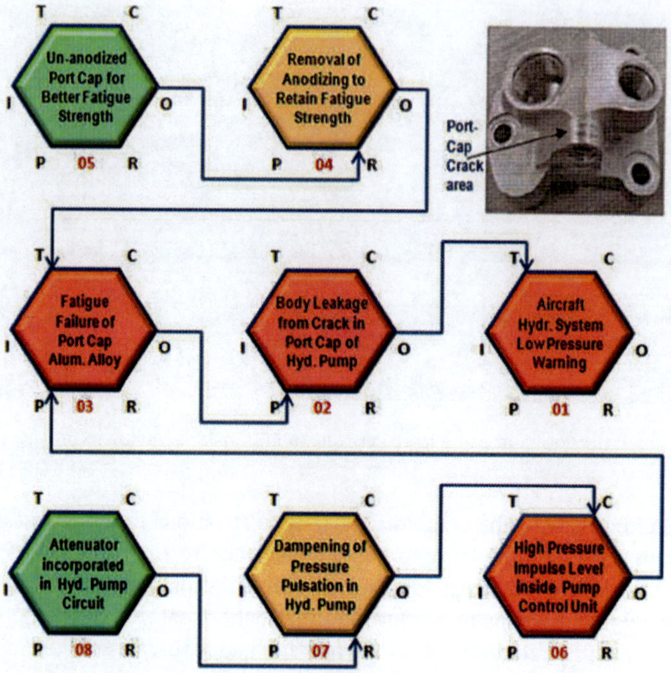

Fig. 29 'FRAMED-IN-FRAM-Diagram' on fatigue failure of hydraulic pump due to pressure pulsation

8 The Defence-in-Depth Principle (S4)

The defence-in-depth principle requires that multiple layers of defences or safety barriers are to be incorporated along the potential path of accident sequence so as to ensure safety protection [6, 12, 16, 17]. The term defence-in-depth came from the multiple defences against enemy provided by a king around his capital by means of forts and moats one after the other. Likewise, multiple safety barriers are to be in the path of accident progression, with successive defensive elements of technical, operational, organizational and systemic nature. Moreover, the defences and safety barriers should not only provide prevention, but also should block further hazardous escalation, contain the damage, and mitigate the potential consequences of the accident or safety event, as shown in Fig. 30.

A case study is now presented here to illustrate how the safety barriers provided along the path of accident causation could mitigate and eliminate the risks associated with a major safety hazard which had resulted in multiple incidents in an aero-engine.

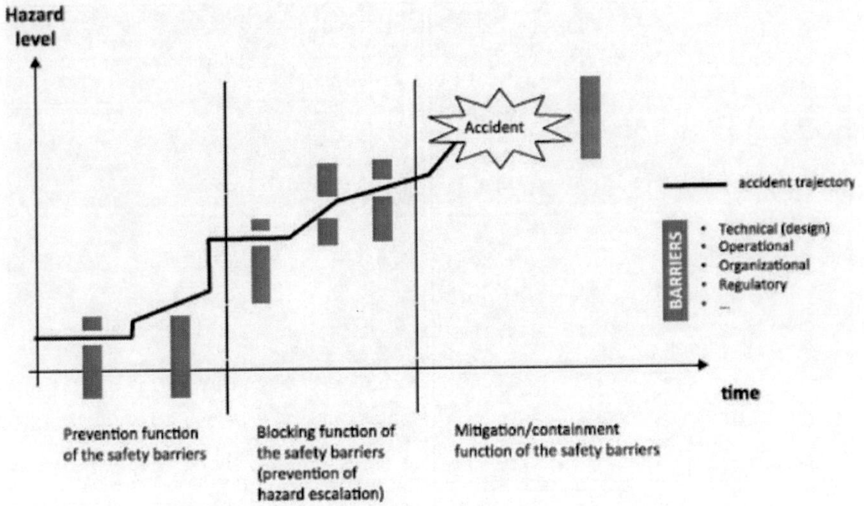

Fig. 30 Defence-in-depth principle: absence, inadequacy, or breach of safety barriers (*Source* [12])

8.1 Fatigue Failure of Compressor Casing Shim—Case Study

Background. In a legacy aero-engine, hot-air leak was reported at the interface between the 2nd stage Compressor Outlet casing and the Intermediate Casing, due to the failure of the aluminum alloy shims, interposed between the Outlet casing and Intermediate casing in order to compensate the loss of height of the casing assembly, since the casings faces are machined flat to correct the distortions after welding done for rectifying cracks.

Strip Examination. On strip examination of one of the aero-engine compressor assemblies, it was found that the Shim had got cracked and a portion of it was blown off. The Retaining nuts at the location of Shim failure were found loose (Fig. 31).

Laboratory Analysis. Investigation into the failed Shim revealed clear fatigue cracks originating from the fretted inner diameter of holes. The fretting of Shim happened against the pulled-out Studs. The failure was attributable to Pulled-out studs in the front flange of the Compressor Outlet casing. The parent material holding the stud got weakened due to repetitive thermal cycles of heat application, which was the root cause, and the stud withdrew from the tapped hole and the clamping loads on the flange were lost. When this occurred over an arc, air passing between the joint faces caused the unclamped portion of the Shim to vibrate and the part failed under High Cycle fatigue.

Remedial Actions. The following improvement actions were introduced based on the recommendations made after the investigations into the failures:

(i) Distortion on the mating faces of the Compressor casings is now being checked and maintained within 0.005″ for both repair and overhaul engines even though the overhaul limit is 0.012″.

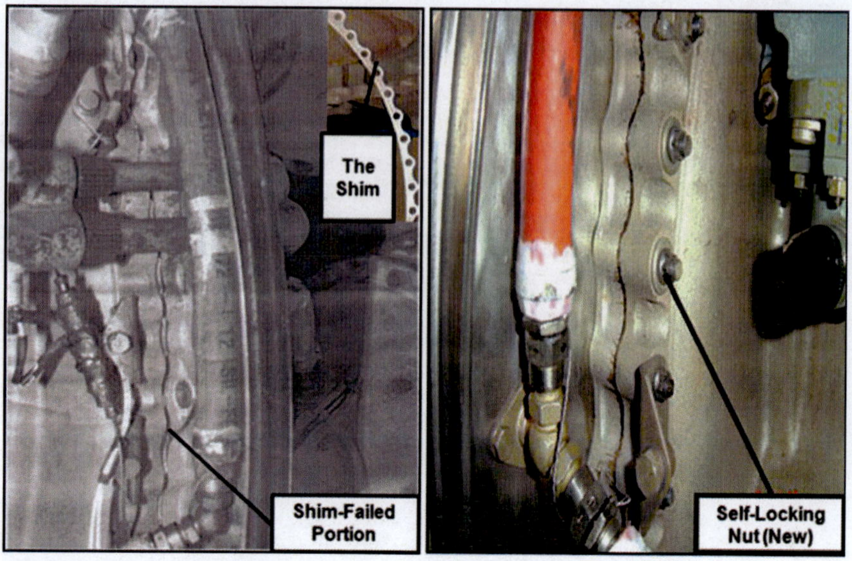

Fig. 31 Zone of fatigue failure of compressor shim, before modification (left) and after modification (right)

(ii) The Studs are checked for Stand-out and torque. Pulled-out studs are replaced with oversize studs during repair and overhaul.
(iii) To arrest the pull out of studs, the 'Loctite 640' Adhesive is applied during replacement of studs on outlet casings through a local modification.
(iv) Self-locking nuts were introduced on the Outlet casing through local modification, instead of the spring washer and hexagon nuts, to retain the torque in service, and all the engines were modified accordingly.
(v) To avoid cracking/shearing of studs on Outlet casing, additional NDT check was introduced on the Studs.
(vi) A temporary rework at Shim-failed location was introduced in service wherein the gap at the Shim-failed location is filled with 'Silcoset' material and retained with L-shaped bracket. This procedure was included in the Engine Maintenance Manual to obviate premature withdrawal of the engine due to hot-air leak.
(vii) A better material used in new aero-engines was introduced through indigenization route.

8.2 Depiction of Accident Sequence and Safety Barriers Using 'FRAMED-IN-FRAM® Diagram'

The FRAMED-IN-FRAM® Diagram given in Fig. 32 shows how the path of Accident sequence (viz., the Initiating Event, the Emerging Event, the Causes, the Symptoms,

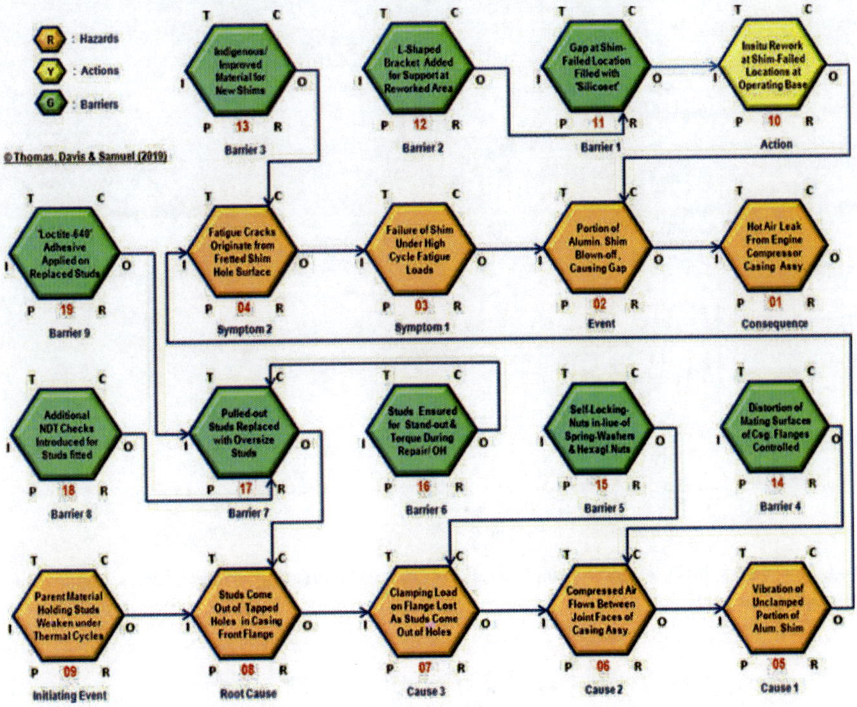

Fig. 32 'FRAMED-IN-FRAM® Diagram' showing accident sequence and safety barriers

the Event and the Consequence), and the Defences or Safety Barriers to prevent propagation of the accident sequence, can be depicted using FRAM Diagram as a new Quality tool. This methodology can be used not only to illustrate the whole sequence of the safety event and the remedial actions and the safety barriers, but also for expanding the horizon of thinking and analysis, to effectively aid in defect investigation and failure analysis.

9 The Observability-in-Depth Principle (S5)

The observability-in-depth principle requires that various features be put in place to observe and monitor for the system state and breaches of any safety barrier, and reliably provide this feedback to the operators, so that all safety-degrading events or states (that the safety barriers are meant to protect against) are observable [6, 12, 17]. Observability-in-depth is characterized by the set of provisions, both technical and operational, designed to enable the scanning and identification of emerging hazardous conditions and accident pathogens and monitoring for hazard escalation and advancement of accident sequences (Fig. 33).

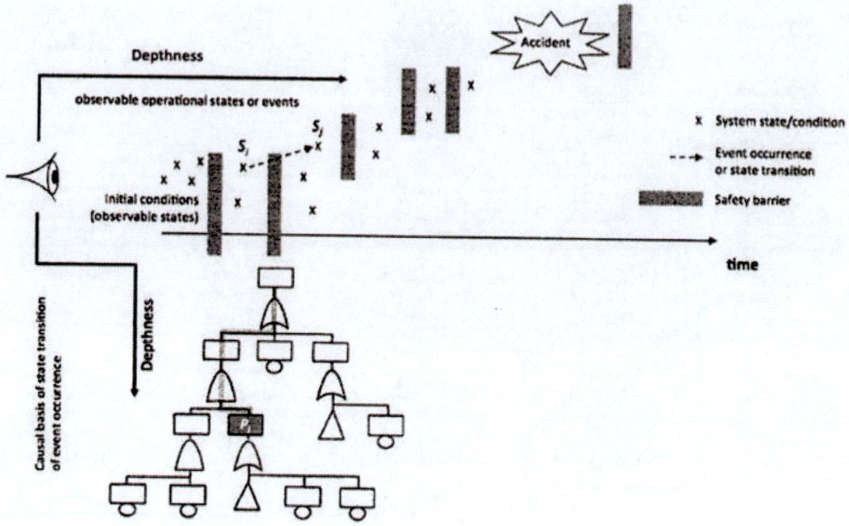

Fig. 33 Observability-in-depth principle: ability to observe breach of safety barriers, and identify accident pathogens (*Source* [12])

9.1 Observability-in-Depth Principle—Case Study on a Safety Event Due to Loss of Spline Drive

Loss of drive between the Internal Gearbox and the External Gearbox of the Turbofan engine of a single-engine trainer aircraft caused engine flame-out and consequent aircraft accident. The Loss of drive was found to be due to failure of the Spline coupling between the starter Drive Shaft and Input Bevel Gear, respectively, of the two Gearboxes. Metallurgical Failure Mode Analysis in the laboratory revealed that the Spline coupling failed because of 'Fretting Wear mechanism'.

There had been a diagnostic mechanism for early detection of wear-out of engine components like gears, splines, bearings, etc. by way of monitoring the count of debris collected at every 25 h in the three Mechanical Chip Detectors (MCDs) mounted in the return lines of the Lubricating channels of the three aero-engine modules, viz., Internal Gearbox, Turbine Bearing Housing and the External Gearbox. As per the methodology followed, the count of Debris collected in each MCD is recorded in the form of a cumulative debris accumulation graph. After an initial period of increased debris count, the debris count stabilizes, whereby the graph becomes a straight line with a slight upward slope, called the Characteristic accumulation gradient, as shown in Fig. 34. If there is a sudden trending up of the cumulative debris graph, that is indicative of an increase in debris generation in the oil system of that particular module, presumably due to deterioration of one of the internal components like splines, gears or bearings along the oil-wetted lubricating-oil channel.

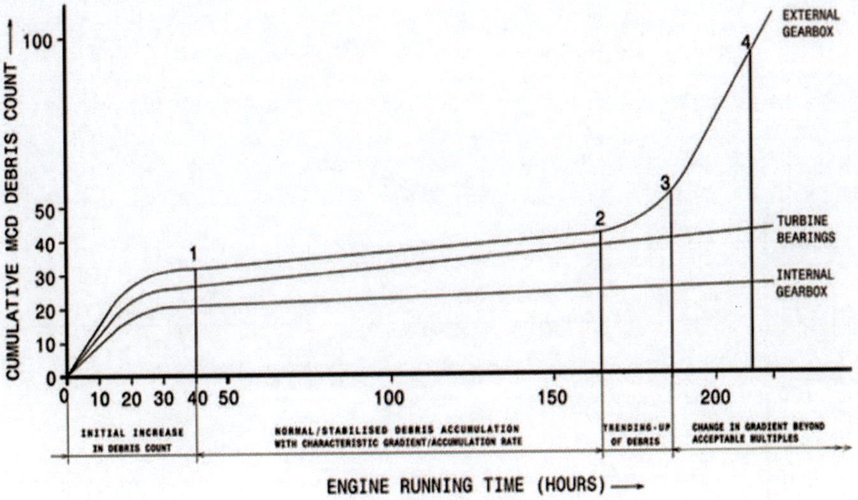

Fig. 34 Illustration of MCD debris monitoring methodology

In case of a trending up of cumulative debris count, as per the MCD Debris monitoring methodology, the MCD Debris Inspection frequency is to be reduced to 10 h from 25 h if the change in graph gradient is 05 times; to 05 h if the gradient change is 10 times; to 02 h if the gradient change is 10 times; until the debris catchment/accumulation rate reverts to the characteristic value. If the change in gradient becomes greater than 15 times, the Debris is to be subjected to microscopic examination for material characterization to decipher the source of debris generation. It is allowed to carry out flushing of the closed-loop engine oil system only once, with subsequent monitoring of MCD Debris accumulation, but the process of Debris monitoring had to be continued. If the trending up of MCD Debris count continues, based on the result of the analysis, the aero-engine is to be withdrawn from service.

In the present case, the aircraft had experienced a bird hit at 670 h of engine operation, after which the Debris accumulation in the MCD attached to the Oil return line of the External Gearbox started trending up. The Inspection frequency was reduced from 25 to 10 to 5 to 2 h, and the trending up persisted. However, when the change in the graph gradient became greater than 15 times, the Engine oil system was flushed, and the Debris monitoring continued. However, instead of continuing the cumulative debris count graph from the place where it was at the time of flushing, the plot was re-started from 'Zero'. Further increase in cumulative debris count was considered to be similar to the initial trending up normally expected at the beginning of engine exploitation, and hence the engine withdrawal as per procedure was not carried out, and improper plotting and wrong monitoring of debris count eventually led to the safety event at 960 h, as shown in Fig. 35.

The FRAMED-IN-FRAM® Diagram for the case is shown in Fig. 36. The diagram depicts the manifestation of the safety event from the instant when the fretting wear

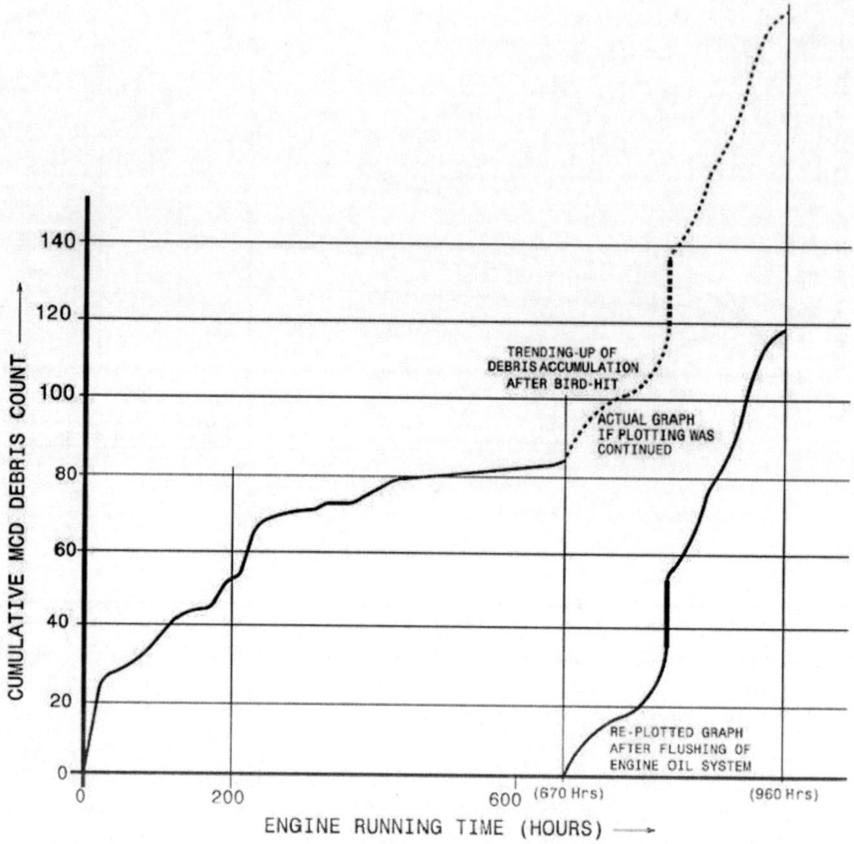

Fig. 35 Non-capturing of trending up of MCD debris accumulation

of the Drive shaft splines started. The generated wear debris got accumulated in the splines, and also got released into the Engine oil system. The debris accumulated in the splines acted as abrasive for aggravation of the fretting wear. The 'Reflexive Resonance Response' here is shown in the form of a cyclic indication with a negative sign at the bottom portion of the diagram.

In the FRAMED-IN-FRAM® Diagram, the two alternate scenarios of the Debris collected at the MCDs (i) not being monitored, and (ii) being monitored, are shown. Non-capturing of the Debris count increase caused further continuation of the wear-phenomenon with additional Negative Reflexive Resonance Response, causing complete wear-out of the splines leading to the catastrophic consequence of 'Complete Loss of Drive' leading to a Class-I failure. Had the MCD data been monitored properly as per the technological procedure, the increase in debris count could have been captured, the periodic monitoring further tightened, and the engine would have been withdrawn in time, in which case the consequence could have been only 'Spline deterioration', a Class-IV failure.

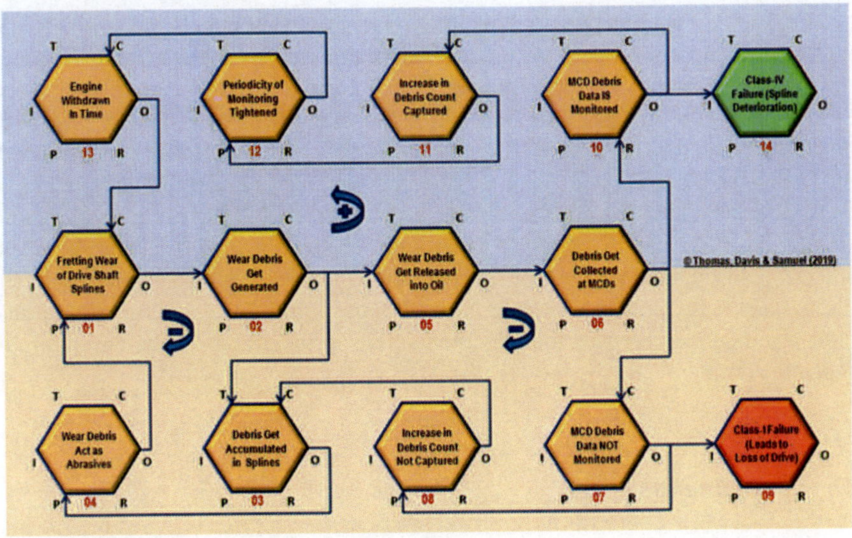

Fig. 36 ‘FRAMED-IN-FRAM® Diagram’ showing negative and positive reflexive response with different failure scenario

10 The Integration-in-Totality Principle (S6) and the Human Factors Principle (S7)

The authors would like to call the afore-discussed system safety principles propounded by J. H. Saleh et al. as the ‘Basic Five’. They are basically dealing only with the physical root causes, and are built on the notion of hazard level and its escalation and the accident sequence.

A safety event, an accident or an incident, can be caused mostly by a combination of multiple causes, which are of three main types of root causes that build upon each other:

(i) Physical root causes: The actual physical mechanism of the failure (for example, manifestation of high/low cycle fatigue as the failure mode.
(ii) Human root causes: The human practices that allowed the physical root causes to exist.
(iii) Systemic/Latent root causes: The way the facility/system is designed and/or managed creating the human roots.

Because the physical root cause leads to the human and systemic/latent root causes, or vice versa, it is necessary to understand the physical roots of a failure to find out the multiple human and latent aspects. To that extent, they are very important from the perspective of system safety analysis [41–43].

In addition to the ‘Basic Five’ system safety principles, the authors would like to add two more system safety principles related to the Systemic and Human root causes of safety events, viz., the ‘Integration-in-Totality principle’ related to the systemic aspects and the 'Human Factors Principle' related to the human aspects, which can be called as the 'Additional Two' system safety principles.

The 'Integration-in-Totality principle' is based on the 'D-C-P Diagram' suggested by Stoop, the 'System-Theoretic Accident Model and Processes (STAMP)' proposed by Leveson and the 'Functional Resonance Analysis Method (FRAM)' propounded by Hollnagel. The 'Human Factors Principle' has been universally accepted as a cardinal principle of System Safety; and International Civil Aviation Organization (ICAO) has mandated that all the organizations engaged in Civil Aviation activities should adopt and train their personnel in the Human Factors Principle [18]. However, discussions in the present article are limited to the 'Basic Five' system safety principles, and deliberations on the 'Additional Two' principles are to be covered in a separate article.

11 Conclusions

This technical article shows how fatigue durability is related to the four most important aspects in the field of aeronautics and aviation, viz., Quality, Reliability, Risk and Safety, which together form a cyclic paradigm. The paper gives a broad view of the 'basic five' system safety principles propounded by J. H. Saleh et al., and also highlights how fatigue fracture failures of aeronautical components can be analyzed in light of these cardinal principles, with the help of a few case studies. The concept of 'Reflexive Resonance Response' has been introduced by the authors to demonstrate the self-reinforcing nature of manifestation of fatigue vulnerability in a cyclic manner leading to safety events, which is shown using the new Quality tool of 'FRAMED-IN-FRAM®', a novel version of the FRAM Diagram, with a few examples from the presented case studies.

Acknowledgements The authors would like to thank the management of HAL for the permission to present this technical article, and express their deep sense of gratitude to all the colleagues who have associated directly or indirectly towards the creation of knowledge in the field.

References

1. Hollnagel E (2011) Introduction to FRAM: the functional resonance analysis method. University of S, Denmark
2. Thomas J et al (2018) Aerospace organisational excellence—quality system standards and global best practices. In: Proceedings of IAF-CSDO golden jubilee seminar on "excellence through maintainability in aviation", 13–14 Dec 2018, Bengaluru
3. Juran JM, Godfrey AB (1999) Juran's quality handbook, 5th edn. McGraw-Hill, pp 2.1–2.2. ISBN 0-07-034003-X
4. Garvin DA (1988) Managing quality: the strategic and competitive edge. Harvard Business School, The Free Press. ISBN 0-02-911380-6

5. Leveson N (2011) Engineering a safer world: systems thinking applied to safety. The MIT Press. ISBN: 978-0-262-01662–9
6. Gnoni MG, Saleh JH (2017) Near-miss management systems and observability-in-depth: handling safety incidents and accident precursors in light of Safety principles. Safety Sci 91:54–167
7. NASA: Accident precursor analysis handbook. NASA/SP-2011-3423, Washington, DC (2011)
8. Bakolas E, Saleh JH (2011) Augmenting defense-in-depth with the concepts of observability and diagnosability from control theory and discrete event systems. Reliab Eng Syst Safe 96:184–193
9. International Civil Aviation Organisation: "Safety Management Manual (SMM)". ICAO Doc. 9854, 3rd edn (2013)
10. Leveson NG (2011) Applying systems thinking to analyze and learn from events. Saf Sci 49:55–64
11. Thomas J, Davis A, Samuel MP (2019) Strategic quality management of aero gas turbine engines, applying functional resonance analysis method. In: Proceedings of national aerospace propulsion conference (NAPC-2018). IIT-Kharagpur, India, 17–19 Dec 2018, under publication by Springer
12. Saleh JH, Marais KB, Favaro FM (2014) System safety principles: a multidisciplinary engineering perspective. J Loss Prevent Process Indust 29:283–294
13. Saleh JH, Marais KB, Bakolas E, Cowlagi RV (2010) Highlights from the literature on accident causation and system safety: review of major ideas, recent contributions, and challenges. Reliab Eng Syst Safe 95(11):1105–1116
14. Favaro FM, Saleh JH (2016) Toward risk assessment 2.0: safety supervisory control and model-based hazard monitoring for risk-informed safety interventions. Reliabi Eng Syst Safe 152:316–330
15. Saleh JH, Geng F, Ku M et al (2017) Electric propulsion reliability: statistical analysis of on-orbit anomalies and comparative analysis of electric versus chemical propulsion failure rates. Acta Astronaut 139:141–156
16. Cowlagi RV, Saleh JH (2013) Co-ordinability and consistency in accident causation and prevention: formal system theoretic concepts for safety in multilevel systems. Risk Anal 33(3):420–433
17. Favaro FM, Saleh JH (2014) Observability-in-depth: an essential complement to the defence-in-depth safety strategy in the nuclear industry. Nuclear Eng Technol 46(6):1–14
18. Shanmugam A, Robert TP (2015) Human factors engineering in aircraft maintenance: a review. J Qual Mainten Eng 21(4):478–505
19. Shanmugam A, Robert TP (2015) Ranking of aircraft maintenance organization based on human factor performance. Comput Ind Eng 88:410–416
20. Stoop J, deKroes J, Hale A (2017) Safety science, a founding fathers' retrospection. Safe Sci 94:103–115
21. Stoop JA, Dechy N, Dien Y, Tulonen T (2016) Past and future in accident prevention and learning: single case or big data? In: Proceedings of ESReDA 50th seminar, Sevilla, Spain, 18–19th May 2016
22. Stoop JA (2016) Safety: a system state or property? J Safe Stud 2(2). ISSN 2377-3219
23. Leveson NG (2014) A system-theoretic, control-inspired view and approach to process safety. AiChE J 60(1) MIT Open Access Articles
24. Hollnagel E (2015) Modelling transport systems with fram: flows or functions? Safety synthesis, information on FRAM. www.functionalresonance.com
25. Hollnagel E (2015) Some myths about industrial safety. Safety synthesis
26. Hollnagel E (2015) Safety-I and safety-II: safety analysis and safety synthesis. Safety synthesis
27. Hollnagel E (2014) The fundamentals of resilient organisational performance. Safety synthesis
28. Gnoni MG, Andriulo S, Maagio G, Nardone P (2013) Lean occupational safety: an application for a near-miss management system design. Safe Sci 53:96–104
29. Teixeira GM (2017) Fatigue of metals: failure and success. Research and development. Dassault Systemes Simulia, Sheffield, UK, Accessed through Researchgate

30. Dang Van K, Papadopoulos IV (1999) High cycle metal fatigue—from theory to applications. Springer
31. Vijayaraju K (2018) Integrated vehicle health management system to improve maintainability. In: Proceedings of IAF-CSDO golden jubilee seminar on "excellence through maintainability in aviation", Bangalore, India, 13–14 Dec 2018
32. Taleb NN (2010) The black swan: the impact of the highly improbable. Penguin Books. ISBN: 978-0-141-03459-1
33. Doorn N, Hansson SO (2011) Should probabilistic design replace safety factors? Philoso Technol 24:151–168
34. John SK, Shetty PB, Mishra RK (2018) Reliability improvement in aero engines during the infant mortality period of operations. J Fail Anal Prevent Springer
35. Jeffcott HH (1919) The lateral vibration of loaded shafts in the neighbourhood of a whirling speed—the effect of want of balance. London, Edinburgh Dublin Philoso Mag J Sci 37(219):304–314
36. Jahromi AF, Bhat RB, Xie W-F (2015) Forward and backward whirling of a rotor with gyroscopic effect. In: Sinha JK (ed) Vibration engineering and technology of machinery, mechanisms and machine science, vol 23, pp 879–887
37. Millsaps KT, Vejvoda CE (1996) Origin of split resonance and backward whirl in a simple rotor. In: Proceedings, international gas turbine and aeroengine congress & exhibition, Birmingham, UK, June 1996, pp 1–6
38. Thomas J (1993) Design-related manufacturing aspects of MiG series aeroengine compressor components. In: Compendium of technical papers, Vol. I, workshop on "MiG engine experience for Kaveri project, pp 105–115
39. Lundberg J, Rollenhagen C, Hollnagel E (2009) What-you-look-for-is-what-you-find—consequences of underlying accident models. Safe Sci. https://doi.org/10.1016/j.ssci.2009.01.004 2009
40. Sujata M et al (2010) Identification of failure mechanisms in nickel base superalloy turbine blades through microstructural study. Eng Fail Anal 17:1436–1446
41. Bhaumik SK (2009) A view on the general practice in engineering failure analysis. J Fail Anal Prev 9:185–192
42. Dennies DP (2005) How to organize and run a failure investigation. ASM Int. ISBN: 0-87170-811-6
43. Sachs NW, Beckman M(ed) (2019) Figuring out why things breakdown. Tribol Lubricat Technol 38–45

Stress Intensity Factor and Damage Index of Reinforced Concrete Beam

Kiran M. Malipatil and S. V. Itti

Abstract Crack is an unavoidable flaw in structures with time, this crack increases and leads to failure of the structure. In this paper, numerical analysis was performed on RCC beam having a particular depth of crack for assessing the damage in the beam. Width as a parameter was used to check the variation in the damage. Variations in Stress Intensity Factor (SIF) for different beam widths were evaluated for opening and closing forces. The force transmitted by the reinforcement bars to the concrete element with the change in cover was evaluated. Additionally analytical studies were carried out on RCC beam strengthened by Fiber Reinforced Polymer (FRP) using ANSYS software. The effect of width of the beam and thickness of FRP laminate were the parameters used to study the behavior of the beam. From the study, it can be concluded that the geometry of the member plays a significant role in deciding the crack properties.

Keywords Stress intensity factor · Damage index · RCC beam · ANSYS · Fracture mechanics · FRP

1 Introduction

Fracture mechanics is the study related to propagation of cracks. Crack in a concrete structure is present in the microform due to shrinkage. This crack on application of load beyond the range may propagate. In some cases even if the load is well below the range, propagation of crack occurs; this is due to fatigue loading. There are various properties related to cracks one among them is stress intensity factor. The stress stage at the tip of the crack is termed to be Stress Intensity Factor (SIF).

K. M. Malipatil (✉)
Department of Civil Engineering, KLE Dr. M. S. Sheshgiri College of Engineering and Technology, Belgaum, India
e-mail: kiranmp@klescet.ac.in

S. V. Itti
Department of Civil Engineering, S. G. Balekundri Institute of Technology College, Belgaum, India
e-mail: ittisv@gmail.com

S. Seetharamu et al. (eds.), *Fatigue, Durability, and Fracture Mechanics*, Lecture Notes in Mechanical Engineering, https://doi.org/10.1007/978-981-15-4779-9_19

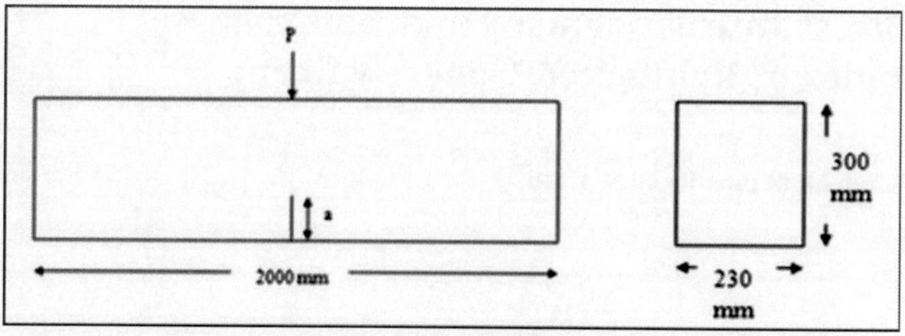

Fig. 1 Reinforced concrete beam with through-thickness crack

SIF related to critical crack depth is termed to be fracture toughness. In order to know whether the crack would propagate and to determine the remaining life of the structure SIF and fracture toughness values play a vital role. The objective of the paper is to numerically evaluate the stress intensity factor for a reinforced concrete beam and to obtain its variation with respect to width of the beam. Study also covers the effect of cover on the crack. To analyze basic three modes of crack propagation are considered. Mode I and Mode II stress intensity factors are considered [1].

A through-thickness edge crack at the tension region of a reinforced concrete beam as shown from Fig. 1 was considered. The sagging moment created by the load produces a tensile force at the bottom of the beam resulting in Mode I crack propagation. The stress intensity due to this moment given by Carpinteri is stated in Eq. 1 [2].

$$K_I = \frac{M}{d^{\frac{3}{2}}B}6(1.99\xi^{1/2} - 2.47\xi^{3/2} + 12.97\xi^{5/2} - 23.17\xi^{7/2} + 24.80\xi^{9/2}) \quad (1)$$

where

$M =$ external moment in N mm,
$d =$ depth in mm,
$B =$ width of the beam in mm.
$\xi =$ relative crack depth (a/d)

Study was carried out to determine the variation of stress intensity factor with respect to the width of the beam. Variation on SIF was obtained for different relative crack depths with different widths of the member given by Fig. 2.

From the figure it is observed that with increase in crack depth SIF values increase positively. For crack depth lesser than the half depth of the beam the curve is approximately horizontal, this indicates there will be no propagation on cracks further. After an increase in the crack depth beyond half of the beam the SIF values increase rapidly. With increase in the width of the member Stress Intensity factor decreases. It could be observed from the figure that increasing the width flattens the curve hence crack propagation could be made stable by increasing the width.

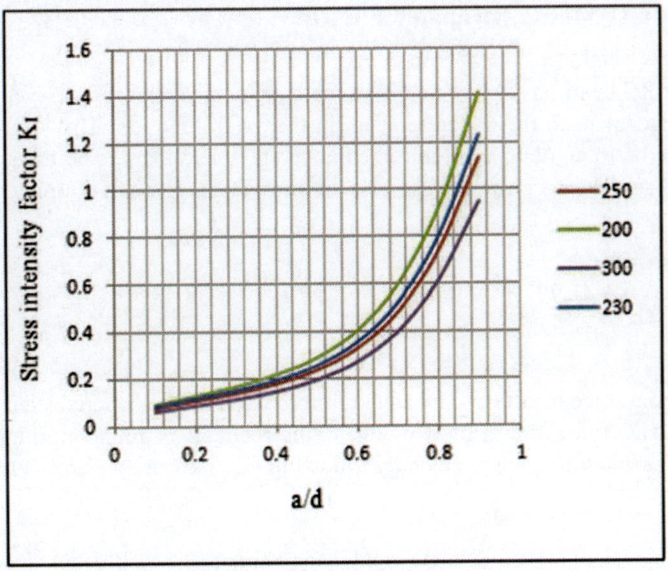

Fig. 2 Mode I stress intensity factor for varying width

Stress Intensity factor obtained found good agreement with the values given in the literature Alberto Carpinteri [2, 3, 4] (Fig. 3).

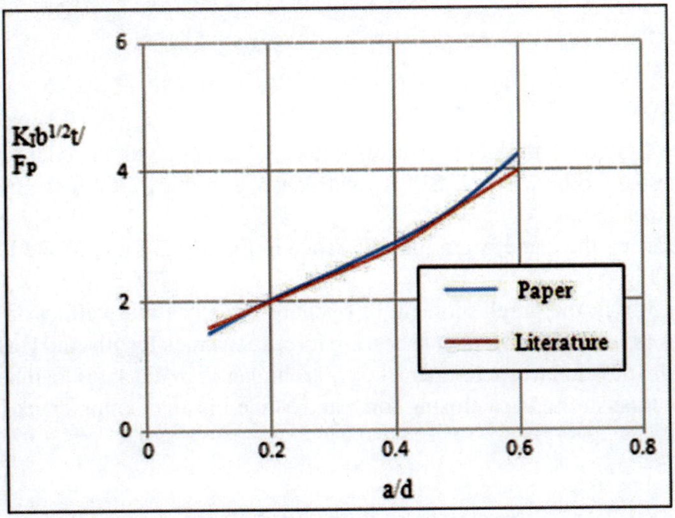

Fig. 3 Validation of SIF

2 Mode II Stress Intensity Factor

When an RC beam is subject to a load there exists a virtual force developed from the reinforcement to the concrete beam in the axial direction. This forms a major difference between plain and reinforced concrete where there is no transfer of load from the bars. This force produces stress intensity factor given by Carpinteri as Eq. 2. [2, 3, 4]

$$K_{Ip} = \frac{F}{d^{\frac{1}{2}}B}(1.99\xi^{1/2} - 0.41\xi^{3/2} + 18.70\xi^{5/2} - 38.48\xi^{7/2} + 53.85\xi^{9/2}) \quad (2)$$

where

F = axial force transferred by steel reinforcement to the concrete element.

This axial force transmitted by the reinforcement is found through rotation congruence condition and is given by following Eq. 3 given by Carpinteri

$$\frac{Fd}{M} = \frac{1}{\left(\frac{1}{2} - \frac{h}{d}\right) + r(\xi)} \quad (3)$$

In which h = cover provided for the reinforcement.

$$r(\xi) = \frac{\int_0^{\xi} Y_{M(\xi)} Y_F(\xi) d\xi}{\int_0^{2} Y_M^2(\xi) d\xi}$$

$$Y_M(\xi) = 6(1.99\xi^{\frac{1}{2}} - 2.47\xi^{\frac{3}{2}} + 12.97\xi^{\frac{5}{2}} - 23.17\xi^{\frac{7}{2}} + 24.80\xi^{9/2}) \quad (4)$$

$$Y_F(\xi) = \left(1.99\xi^{\frac{1}{2}} - 0.41\xi^{\frac{3}{2}} + 18.70\xi^{\frac{5}{2}} - 38.48\xi^{\frac{7}{2}} + 53.85\xi^{\frac{9}{2}}\right) \quad (5)$$

Figure 4 gives the variation of stress intensity factor due to the axial force which may be termed to be Mode II SIF K_{II} with change in the width of the member for varying crack depth.

Considering the same beam the difference in the Mode I and Mode II SIF is as plotted in Fig. 5.

As seen from the graph moment or opening force has more effect on the crack properties as compared to axial or shearing force. At a crack length equal to half depth of the beam SIF due to opening crack is 50% higher than the axial load because the axial force transmitted is a closing force and Mode I is an opening crack force.

3 Variation of Force Induced in Steel Bar

The axial force transmitted by the reinforcement bars to the concrete element as given by Eq. 3 is an important parameter related to stability of the crack. Figure 6

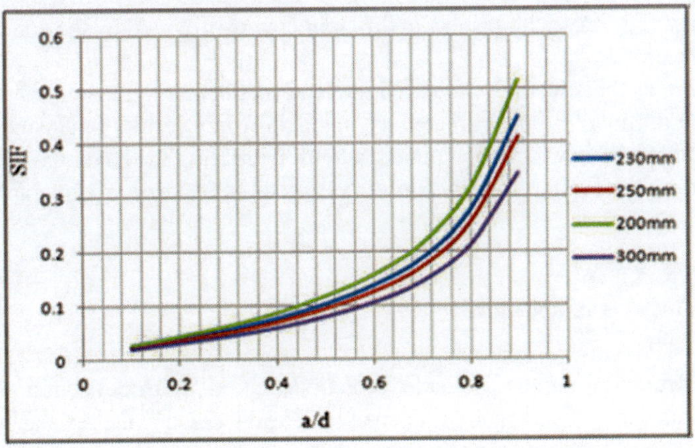

Fig. 4 Mode II SIF with width change

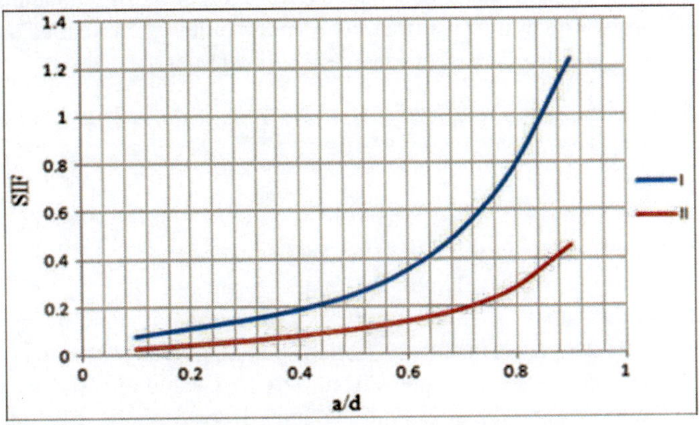

Fig. 5 Difference between Mode I and Axial SIF

Fig. 6 Reaction from reinforcement with cover change

states this axial force with respect to different covers provided by the concrete to the reinforcement.

As seen, as the cover has increased the load that is being transmitted to concrete is increased. Since moment created by this axial force results in closing of crack, from obtained value we can conclude that with increasing the cover the moment due to axial force gets reduced. Hence lower values of axial force are favorable.

4 Damage Assessment

When a structural member is loaded it gets deflected. With deflection as the load increases its resistance capacity is decreased. The decrease in the resistance is due to the formation of crack and propagation of the crack. Hence when a structural member is to be repaired the percentage of damage in that member has to be evaluated. This assessment provides what the residual strength is remaining in the member. In this paper, the damage is quantified using a fraction number less than 1 called the Damage Index. Damage index is given by the formula below [5–8]

$$D = 1 - \frac{\lambda_{\text{mind}}}{\lambda_{\text{minu}}} \tag{6}$$

where

λ_{mind} = minimum eigen value of damaged beam and

λ_{minu} = minimum eigen value of undamaged beam.

Hence Damage index is the measure of stiffness of damaged beam to that of undamaged beam. Damaged beam is a cracked beam. Undamaged beam is uncracked beam. Damage index (D) value ranges from 0 to 1. A value of 0 for damage index corresponds to there is no damage. And D equals 1 corresponds to critical damage of the member.

Stiffness matrix of undamaged RCC beam

Stiffness of undamaged beam depends on various parameters like area of concrete, area of steel Young's modulus and the geometry of the beam. Stiffness matrix as per Chajes (1974) is given by [9] (Table 1)

Table 1 Details of RCC beam

Parameter	Values
Steel area (As, mm^2)	201.06
Es (N/mm^2)	200,000
Ec (N/mm^2)	25,000
a (mm)	90

$$[k]_{\text{beam}} = [k]_{\text{std}} - F[k]_{\text{axial}} \tag{7}$$

$$[\text{k}]_{\text{std}} = \begin{bmatrix} (\frac{\text{AcEc}}{1} + \frac{\text{AsEs}}{1}) & 0 & 0 & -(\frac{\text{AcEc}}{1} + \frac{\text{AsEs}}{1}) & 0 & 0 \\ 0 & \frac{12\text{EcIc}}{1^3} & \frac{-6\text{EcIc}}{1^2} & 0 & \frac{-12\text{EcIc}}{1^3} & \frac{-6\text{EcIc}}{1^2} \\ 0 & \frac{-6\text{EcIc}}{1^2} & \frac{4\text{EcIc}}{1} & 0 & \frac{6\text{EcIc}}{1^2} & \frac{2\text{EcIc}}{1} \\ -(\frac{\text{AcEc}}{1} + \frac{\text{AsEs}}{1}) & 0 & 0 & (\frac{\text{AcEc}}{1} + \frac{\text{AsEs}}{1}) & 0 & 0 \\ 0 & -\frac{12\text{EcIc}}{1^3} & \frac{6\text{EcIc}}{1^2} & 0 & \frac{12\text{EcIc}}{1^3} & \frac{-6\text{EcIc}}{1^2} \\ 0 & \frac{-6\text{EcIc}}{1^2} & \frac{2\text{EcIc}}{1} & 0 & \frac{6\text{EcIc}}{1^2} & \frac{4\text{EcIc}}{1} \end{bmatrix}$$

In the above equation, F is the axial force induced in steel rod. It is an eccentric axial force that is transmitted by reinforcement to concrete beam. The value is given by Carpinteri Eq. 3.

$$[k]_{\text{axial}} = \begin{bmatrix} 0 & 0 & 0 & 0 & 0 & 0 \\ 0 & \frac{6}{5l} & \frac{-1}{10} & 0 & \frac{6}{5l} & \frac{-1}{10} \\ 0 & \frac{-1}{10} & \frac{2l}{15} & 0 & \frac{1}{10} & \frac{-l}{30} \\ 0 & 0 & 0 & 0 & 0 & 0 \\ 0 & \frac{6}{5l} & \frac{1}{10} & 0 & \frac{6}{5l} & \frac{1}{10} \\ 0 & \frac{-1}{10} & \frac{-l}{30} & 0 & \frac{1}{10} & \frac{2l}{15} \end{bmatrix}$$

Stiffness matrix of cracked beam [10]

$$[k]_{\text{crack}} = \begin{bmatrix} \frac{\lambda_{mm}}{C} & 0 & \frac{-\lambda_{mf}}{C} & \frac{-\lambda_{mm}}{C} & 0 & \frac{\lambda_{mf}}{C} \\ 0 & \frac{1}{\lambda_{vv}} & 0 & 0 & \frac{-1}{\lambda_{vv}} & 0 \\ \frac{-\lambda_{mf}}{C} & 0 & \frac{\lambda_{ff}}{C} & \frac{\lambda_{mf}}{C} & 0 & \frac{-\lambda_{ff}}{C} \\ \frac{-\lambda_{mm}}{C} & 0 & \frac{\lambda_{mf}}{C} & \frac{\lambda_{mm}}{C} & 0 & \frac{-\lambda_{mf}}{C} \\ 0 & \frac{-1}{\lambda_{vv}} & 0 & 0 & \frac{1}{\lambda_{vv}} & 0 \\ \frac{\lambda_{mf}}{C} & 0 & \frac{-\lambda_{ff}}{C} & \frac{-\lambda_{mf}}{C} & 0 & \frac{\lambda_{ff}}{C} \end{bmatrix}$$

λ_{mm}, λ_{mp}, and λ_{pp} are the compliances as given below

$$\lambda_{mm} = \frac{2B(1-\gamma^2)}{E} \int_0^A (\frac{K_{IM}}{M})^2 \text{d}A$$

$$\lambda_{mf} = \frac{2B(1-\gamma^2)}{E} \int_0^A \left(\frac{K_{IM}}{M}\right)(\frac{K_{IF}}{F}) \text{d}A$$

$$\lambda_{ff} = \frac{2B(1-\gamma^2)}{E} \int_0^A (\frac{K_{IP}}{P})^2 \text{d}A$$

$$\lambda_{vv} = \frac{2B(1-\gamma^2)}{E} \int_0^A (\frac{K_{IIv}}{V})^2 \text{d}A$$

where A is the area of crack and dA is incremental area which is d X da.

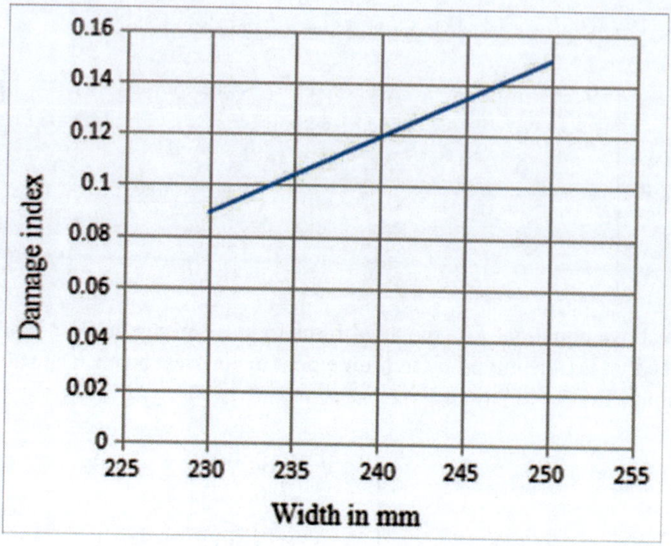

Fig. 7 Effect of width on damage index

$$C = \lambda_{mm}\lambda_{ff} - \lambda_{mp}^2$$

Study was carried on the beam as shown in Fig. 1 to determine the Damage Index with variation in the width of the member. Figure 7 gives the results of the study for two different widths in a form of graph.

As seen from the graph with increase in the width of the member damage index has been increased. The reason for increase in the index is due to increase in the crack area. For 230 mm width of the member, the damage is around 9%. With increases in the width, the damage increased to 15%. Hence damage index is directly proportional to the width of the member.

5 Strengthening of Reinforced Concrete Beam

Every structure is designed for a design period, i.e., called the service life of the structure and with certain factor of safety. During the service period the resistance of the structure that's the load-carrying capacity of the structure is reduced thereby the remaining life or residual strength is decreased. In order to extend the life of the structure FRP like composite was introduced. FRP is a material in which fibers are embedded in polymer matrix.

Many investigations have been carried on strengthening of beam using FRP. Thermal, torsion, shear behavior of the material has been investigated. Along with this test fatigue analysis is also carried out. Fatigue behavior on FRP has shown that

Table 2 Numerical validation RCC [12]

Load required for first crack to appear (KN)	
ANSYS	45.89
Numerical	41.52

the behavior is same as RC beam. As per the literature it is found out that FRP is more effective on damaged beam than on undamaged beam. FRP laminates are in form of mat. They can be single, double woven wrap. The applications of FRP laminate are mainly in the tension region of the beam. In this paper, FRP laminate is applied in the tension region, i.e., at the bottom as bottom single strip. The main concern regarding FRP is debonding of the laminate. Hence proper care must be taken in this direction [11].

Different models are used for the study. RCC beam (Fig. 1) strengthen by 25 mm FRP laminate, and RCC beam strengthen by 25 mm FRP laminate with change in the width.

5.1 Structural Modeling and Analysis

The specimen to be analyzed was modeled in ANSYS 14. Nonlinear analysis was carried out. In order to verify the results from the model from the ANSYS it is necessary to validate the results. Validation can be done numerically and also experimentally cross verifying the results. In this paper validation is conducted both numerically with respect to load required for First crack to appear and with the experimental results given in literature [11] (Table 2).

Experimental results from literature for both RCC and FRP beam as given in Figs. 8 and 9 along with experimental results, the RCC model is validated with Load required for first crack to appear. Variation in results is about 9.5%.

6 Effect of Width on Behavior of FRP Beam

One beam model with width 250 was modeled to compare with the existing results of model having 230 mm width. The variation in the behavior was captured with the increase in the width of the beam. Figure 10 shows the comparison between the two models. As seen from the graph deflections for the beam having 250 mm width have been decreased.

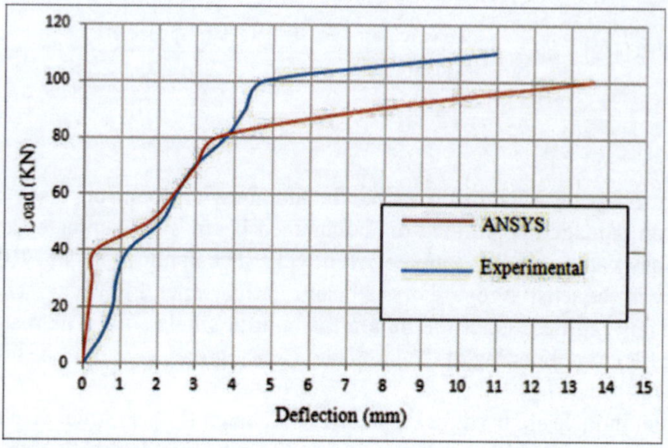

Fig. 8 Validation for RCC

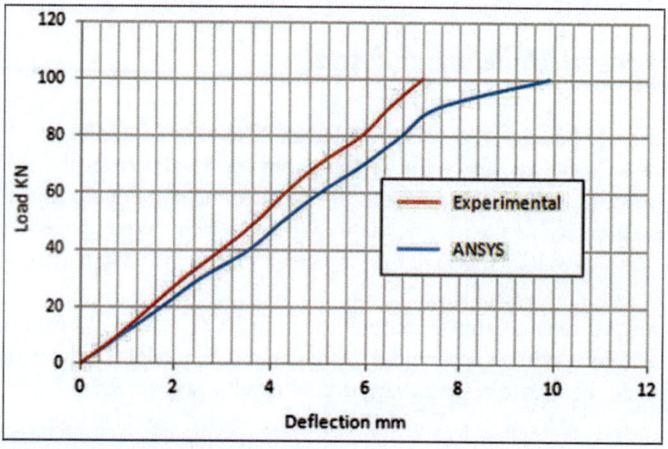

Fig. 9 RCC beam strengthen by FRP laminate (validation)

7 Stress Intensity Variation

Variation of stress intensity along the depth and its variation with respect to load have been plotted in this section. The test is conducted on RCC beam and RCC beam with FRP. Stress intensity values are maximum at the application of the load in case of RCC beam. In case of FRP beam maximum stress intensity occurs in the plate. Both beams were tested at 48 KN to check the variation along the depth. As seen from the figure the nature of graph is similar for both the beams. It is maximum at bottoms thereafter decreases and approaches nearby to zero at the center thereby increases linearly with higher value at the top. Values at the bottom are higher in

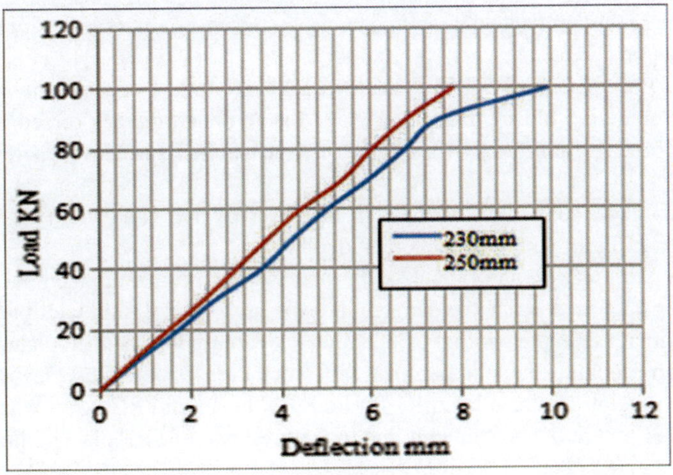

Fig. 10 Effect of width on FRP

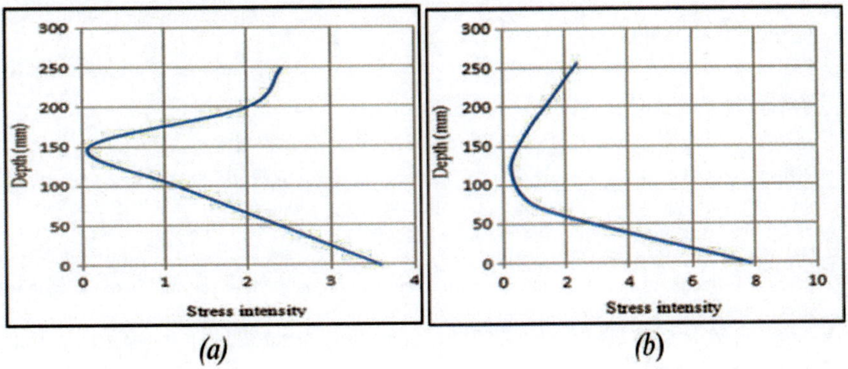

Fig. 11 Stress intensity along the depth. **a** RCC, **b** FRP

FRP composite beam due to the presence of FRP plate at the bottom. There is very little difference in the value along the top half part as compared to bottom half part in RCC and FRP composite beam.

As seen from Fig. 11 stress intensity variation for both RCC and FRP is almost same at top half portion. At bottom due to FRP plate stress intensity is high.

8 Conclusions

1. SIF values are increasing with the positive value; it concludes that the fracture process is unstable.

2. For a particular crack depth with increase in the width Stress Intensity Factor has decreased.
3. The value of axial SIF is less with respect to Mode I because the axial force transmitted is a closing force and Mode I is an opening crack force.
4. The force transmitted by the reinforcement has increased with increase in the cover.
5. The damage index of the RC beam is increased by 6% when width changed from 230 to 250 mm with constant crack depth.
6. For a particular depth of crack, with increase in width of the member the crack area is increased hence the damage index is high for 250 mm width RC beam.
7. With increase in the width the deflection is decreased in the FRP plate. With this we can conclude that load-carrying capacity increases with increase in the width.
8. The variation of stress intensity for both RC beam and RC beam with FRP are same expect at the bottom half due to the presence of FRP plate at the bottom.

Acknowledgements The Department of Civil Engineering, KLE DR. MSSCET Belgaum, Staff and Management is greatly fully acknowledged by the authors.

References

1. Murthy RC, Palani GS, Nagesh RI, Gopinath S, Raghu Prasad BK (2012) Residual strength evaluation of concrete structural components under fatigue loading. Sadhana 37, Part 1:133–147
2. Carpinteri A (1984) Stability of fracturing process in RC beams. J Eng Mech 110(3) ASCE
3. Jakubczak H, Glinka G, Calculation of stress intensity factors for cracks subjected to arbitrary non-linear stress fields. In: 5th international conference on biaxial/multiaxial fatigue and fracture, Cracow '97, Poland, pp 431–447
4. Murti V, Valliappan SF, Lee IK (1985) Stress intensity factor using quarter point element. J Eng Mech Ill(2), ASCE
5. Chandra Kishen JM (2012) Quantification of damage for residual life assessment of reinforced concrete infrastructures. Department of Civil Engineering, Indian Institute of Science
6. Lakshmikandhan KN (2013) Damage assessment and strengthening of reinforced concrete beams. Int J Mater Mech Eng 2(2):34–42, IJMME
7. Sain T, Chandra Kishen JM, Residual fatigue of RCC: deterministic & probabilistic approach. Indian Institute of Science
8. Sain T, Chandra Kishen JM (2013) Damage and residual life assessment of structures using fracture mechanics. In: 16th ASCE engineering mechanics conference, July 16–18
9. Chaje A (1974) Principles of structural stability theory
10. Tharp TM (1987) A finite element for edge-cracked beam columns. Int J Numer Meth Eng 24(10):1941–1950
11. Shaishav TPV, Viradiya R (2014) Comparative study of experimental and analytical results of FRP strengthened beam in flexure. Int J Res Eng Tech 3(4)
12. IS 456: 2000, Code of practice for plain and reinforced concrete. Bureau of Indian Std

Failure Investigation of HSS Counter Punch Tool

G. J. Naveen, L. Deeksha, P. Sampathkumaran, P. Rangasamy, B. Jayadev, S. Seetharamu, and A. Sathyanarayana Swamy

Abstract The manufacturing processes such as shaping, forming, cutting, heat treatment and joining processes are being adopted in engineering applications, and various types of tools are necessarily being deployed during these operations to produce components. In this context, the failure of HSS counter punch tool used in cold forging is reported from an industry, as cold forging is a part of forming processes. The problem has been entrusted to investigate the cause of failure taking into account the selection of input parameters considered during cold forging and also the steps involved in the tool making. The material M35 HSS tool had developed a long crack along the periphery passing through the central axis of the counter punch tool. The mechanical and metallurgical investigations were done involving hardness, microstructure, retained austenite measurement, fractography as well as the non-destructive tests, viz., visual inspection, dye penetrant test and ultrasonic detection as a part of this study. The investigation revealed fatigue-induced damage due to the application of one-time load during forging caused by the microstructural changes observed during the post processes. The root cause of failure of the tool is established based on the analysis and interpretation of the experimental data. Further, the reason of failure is communicated to the industry along with the possible solutions from the point of avoiding failure occurrence in future.

Keywords Counter punch HSS tool · Cold forging · Fractography · Fatigue · Failure investigation · Mechanical and metallurgical tests · Non-destructive tests

G. J. Naveen (✉) · L. Deeksha · P. Sampathkumaran · A. S. Swamy
Department of Mechanical Engineering, Sambhram Institution of Technology, Bengaluru 560097, India
e-mail: gj_naveen@yahoo.co.in

P. Rangasamy · B. Jayadev
Essae Digitronics Pvt Ltd, Bengaluru 560099, India

S. Seetharamu
Department of Mechanical Engineering, Nitte Meenakshi Institute of Technology, Bengaluru 560064, India

S. Seetharamu et al. (eds.), *Fatigue, Durability, and Fracture Mechanics*, Lecture Notes in Mechanical Engineering, https://doi.org/10.1007/978-981-15-4779-9_20

1 Introduction

There are several manufacturing methods practised such as shaping, milling, forming, cutting, heat treatment and welding. The life assessment of a tool is important from the point of superior performance as several tools fail at an early stage. There are various reasons attributed to the failure such as processing, heat treatment and other metallurgical issues. In this context, a failure of HSS counter punch tool has been reported by an industry, and thc work has been assigned to solve the problem. This would help the industry to take remedial actions and avoid future reoccurrence. It is well known that, in forging, hot, warm and cold forging are the various kinds. Cold forging has several distinct advantages such as the operation is done at room temperature, near net shape of a component is produced and the inventory required and risk involved are less. Thus, cold forging process has unique features compared to other forging processes [1].

The documentation of tool steels regarding the failure causes and preventions done by Jung et al. [2] discusses classification and type of failures, its assessment and giving practical solutions to the problems. It is reported that the defects originating during manufacturing, operating conditions, post processing have bearing on the tool life. Further, it is reported that many defects appear after the heat treatment stage may be due to either higher retained austenite content or precipitation of carbides along the grain boundaries or bonding and coarseness seen in the microstructure, etc., thus resulting in component failure.

The materials subjected to load, temperature and environmental factors result change in the behavioural aspects, which in turn depends upon the mechanical, physical, chemical and thermal properties. The effect of load, temperature, speed, microstructure, retained austenite content and other NDT parameters on the material properties has been extensively studied by other researchers and reported [3, 4]. Among them, the characteristics, viz., mechanical properties, Elasticity, Yield point, Stiffness, Ductility, Hardness, Plasticity, Brittleness, Creep, Toughness [5], Impact energy and Wear [6], are important from the point of performance and applications. These properties have to be considered appropriately depending upon their requirement and applicability. If one considers recovery, recrystallization and grain growth [7] in any material due to the cold working or hot working, it has deleterious effect on the crystal structure and in turn property change. Generally, deformed materials are in non-equilibrium state as well as thermodynamically unstable. All these aspects contribute to the healthiness of the component. When material/component/part fails, one should ascertain the origin and should find out why it has occurred. The premature failure of component during service condition needs to be assessed to provide possible solutions. The failure of any part may be occurring either due to operational problems or the post processes employed or design aspects. Thus, it is very much necessary to follow the design practice and carry out the post processes such as heat treatment, cold treatment, tempering, stress reliving as per the laid down procedures provided in the literature [7].

In the present work, the material M35 HSS grade steel used in the counter punch has suffered from a damage having long cracks along the periphery passing through the central axis. Hence, a thorough understanding and in-depth study is required to be carried out. For this purpose, the evaluation involving visual examination, magnetic particle inspection, ultrasonic detection as well as mechanical tests such as hardness, retained austenite measurement, microstructural analysis and fractographic analysis are taken up.

1.1 Cold Forging as a System

It is reported [8] that cold working of a metal is carried out below its recrystallization temperature. Although normal room temperatures are generally used for cold working of various types of steel, temperatures up to the recrystallization range are sometimes used for improved performance. In cold working, as reported, the recovery processes are not that effective. The input variables considered for cold forging process [9], shown in Fig. 1, are billet or blank (geometry or material), tooling (the geometry and material) material used for the tool/material interface, the mechanism involved, the type of equipment used, final product characteristics and the surrounding environment where the process is being done. The study of the input/output relationships and the effect of process variables on the product quality and processing methods forms a part of system approach in cold forging. The important aspects to be followed in metal forming operation are to yield desired shape and size, specific properties and finally understanding to control metal flow. The final

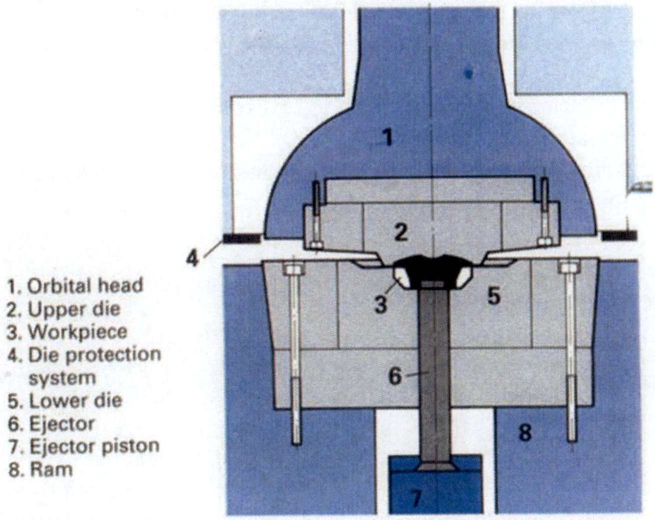

Fig. 1 Cold forging system

component characteristics are very much dependent on the direction and magnitude of deformation during metal flow, temperature and in turn the mechanical properties relating to local deformation and the formation of defects such as cracks below the surface.

The metals such as steels and aluminium alloys possessing good ductility may be cold forged at room temperature. The alloys such as copper, zinc, tin, titanium, beryllium and nickel may be cold forged depending on the applications. The cold forging of stainless steels are not easy, as high forces and tool pressures are required to be exercised during the process, especially, for austenitic or austenitic-ferrite steels (work-hardening ability is very good). The materials used for cold forging are rolled or drawn rod or wire as well as in the form of sheared or sawed-off billets. In order to maintain dimensional tolerances of the cold forged component and to avoid excessive loading of the press and tool, weight and surface finish of the sheared (or sawed) billet or preform are required to be controlled pretty closely. The key issues which are required to be controlled and their requirements during cold forging are detailed below.

1. Mechanical design: Compatible with the steel grade chosen and the method to produce tool or die for its use.
2. Grade selection: Based on the design, the composition to produce tool or die needs to be optimized for the intended service life.
3. Steel quality: It should be structurally sound, free from harmful inclusions and defects to the degree required for the application.
4. Machining processes: The surface microstructure or surface finish should not get altered due to excessive residual stresses induced during fabrication, machining, heat treatment or service conditions.
5. Heat treatment operation: The desired microstructure, hardness, toughness and hardenability at the surface and the interior to be obtained.
6. Grinding and finishing operations: The surface integrity of the component should not get impaired.
7. Tool and die setup: Alignment is ensured to avoid irregular, excessive load which may accelerate wear or cause cracks.
8. Tool and die operation: The desired component life to be achieved not by over loading them during processing.

2 Materials and Methods

2.1 *Materials*

The M35 HSS grade steel [10] is conventionally manufactured by cobalt-alloyed high-speed steel, and it is used as a punch tool during cold forging process. This material is suitable for cold work applications with the constraints imposed on wear resistance and service life [6]. The tool steels possess good wear resistance as well as reasonable toughness characteristics compared to high-alloyed cold work steels. The manufacturing processes at every stage are well defined and controlled so that the final product obtained yields good microstructure with uniform distribution of carbides and good mechanical properties.

2.2 *Processing of the Tool*

The processing method of the counter punch tool is as follows; the raw material of size diameter 85 mm × 146 mm is turned into diameter 82.5 mm followed by machining process where the tool is finished to its nominal diameter 82 mm.

2.3 *Heat Treatment*

The tool is preheated and stress relieved before it is hardened to a temperature of 1160 °C after which it is quenched in oil medium following with tempering reaching temperatures of 560 °C for 3 times. A sub-zero treatment is carried out at −72 °C after first tempering. The hardening characteristics [11] are shown in Fig. 4, wherein the influence of hardening and tempering temperatures on the hardness level is very well seen. To get the highest hardness level, it is preferred to choose temperature in the range 520–540 °C. Following this, the tool is ground to about 30 μm and polished (Fig. 2).

2.4 *Cold Forging Process*

The working condition of the tool can be described as the counter punch tool is held in the vertical axis of the cold forging machine, the work is placed on the counter punch tool and an average force of 45 MN is applied by a hydraulic press and is transmitted through a punch tool on to the work (Fig. 3).

2.5 *Characterization Methods*

It is reported that the tool has failed after producing only 292 parts whereas the desired tool life is to produce minimum 500 parts. The assessment of failure involved visual inspection, composition analysis, hardness, microstructure (Light Microscope), retained austenite (XRD), non-destructive testing (NDT) and fractography using scanning electron microscope (SEM).

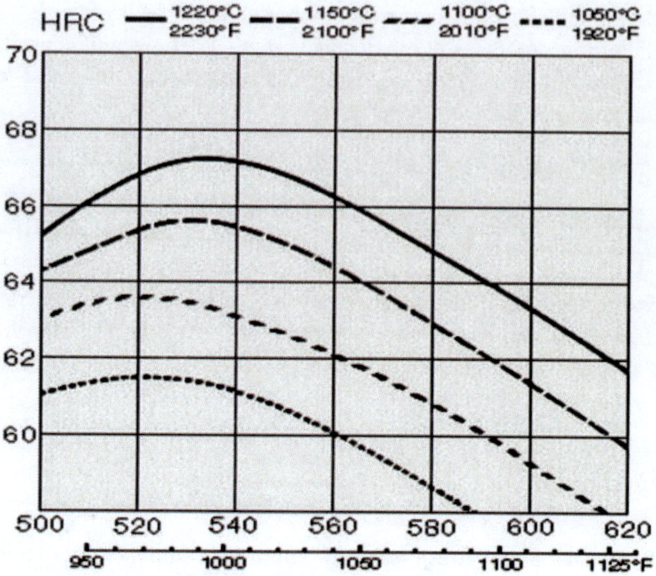

Fig. 2 Hardness after hardening, step quenching and tempering 2 * 1 h of sample size 25 mm

Fig. 3 Cold forging process of counter punch tool

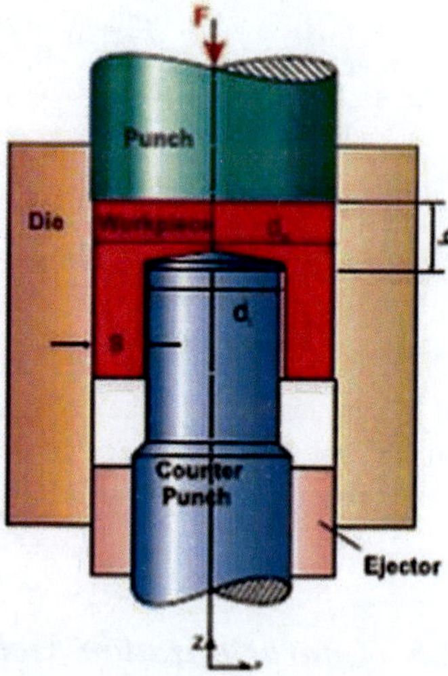

The chemical composition was carried out using optical emission spectrophotometer. The hardness measurement is done using Rockwell C scale using diamond cone indenter (angle 120 degree) at a standard load of 150 kg. The metallographic examination was done on the tool steel by grinding the surface using coarse/fine emery sheets and final polishing using a velvet cloth till a smooth scratch-free surface is obtained. This is followed by chemical etching process using 3% nital solution. The x-ray diffraction technique was employed to measure the retained austenite content in the sample. The non-destructive tests (NDT), viz., dye penetrant examination (DP), magnetic particle inspection (MPI) and ultrasonic detection (UT) to characterize and quantify the defects in the material.

a. **Dye Penetrant**: Penetrant solution containing visible dye or fluorescent particles is drawn into surface-breaking cracks and voids by capillary action. A developer is applied to draw the penetrant out of the flaw, giving a visible indication.
b. **Magnetic Particle Inspection**: A magnet or electrical current creates magnetic fields in a specimen. Magnetic particles applied to the surface indicate the location of a field leakage, signalling a discontinuity.
c. **Ultrasonic Testing**: Ultrasonic waves are injected into material. Measurements of the time or frequency response allow detection of defects and changes in material properties. A transmitter cum receiver is used to transmit and receive the ultrasonic signal, and the data seen on the CRT with the two ends (bottom and top) of the sample undergoing ultrasonic testing and representing in the form of signals on the left and right side corners with the defects, if any, are projected in the form of an echo pattern.

2.6 Fractographic Analysis Using SEM

The SEM is sophisticated equipment, which is mainly used for surface topographical studies as well as for fractographic analysis. Under electron bombardment, a variety of different signals is generated (including secondary electrons, backscattered electrons, characteristic x-rays, and long-wave radiation in the ultraviolet and visible region of the spectrum) and detected from the point of characterizing the material.

3 Results and Discussion

The chemical composition of the counter punch tool has been analysed, and it is shown in Table 1.

The visual inspection of the failed punch tool shown in Fig. 4 gives the fractured features after the counter punch tool is made to cut into two halves.

Table 1 Composition analysis of the tool

Element	C	Si	Mn	S	P	Cr	Mo	Co	V	W
Composition (wt%)	0.85	0.28	0.25	0.04	0.04	4.06	5.01	4.92	1.88	6.4

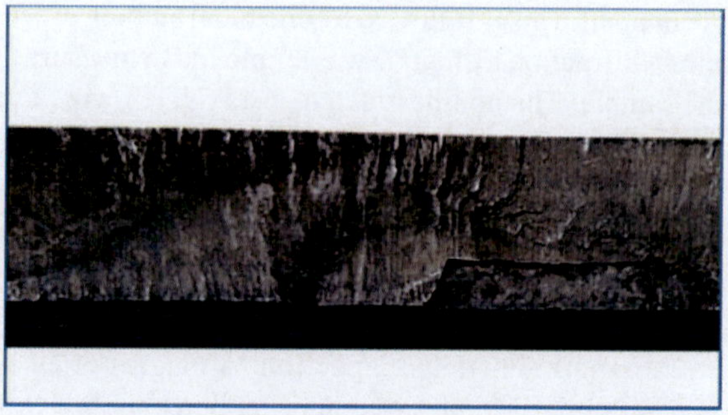

Fig. 4 Fractured surface features

The beach marks are noticed on the fractured surfaces as shown in Fig. 4. The tool steel has shown long cracks (refer Fig. 5) along the periphery of the tool passing through the central axis.

The hardness and the retained austenite content of the failed tool are shown in Table 2, respectively, wherein these measurements have been done at three regions

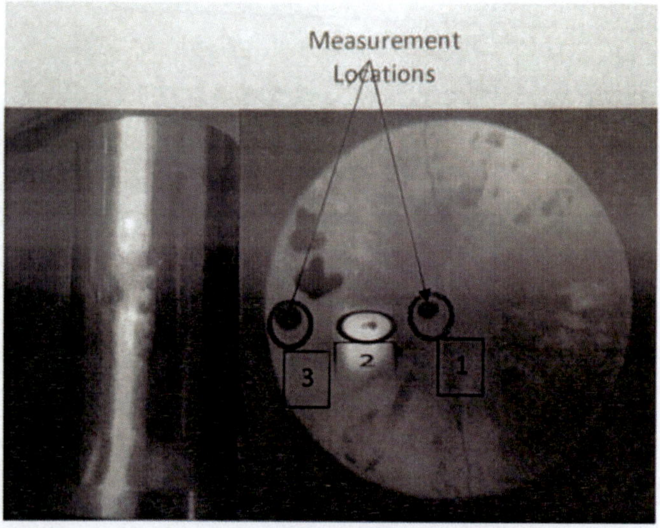

Fig. 5 Photograph of the counter punch tool cracked along the diameter

Table 2 Hardness and retained austenite content data of the counter punch tool

Sl. No.	Region	HRc	RA (%)
1	R1	57	5.8
2	R2	62	4.2
3	R3	63	3.9

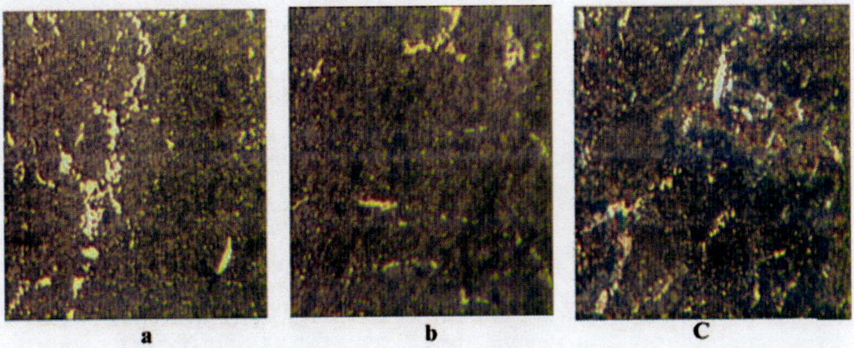

Fig. 6 Light microscopic features of the counter punch tool for the regions R1, R2 and R3

(R1, R2 and R3) as shown in Fig. 5. The region R1 is very close to the crack, whereas the region R3 is almost at the edge of the punch tool. R2 is in between them.

The microstructural features of the tool pertaining to the region R1 is shown in Fig. 6a. For the regions R2 and R3, the microstructures are shown in Fig. 6b and c, respectively.

The scanning electron microscopic features are shown in Fig. 7 a, b, c and d at magnifications 2k, 1k, 500x and 200x, respectively.

The non-destructive test namely dye penetrant test and magnetic particle inspection (DPT, MPI) have been carried out and shown in Fig. 8 a and b, respectively. The ultrasonic detection was carried out on the failed punch tool and recorded.

As regards the composition of the HSS tool steel, it is confirming to the standard grade of M35 steel sample. Thus, there is no discrepancy with regard to its composition.

There is a variation in the hardness value measured in the regions R1, R2 and R3, wherein R1 is showing lower hardness value (57 HRc); on the other hand, the regions R2 and R3 are showing an average value of 62.5 HRc which is in close agreement with the HRc reported for the M35 grade steel (62–64 HRc). Thus, there is a hardness gradient observed along the radial direction when one assesses from the central region to the periphery of the punch tool. Similarly, the RA content when measured along the radial direction reveals a variation in its value at the centre of the punch tool, (R1) close to the crack; the RA% is much higher (5.8%) compared to regions R2 and R3, which show in the range of 3.9–4.2%. Thus, both hardness and RA% are giving good indications of probable reason of the failure of the tool.

The light microscopic features of the three regions R1, R2 and R3 when analysed reveal that, in region R1, the segregation of carbides in a tempered martensitic structure could be observed. In regions R2 and R3, network of carbides delineating the arms of dendritic structure uniformly distributed in a tempered martensitic matrix is noticed. The carbide size in region R1 does seem to be slightly bigger than in the regions R2 and R3.

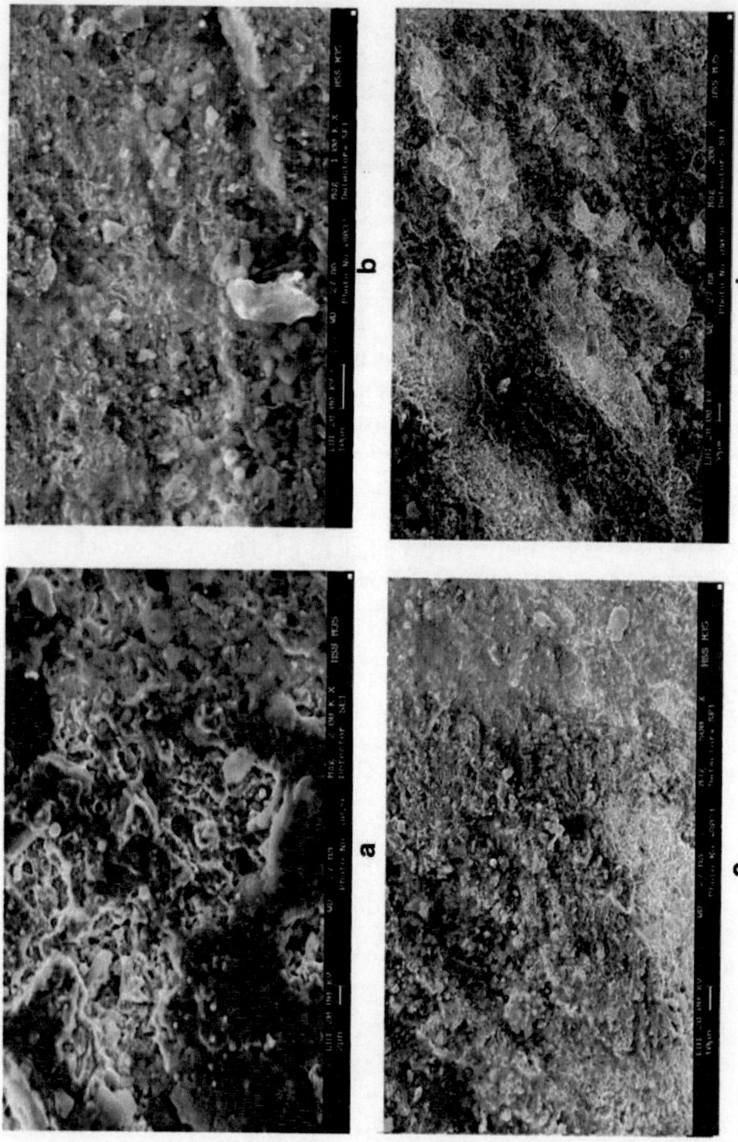

Fig. 7 Scanning electron microscopic pictures of the fractured surface taken at **a** 2000X, **b** 1000X, **c** 500X and **d** 200X

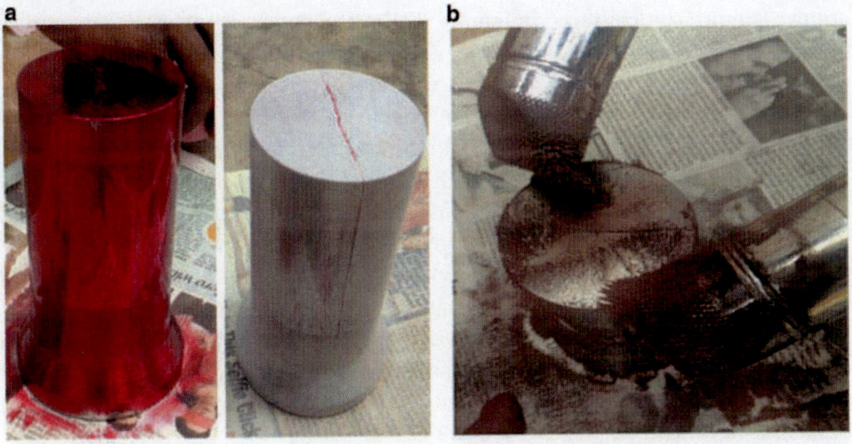

Fig. 8 Photographs showing **a** Application of die and developer for Die Penetrant (DP) test **b** DP revealing crack along the central axis of the punch tool and **c** MPI showing magnetic particles aligned along the length of the crack

The region R1 clearly shows banding of carbides and the probable reason for this may be attributed to the heat treatment process, which may not have been adhered to properly resulting in concentration of carbides in one particular region of the matrix. Finally, the banding structure has caused cracking along the centre line, and with the application of sudden one-time load, consequently, it is damaged and failed. The heat treatment procedure given in Sect. 2.3 needs to be looked into and ascertained from the point of failure. This needs further investigation.

To get further support on the crack formation in the counter punch tool, the NDT measurements were employed and revealed the following points.

- The DPT shows linear indications of the crack in diametrically opposite side of the OD (Fig. 8a).
- The MPI (Fig. 8b) also reveals the linear indication of the crack formed during the process.
- The Ultrasonic detection done on the failed sample clearly quantifies the defect. The length of about 130 mm and depth of about 30 mm are the dimensions of the crack obtained. Further, the crack size is more than the accepted level that it has crossed the critical length, and hence, the counter punch tool has failed.

The scanning electron microscopic features show the presence of beach marks (Fig. 7c and d) with deformed zones having deep grooves and craters observed (Fig. 8a and b) in the matrix. Figures b and c show the matrix in the highly disturbed condition perhaps attributed to ploughing action with visible cracks together with debris formation. Thus, the SEM pictures are giving good indications of fatigue type of failure of the punch tool.

Thus, the mechanical and metallurgical tests clearly support the fact that failure may be attributed to the fatigue-induced damage with the crack size crossing the critical length. The cause of fatigue is due to the structural changes taken place during the processing stage, especially, the coarse carbide concentration at the mid portion of the punch tool and resulting in banding (segregation).

The investigation conducted by Beth Brazitus and Paul Brda has [12] reported the decrease in fatigue life of bearing steel during experimentation because of carbide segregation in the matrix. The in-depth study reported by them revealed that the micro-cracks formed due to segregation beneath the surface tend to grow and form network of cracks, and finally, due to cyclic stress induced, the component reported to have been failed. Further, it is also reported how to minimize carbide segregation by introducing faster cooling of ingots, increasing the annealing time, etc.

In the present case also, one could see, the carbides are concentrated in a specific region and not well distributed in the matrix; crack formation and propagation due to load application as well as minute misalignment does seem to be the root cause of failure.

4 Conclusions

The important findings of the failure investigation carried out on the counter punch HSS tool steel are given below:

- The chemical composition analysis is conforming to the standard grade.
- The results on hardness data show much lower in value in the vicinity of the crack compared to the standard value of the hardness of M35 grade HSS steel.
- Like in the case of hardness, the RA% is higher and very close to the crack compared to the other regions.
- The microstructure of the sample shows clusters of carbides near the cracked region in the form of bands, which would have originated due to variation in the processing condition, especially, the heat treatment procedure with respect to tempering cycles.
- The examination of SEM photographs reveals the appearance of beach marks on the fractured surface indicative of fatigue type of failure.
- The fatigue would have taken place due to the sudden application of one-time load, and in the process, the region where non-uniform carbide dispersion is present would be brittle and get fractured easily, thus leading to cracks formation. The crack has propagated further along the periphery of the tool passing through the central axis leading to failure. Even any minute misalignment of the tool also would have contributed to the failure.
- The heat treatment process needs to be looked into very carefully, particularly, the tempering procedures from the point of avoiding failures in future.

Acknowledgements The authors wish to acknowledge with thanks the management of Sambhram Institute of Technology, Bengaluru and Essae Digitronics Pvt. Ltd., Bommanahalli Industrial Area,

Bengaluru for their kind support and help rendered as well as encouragement given to present and publish this paper.

References

1. Rathi MG, Jakhad NA (2014) An overview of forging processes with their defects. Int J Sci Res Publ 4–6 (2014)
2. Jung I, Lubich V, Wielaud HJ (2010) Tool failures-causes and prevention. In: 6th international tooling conference, pp 1343–1363
3. Schijve J (eds) (2009) High temperature and low temperature fatigue. In: Fatigue of structures and materials. Springer, Dordrecht, pp 481–495
4. Zohu SQ, Liu XY, Xu Y (2018) Microstructure, mechanical properties and tribological properties of composites reinforced by Cr Mn Fe Co Ni high entropy alloy. Materials (Basel), 11–10:1850. Multidisciplinary Publishing Institute, MDPI (2018)
5. Molinari AJ, Pellizzari M, Gialanella S, Straffelini G, Stiasny K (2018) Effect of deep cryogenic treatment on mechanical properties and microstructure of the tool steel CR7V for hot stamping. J Mater Eng Perform 27(1–3)
6. Tobola D, Brostow W, Czechowski K, Rusek P (2017) Improvement of wear resistance of some cold working tool steels. Wear 382–383:29
7. ASM Hand Book, 2. Heat treating and cleaning of metals
8. Campbell FC (2008) Recovery, recrystallization and grain growth. Book on Elements of metallurgy and engineering, ASM International (2008)
9. Groover MP (2009) Fundamentals of modern manufacturing, materials-processes-systems, 4th edn. John, USA
10. Bayer AM, Becherer BA, Vasco T (1989) High speed steels. ASM Hand Book, 16:51–59
11. Taylor special steels, M35 High speed steel. www.taylorspecialsteels.co
12. Brazitis B, Brda P (2018) Bearing fatigue life: the detrimental effect of carbide segregation. Technical Article, Iron & Steel Technology, AIST.ORG, September

Experimental and Numerical Prediction of Fatigue Behavior of Ring Type Specimen

V. Raja Prabhu, GT Shaji Ebron, V. Gautham, and PR Thyla

Abstract Nuclear pressure vessels are generally made of butt weld joints which are larger in size. Since it is difficult to perform experimental investigation on large structures with weld defects, a ring specimen of simpler geometry is considered for investigation. An experimental setup was fabricated to apply repeated loading on ring specimen, and the fatigue life of ring specimen with various weld defects like lack of penetration, lack of fusion, porosity, and undercut was recorded. Also, finite element models of ring specimen incorporated with weld defects are used to predict the fatigue life, and the same is compared with experimental results. The methodology adopted in the present work can be extended to investigate the influence of weld defects on fatigue life of roof slab.

Keywords Structural integrity · Fatigue life · Weld defects · Ring type specimen · Nuclear reactor · Roof slab

1 Introduction

Structural integrity has been defined as the science and technology of the margin between safety and disaster [1]. Systematic evaluation of structural integrity of the nuclear structure and estimation of the remaining life of structures are very crucial

V. R. Prabhu (✉) · P. Thyla
Professor, Department of Mechanical Engineering, PSG College of Technology, Coimbatore, India
e-mail: vpr.mech@psgtech.ac.in

P. Thyla
e-mail: prt.mech@psgtech.ac.in

G. S. Ebron · V. Gautham
Research Scholar, Department of Mechanical Engineering, PSG College of Technology, Coimbatore, India
e-mail: ebronshaji@gmail.com

V. Gautham
e-mail: gauthamvelayudham@gmail.com

S. Seetharamu et al. (eds.), *Fatigue, Durability, and Fracture Mechanics*, Lecture Notes in Mechanical Engineering, https://doi.org/10.1007/978-981-15-4779-9_21

to ensure human safety, environmental protection, and the economical consideration for developing new structures or maintaining them [2].

Structural integrity of nuclear reactor roof structure presents a unique challenge in the production of safe and cost-effective structures by means of analysis, inspection, and rehabilitation. IGCAR (Indira Gandhi Center for Atomic Research) planned to start the 500 MWe Prototype Fast Breeder Reactor (PFBR) which is a sodium-cooled fast reactor type in India. The schematic diagram of the reactor assembly is as shown in Fig. 1. The dome-shaped roof slab with conical skirt under compression is an important component in a nuclear reactor fitted with the main vessel in order to hold all other components like control plug, sodium pump, heat exchanger, etc. The total weight of the dome-shaped roof slab is approximately 100 tons, and the inner radius of the roof slab is around 6050 mm as shown in Fig. 2.

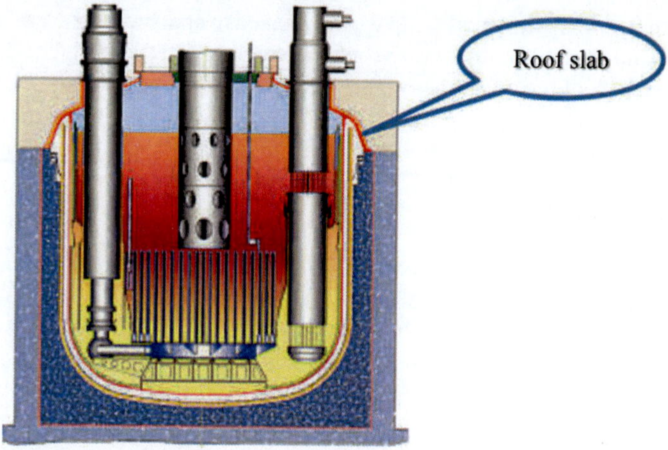

Fig. 1 Schematic of PFBR reactor assembly [9]

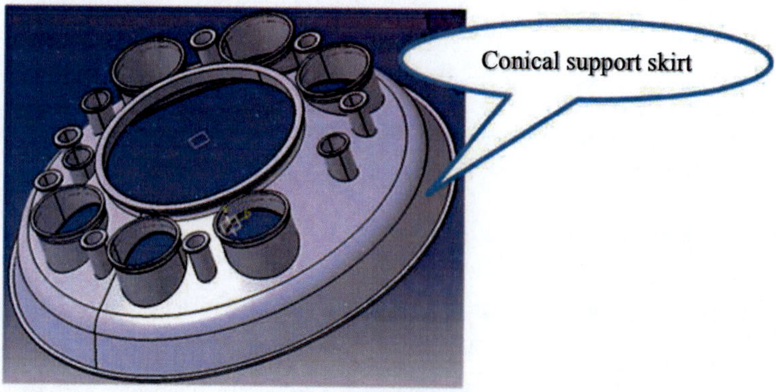

Fig. 2 Dome-shaped roof slab with conical support skirt for PFBR [10]

One of the technological challenges in the construction of nuclear reactor is welding of roof slab with reactor main vessel [3]. Any inclusion of weld slag or presence of weld defects will cause a reduction in structural strength which, in turn, affects the load-carrying capacity of the roof slab. In order to ensure the structural integrity of roof slab, it is essential to investigate the fatigue behavior of welded joints with defects [4].

2 Design of Ring Specimen

In order to carry out experimental and numerical investigation, it is required to incorporate the weld defects like lack of fusion, lack of penetration, and porosity in the weld joints. It will be cumbersome to incorporate all the weld defects in a huge structure like roof slab and subsequently carry out fatigue testing. Hence, a simplified specimen model needs to be developed to match the stress pattern of roof slab [5].

The tri-joint weld zone which is in the meridional plane of a dome will be subjected to membrane compressive stress of 100 MPa [6]. Based on the numerical prediction of stress behavior of dome type roof slab, a cylindrical ring type specimen configuration is found to be more suitable for inducing idealized membrane compressive stresses along the meridians [7]. A circular ring specimen of mean radius 200 mm with shell thickness of 6 mm to a width of 12 mm was finalized by applying analytical equations and verified by numerical analysis [8]. A weld joint shall be introduced in the ring specimen to represent tri-joint weld area.

An elementary strip of axial length ‘b’ subjected to a concentrated radial load ‘W’ acting in-plane at point ‘A’ is shown in Fig. 3. The following assumptions and factors are considered for arriving at the specimen dimensions:

1. The material is homogeneous and it obeys Hooke’s Law.
2. The modulus of elasticity in tension and compression is equal.

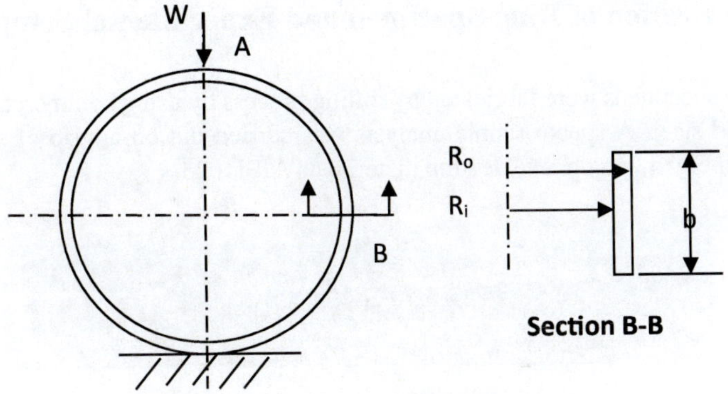

b= axial length; Ro =outer radius; Ri =Inner radius

Fig. 3 Ring specimen with radial load

Table 1 Specifications of the ring type specimen

Parameter/Type		Dimensions
Inner radius, R_i	(mm)	197
Outer radius, R_o	(mm)	203
Radial width, h	(mm)	6
Axial length, b	(mm)	12
Mean radius, R_m	(mm)	200
Ratio of mean radius to thickness		33.33
Moment of inertia, I	(mm^4)	216
Load estimated using Roark's formula	(N)	194
Load estimated using curved beam theory	(N)	200
Deflection, δ	(mm)	5.1

3. The ratio of $R_i/(R_o - R_i) > 10$ satisfying thin wall cylinder condition.
4. The section 'B-B' under consideration experiences a stress of 100 MPa that exists in a portion of roof slab.
5. Feasibility for manufacturing and conducting tests.
6. The ring is of uniform cross section and is symmetrical about the plane of curvature.
7. All the loadings are applied at the radial position of the centroid of the cross section.
8. The deflection is primarily due to bending of the thin ring.

b = axial length; Ro = outer radius; Ri = Inner radius.

The specimen dimensions are arrived in such a way that the section B-B will be subjected to a stress of 100 MPa, and the load required for inducing the above stress is also predicted. The specifications of the ring type specimen are as shown in Table 1.

3 Fabrication of Ring Specimen and Experimental Setup

The ring specimens were fabricated by rolling process by using commercially available mild steel. A spectroscopic analysis was carried out on a strip of sample as shown in Fig. 4, which revealed the material as AISI1023.

Fig. 4 Strip of ring specimen for spectroscopic analysis

Table 2 Chemical composition and mechanical properties

Properties	Base metal	Filler metal
Chemical composition (%)		
C	0.237	0.071
Si	0.125	0.46
Mn	0.426	1.21
Fe	99	98
Others	0.212	0.25
Mechanical properties		
Yield strength (MPa)	360	512
Tensile strength (MPa)	425	577
Elongation %	15	29

The chemical composition and mechanical properties of the material under consideration are given in Table 2.

A flat plate of cross section 12 mm × 6 mm was rolled to obtain the test specimen of mean diameter 400 mm.

A special fixture was developed taking into consideration the following aspects:

- The test specimen developed is to be within the limit of cylindricity.
- To hold and mount the specimen during the edge preparation and welding process.
- To avoid warping after welding the specimen.

The fixture made of mild steel was machined with an annular groove to accommodate the ring specimen of cross section 20 mm × 7 mm. A rectangular slot was made to enable easy access to the weld operator as shown in Fig. 5.

The ring specimen was inserted in the annular machined slot, and bevel edge preparation was carried out on the end surface. The region to be welded was cleaned with acetone to ensure that no oil, dirt, or foreign particles were present.

The gas tungsten arc welding was used considering the parent and filler material as AISI1023 and AWS:A5.1:E:7018, respectively. The electrode used in the welding

Fig. 5 Fixture for checking cylindricity

Fig. 6 Experimental test setup

process was made of tungsten material of 2.5 mm rod diameter. Since the focus of this study is on the structural integrity of the weld specimen, the filler material plays a major role. The chemical and mechanical properties of the filler material E7018 without any flux coating are given in Table 2.

Experimental setup for prediction of fatigue life of ring specimen was designed and developed for performing fracture analysis on the ring specimens as shown in Fig. 6.

Vibration mounts were installed to absorb vibration, and vibration arrester pads were also installed at the base of the motor on each of its legs. The test setup was assembled by fastening instead of welding, so that it can accommodate any specimen of this category and facilitate easy transportation. Spring washers of IS 6735:1994 standards were used to give pretension to the bolts, and all the components were coated to avoid corrosion.

4 Incorporation of Weld Defects in Ring Specimen

The major defects observed in welding are undercut, lack of penetration, porosity, and lack of fusion. Lack of penetration occurs when weld material has not reached the root leaving a narrow gap as shown in Fig. 7a. Porosity in weldments can occur as a result of momentary loss of shielding and also from the use of welding wire having a hydrated oxide or oily surface finish which causes hydrogen contamination in the weld pool as shown in Fig. 7b. Lack of fusion can occur due to improper welding

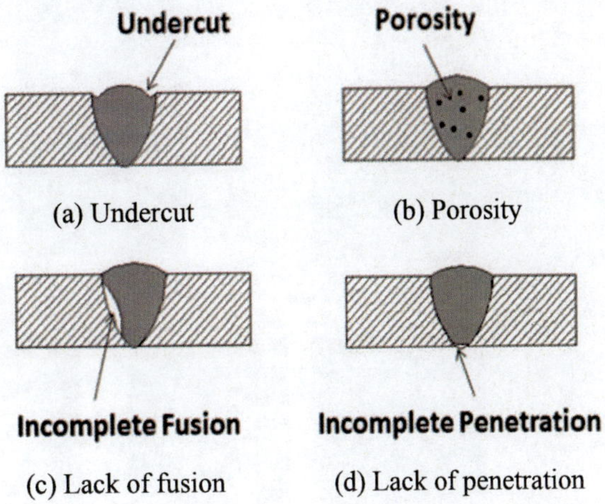

Fig. 7 Types of weld defects [11] **a** undercut **b** porosity **c** lack of fusion **d** lack of penetration

procedure or inappropriate joint design, spatter ahead of the weld, etc., as shown in Fig. 7c. Undercut can occur as shown in Fig. 7d when the arc voltage and travel speed are high.

The radiographic test images of the weld defects such as lack of penetration, porosity, lack of fusion, and undercut introduced into the ring specimen are shown in Fig. 8.

5 Experimental Determination of Fatigue Life of Ring Specimen

The ring specimen was mounted on the test setup as shown in Fig. 6. Experiments were carried out using the setup, and the specimen was monitored for every 20,000 cycles. Considering specimen with lack of fusion, cracks were observed at 80,000 cycles. The crack was periodically monitored until the specimen ruptured as shown in Fig. 9, and the corresponding fatigue life was recorded for ranking the severity.

The same procedure was followed for testing specimens with other weld defects for three trials. Fatigue life was noted after the appearance of crack, and after complete rupture, the fracture life was recorded. Similarly, for all trial specimens, experimental fatigue life was determined, and the results are given in Table 4.

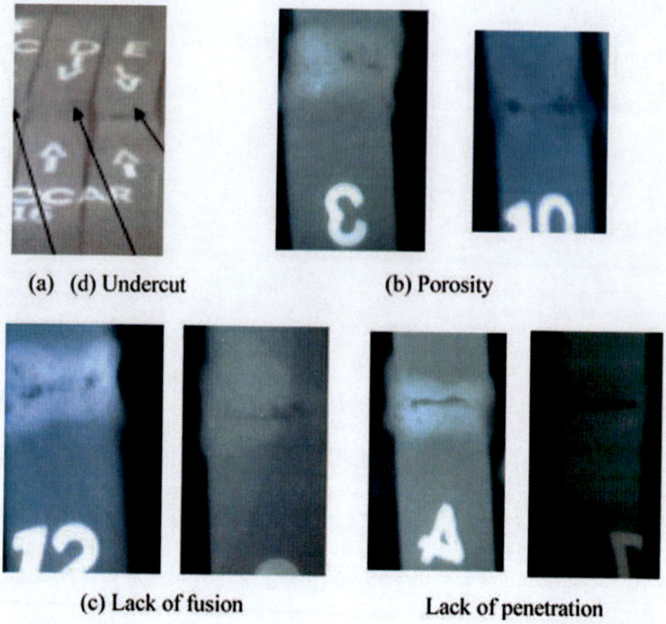

Fig. 8 Radiographic images of ring specimens with weld defects **a** undercut **b** porosity **c** lack of fusion **d** lack of penetration

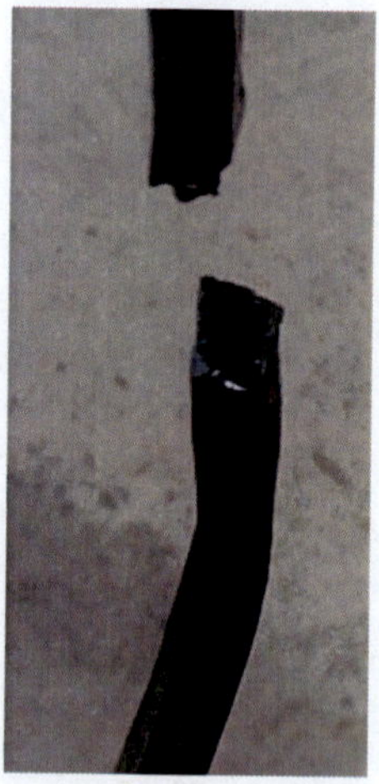

Fig. 9 Ruptured ring specimen with lack of fusion

6 Numerical Prediction of Fatigue Life of Weld Defects in Ring Specimen

Methods used in the prediction of fatigue life of welded joints and their implementation are thoroughly described in the International Institute of Welding (IIW) recommendations. The methods can be divided into local and global approaches. The nominal stress method and structural hot-spot method are both global approaches where the fatigue life is calculated based on stresses that are unaffected by the weld geometry. The effective notch stress method and the fracture mechanics approach are local approaches. Fatigue life is calculated using stresses obtained on the local weld geometry most often by using FEA. Fracture mechanics approach is used to predict crack growth in structures subjected to fatigue loading. The FRANC3D is a FE-based simulator for propagating curvilinear crack propagation in 3D components. This program has the ability to analyze a cracked body using special isoparametric crack tip elements to describe the singularity ahead of the crack tip.

Before performing the numerical analysis in ring specimen, a validation study is carried out using Newman–Raju equation on a finite body with surface crack as shown in Fig. 10, and a close validation with FRANC3D was obtained. The following condition was considered for validation:

$0 < a/c < 2$ and $c/b < 0.5$ provided that $a/t < 1$.

The semi-elliptical crack of $a = c = 1$ mm is modeled on a finite plate of $100 \times 100 \times 10$ mm and is propagated to 60% of total thickness of the plate.

$$K = \sigma \times \left(\sqrt{\frac{\pi \times a}{Q}}\right) \times F_S\left(\frac{a}{c}, \frac{a}{t}, \frac{c}{b}, \emptyset\right) \quad (1)$$

Crack propagation on a finite plate predicts a closer agreement with numerical and analytical solution as shown in Fig. 11.

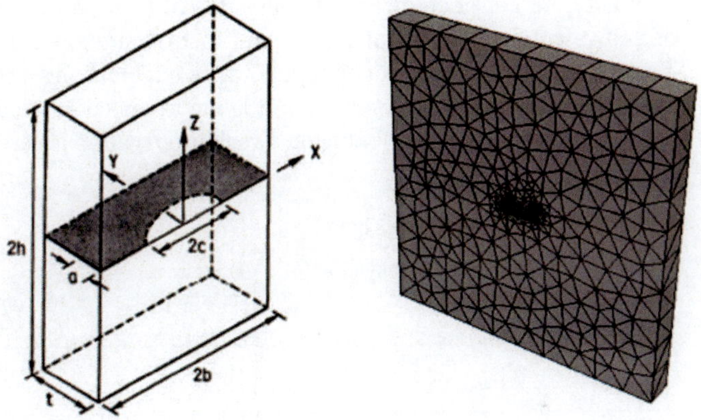

Fig. 10 Finite plate with surface crack

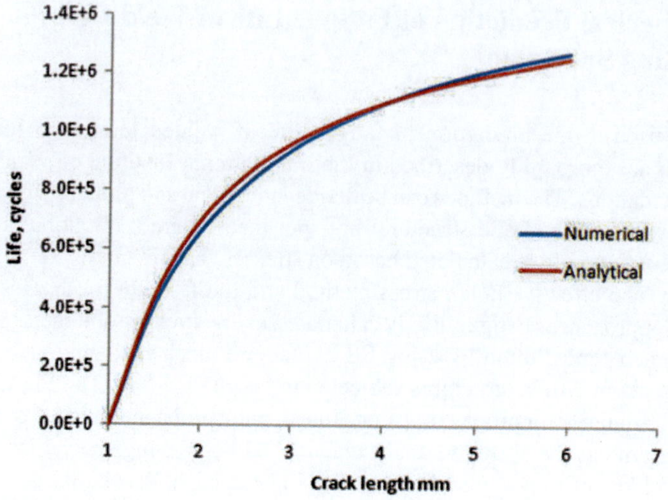

Fig. 11 Fatigue life comparison between numerical and analytical solutions

The ring specimen is modeled with mean radius 200 mm and thickness 6 mm with 12 mm axial length, and the material properties were assigned considering the ring as mild steel. The maximum stress corresponding to a load of 194 N is found to be 100 MPa at the welded portion with the maximum deflection found to be 5.4 mm. Initial crack dimensions for the corresponding weld defects are considered from the radiographic images. The weld defect is modeled as a semi-elliptical crack and incorporated into the ring specimen considering the material propetires given in Table 3.

The sub modeling concept is adopted to reduce the computational time. The weld zone of the ring specimen is considered as a local model as shown in Fig. 12b where the crack is incorporated. Weld defects like undercut, lack of penetration, lack of fusion, and porosity are considered for the analysis. All the defects are modeled by means of equivalent semi-elliptical cracks and incorporated at corresponding location of ring specimen. Initial crack corresponding to undercut and lack of fusion are incorporated at weld toe, lack of penetration is incorporated at weld root, and the porosity is incorporated into the weldment. Crack propagation is carried out till

Table 3 Properties of parent and filler material

Materials	Yield strength (MPa)	Tensile strength (MPa)	Elongation %	Fracture threshold K_{th} MPa$\sqrt{}$mm	Fracture toughness K_{1c} MPa$\sqrt{}$mm
Parent material	360	425	15	160	1600
Filler material	512	577	29	180	1750

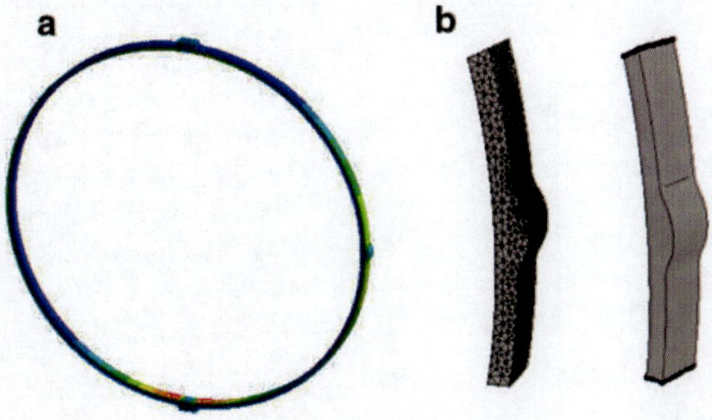

Fig. 12 **a** Ring Specimen—global model **b** ring Specimen—local model

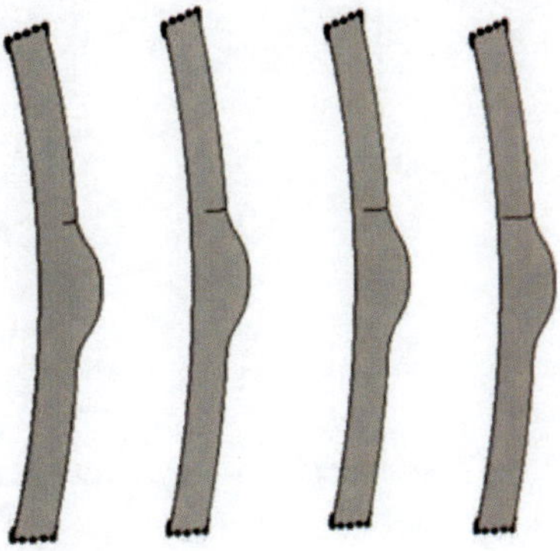

Fig. 13 Crack propagation in ring specimen

it reaches critical state as shown in Fig. 13 for all the defects, and the corresponding fatigue life is determined as shown in Fig. 14.

Stress intensity factor of the crack increases as it propagates toward the thickness of ring specimen. Crack propagation analysis is carried out until the slit reaches the fracture toughness of 1600 MPa$\sqrt{mm}$ as shown in Fig. 15, and the corresponding fatigue life is considered as the failure of ring specimen.

Similarly, experimental and numerical analysis were carried out in the ring specimen with other weld defects using fatigue testing setup and the results are shown in Table 4. The infinite life given in Table 4 corresponds to more than 1×10^6 cycles. The size of initial crack is defined in terms of 2c and shown in Fig. 10.

It is difficult to monitor the internal crack formation when a body is subjected to fatigue loading. Numerical analysis predicts only the fatigue life corresponding to the propagation of cracks representing weld defects. Even though there is considerable error between experimental and numerical prediction of fatigue life, the ranking of severity of weld defects on fatigue life is found to be the same.

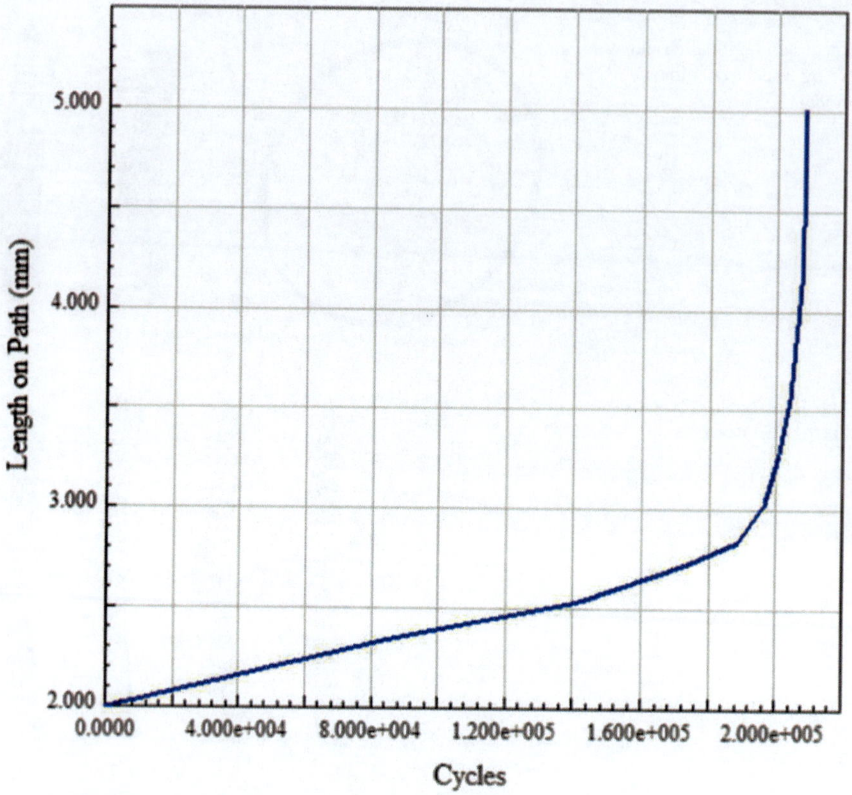

Fig. 14 Fatigue life corresponding to the propagation of undercut

7 Conclusion

Experimental and numerical investigations were carried out to establish the severity of weld defects on the fatigue life considering ring specimen. It is found from the investigation that undercut and lack of fusion have significant influence on the fatigue life of the ring specimen compared to porosity and lack of penetration. The methodology adopted in the present work can be extended to investigate the influence of weld defects on fatigue life of roof slab.

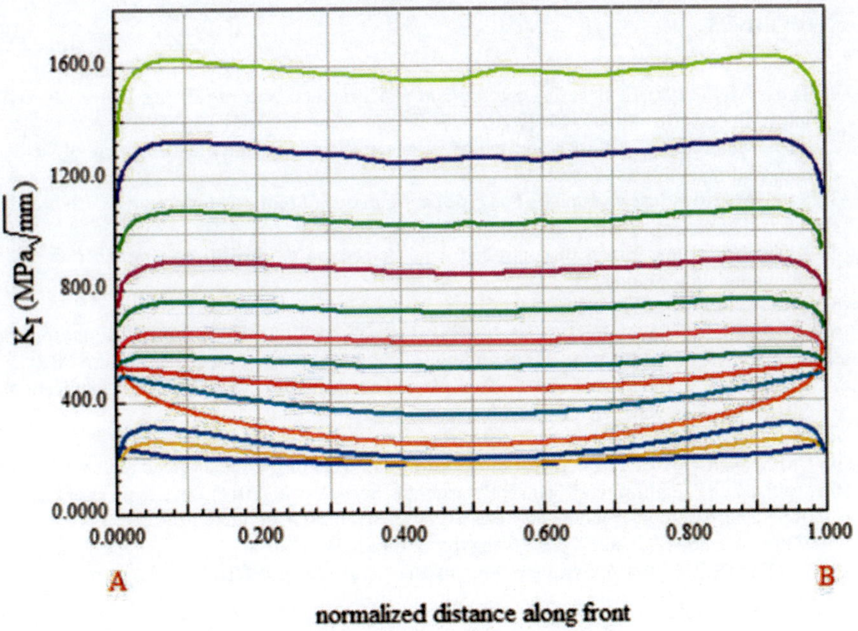

Fig. 15 Stepwise mode 1 stress intensity factor of undercut

Table 4 Comparison of numerical and experimental results

Description	Size of initial crack (mm)	Fatigue life, cycles		Error
		Experimental	Numerical	
Lack of fusion	2c—6 a—2 33% of cross section	426,666	356,874	17%
Undercut	2c—8 a—2 33% of cross section	324,896	208,247	36%
Lack of penetration	2c—8 a—2 33% of cross section	880,000	690,843	22%
Porosity	2c—4 a—2 33% of cross section	Infinite life	Infinite life	–

Acknowledgements The outcome of research described in this paper was financially supported by the Board of Research in Nuclear Sciences (BRNS), Mumbai, and the authors are indebted to Board of Research in Nuclear Sciences (BRNS), Mumbai and Indira Gandhi Center for Atomic Research (IGCAR), Kalpakkam, Tamil Nadu for their support throughout the investigation.

References

1. James MN (1998) Engineering materialism and structural integrity. J Eng Des 9:329–342. https://doi.org/10.1080/095448298261471
2. Alam MS (2005) Structural integrity and fatigue crack propagation life assessment of welded and weld-repaired structures 1–191
3. Ramu M (2013) Investigation on box type configuration of roof slab for prototype fast breeder reactor. Anna University, Chennai
4. Kasen MB (1961) Effect of controlled weld defects on the ballistic performance of 5083 aluminum alloy weldments
5. Isler H, Balz M (2011) Balz House
6. IGCAR Tender NO. MRPU/IGCAR/FAB/1155/TPT-807 Technical specification for the development of critical tri-junction forging for the dome shaped roof slab for future FBRs
7. Ventsel E, Krauthammer T (2001) Thin plates and shells theory, analysis, and applications. CRC Press, Newyork
8. Roark RJ, Young WC, Plunkett R (1976) Formulas for stress and strain
9. Chetal SC, Chellapandi P (2013) Indian fast reactor technology. Indian Acad Sci 38:795–815
10. Chellapandi P, Puthiyavinayagam P, Balasubramaniyan V et al (2011) Development of innovative reactor assembly components towards commercialization of future FBRs. Energy Procedia 7:359–366. https://doi.org/10.1016/j.egypro.2011.06.047
11. Welding Defect. https://www.theweldingmaster.com/welding-defects/

Fatigue Validation of Iron Casting Components by Strain-Life Principle

Sarvesh Ghongade and Shanmukha Rao Damu

Abstract Air Disc Brake is a safety critical system in a heavy commercial vehicle, and its components are subjected to various types of loads during braking. Fatigue damage of brake components due to cyclic loading is the most critical and common component failure among other failures. Fatigue is the most widely recognized reason for failure in metal components/structures, and the main aim of this study is to predict fatigue life of casting components with Finite Element Method (FEM) simulation and correlate/validate with experimental results. Fatigue analysis of Brake Caliper of Air Disc Brake which is manufactured by sand casting is carried out using strain-life method. This study considers only mechanical fatigue of strain-life theory with plastic stress and strain. Softwares, ANSYS 18.2 and nCodeDesignLife, are used to calculate component fatigue life under static loading condition. The influence of the various parameters like Surface Roughness, different Mean Stress Correction Theories, Stress Combination Method, and Stress Gradient on the simulated results were studied in detail to correlate/validate with experimental results. It is observed that fatigue life by nCode DesignLife strain base method gives good result in comparison with experimental result.

Keywords Brake caliper · Static loading and simulation · Fatigue analysis · Mean stress correction · ANSYS · nCode DesignLife

1 Introduction

Brake is a mechanical device by means of which frictional resistance is applied to a moving member, to retard or stop the motion of a vehicle. Air brake system is used in heavy commercial vehicle, and it is one of the most important safety systems in the vehicle. Air brake systems are carefully designed with valves, tubes,

S. Ghongade (✉) · S. R. Damu
Knorr-Bremse Technology Center India Private Limited, Pune, India
e-mail: Sarvesh.Ghongade@knorr-bremse.com

S. R. Damu
e-mail: Shanmukharao.Damu@knorr-bremse.com

S. Seetharamu et al. (eds.), *Fatigue, Durability, and Fracture Mechanics*, Lecture Notes in Mechanical Engineering, https://doi.org/10.1007/978-981-15-4779-9_22

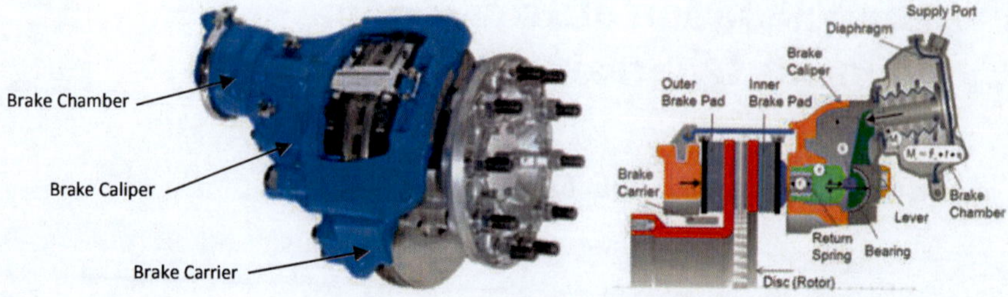

Fig. 1 Air disc brake

ABS, Air disc brake, etc., and properly selected to result in balanced vehicle braking performance. Brake Caliper is the most important component of Air disc brake which is manufactured by sand casting method with material EN GJS 600 (Fig. 1).

Brake Caliper is subjected to cyclic loading over brake/vehicle life and results in fatigue failure. Particularly when designing brake systems, the manufacturer must take great care of component/structural strength as well as fatigue life as a part of design validation before releasing designs for manufacturing samples/series production.

In many fatigue situations, a mean strain can be present on the component/system due to static/dynamic load. Strain-controlled cycling with a mean strain results in a mean stress. This relaxation is because of plastic deformation in the body, and therefore, the amount of relaxation depends on the magnitude of the plastic strain amplitude. There is more mean stress relaxation at larger strain amplitudes due to larger plastic strains. Mean stress influences fatigue life. There are different mean stress correction theories derived by strain base fatigue life principal.

In this work, static analysis of Brake Caliper carried out in FEA software tool ANSYS and fatigue life determined nCode DesignLife by strain base method with Linear and Non-Linear material properties. nCode DesignLife strain base module performs analysis using both Linear and Non-Linear FE solutions. Fatigue life has been evaluated by considering strain base parameters like strain base theory, for example, Smith–Watson–Topper, Morrow's theory, and stress combination method.

2 Methodology

To reduce the computational time, simplified models were developed for the analysis of Brake Caliper subjected to load and boundary condition. While modeling of different models, many geometrical features are neglected to avoid geometrical singularities. Modeling in FEA tool includes all the contact, mesh control, and loading condition on the caliper geometry.

Static analysis results were used as input to fatigue life calculations in nCode DesignLife strain base module. Hence, fatigue analysis of caliper is calculated using different parameters in nCode DesignLife to set process for caliper fatigue calculation (Figs. 2 and 3).

Following are the main analysis steps.

1. CAD modeling in Creo
2. Static Analysis in ANSYS Workbench
3. Determination of deformation and stress
4. Fatigue analysis in nCode DesignLife by strain base method
5. Comparison with experimental result.

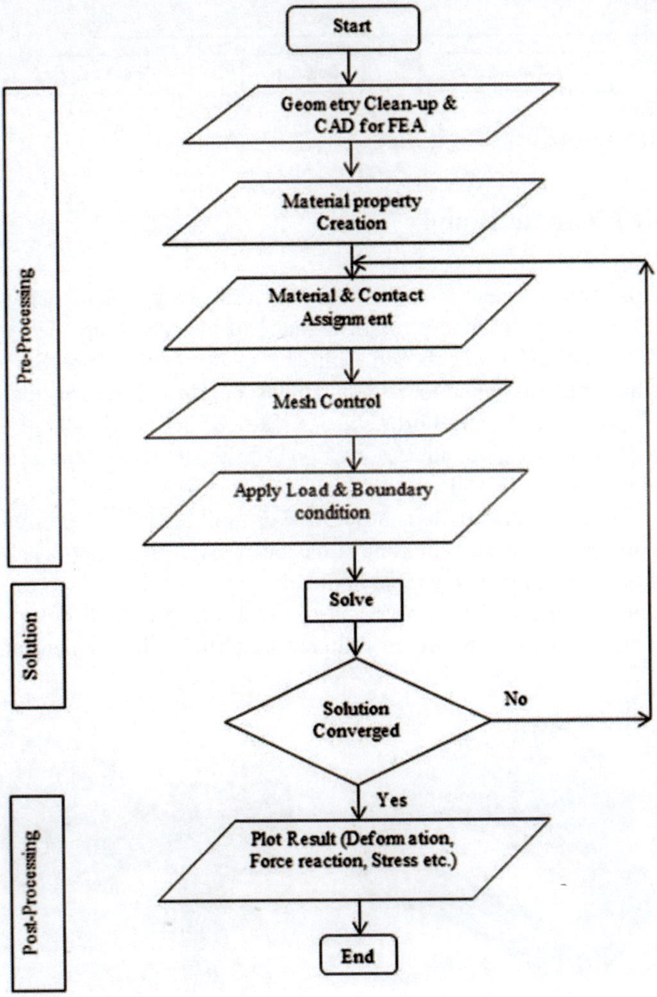

Fig. 2 Static analysis flow chart

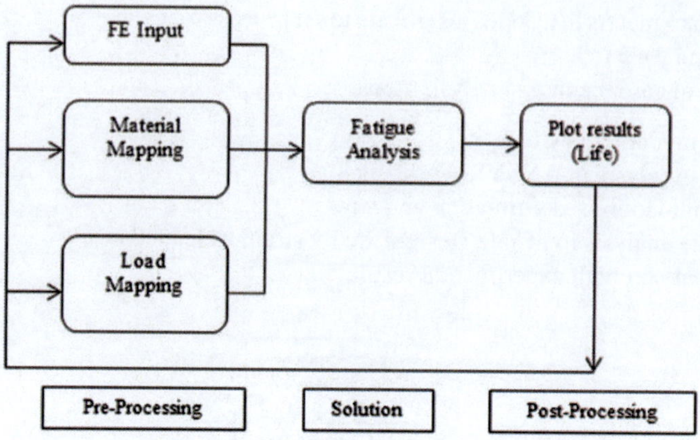

Fig. 3 Fatigue analysis flow chart

3 Finite Element Model

Brake Caliper finite element model for the structural analysis is modeled by simplifying CAD geometry; it reduces total calculation time. The finite element analysis has been done in ANSYS Workbench. Caliper is made of cast iron; GGG 60 material properties have been assigned to caliper. Analysis is carried out considering Linear as well as Non-Linear material properties. Brake Caliper analysis has been carried out for the given load and boundary condition. Mesh pattern and contact setting have been selected properly (Fig. 4).

Brake force of 275 kN is applied through the piston. Boundary condition of various fixed supports and displacement control has been given to the model. Brake force for caliper is given in the below table in detail.

The main purpose of FE analysis is to determine the total deformation and maximum principal stress of Brake Caliper. Result of finite element analysis has

Fig. 4 Brake caliper

been used for fatigue life evaluation in nCode. Fatigue life is determined in nCode DesignLife by strain base method.

4 Strain Base Fatigue Theory

It is observed that more than 50% of structural failures are because of fatigue failure only. Whenever a structure is subjected to time varying loads, fatigue must be considered.

Fatigue approach: Fatigue analysis methods are based on stress-life which is also called SN approach and is appropriate for high-cycle fatigue and strain-life, also known as εN approach.

4.1 Strain-Life

It uses elastic–plastic strains and assesses fatigue damage using cyclic strain ranges and strain-life relationship equation. It is applicable to both low-cycle and high-cycle applications. It uses local elastic–plastic strains.

The following equation is for strain-life data of small smooth axial specimens:

$$\frac{\Delta\varepsilon}{2} = \varepsilon_a = \frac{\Delta\varepsilon_a}{2} + \frac{\Delta\varepsilon_p}{2} = \frac{\sigma'_f}{2}(2N_f)^b + \varepsilon'_f(2N_f)^c$$

where

$\Delta\varepsilon/2 =$ total strain amplitude $= \varepsilon_a$.
$\Delta\varepsilon_e/2 =$ elastic strain amplitude $= \Delta\sigma/2E = \Delta\sigma_a/E$.
$\Delta\varepsilon p/2 =$ plastic strain amplitude $= \Delta\varepsilon/2$—$\Delta\varepsilon_e/2$.
$\varepsilon'_f =$ fatigue ductility coefficient.
$c =$ slopes of plastic lines (fatigue ductility exponent).
$\varepsilon'_f =$ fatigue strength coefficient.
$b =$ slopes of elastic line (fatigue strength exponent).
$E =$ modulus of elasticity.
$\Delta\sigma/2 =$ stress amplitude $= \sigma_a$.

4.2 Mean Stress Correction

While designing fatigue durability, the presence of non-zero mean stress can influence fatigue behavior of the material because of the compressive or tensile nature. Mean strain can be present. Strain-controlled cycling with mean strain usually results in a mean stress which may relax fully or partially.

The main aspect of a strain cycle that affects fatigue damage is its range. Fatigue damage is also influenced by the mean stress of each cycle. Mean stress correction methods allow the effect of mean stress to be modeled and considered in the life prediction. Several theories are developed to consider the effect of mean stress. Morrow's mean stress method replaces σ'_f with σ_m in strain-life equation, where σ_m is the mean stress, such that

$$\frac{\Delta\varepsilon}{2} = \varepsilon_a = \frac{\sigma'_f - \sigma_m}{E}(2N_f)^b + \varepsilon'_f(2N_f)^c$$

σ_m is taken to be positive for tensile values and negative for compressive values.

The Morrow mean stress correction adjusts the elastic part of the strain-life curve before looking up the life/damage. Morrow's mean stress theory is used for steel and for components which are having long life and less plastic strain significance.

Smith–Watson–Topper method considers the product of strain amplitude and maximum stress of each cycle. Smith, Watson, and Topper (often called the "SWT") is

$$\sigma \text{max} \varepsilon_a E = (\sigma'_f)^2(2N_f)^{2b} + \sigma'_f\varepsilon'_f E(2N_f)^{b+c}$$

where $\sigma_{\max} = \sigma_m + \sigma_a$ and ε_a is alternating strain. This equation assumes that for different combinations of strain amplitude ε_a and mean stress σ_m, the product is σmax. ε_a remains constant for given life. Smith–Watson–Topper theory is suitable for grey cast iron, hardened carbon steel, and micro-alloyed steel. Smith–Watson–Topper theory is suitable for Brake Caliper analysis.

Elastic–plastic corrections are often required because the local strain approach requires total (elastic–plastic) strain as well as stress; yet for efficiency reasons, calculations often must be based on Linear elastic FE calculations. Notch corrections allow elastic–plastic strains and stresses to be estimated based on elastic FE results.

(1) Neuber: Suitable when structure is under uniaxial loading.
(2) Hoffmann–Seeger: Suitable when structure is under multi-axial loading.

This correction is also suitable for structure with notch having nominal elastic stresses; local plastic strains can occur near notches (Fig. 5).

5 nCode DesignLife Modeling

nCode is a tool of HBM. It is specially used for test measurement as well as various fatigue calculations, e.g., stress-life, strain-life, crack growth, etc. Brake Caliper fatigue life is evaluated by strain base (E-N) module. Strain-life material properties have been assigned to Brake Caliper.

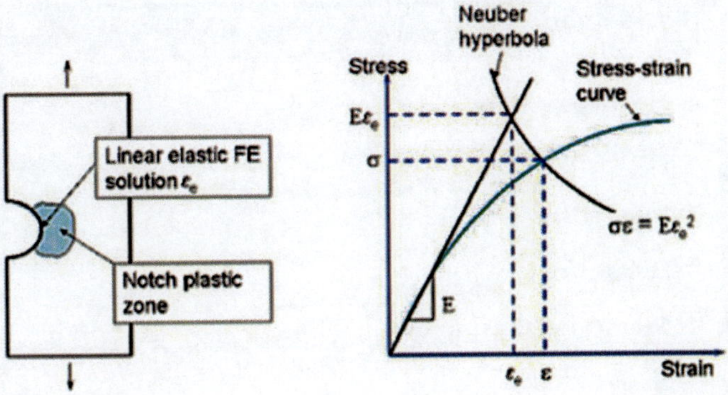

Fig. 5 Neuber correction

5.1 Material Mapping

The most important parameter which affects fatigue life of component is Surface Roughness. nCode DesignLife allows to consider roughness parameter through material mapping. nCode DesignLife has three options to enter surface parameter.

(1) Surface treatment factor ($K_{treatment}$): It considers the type of surface treatment and their effects.
(2) User surface factor (K_{user}): This factor allows user to adjust the fatigue life, based on the past experience and actual experimentation result.
(3) Surface roughness ($K_{roughness}$): It defines the type of surface roughness. It is defined by various options like selecting manufacturing operation, entering surface roughness value in micron, and entering surface roughness factor (Fig. 6).

nCode DesignLife combines effect of all three parameters as one surface factor.

$K = K_{treatment} * K_{user} * K_{roughness}$

These parameters are selected from FKM guideline for Brake Caliper analysis. Caliper is having shot blasting surface treatment. 0.8 $K_{roughness}$ factor and 1.1 $K_{treatment}$ factor has selected from FKM guideline (Fig. 7).

5.2 Load Mapping

nCode has different type of load mapping consideration, it has been used according to the application.

Constant amplitude assumes FE stress cycle between minimum and maximum. Time step load mapping directly uses the FE stress and strain result. In constant amplitude, minimum stress is zero; there is no effect of unloading. It allows to consider stress ratio as $R = 0$ or -1.

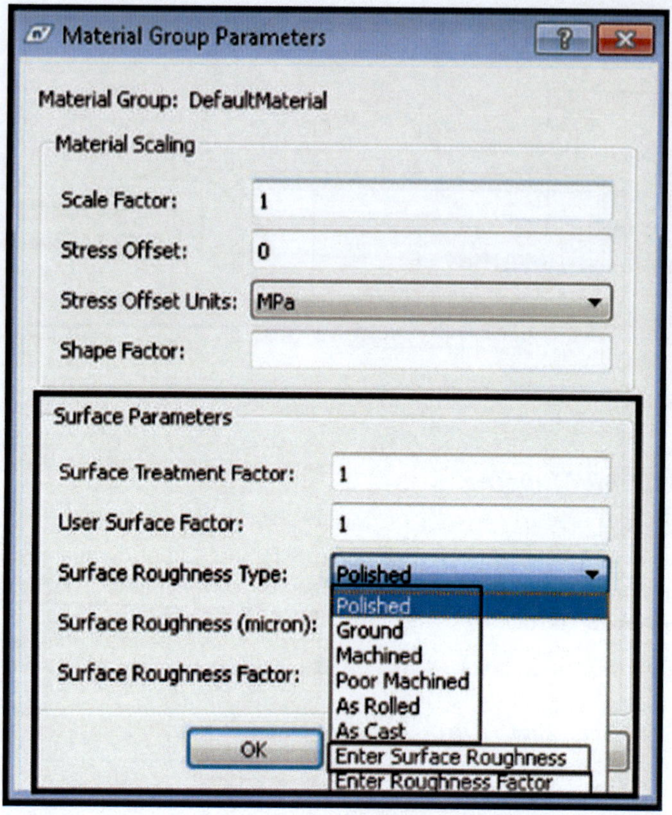

Fig. 6 nCode Material parameters

For caliper analysis, time step method is used by considering effect of loading and unloading (Fig. 8).

5.3 *Fatigue Analysis Input*

nCode analysis carried out by assigning material properties, load and different fatigue properties. FE Analysis carried out with considering Linear material properties. nCode Fatigue analysis carried out with considering Smith Watson Topper Mean Stress correction theory and Hoffmann-Seeger Elastic-Plastic Correction factor.

nCode DesignLife has elastic–plastic correction consideration when FE results are with Linear material properties. This factor converts Linear FE result into Non-Linear and computes fatigue life. There is no need to consider this factor for non-Linear FE results. All parameters are defined in nCode and analysis is performed (Fig. 9).

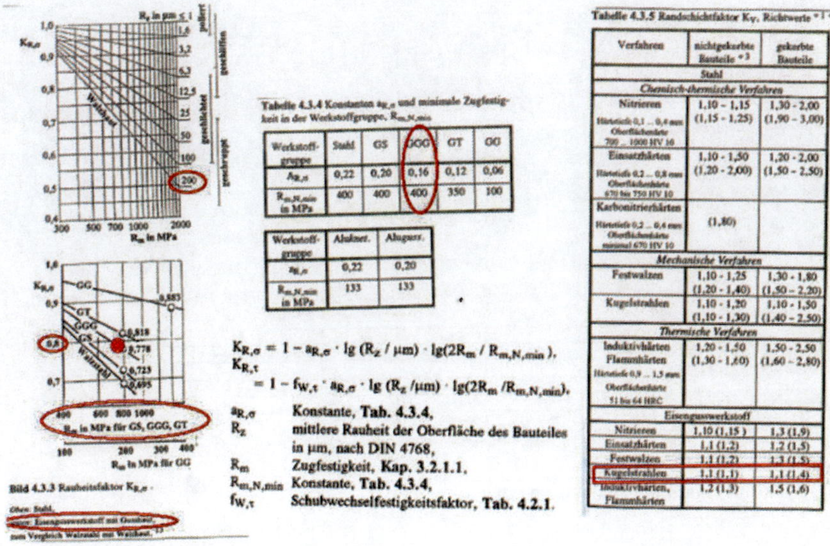

Fig. 7 FKM Surface Roughness and Surface Treatment Factor

Fig. 8 nCode Load mapping

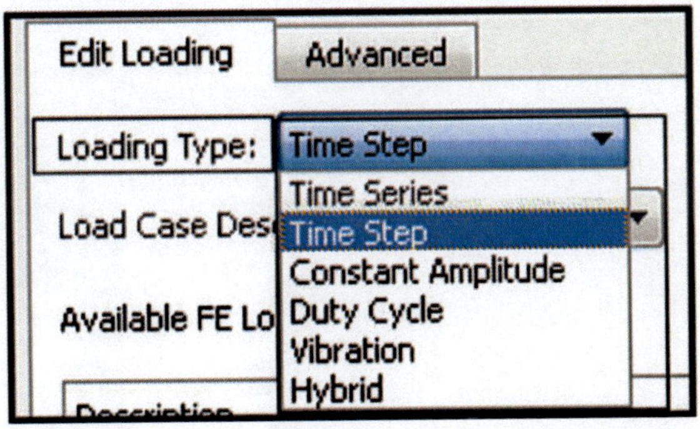

6 Result

Total deformation and maximum principal stress are evaluated for Brake Caliper. The results of different Brake Caliper models are given in the below table. Caliper deformation and stresses are observed.

Fatigue life estimated by Smith–Watson–Topper (SWT) mean stress correction theory. Fatigue life detail is given in the below table (Table 1 and Figs. 10 and 11).

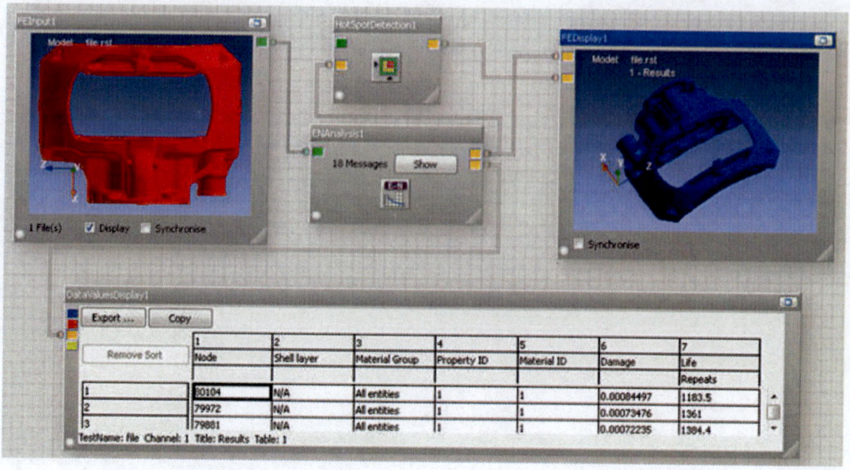

Fig. 9 nCode Strain life analysis layout

Table 1 nCode Fatigue life

Model no	Caliper 1	Caliper 2
nCode	31220	24490

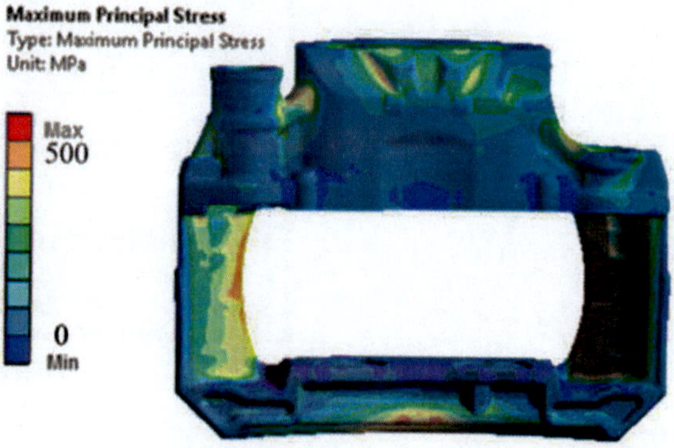

Fig. 10 Caliper maximum principal stress

7 Experimentation

Brake Caliper test has been carried on special test rig. Test has been carried out with complete Air Disc Brake model. Initially, air pressure has been adjusted to get constant brake force on Brake Caliper. Force has been measured by various load cells. Brake Caliper is subjected to repetitive loading for over period. Few numbers of sample get tested. Results were recorded.

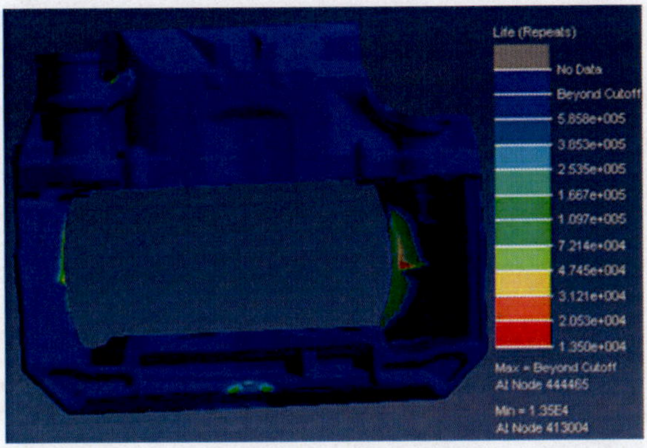

Fig. 11 nCode caliper fatigue plot

Table 2 Experimental Fatigue life

Model no	Average fatigue life
Caliper 1	32600
Caliper 2	32400

Experiment test has been carried out for different brake model sample. The experimental results of different Brake Caliper models are given in the below table (Table 2).

Experimental failure location is similar like failure location determined by finite element analysis (Fig. 12).

Fig. 12 Test caliper

Table 3 Fatigue life comparison

Model no	Caliper 1	Caliper 2
ncode	31220	24490
Exp. result	32600	32400

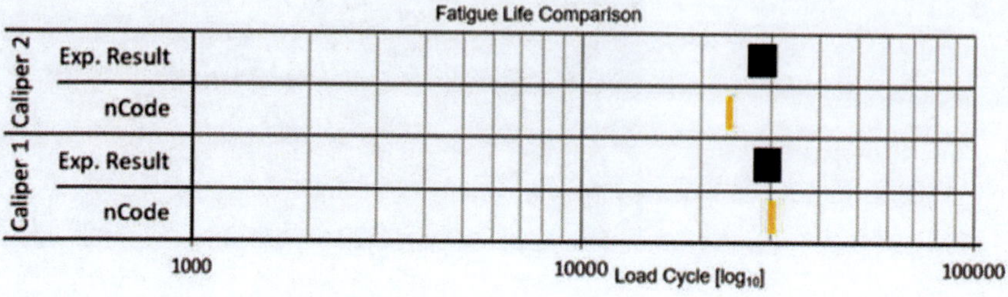

Fig. 13 Fatigue life result comparison

8 Result Comparison

Experimental fatigue life results are compared with analytical fatigue life (Table 3).

From fatigue life comparison, it is found that Smith–Watson–Topper is giving good fatigue life result.

Fatigue life comparison result is shown in the below graph (Fig. 13).

9 Conclusion

From static analysis, it is found that caliper failure location determined by finite element analysis is similar to caliper failure location determined by experiment. In this work, different nCode parameter has been used for strain base fatigue analysis of Brake Caliper. Strain base fatigue life has been estimated and validated with experimental result. It is found that Brake Caliper fatigue life estimated by nCode strain base method is giving good result in relation with experimental result.

References

1. Schlitz W (1996) A history of fatigue. Eng Fract Mech 54(2):263–300, Elsevier
2. Aghav HL, Walame MV (June 2016) Stress analysis and fatigue analysis of front axle of heavy-duty truck using ANSYS ncode design life for different loading cases ISSN: 2248–9622 6(6):78–82
3. Shivakumar MM, Nirmala L (August 2014) Fatigue life estimation of chassis frame FESM bracket for commercial vehicle. (IJSR) ISSN (Online): 2319–7064

4. Stiffness calculation of car camber link to improve the buckling strength. Int J Sci Res (IJSR) ISSN (Online): 2319–7064
5. Sachin R, Rakesh C (2016) A study on fatigue life optimization and performance analysis of wheel disc. IJEDR 4(4) ISSN: 2321–9939
6. Vikram A, Shedge, Munde KH (May 2015) Fatigue analysis correlation and failure analysis of optimized connecting rod. Int J Mech Eng ISSN 2321–6441
7. Finite element method. Int J Recent Adv Mech Eng (IJMECH) 4(4) (2015)
8. Stephens, Fatemi, Fuchs. Chapter 1–5 from "Metal fatigue in engineering". Willy Interscience publication
9. Chapter 1 and 3 from nCode glyphworks fatigue theory guide

Numerical Simulation of Fatigue Crack Growth in A Pre-Cracked Thin Plate Repaired With A FRP Patch

Hanmant Shinde, Prashant Kumar, Madhuri Karnik, Prakash Shinde, and Ghanshyam Prajapati

Abstract A numerical simulation was performed to study the fatigue behavior of a center pre-cracked thin panel of an aluminum alloy 6061-T6 repaired with a polymer composite patch. The patch consists of a thin separating woven GFRP (glass-fiber-reinforced plastics) ply and load-bearing UD-CFRP (carbon-fiber-reinforced plastics) plies. The model was simulated using the finite element package ANSYS 15.0. The cohesive zone material model (CZM) was employed to simulate the behavior of the interface between the skin and the patch. The numerical model evaluated J-integral for different crack lengths ($2a = 20$ mm, 22 mm, …., 32 mm). The value of stress intensity factor (K_I) was evaluated for each crack length using equivalence relation between the J-integral and K_I. A second-order polynomial curve was fitted to obtain the algebraic relationship between ΔK_I and crack length. The Paris law was invoked to get the relation between the crack growth and the number of cycles. The relationship, predicted by the numerical simulation, matched well with those obtained from experimental studies.

Keywords Repair patch · Fatigue life · Fatigue crack growth · J-integral · Stress intensity factor · UD-CFRP

1 Introduction

The replacement of damaged aerospace structures with a new one is often expensive. Thus, the repairing of the damaged component is a cost-effective method. The repair of metal structures with composite materials was first introduced in Australia in the early 1970s and later in the USA in the early 1980s [1–3]. The bonding of composite patches for repairing components with cracks and defects may have advantages over the riveting and the welding process from a stress concentration point of view [4]. A CFRP patch reduces stress in the skin and thus decreases stress intensity factor which substantially increases fatigue life. Repair of cracked or corroded structures using

H. Shinde (✉) · P. Kumar · M. Karnik · P. Shinde · G. Prajapati
Department of Mechanical Engineering, College of Engineering Pune, Pune 411005, India
e-mail: hps.mech@coep.ac.in

S. Seetharamu et al. (eds.), *Fatigue, Durability, and Fracture Mechanics*, Lecture Notes in Mechanical Engineering, https://doi.org/10.1007/978-981-15-4779-9_23

composite patches resulted in a cost-effective method for extending the service life [5]. Researchers have published test results at different fatigue loading conditions to assess the viability of repaired steel and aluminum specimens [6–11]. Fatigue tests of repaired specimens in different environmental conditions are performed by many researchers [12–18]. The effect of change in temperature on fatigue life was studied. Results showed that patch–skin interface properties affected fatigue life at different environment temperatures. The behavior of the epoxy interface is sensitive to temperature compared to CFRP stiffness. It was observed that repair efficiency with the CFRP patch improves from cold to high temperature [18]. Results were different for tests at low and high temperatures due to a shift in failure mechanisms.

In this study, the experimental fatigue testing of a pre-cracked aluminum alloy 6061-T6, repaired with the asymmetric CFRP patch, was performed at room temperature of 32 °C and at the high temperature of 80 °C. The relation between the applied stress and the number of cycles required for the crack growth was obtained through experiments. Numerical simulation of the specimen was performed at the fatigue load conditions for both temperatures using the cohesive zone material (CZM) model in ANSYS software and results thus obtained were compared with experimental results.

2 Experimental Setup

A thin sheet of pre-cracked aluminum alloy 6061-T6 was repaired with the 6-ply asymmetric CFRP patch. CFRP patch is bonded to aluminum alloy using VARTM (vacuum-assisted resin transfer molding) technique. The tension-tension fatigue test was conducted at the room and the elevated temperature. The heating chamber was made designed and fabricated to maintain 80 °C temperature of its inside air using heaters controlled by a PID controller. The relation between the crack growth and the number of cycles was obtained through experiments. Crack growth with respect to the number of fatigue cycles was observed using a digital microscope during fatigue testing of specimens and was recorded for further analysis.

2.1 Fatigue Testing Facility

The tension-tension fatigue testing was done on a uniaxial fatigue testing machine with a maximum load capacity of 40 KN and up to 10 Hz frequency. The specimen was loaded by the computerized axial fatigue test machine as shown in Fig. 1a at room temperature. The specimen was loaded between the upper clamp and the lower clamp of the fatigue machine. The fatigue load was applied to the lower clamp while the upper clamp remains stationary. A digital microscope was used to observe crack growth at magnification of 50x. Figure 1b shows specimen mounted within the high-temperature environment chamber and loaded between two clamps of the fatigue machine.

Fig. 1 Tension-tension fatigue testing with arrangement for crack observation: **a** at room temperature and **b** the specimen mounted within the high-temperature environment chamber

2.2 *Specimen Preparation*

Figure 2a and b shows the configuration of the 6-ply specimen and prepared FRP patch on skin, respectively. The skin of the specimen was cut in the rolling direction of aluminum alloy sheet of 1 mm thick. The length and width of the skin were kept 335 and 50 mm, respectively, as per the ASTM E-647 standard. The mechanical properties of the aluminum 6061-T6 sheet were determined through the standard tensile test coupons using a 10 ton UTM as yield stress = 277 ± 3 MPa and ultimate strength = 318 ± 4 MPa. A center pre-crack of 18 mm length and 2 mm width was cut with a wire EDM in aluminum alloy plate. The crack length on each crack tip was extended using surgeon blade and fresh razor blade by 0.5 and 0.25 mm, respectively. Further, 0.25 crack length was extended by applying 40% of yield stress as fatigue load, providing an overall crack length of $2a = 20$ mm.

The center cracked skin was repaired by bonding a single-sided UD-CFRP patch. The 6-ply patch specimens were prepared with different lengths of each ply to achieve a ply drop of 4 mm between the neighboring plies on the leading edges of the patch. The shortest and longest CFRP plies were 24 and 64 mm, respectively. The thicknesses of GFRP and CFRP plies were 0.08 mm and 0.17 mm, respectively. Unidirectional (UD) carbon fiber sheet of 160 gsm was chosen as reinforcement of the FRP patch. The mechanical properties of carbon fiber were $E = 230$ GPa and $\nu = 0.2$. Between the aluminum alloy panel and the CFRP plies, the thin GFRP ply, reinforced with a glass fiber sheet, was placed to avoid corrosion with aluminum alloy panel. The mechanical properties of epoxy were $E = 3.8$ GPa, $\nu = 0.4$, and ρ

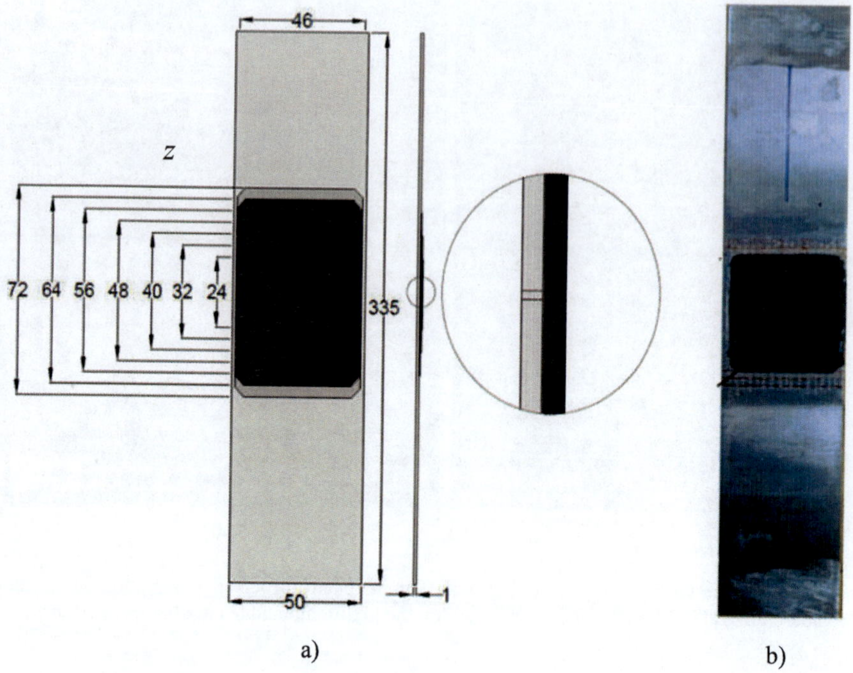

Fig. 2 The 6-ply patch specimen: **a** schematic view and **b** FRP patch on the skin

= 1200 kg/m^3. A silane material, 3-aminopropyltriethoxysilane, was mixed in the epoxy resin, 0.5% of resin mass, to improve the bonding strength between aluminum alloy and FRP patch. CFRP patch is bonded to aluminum alloy using the VARTM technique. The specimen was allowed to be cured completely within the bag for at least 12 more hours. Post curing of the specimen was done in an oven for 5 h at 80 °C, as per the specifications of the epoxy supplier so that the patch was cured completely and the bonding between patch and skin was improved. The volume fraction of GFRP and of CFRP plies was approximately 50%.

The thickness of the CFRP patch was determined using stiffness ratio considerations required for sufficient life of specimen under fatigue loading. Stiffness ratio of the repaired section of the skin is defined as

$$S = \frac{E_p t_p}{E_s t_s} \quad (1)$$

where E_p and E_s are the moduli of the patch and the skin, and t_p and t_s are the thicknesses of the patch and the skin. The stiffness ratio of the 6-ply patch specimen is 1.7. Several investigations are reported in the literature, the patch is designed to have "S" close to 1.6 for static loading [17].

2.3 Fatigue Testing

The specimen was loaded in the fatigue testing machine under tension-tension loading condition at room temperature. Specimen is loaded in a universal joint clamping fixture which allows two degrees of freedom about x- and y-directions of the specimen. For fatigue tests to be conducted at high temperature, heating chamber will be clamped on the specimen and then this assembly will be mounted on the fatigue testing machine. The upper limit of the load was set, and the lower limit was kept 10% of the upper limit, thus maintaining the stress ratio, $R = 0.1$. All the tests were carried out at a frequency, $f = 9$ Hz. The far-field stress $\sigma_{\infty} = \sigma_{\max}$ applied to the specimen was chosen as 50% of the yield stress of the skin material, i.e., 145 MPa. Three similar specimens were tested at the same fatigue load.

The specimen was continuously monitored during the fatigue test by illuminating the area near the pre-crack on the unpatched side. Crack length (a) versus the number of cycles (N) was recorded for each 0.25 mm increase in crack length. Crack propagation was monitored until the final failure of the skin. Number of cycles for final failure, N_f were also recorded.

3 Numerical Technique

A numerical model of the specimen was prepared in ANSYS. The study was carried out in two phases: (i) to determine the stress intensity factor (K_I)-based J-integral values obtained in numerical simulation and (ii) invoking the Paris law to obtain the relation between crack growth and cycles.

3.1 Geometry Modeling

The specimen was symmetric about x- and y-axes. Therefore, only a quarter model ABCD was considered for analysis, which saves computational time and computational memory. Figure 3b shows the quarter model of the repaired specimen with the crack tip at point T.

An aluminum skin of dimensions 190 mm × 50 mm with a thickness of 1 mm was used for the numerical purpose because the actual length, i.e., 335 mm would unnecessarily increase the number of nodes and number of elements leading to increase in computational time. It was modeled in such a way that it incorporates the initial crack length $a = 10$ mm. The crack was introduced on AT portion of AB (Fig. 3b). The polymer composite patch consists of GFRP ply and CFRP plies were also modeled.

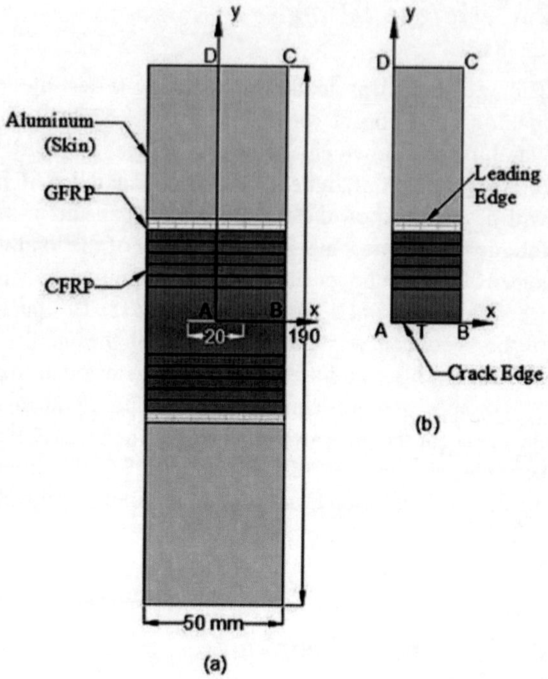

Fig. 3 Specimen geometry for numerical simulation: **a** specimen repaired with a polymer composite patch of the 6-ply thickness and **b** a quarter model

3.2 Meshing

External tensile stress (σ^{∞}) is applied to the end face along the width of the specimen skin. A part of the load is transferred to the patch and thus remaining stress in the skin is reduced which in turn develops smaller stress intensity factor at the crack tips. Due to asymmetrically applied patch, the neutral axis of the bare portion of the skin shifts from its middle plane toward the patch in the patched area of the specimen. Consequently, a bending moment is induced in the patched portion of the specimen. The induced bending moment develops normal stress σ_{zz} at the interface, which is known as peel stress. Fine meshing was done near the crack tip and at the interface while coarser meshing was employed away from the crack tip. Figure 4 shows meshing on the quarter specimen: (a) meshing in the xy-plane and (b) meshing in the thickness direction.

The skin was meshed with the help of solid 186 elements which is 20-noded isoparametric element. The element size of 0.2 mm was uniform along the x-direction of the skin. The small element size of 0.25 mm was chosen up to 12 mm length of skin along y-direction from the crack plane to improve the accuracy of FEM analysis. The element size of 1mm was used from 12 mm (smallest CFRP ply) to 36 mm (GFRP ply) length along the y-direction, while the element size of 2 mm was used beyond 36 mm length of the. From the interface of the skin and FRP patch, the element size chosen was 0.03, 0.07, 0.15 mm in sequence toward the free surface of the

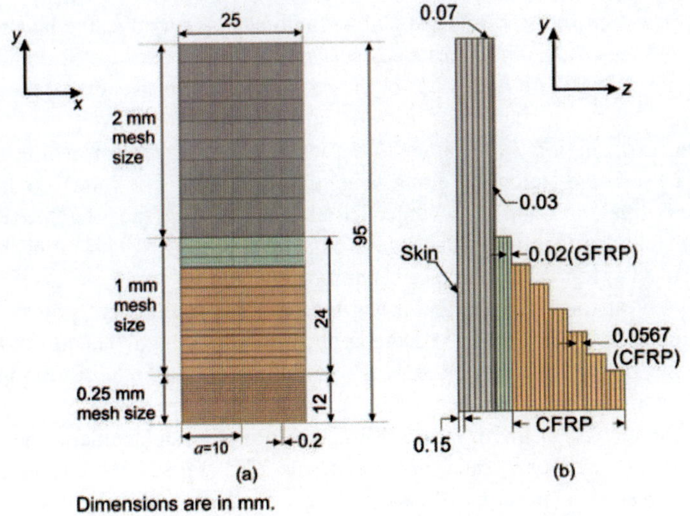

Fig. 4 Meshing on the quarter specimen: **a** meshing in xy-plane and **b** meshing in the thickness direction

skin (unpatched side) to increase the accuracy. The crack was introduced in the skin with the help of the fracture mechanics module provided by ANSYS. The crack was introduced in "AT" portion of "AB" by defining the crack edge and crack propagation direction.

The cohesive zone material (CZM) model was used to simulate the interface between the skin and the patch. In the CZM model, the node matching technique was used. Element size along x- and y-directions of GFRP ply and CFRP plies was kept the same as that of skin to match the nodes on the skin and FRP patch. The interface between the skin and the patch was meshed by using special interface elements INTER 204 which is 3D 16-node quadratic elements having 3 DOF at each node.

Separate numerical simulation was carried out for crack length, i.e., $2a = 20$, 22, 24, 26, 28, and 32 mm. Therefore, meshing needs to be repeated as per the above-discussed strategy for each different crack length.

3.3 *Cohesive Zone Material Model*

The portion of the patch may separate from the skin due to the applied fatigue load. For example, some separation exists around the crack tip and near the crack edge. Also, some separation may be there close to the leading edge [19]. To simulate the failure of the patch, such separation between the skin and the patch should be accounted for. In the current study, this is carried out by considering a cohesive zone material layer at the interface. In the CZM model, there is a single layer of interface elements between the skin and the patch which acts as an adhesive layer between the skin and the patch. The interface is susceptible to fail because of the difference in material properties on both sides of the interface and the strength of the adhesive is not high.

The normal traction T_n separates interface in normal direction while tangential traction T_t separates interface along tangential direction. The interface separation along the normal direction is accounted by distance δ_n, while the interface separation along tangential direction by distance δ_t is shown in Fig. 5. The bilinear traction model is used to represent interface separation.

At a point at the crack tip and along the interface, coincident points A and B are considered as an initial state before application of the load. The normal traction T_n increases linearly with an increase in separation distance δ_n until point A where normal traction is $T_n^{\max}$ as shown in Fig. 5a. It then decreases to zero value due to the gradual degradation of the interface. Similar to normal displacement, the interface starts becoming weaker in shear once it attends $T_t^{\max}$ (Fig. 5b). It causes complete separation in shear at point D, when δ_t becomes δ_t^c. The maximum normal traction, $T_n^{\max} = 35$ MPa, and the maximum shear traction, $T_t^{\max} = 25$ MPa, were the yield stresses of the epoxy adhesive and they were taken as the cohesive strength of the interface [20].

For mode I and II, the area under ΔOAC and ΔOBD is equal to the critical energy release rate required for patch separation, G_{Ic} and G_{IIc}, respectively. Thus,

$$G_{IC} = \left(T_n^{\max}\right) x \left(\delta_n^c\right) \tag{2}$$

$$G_{IIC} = (T_t^{\max}) x \left(\delta_t^c\right) \tag{3}$$

Critical energy release rates of an interface between aluminum alloy and FRP patch were determined experimentally as Mode I, G_{Ic}= 87.3 J/m^2 and Mode II, G_{IIc} = 771 J/m^2 [21]. Therefore, by using Eqs. 2 and 3, critical separation distance was found to be $\delta_n^c = 0.0049$ mm and $\delta_t^c = 0.062$ mm. In this case of mixed-mode fatigue, under the combined action of mode I and mode II, a failure criterion is needed. ANSYS uses the following criterion:

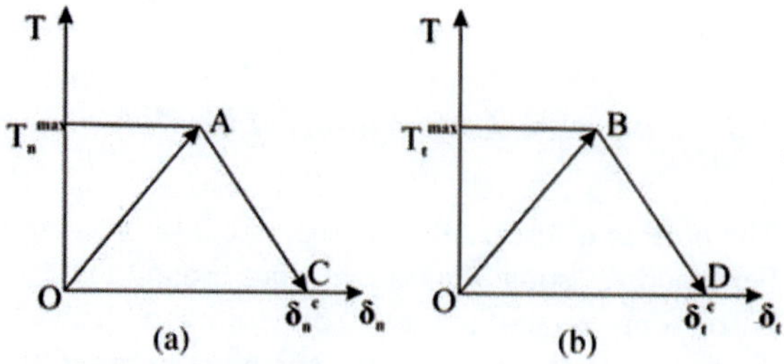

Fig. 5 Traction–separation curve in **a** normal and **b** tangential directions

Table 1 Stress–strain data input to numerical analysis of aluminum alloy 6061-T6

True stress (MPa)	0	290	302	320	331	342	348	354	358	362
Plastic strain (mm/mm)	0	0	0.004	0.0096	0.015	0.023	0.03	0.04	0.061	0.09

$$\left(\frac{G_I}{G_{Ic}}\right)^2 + \left(\frac{G_{II}}{G_{IIc}}\right)^2 \geq 1 \tag{4}$$

3.4 *Material Properties*

The elastic–plastic behavior of the skin material was given as input to the analysis through the multilinear isotropic hardening model. The stress–strain relation was obtained by testing the specimen on the Universal Testing Machine (UTM). This model considers multilinear true stress versus plastic strain relation in which the elastic–plastic behavior was characterized by multiple small linear segments defined by values given in Table 1.

Linear orthotropic material properties were assigned for woven GFRP and CFRP patches. The elastic constants of the UD patches were obtained using the rule of mixture and Halpin–Tsai relations from the properties of constituent materials, carbon fibers, and epoxy [22]. Properties of GFRP and CFRP are mentioned in Table 2.

3.5 *Solving the Material Model*

Figure 6 shows the symmetric boundary conditions which were applied on the face marked as AD and BT. In the fatigue experiments, the specimen was pulled as a two force body with a hinge joint at each end. Thus, in the numerical analysis, uniform load was applied in y-direction at the nodes located on the end face CD. Once the physical model is converted into a virtual mathematical model, it was solved to obtain an elemental and nodal solution. The iterative solver was used for numerical analysis.

The model is solved to determine the stress field and J-integral values for different initial crack lengths ($2a = 20, 22, 24, 26, 28, 30, 32$ mm). The above-described procedure was repeated for each crack length to determine the stress field and J-integral. The above numerical simulation procedure was repeated for analysis at elevated temperature. The thermal load was defined as 80 °C temperature on the specimen.

Table 2 Orthotropic elastic constants of GFRP and CFRP

Material property	Woven glass/epoxy (GFRP) composite	Unidirectional carbon/epoxy (CFRP) composite
Longitudinal modulus, E_1 (GPa)	17	117
Transverse in-plane modulus, E_2 (GPa)	17	14.16
Transverse in-plane modulus, E_3 (GPa)	9.75	14.16
Major in-plane Poisson's ratio, υ_{12}	0.2	0.28
Major out-of-plane Poisson's ratio, υ_{23}	0.4	0.37
Major out-of-plane Poisson's ratio, υ_{13}	0.4	0.28
In-plane shear modulus, G_{12} (GPa)	3.5	4.03
Out-of-plane shear modulus, G_{23} (GPa)	2.98	5.147
Out-of-plane shear modulus, G_{13} (GPa)	2.98	4.03

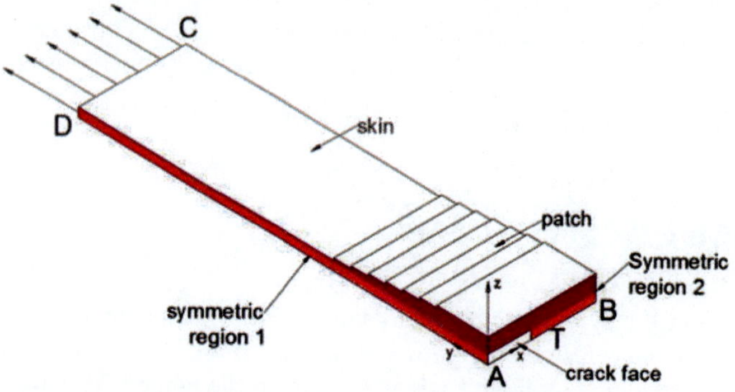

Fig. 6 Boundary conditions for the quarter model of the specimen

4 Results and Discussion

The fatigue tests were conducted at room temperature (RT) and elevated temperature (ET) and crack length (*a*) versus number of cycles (*N*) for the 6-ply patch specimen were monitored. Results of fatigue testing at both environment temperatures were compared with each other and the results of numerical simulation.

Table 3 Applied fatigue load versus failure cycle for the bare pre-cracked specimen at room temperature (R.T.) and 80 °C temperature

Exp. no	Applied maximum stress (MPa)	Failure cycles	
		At 32 °C	At 80 °C
1	120	3203	3151
2	115	4512	4293
3	110	6573	5971
4	105	8147	8265

4.1 Experimental Results for Bare Specimen

The bare (without patch) specimens with pre-crack were tested at room and elevated temperatures at stress ratio 0.1 and frequency 9. The fatigue load σ_{max} stress varied between 105 and 120 MPa. Table 3 shows the applied fatigue load versus failure cycle for the bare pre-cracked specimen at room temperature and 80 °C temperature. The fatigue cycles required to complete failure increased with decreasing applied maximum stress. The result shows that there was a very minor effect of the increase in environmental temperature on the fatigue life of the bare specimen. This was expected as properties of the aluminum skin do not change significantly at 80 °C.

4.2 Experimental Results for the Patched Specimen

Three specimens were tested under tension-tension fatigue loading at room temperature. Similarly, another three specimens were tested at elevated temperature. The maximum stress, σ_{max}, was chosen 50% of the yield stress of the aluminum alloy. Table 4 shows specimen (Skin) failure cycles for three specimens at room and elevated temperature.

The crack tips grew under the fatigue cycles till the skin was broken into two parts. Then the entire load was taken by the patch requiring few additional fatigue cycles for the complete failure. The specimens tested at high temperature showed the same failure behavior as that of corresponding specimens tested at room temperature.

Table 4 Number of cycles (*N*) for the patched specimen at room at elevated temperature

Exp. no	Failure cycles	
	At 32 °C	At 80 °C
	Skin failure	Skin failure
1	114,803	315,166
2	123,712	297,587
3	107,849	323,322
Average	115,455	312,025

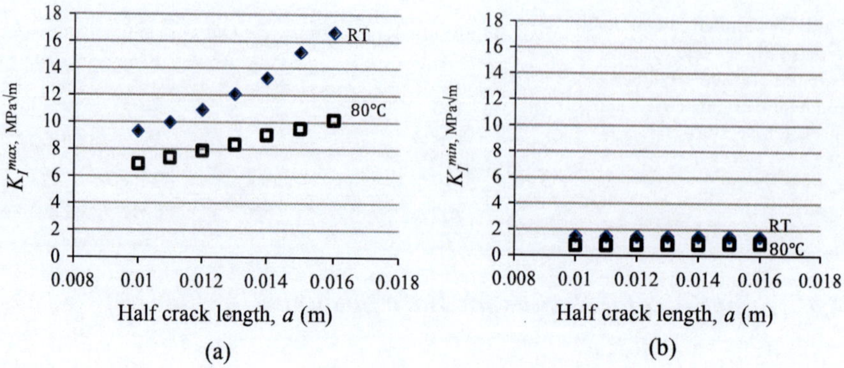

Fig. 7 Comparison of stress intensity factor (K_I) vs. crack length (a) for patched specimen at room temperature and at 80 °C, **a** $K_I^{\max}$ versus crack length, and **b** $K_I^{\min}$ versus crack length

4.3 Numerical Simulation

The specimen was numerically analyzed at room and elevated temperature. The maximum and minimum stresses applied were $\sigma_{\max} = 145$ MPa and $\sigma_{\min} = 14.5$ MPa, respectively. Stress $\sigma_{\max} = 145MPa$ was elastic almost in the entire skin except with a small plastic zone at the crack tip, and therefore the linear elastic fracture mechanics (LEFM) was invoked to determine SIF at the crack tip of skin. The stress intensity factor K_I was evaluated using the following relation [23]:

$$K_I = \sqrt{J_I E} \tag{5}$$

J-integral for different initial crack lengths at a maximum load was determined through numerical simulation. $K_I^{\max}$ and $K_I^{\min}$ were evaluated for maximum and minimum applied loads using Eq. 5 for all J-integral values.

Figure 7a shows a plot of $K_I^{\max}$ versus a. It can be observed that the stress intensity factor ($K_I^{\max}$) for elevated temperature is less than the stress intensity factor at room temperature. This is due to the large difference in the coefficient of thermal expansion ($\propto$) values of the CFRP plies and the aluminum alloy skin. The $\propto$ of CFRP plies in the longitudinal direction is $1.25 \times 10^{-6}/$ °C and in the transverse direction $\propto$ is $20.2 \times 10^{-6}/$ °C [22]. The $\propto$ of skin (Aluminum 6061-T6) is $23 \times 10^{-6}/$ °C [24].

At elevated temperature, the skin tends to expand while the CFRP patch with much lower $\propto$ restricts its expansion. Therefore, compressive in-plane thermal stress was developed in the skin. In short, the in-plane stress σ_{yy} in the skin is the superposition of two kinds of stresses: (i) compressive stress through temperature increase and (ii) tensile stress developed in the skin through applied stress σ^{∞}. This results in the reduction of σ_{yy} in the skin and corresponding reduction in $K_I^{\max}$. Similarly, the stress intensity factor ($K_{\min}$) for $\sigma_{\min} = 14.5MPa$ is less at elevated temperature than the stress intensity factor at room temperature (Fig. 7b).

Table 5 Crack length ($2a$) and stress intensity range (ΔK_I) for 6-ply patch specimen at room and elevated temperatures

Crack length, $2a$ (mm)	ΔK_I (MPa$\sqrt{m}$)		
	At 32 °C	At 80 °C	% change
20	8.542	5.455	36
22	9.180	5.904	36
24	10.06	6.428	36
26	11.22	6.906	38
28	12.38	7.608	38
30	14.26	8.101	43
32	15.70	8.726	44

The stress intensity range, $\Delta K_I = K_I^{\max} - K_I^{\min}$, was evaluated using values of $K_I^{\max}$ and $K_I^{\min}$. Table 5 shows crack length (2a) and stress intensity range (ΔK_I) for 6-ply patch specimen at room and elevated temperatures. The second-order curve was fitted to obtain the algebraic relation between ΔK_I and a for stresses in the skin adjacent to the patch–skin interface as

$$\Delta K_I = 106867a^2 - 1565.7a + 13.489 \quad (6)$$

where ΔK_I is in MPa $\sqrt{m}$ and crack length, a unit is the meter.

Once the algebraic relation between the ΔK_I and a was known, the Paris law was invoked. Experiments were conducted to evaluate the values of material constants, C and m, in the Paris law, $\frac{da}{dN} = C(\Delta K_I)^m$ for the skin used for specimens. Paris law for aluminum 6061-T6 has already been determined as

$$da/dN = 6.40 \times 10^{(-10)}(\Delta K_I)^{1.960} \quad (7)$$

Using Eqs. (6), (7) can be rewritten and integrated for crack growth from a_0 to a to yield equation as

$$N = \int_{a_0}^{a} \frac{da}{6.40x10^{-10}(106867a^2 - 1565.7a + 13.489)^{1.960}} \quad (8)$$

Equation 8 was integrated to predict fatigue life. Fatigue life determined in numerical simulation for ΔK_I versus a relations obtained at interface nodes matched well with experimental results.

Figure 8 shows half crack length (a) versus number of cycles (N) plot obtained through numerical and experimental study for 6-ply specimen at room and elevated temperatures. Results have shown that numerical results were in a close match with experimental results. As discussed in the previous section, $K_I^{\max}$ at elevated temperature is less compared to $K_I^{\max}$ at room temperature. Therefore, there is an increase in the number of cycles at elevated temperature for particular crack length compared to

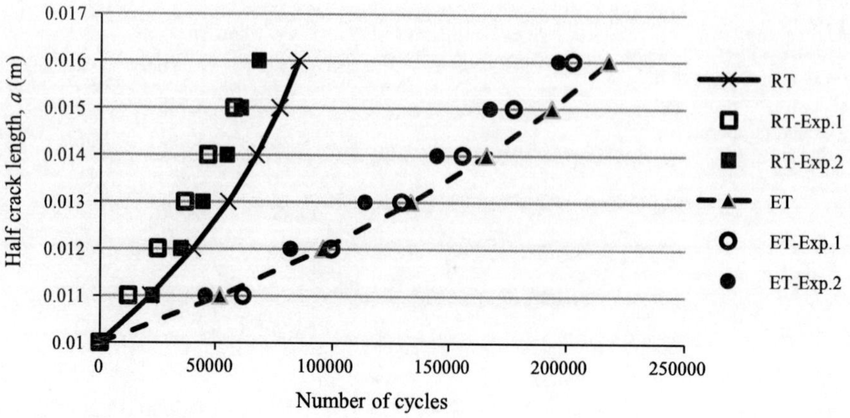

Fig. 8 Half crack growth (*a*) and the number of fatigue cycles (*N*) for the patched specimen at elevated temperature

the number of cycles at room temperature. Thus, bonding a patch at room temperature is advantageous to improve performance at elevated temperature.

5 Conclusion

Thin plate with a pre-crack was repaired with CFRP patch fatigue tested with stress ratio 0.1 and frequency 9 Hz at room and elevated temperatures. Repair of pre-crack using 6-ply patch had increased the fatigue life of the specimen considerably. Specimen failed by the extension of the crack in the skin all the way to the specimen edges. Fatigue behavior observed at elevated temperature was different than that at room temperature. Higher fatigue life was observed for specimens tested at a higher temperature for the same fatigue loading conditions. The increase in fatigue life is attributed to the reduction in stress intensity factor at elevated temperature. Stress intensity factor at elevated temperature has reduced due to a decrease in stresses in the skin.

The numerical results of the crack growth versus number of fatigue cycles were predicted using stress intensity range (ΔK_I) versus crack length (a) and Paris law. Crack growth with fatigue cycles determined using linear elastic fracture mechanics approach is working satisfactory and is in good match with experimental results.

Acknowledgements This work is carried out with the help of a research grant from ARDB (Aeronautical Research and Development Board), DRDO, New Delhi for the project "Fatigue growth characterization at cold and elevated temperatures of thin aluminium-alloy panel repaired with polymer composite patches".

References

1. Baker AA, Callinan RJ, Davis MJ, Jones R, Williams JG (1984) Repair of MIRAGE III Aircraft using the BFRP crack patching technique. Theor Appl Frac Mech 2:1–15
2. Baker AA (1987) Fibre composite repair of cracked metallic aircraft components-practical and basic aspects. Composites 18(4):293–308
3. Baker Alan (1999) Bonded composite repair of fatigue-cracked primary aircraft structure. Compos Struct 47:431–443
4. Domazet Z (1996) Comparison of fatigue crack retardation methods. Eng Fail Anal 3(2):137–147
5. Shinde PS, Kumar P, Singh KK, Tripathi VK, Sarkar PK (2015) Experimental Study of CFRP patches bonded on a cracked aluminum alloy panel. Compos Interfaces 22(4):233–248
6. Pascoe JA, Alderliesten RC, Benedictus R (2013) Methods for the prediction of fatigue delamination growth in composites and adhesive bonds—a critical review. Eng Fract Mech 72–96
7. Benyahia F, Aminallah L, Albedah A, Bouiadjra BB, Achour T (2015) Experimental and numerical analysis of bonded composite patch repair in aluminum alloy 7075 T6. Mater Des 73:67–73
8. Bouiadjra BB, Benyahia F, Albedah A, Bouiadjra BAB, Khan SMA (2015) Comparison between composite and metallic patches for repairing aircraft structures of aluminum alloy 7075 T6. Int J Fatigue 80:128–135
9. Aakkula Jarkko, Saarela Olli (2014) An experimental study on the fatigue performance of CFRP and BFRP repaired aluminium plate. Compos Struct 118:589–599
10. Albedah A, Khan SMA, Benyahia F, Bouiadjra BB (2015) Experimental analysis of the fatigue life of repaired cracked plate in aluminum alloy 7075 with bonded composite patch. Eng Fract Mech 145:210–220
11. Albedah A, Khan SMA, Benyahia F, Bouiadjra BB (2016) Effect of load amplitude change on the fatigue life of cracked Al plate repaired with composite patch. Int J Fatigue 88:1–9
12. Usman M, Pascoe JA, Alderliesten RC, Benedictus R (2017) The effect of temperature on fatigue crack growth in FM94 epoxy adhesive bonds investigated by means of energy dissipation. Eng Fract Mech 189:98–109
13. Syed AK, Zhang X, Moffatt JE, Fitzpatrick ME (2017) Effect of temperature and thermal cycling on fatigue crack growth in aluminium reinforced with GLARE bonded crack retarders. Int J Fatigue 98:53–61
14. Hosseini-Toudeshky Hossein, Sadighi Mojtaba, Vojdani Ali (2013) Effects of curing thermal residual stresses on fatigue crack propagation of aluminum plates repaired by FML patches. Compos Struct 100:154–162
15. Ergun Emin, Tasgetiren Suleyman, Topcu Muzaffer (2010) Fatigue and fracture analysis of aluminum plate with composite patches under the hygrothermal effect. Compos Struct 92:2622–2631
16. Feng Peng, Lili Hu, Xiao-Ling Zhao Lu, Cheng Shanhua Xu (2014) Study on thermal effects on fatigue behavior of cracked steel plates strengthened by CFRP sheets. Thin-Walled Struct 82:311–320
17. Vlot A, Massar JMA, Guijt CB, Verhoeven S (2000) Bonded aircraft repairs under variable amplitude fatigue loading and at low temperatures. Fatigue Fract Eng Mater Struct 23:9–18
18. Okamoto Y, Kamizono T, Namiki N, Cai H (2007) Environmental fatigue test of bonded composite patch repaired metallic structure. In Proceedings of Asia-Pacific conference on FRP in structures (APFIS 2007), pp 1025–1031
19. Shinde PS, Kumar P, Tripathi VK (2018) Dependence of repair strength on the size of FRP patch bonded to a cracked aluminum alloy panel. Thin-Walled struct 124:303–311
20. Gilat A, Goldberg RK, Roberts GD (2007) Strain rate sensitivity of epoxy resin in tensile and shear loading. J Aerosp Eng 20(2):75–89

21. Kumar Prashant (2018) Prakash Sonyabapu Shinde: fracture toughness and shear strength of the bonded interface between an aluminium alloy skin and a FRP patch. J Inst Eng India Ser. https://doi.org/10.1007/s40032-018-0467-1
22. Agarwal BD, Broutman LW (1990) Analysis and performance of fiber composites. Wiley, New York. Chapter 3, pp 76–79, 101–106
23. Kumar P (2014) Elements of fracture mechanics. Tata McGraw-Hill, New Delhi
24. Daniel IM, Ishai O (1994) Engineering mechanics of composite materials. Oxford university press, 2nd edn

Correlating Stress Ratio Effects on the Fatigue Crack Growth Rate of a Nickel Base Superalloy IN718

Sharanagouda G. Malipatil, Anuradha N. Majila, D. Chandru Fernando, and C. M. Manjunatha

Abstract Constant amplitude fatigue crack growth rate tests were conducted on a nickel base superalloy IN718 at various stress ratios, R ranging from $R = 0.1$ to 0.7. Tests were conducted at room temperature and in lab air atmosphere. Tests were performed in a 100 KN computer-controlled servo-hydraulic test machine using compact tension specimens with sinusoidal waveform at 10 Hz. Crack length was monitored by compliance technique using COD gage. Increasing stress ratio was observed to increase crack growth rates and also decrease threshold stress intensity factor range, ΔKth. Stress ratio effects on crack growth rates were correlated by using a two-parameter crack driving force, $\Delta K*$. This approach was observed to provide a reasonably good correlation which can further be employed in modeling crack growth behavior under service loads.

Keywords Fatigue crack growth · Ni base superalloy · Stress ratio · Crack driving force

1 Introduction

Nickel base superalloys are widely used in gas turbine engines for engine disk and blade constructions. They possess high strength, stiffness, corrosion resistance, and also they retain high strength at elevated temperatures. The disk material experience fatigue loads during operation of the engine. These service loads initiate cracks at stress concentration locations and further this crack starts growing under the fatigue loads and leads to fracture on attaining the fracture toughness of the material. As a part of damage tolerance evaluation in these materials, it is required to study and

S. G. Malipatil (✉) · C. M. Manjunatha
Academy of Scientific and Innovative Research (AcSIR), Ghaziabad 201002, India
e-mail: sgmalipatil@nal.res.in

A. N. Majila · D. Chandru Fernando
Materials Group, Gas Turbine Research Establishment, DRDO, Bengaluru 560093, India

S. G. Malipatil · C. M. Manjunatha
Structural Integrity Division, CSIR-National Aerospace Laboratories, Bengaluru 560017, India

S. Seetharamu et al. (eds.), *Fatigue, Durability, and Fracture Mechanics*, Lecture Notes in Mechanical Engineering, https://doi.org/10.1007/978-981-15-4779-9_24

understand the fatigue crack growth behavior of Ni base superalloys under spectrum loads.

Fatigue crack growth rate (FCGR) of a material depends on various parameters such as stress or load ratio, microstructure, loading waveform, frequency, temperature, environment, etc. [1–5]. Several studies have been made [1, 3, 6–8] on the effect of stress ratio, $R = \sigma \min/\sigma \max$ on FCGR behavior of structural alloys. The constant amplitude FCGR data shows that stress ratio does influence crack growth rate significantly [6–8]. While estimating the damage tolerance behavior of such alloys under service loads, it is necessary to merge all FCGR curves at different stress ratios and find a general correlation between crack speed and crack driving force, eliminating stress ratio parameter in the empirical equation [6–11].

Many efforts have been made to correlate stress ratio effects on the FCGR behavior in various materials [6–12]. Walker's modified equation has been used to correlate R ratio effects quite well in Al and Ti alloys [13]. However, this approach has been found to be suitable for Paris regime of crack growth only and not applicable to all materials. Crack closure effects have been used by several authors to collate FCGR curves into a single curve [8–11]. However, accurate crack closure estimation under service loads in real components is quite difficult and may lead to significant errors in prediction of fatigue life under such cases. Recently, Kujawski and co-workers [6] have demonstrated that use of two-parameter crack driving force parameter, ΔK^*, provides a better approach for correlating stress ratio effects in many materials.

As a part of detailed investigations on the damage tolerance behavior of Inconel 718 under service loads, in this investigation, an attempt is made to generate constant amplitude FCGR curves at various stress ratios and determine the correlation between stress ratio and crack speed which will be used further for crack propagation life under service loads.

2 Experimental

2.1 Material and Specimen

Nickel base superalloy Inconel 718 was used in this investigation. The chemical composition depicting major alloying elements is shown in Table 1. The material was obtained as blanks extracted from a forged disk. The mechanical properties of this material are shown in Table 2. The initial microstructure of the material is shown

Table 1 Composition (wt %) showing major alloying elements in Inconel 718

C	Si	Mn	Nb	P	Ni	Cr	Mo	W
0.025	0.07	0.06	5.39	0.01	53.80	17.9	2.93	0.04
Al	Co	Cu	Ti	Fe	B	N	Mg	Ag
0.54	0.18	0.03	0.99	18.28	0.004	0.0078	0.0015	0.0001

Table 2 The mechanical properties of Inconel 718 used in this investigation

UTS (MPa)	YS (MPa)	Poisson's ratio	Fracture toughness (MPa$\sqrt{m}$)
1241	1034	0.33	85

Fig. 1 Initial microstructure of Inconel 718 nickel base superalloy

in Fig. 1. Standard compact tension, CT specimens were machined from the blanks. Dimensions of the CT test specimen are shown in Fig. 2.

The stress intensity factor range (ΔK) for a crack in C(T) specimen was determined as [14]

$$\Delta K = \frac{\Delta P}{B\sqrt{W}} f(a/w) \quad (1)$$

where $f(a/w) = \frac{\left(2+\frac{a}{w}\right)\left(0.886+4.64*\left(\frac{a}{w}\right)-13.32*\left(\frac{a}{w}\right)^2+14.72*\left(\frac{a}{w}\right)^3+5.6*\left(\frac{a}{w}\right)^4\right)}{\left(1-\frac{a}{w}\right)^{3/2}}$ (2)

2.2 Test Equipment and Procedure

Fatigue crack growth tests were performed in a 100 KN computer-controlled servo-hydraulic test machine at room temperature (RT) and in lab air atmosphere. A crack opening displacement (COD) gage with 5 mm gage length and $\pm$ 2 mm travel was fixed at the front face of the specimen for online automated crack length measurement. All the tests were performed following ASTM E647 test standard specifications [14]. Initially, specimen was pre-cracked under cyclic loads to produce a sharp crack of

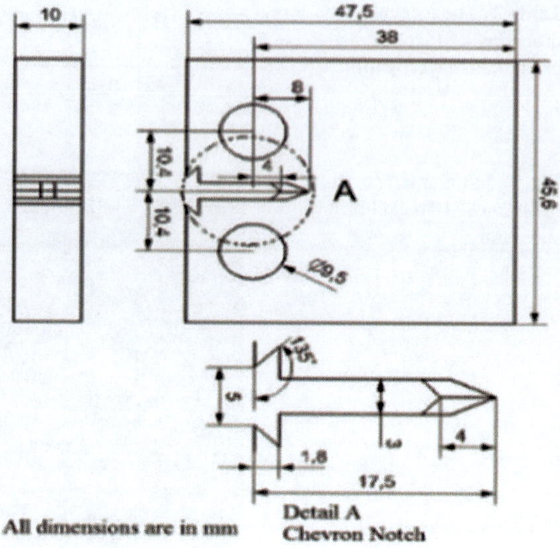

Fig. 2 Dimensions and notch details of the CT specimen geometry

about 2 mm in length from the notch root, using decreasing ΔK test procedure. Then, near-threshold fatigue crack growth rates were obtained by conducting decreasing ΔK test. The K-gradient parameter used was −0.08 [14]. Further, constant load amplitude (increasing ΔK) test was continued on the same specimen to obtain FCGR data in the Paris and high ΔK regimes. The loads were chosen in such a way that at least some overlapping FCGR data was obtained between decreasing and increasing ΔK test in the Paris regime. The tests were conducted at various stress ratio, R ranging from 0.1 to 0.7. Two tests were conducted for each R ratio and average crack growth rates were determined.

3 Results and Discussion

Constant amplitude (CA) fatigue crack growth rate (FCGR) curves at various stress ratios determined in nickel base superalloy IN718 are shown in Fig. 3. Increasing the stress ratio was observed to increase the crack growth rate as well as decrease threshold stress intensity factor range, ΔKth. Similar trend in the constant amplitude FCGR behavior of various alloys has been observed by many investigators [6–8].

An effort was made to correlate stress ratio effects using empirical equation and nullify the effect of stress ratio on the FCGR behavior. Kujawski and co-workers defined a two-parameter crack driving force, K^* as [15]

$$K^* = (\Delta K^+)^\alpha (K_{max})^{(1-\alpha)} \tag{3}$$

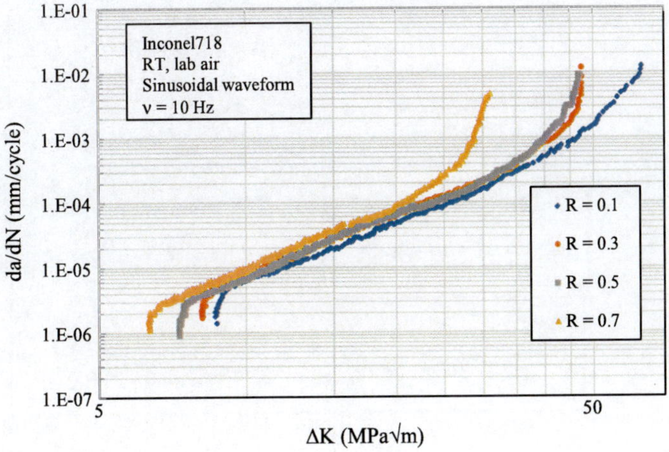

Fig. 3 Constant amplitude FCGR curves for nickel base superalloy IN718 at various stress ratios

where $\Delta K+$ = positive part of applied stress intensity factor range, $K_{\max}$ = maximum stress intensity factor range, and α = material constant. Further studies [15] have refined the crack driving force and recently [6] it is shown that the two-parameter crack driving force ΔK^* could be expressed as

$$\Delta K^* = \Delta K\left(\frac{1-R_t}{1-R}\right)^{\gamma} \tag{4}$$

where ΔK = applied SIF range, R = load ratio, R_t = transition load ratio, and γ = constant. The value of γ depends on R_t. Sree and Kujawski [6] have provided a procedure for estimating the values of the constants in Eq. (4). The same procedure was used as shown in Fig. 4 to determine these constants for Inconel 718 as follows.

Now, the da/dN is replotted as a function of ΔK* as defined in Eq. (4) in Fig. 5. It is clear that use of two-parameter crack driving force, ΔK^*, appears to provide a better correlation for stress ratio effects in Inconel 718 alloy. This collapsed data can be fitted to an empirical equation and used further in crack propagation studied under service loads.

4 Conclusions

Following conclusions may be drawn based on the results obtained in this investigation:

1. Increasing stress ratio increases the fatigue crack growth rate in IN718 alloy.
2. Increasing stress ratio decreases threshold stress intensity factor range.

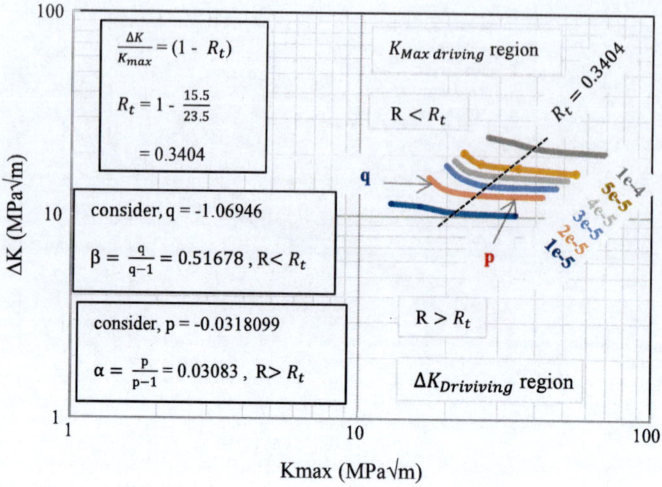

Fig. 4 Determined α, β, and R_t using FCGR Fig. 3 of Inconel 718 alloy

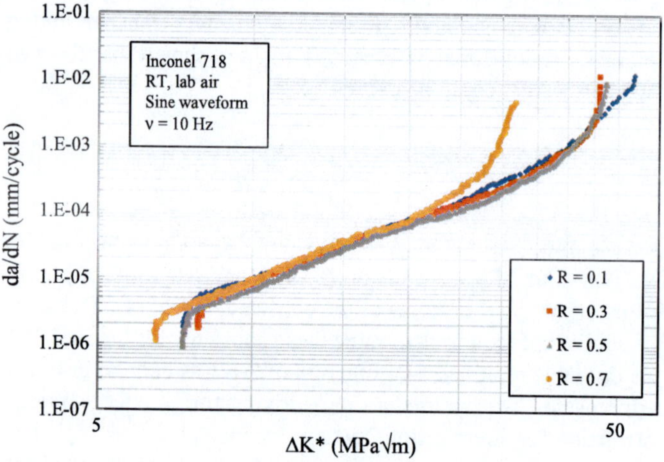

Fig. 5 Correlation of stress ratio effects with crack driving force parameter ΔK^*

3. Two-parameter crack driving force, ΔK^*, appears to correlate stress ratio effects reasonably well which could further be employed in crack propagation studied under service loads.

Acknowledgements Authors would like to thank the Director, GTRE and the Director, CSIR-NAL, for their encouragement and support during this work. They also would like to acknowledge the technical support rendered by the scientists and technical support staff members of SID, CSIR-NAL.

References

1. Padula Ii SA, Shyam A, Ritchie RO, Milligan WW (1999) High frequency fatigue crack propagation behavior of a nickel-base turbine disk alloy. Int J Fatigue 21(7):725–731
2. Li HY, Sun JF, Hardy MC, Evans HE, Williams SJ, Doel TJA, Bowen P (2015) Effects of microstructure on high temperature dwell fatigue crack growth in a coarse grain PM nickel based superalloy. Acta Mater 90:355–369
3. King JE (1987) Fatigue crack propagation in nickel-base superalloys–effects of microstructure, load ratio, and temperature. Mater Sci Technol 3(9):750–764
4. Dahal J, Maciejewski K, Ghonem H (2013) Loading frequency and microstructure interactions in intergranular fatigue crack growth in a disk Ni-based superalloy. Int J Fatigue 57:93–102
5. Gustafsson D, Moverare J, Simonsson K, Johansson S, Hörnqvist M, Månsson T, Sjöström S (2011) Fatigue crack growth behaviour of Inconel 718–the concept of a damaged zone caused by high temperature hold times. Procedia Eng 10:2821–2826
6. Sree PC, Kujawski D (2014) A two-parameter fatigue crack growth correlation using ΔK and KMAX parameters. In: ASME 2014 international mechanical engineering congress and exposition. American Society of Mechanical Engineers, pp V009T12A033–V009T12A033
7. Manjunatha CM (2008) Fatigue crack growth prediction under spectrum load sequence in an aluminum alloy by K^*-RMS approach. Int J Damage Mech 17(6):477–492
8. Malipatil SG, Majila AN, Fernando DC, Manjunatha CM (2019) Influence of crack driving force on correlating stress ratio effects in fatigue crack growth rate of a nickel base super alloy IN720. In: 4th Indian conference on applied mechanics, at indian institute of science (IISc). Bengaluru, pp 1–4
9. Kujawski D (2001) A fatigue crack driving force parameter with load ratio effects. Int J Fatigue 23:239–246
10. Newman JC Jr, Kota K, Lacy TE (2018) Fatigue and crack-growth behavior in a titanium alloy under constant-amplitude and spectrum loading. Eng Fract Mech 187:211–224
11. Elber W (1971) The significance of fatigue crack closure. Damage tolerance in aircraft structures. ASTM International
12. Tzamtzis A, Kermanidis AT (2014) Improvement of fatigue crack growth resistance by controlled overaging in 2024-T3 aluminium alloy. Fatigue Fract Eng Mater Struct 37(7):751–763
13. Li Y, Wang H, Gong D (2012) The interrelation of the parameters in the Paris equation of fatigue crack growth. Eng Fract Mech 96:500–509
14. ASTM International (2011) Standard test method for measurement of fatigue crack growth rates. ASTM E647 International
15. Kujawski D (2001) A new $(\Delta K + K_{\max})^{0.5}$ driving force parameter for crack growth in aluminum alloys. Int J Fatigue, 23(8), pp 733–740

Design and Development of Plane Bending Fatigue Testing Machine

Virendra Patil, Pravin Kulkarni, and Vivek Kulkarni

Abstract Fatigue failure in dynamic structures occurs because of continuous cyclic loading; especially bending fatigue often takes place depending upon the loading under yield and ultimate strength of the material. Generally, fatigue data of material is not commonly accessible because of lengthy experimentation and unavailability of testing machines. There are some fatigue testing machines that are available with complex mechanism and data acquisition system at high operating cost. The research work presents design and development of a plane bending fatigue testing machine with simple operating mechanism and data acquisition unit to obtain fatigue life of isotropic and orthotropic material at comparatively low operating and manufacturing cost. The machine is designed with the application of DFMA principles, manufactured, assembled, and tested for design specifications and performance which includes measurement of angular displacement, speed setting, load measurement, and vibrations generated. At the end, the performance of the machine is tested with the experimental work on fatigue testing of Al alloy for plane bending. The testing performance is considered satisfactory, and an economical and accurate testing setup is developed.

Keywords Fatigue testing · DFMA · Plane bending · LCF · Aluminum

V. Patil (✉)
Department of Mechanical Engineering, Sanjay Ghodawat University, Kolhapur, Maharashtra, India
e-mail: Patilvirendra33@gmail.com

P. Kulkarni · V. Kulkarni
Department of Mechanical Engineering, Sanjay Ghodawat Group of Institutions, Kolhapur, Maharashtra, India
e-mail: Sagitaria0460@gmail.com

V. Kulkarni
e-mail: Kulkarni.vv@sginstitute.in

S. Seetharamu et al. (eds.), *Fatigue, Durability, and Fracture Mechanics*, Lecture Notes in Mechanical Engineering, https://doi.org/10.1007/978-981-15-4779-9_25

1 Introduction

Generally fatigue failure is the weakening or breaking of material due to cyclic loading which may be either unidirectional ($R = 0$) or fully reversible ($R = -1$). The failure of materials under cyclic loading is called fatigue failure, and techniques for measuring susceptibility to fatigue failure are called fatigue testing. The total number of cycles of those specimens that can sustain before failure is called as fatigue life. The high-cycle fatigue (HCF) and low-cycle fatigue (LCF) are the types of fatigue lives depending upon the loading under yield and ultimate strength of the material, respectively. Especially, fatigue occurs before ultimate strength of material. The fatigue life is the function of many variables, including stress level, stresses state, cyclic waveform, fatigue environment, and the metallurgical condition of the material. The small geometrical changes in the specimen or test conditions can significantly affect fatigue behavior, making analytical prediction of fatigue life difficult. In recent years, orthotropic materials have developed more rapidly than isotropic materials in structural applications. It is very difficult to find out the specifications of the fatigue testing machines for orthotropic materials. Also, it is specifically not available for the plane bending fatigue testing machine. The designers have assumed as the performance of composites in fatigue loading is similar to metals. Also, the composites are very good in fatigue life that has not completely tested the applications in high stresses. This leads to overdesigning and using them uneconomically.

Ansar et al. [1] presented work on fatigue analysis of glass fiber reinforced composite material, when compared to aluminum subjected to vibrating or oscillating force applied. Wood et al. [2] presented application and modification of design principle for the development of new products. Therefore, modification and clarification are suggested in DFMA principles. Bayoumi et al. [3] explained the design for manufacturing and assembly concepts, manufacturing improvements, the early effects of design, benefits of DFM, design philosophy, DFMA versus design freedom and design time, the system approach to DFMA, creativity and brainstorming, design for assembly, and design for repair and maintenance. Harris et al. [4] presented the fatigues test methods, standards, and issues related to the fatigue under multi-axial loading, life prediction methods, and fatigue failure criteria. Different methods of measurement of fatigue data including NDT are also described. Surati et al. [5] presented design and development of the fatigue testing machine. In these, they have explained the following points such as fatigue failure, factors affecting fatigue life, and stages of fatigue failure they consist of crack initiation, crack propagation, and final fracture. Also, the work is done by analyzing the fatigue life using stress life-based approach for S-N method. Gerhardt et al. [6] presented work on the implementation of DFMA procedure in combination with simultaneous engineering in their industry (Ingersoll-Rand) for a new radiator and oil-cooler design. Di Franco et al. [7] presented the details of the setup required for the experimental work of fatigue testing machine in plane bending for composite specimens and bolted joint. Barbose et al. [8] presented work on design for manufacturing and assembly methodology for application on aircrafts design and manufacturing. It involves the guideline for the evaluation of manufacturing processes, assembly, maintenance, and human factors. Thompson et al. [9] worked on DFMA key performance indicators to measure and improve manufacturability and product quality throughout the high-speed product development process. Abo-Elkhier et al. [10] presented the work on investigation of the capability of the non-destructive tool to characterize and quantify the fatigue limit of the specimen. Kulkarni et al. [11] presented work on the development of fatigue testing setup for the fatigue life estimation of CFRP in plane bending with different orientations (0°, 45°, and 90°) and suggested modal testing (NDT) for estimation of fatigue life. Landge et al. [12] presented the work on design and development of the bending fatigue testing machine for glass fiber composite material at low frequency. Paepegem and Degrieck et al. [13] developed the experimental setup required for plane bending fatigue testing of glass/epoxy composites. The damage behavior and the stiffness degradation are studied for two different configurations as 0° and 45° orientations of fibers. A damage accumulation model was proposed to estimate the fatigue life. It is found that the stresses are redistributed across the structure during fatigue life.

In the present work, the machine is designed with application of DFMA principles, manufactured, assembled, and tested for design specifications as well as performance by considering angular displacement, speed setting, load measurement, and vibrations generated. At the end, the performance of the machine is tested with the experimental work on fatigue testing of Al alloy for fully reversed bending ($R = -1$). The testing performance is considered satisfactory, and an economical and accurate testing setup is developed.

2 Working Principle of Fatigue Testing Machine

The block diagram of mechanism of fatigue testing machine and data acquisition system is shown in Fig. 1. The fatigue testing machine shown in Fig. 2 consists of a reaction bracket that can transfer applied load at free end of cantilever structure to fix end when test specimen is attached. The reaction bracket is attached to variable eccentric flange connected to load applying mechanism on test specimen by means of connecting rod driven by 2.2 kW-1440 rpm motor which gives different angles of deflection such as 6°, 80°, 10°, and 12° on half of the rotation of flange. The compressive button type load cell and proximity sensor are used to obtain applied load and number of cycles respectively. The Human Machine Interface (HMI) is used to control the machine at different rpms as well as to obtain the results from machine.

The fatigue testing machine is developed for plane bending with specification given in Table 1.

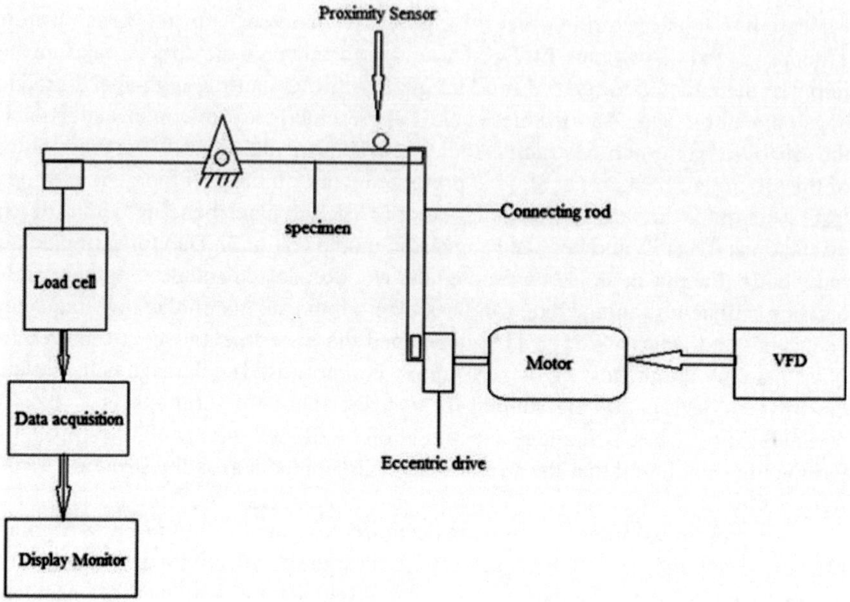

Fig. 1 Block diagram of experimental setup

3 Testing of Machine

The testing of machine for the specification includes the measurement of angular deflection, speed, load, and vibrations of machines.

3.1 Angle Measurement

The angle measurement is required for the application of bending moment to the test specimen. The angular deflection depends upon the eccentricity applied by the eccentric flange. There are three readings taken for measurement of angle of deflection at each eccentricity, and the results are shown in Table 2.

From the calculations, it has seen that there is 5% deviation in the angular displacement for the designed values which is within acceptable limit.

3.2 Speed Measurement

Setting up the rotation or working frequency to the test specimen and to obtain fatigue life in the form of number of cycles is a very important task in the experimentation

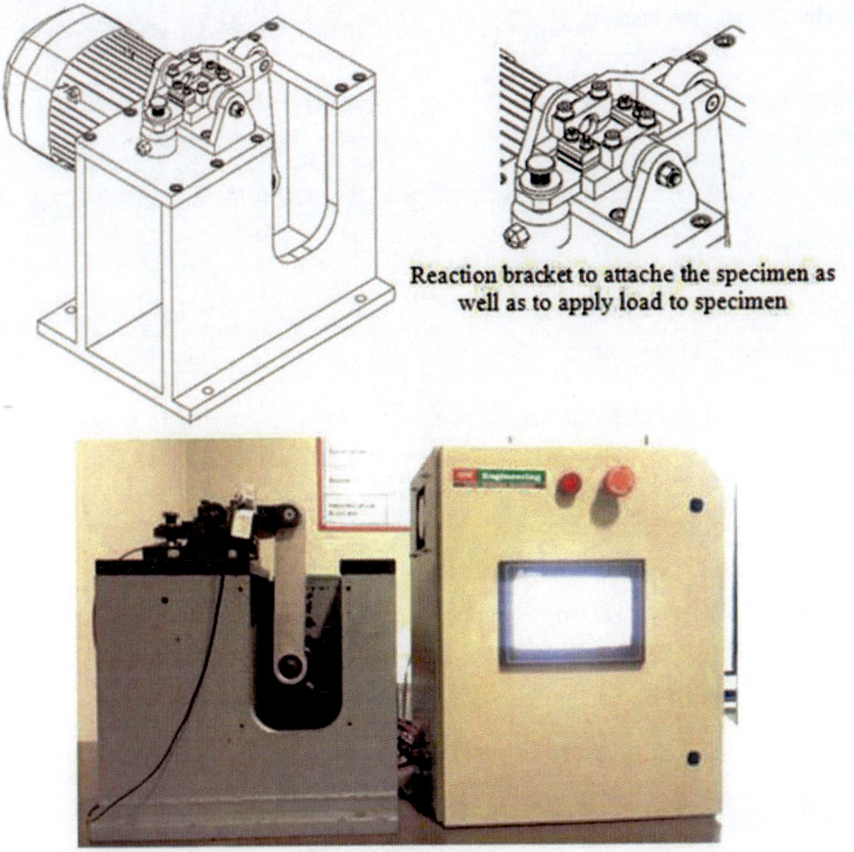

Fig. 2 Fatigue testing machine

Table 1 Machine specifications

Bending moment	−30 to +60 nm
Repeated bending moment	0 to ± 30 nm
Mean bending moment	0 to 30 nm
Static bending angle	(−12) to +24
Dynamic bending angle	±12
Frequency	300−1500 cpm
External signal output for bending moment	DC ± 10 V/F.S
Power supply 220/200v 50/60 Hz	Single-phase 2KVA

Table 2 Angle measurement

Eccentricity (mm)	Designed angle	Angle measurement	
		Readings	Average
10.5	60	5.810	5.810
		5.940	
		5.680	
13.9	80	7.690	7.700
		7.770	
		7.650	
17.4	100	9.380	9.420
		9.420	
		9.460	
20.8	120	11.30	11.370
		11.430	
		11.390	

of fatigue testing. The speed measurement is carried out by setting rpm on machine by means of HMI as shown in Fig. 3 and checked by contact-type tachometer at different rpms as shown in Table 3.

The speed deviation is within 0.02–0.06% that is within acceptable limits.

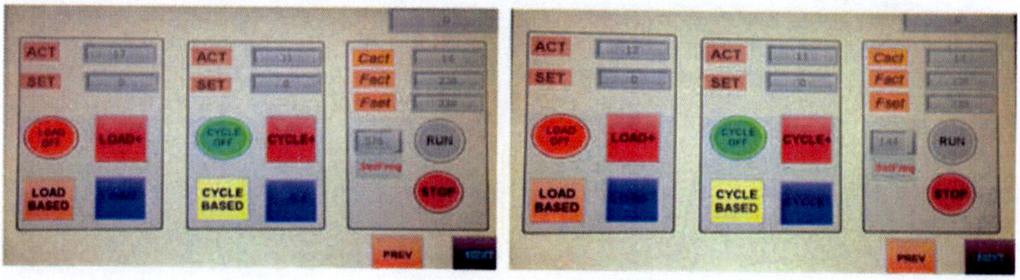

Fig. 3 Setup of rpm in HMI

Table 3 Speed reading

Set value (rpm)	Tachometer reading (rpm)				% error
	Reading-1	Reading-2	Reading-3	Average	
144	144.8	144	144.6	144.4	0.02
288	288.4	288.7	288.9	288.6	0.02
432	432.4	432	432.2	432.2	0.04
576	576.1	576.2	576.8	576.3	0.06

Fig. 4 Vibration measurement by using vibrometer

3.3 Vibration Measurement

The vibration is measured at five different locations on machine by using vibrometer as shown in Fig. 4. The vibration machine is measured by using vibrometer with three different speeds at five different locations. The readings are presented in Table 4.

The vibrations are within the acceptable limit, but the use of vibration-absorbing pads is required to reduce vibrations at high operating speed.

4 Fatigue Test on Machine

The aluminum alloy with transverse and longitudinal rolling direction is used to prepare test specimens. The chemical composition and tensile properties are shown in Tables 5 and 6. Total of 24 test specimens (12 for one orientation) were prepared for both cases as per the standard (JIS Z2275). The dimension of the specimen is as shown in Fig. 5.

While experimentation, the machine set on 400 rpm and crack initiation, propagation, and final failure is examined as shown in Table 7. The fatigue life of material obtained at different angles of deflections is given in Table 8. The S-N_f curve is shown in Fig. 6. The specimens tested to fatigue life are as shown in Fig. 7.

The power equations obtained by results are as follows:

For transverse rolling direction,

Table 4 Vibration measurement

Speed (rpm)	Location	Vibrometer reading (m/s)			
		Reading-1	Reading-2	Reading-3	Average
540	1	2.7	3.8	4	3.5
	2	2.8	3.8	4.3	3.6
	3	6.6	7.4	8.1	7.4
	4	6.2	8.4	8.5	7.7
	5	3.7	2.9	3.7	3.4
852	1	3.8	3.7	4.2	3.9
	2	3.8	3.8	4	3.9
	3	4.4	3.8	4.5	4.2
	4	7.5	4.3	4.2	5.3
	5	3.6	3.7	3.9	3.7
1084	1	5.6	6.3	5.7	5.9
	2	4.3	4.6	4.6	4.5
	3	4	4	4	4.0
	4	3.5	9.7	9.2	7.5
	5	4.1	4.1	4.3	4.2

Table 5 Chemical composition of aluminum alloy

Mn	0.01
Si	0.782
Ni	0.009
Cu	0.017
Al	98.492
Ti	0.012
Sn	0.032
Mg	0.156
Pb	0.129
Fe	0.378
Zn	0.045

Table 6 Tensile property of aluminum alloy

Aluminum alloy (Al)	Al-(T)	Al-(L)
Tensile strength (MPa)	124	132
Modulus of elasticity (GPa)	32	32

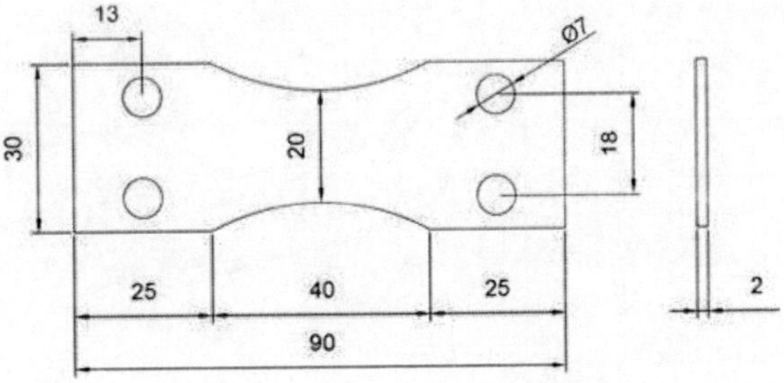

Fig. 5 Dimension of test specimen for fatigue testing (mm)

Table 7 Sample of readings

Angle of deflection		N_f (no of cycles)			
		Reading-1	Reading-2	Reading-3	Average
60	Crack initiation	6600	7100	7200	6967
	Crack propagation	8340	8800	9010	8717
	Final crack	10397	11204	11427	11009
80	Crack initiation	2350	2947	2974	2757
	Crack propagation	5230	5412	5456	5366
	Final crack	7410	7218	7711	7446
100	Crack initiation	1493	1379	1576	1483
	Crack propagation	1901	1896	1630	1809
	Final crack	2190	2103	1999	2097
120	Crack initiation	1254	1640	1709	1534
	Crack propagation	1301	1705	1863	1623
	Final crack	1375	1803	2006	1728

Table 8 Result of fatigue test

Angle of deflection	Applied force (N)	Stress level (kN/m^2)	Fatigue life (no of cycle)	
			Al-(T)	Al-(L)
60	28	700	11009	5534
80	37	927	7446	1731
100	46	1161	2097	1024
120	55	1387	402	671

$$\sigma = 11089 \times (N_f)^{-0.289} \tag{1}$$

For longitudinal rolling direction

$$\sigma = 10686 \times (N_f)^{-0.32} \tag{2}$$

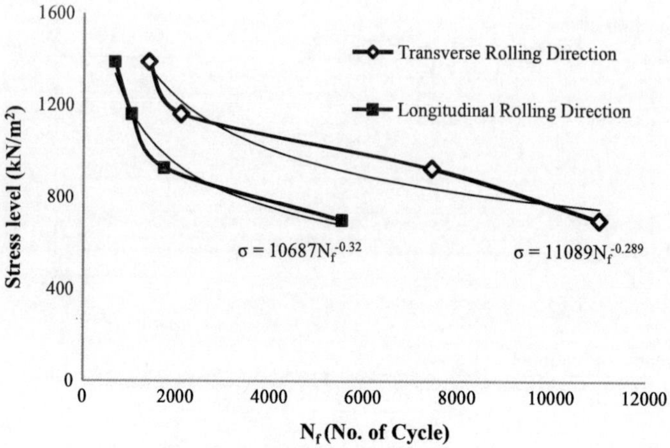

Fig. 6 Stress level V/S no. of cycles

Fig. 7 Failure of test specimen

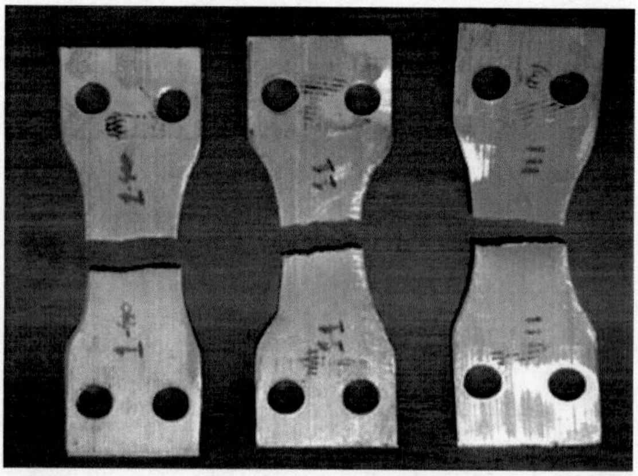

The results show that fatigue life of aluminum is higher when rolling direction is perpendicular to the applied load.

5 Conclusions

Fatigue testing machine for plane bending is designed, developed, and tested for design specifications and performance. The DFMA principles applied at the design stage has reduced the complexity of the design, and the assembly is made easy and cost-effective. The design, manufacturing, and assembly of the setup are carried out as per the design specifications, and the performance is satisfactory. The fatigue life testing of Al alloy is indicating the expected trend in the results, and the performance testing for fatigue testing is satisfactory. The fatigue life of Al alloy is greater in direction of rolling than in the perpendicular direction.

References

1. Ansar AM, Singh D, Balaji D (2013) Fatigue analysis of glass fiber reinforced composites. Int J Eng Res Appl
2. Wood AE, Wood CD, Mattson CA (2014) Application and modification of design for manufacturing and assembly principles for the developing word. In: IEEE global humanitarian technology conference 2014
3. Bayoumi A.E (2000) Design for manufacture and assembly (DFMA): concepts, benefits and applications. In: Current advances in mechanical design and production seventh Cairo University international MDP conference Cairo, 15–17 Feb 2000
4. Harris Bryan (2003) Fatigue in composites, science and technology of the fatigue response of fibre-reinforced plastics. Woodhead Publishing Limited, England
5. Surati C, Hiral P (2016) Design and development of fatigue testing machine. Int J Eng Res Gen Sci 4(2), March–April, ISSN 2091–2730
6. Gerhardt DJ, Hutchinson WR, Mistry DK (1991) Design for manufacturing and assembly: case studies in its implementation. Int J Adv Manuf Technol
7. Di Franco G, Marannano G, Pasta A. Virzì Mariotti G (2011) Design and use of a fatigue test machine in plain bending for composite specimens and bonded joints. http://www.intechopen.com
8. Barbosa GF, Carvalho J (2013) Guideline tool based on design for manufacturing and assembly (DFMA) methodology for application on design and manufacturing of aircrafts. Springer
9. Thompson MK, Jespersen IKJ, Kjergaard T (2018) Design for manufacturing and assembly key performance indicators to support high-speed product development. In: 28th CIRP design conference. Nantes, France, May 2018
10. Abo-Elkhier M, Ahamada AA, El-Deen AB (2014) Prediction of fatigue life of glass fiber reinforced polyester composite using model testing. Int J Fatigue 69:28–35
11. Kulkarni PV, Sawant PJ, Kulkarni VV (2017) Design and development of plane bending fatigue testing machine for composite material. In: Proceedings of international conference on materials manufacturing and modelling
12. Landge SM, Patil DN, Talekar AV, Ghosh AA, Bhaumik M (2018) Design and development of bending fatigue testing machine for composite materials. Int Res J Eng Technol (IRJET) 05(04) e-ISSN: 0056–2395
13. Van Paepegem W, Degriek J (2001) Experimental setup for and numerical modelling of bending fatigue experiments on plain woven glass/epoxy composites. Compos Struct 51(1):1–8

Fatigue Life Analysis of Glass Fiber Reinforced Epoxy Composite Using Rotating Bending Fatigue Test Machine

Ravindra R. Malagi and Venkatesh K. Havanur

Abstract Fatigue load is the most major type of load acting on the structural components. About 90% of the components fail due to fatigue. Hence, determining the fatigue life of a material is an important aspect of engineering applications. The fatigue life is estimated as number of cycles to failure. In this paper, fatigue life of polymer matrix composite material is determined using rotating bending fatigue test machine. Here, the composition of composite material is bidirectional fiberglass as a fiber and epoxy resin in the form of matrix. The volume fraction for epoxy is 40% and fiber is 60%. Hand layup technique is used as a method to assemble fiber and matrix. This sandwich material is cut in the form of dog bone shape, and specimens are tested on the machine. In this present work, experimental and simulation tests were carried out. Results obtained are within the acceptable limits. The bending moment applied to the specimen varied from 50 to 80 kg-cm. The experimental results have been compared with the simulation results. The maximum stress or bending stress obtained by manual calculation for 75 kg-cm bending moment is 377.1 MPa, and by simulation it is 381.41 MPa with an error of 1.13%. The number of cycles to failure for 55 kg-cm bending moment obtained by experiment is 7856 cycles, and by simulation it is 7950.5 cycles with an error of 1.19%. The range of values obtained by experiment and simulation are plotted as stress(S) versus number of cycles to failure (N) (S-N) graph.

Keywords Fatigue · Bending stress · Number of cycles to failure · Fatigue life · Composite · Rotating bending machine

R. R. Malagi
Visvesraya Technological University, Belgavi, India
e-mail: rrmalagi@rediffmail.com

V. K. Havanur (✉)
S.D.M. College of Engineering and Technology, Dharwad, India
e-mail: venki.vkh@gmail.com

S. Seetharamu et al. (eds.), *Fatigue, Durability, and Fracture Mechanics*, Lecture Notes in Mechanical Engineering, https://doi.org/10.1007/978-981-15-4779-9_26

1 Introduction

Fatigue accounts for 90% of service failures. Fatigue analysis helps in predicting the life of the component in design phase itself. Static or dynamic analysis can tell about stress, displacement, acceleration, etc. but not how long the component will survive. Failures or crack usually initiates at surface. Life of the component depends on surface conditions like grinding, induction hardening, shot peening, etc. Static or dynamic analysis cannot take into account these details, while fatigue can. Fatigue is the life of structure when it is subjected to repetitive load. The fatigue is divided into two categories. One is low-cycle fatigue in which life of component is less than 10^5 cycles, applicable for heavy-duty application loading, and another is high-cycle fatigue in which component is subjected to less severe loads and life is greater than 10^5 cycles. Stress level below which material never fails is known as endurance limit or fatigue limit. This is called infinite life. This is the point where S-N curve slope changes, and it becomes parallel to x-axis. S-N curve can be plotted for constant amplitude rotating bending test, i.e., shaft subjected to pure alternating bending stress. Similar test could be conducted for tension, compression, shear, and torsion stress. Bending fatigue strength is higher than tension/compression, and torsional fatigue strength is the lowest. In this project, stress life approach is adopted. It is the first method used for fatigue calculation. This approach is simple based on S-N curve also known as Wohler diagram [1]. S-N curve is generated by conducting rotating bending test. Fatigue design data can be generated using stress life fatigue tests. Tests are usually carried out on several specimens at different bending stress amplitudes over a range of fatigue lives for identically prepared specimens.

Fatigue life estimation of polymer matrix and glass fiber combination of composite material is the scope of this project work carried out. Composite materials are used in many applications such as aerospace, automotive, space, etc. because of their high strength to low weight ratio. Glass fiber reinforced polymer composites are widely used as structural load carrying composites. Their durability is affected by the onset and growth of fatigue damage under cyclic loading, leading ultimately to catastrophic failure. Dominant fatigue mechanisms in composite include fiber breakage, fiber pullout (separation from matrix), matrix delamination [2], and matrix cracking. In the absence of tools to track fatigue damage in composites, a fatigue test on composites has for decades been reduced to just one number—cycles to failure.

2 Literature Review

Ueki [3] in his paper on high-speed bending fatigue testing of composite materials (2018) devised and evaluated a methodology for fatigue testing of fiber reinforced composites. In this experiment, specimen was held as a cantilever and given a periodic stress at a frequency of 230 Hz. The specimen was kept at a room temperature by external cooling since the rise in temperature was observed during fatigue testing. A reversed bending test at 1 Hz was also performed with identical specimens to confirm the validity of the test method. Then the S-N plot was plotted.

Wilson and Heyes [4] in their paper on fatigue of short fiber reinforced polymers: from material process to fatigue life of industrial components (2018) devised a simulation strategy to be adopted by the automotive industry. Then based on numerical microstructure, integrative simulation is used to analyze and predict mechanical behavior of fatigue specimens.

The paper by Khalid, Mutasher, Sahari, and Hamouda [5] deals with bending fatigue behavior of hybrid aluminum/composite drive shafts (2005). Filament winding technique was used to prepare the hybrid shafts. External composite layers were constructed using glass fiber with a matrix of epoxy resin and hardener. In this paper, four cases were studied. The usage of aluminum tube wounded by different layers of composite materials and at different fiber orientation angles was done. It is observed that there is no fiber breakage from the rotating bending fatigue test.

In the paper of Khashaba [6] on fatigue and reliability analysis of unidirectional GFRP composites under rotating bending loads (2003), rotating bending fatigue tests have been conducted using circular-shaped specimens with having material composition of unidirectional glass fiber reinforced polyester composite of various fiber volume fraction ratios. Then the test results were plotted on S-N graph.

3 Methodology

Flowchart of the methodology is shown in Fig. 1. The first step is the selection of material. The material selected was based on the literature review. Bidirectional glass fiber fabric was selected as fiber, and epoxy resin was selected as matrix. The volume

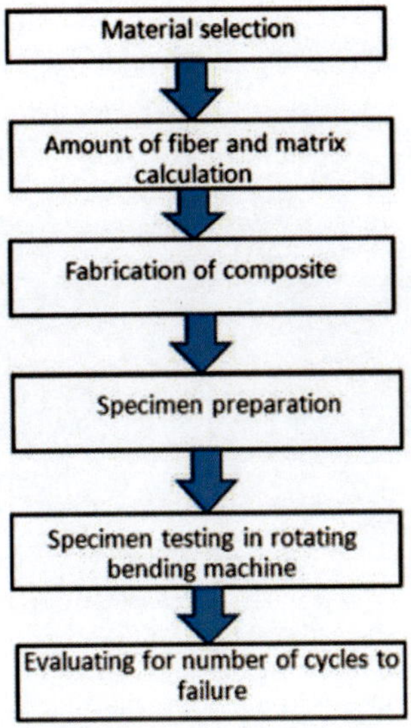

Fig. 1 Methodology flowchart

fraction of glass fiber taken was 60% and that of epoxy resin was 40%. Using hand layup technique, composite material was fabricated. From the sandwiched composite sheet, dog-bone-shaped specimens were cut according to ISO 527B standard. Then, the specimens were loaded on the rotating bending fatigue test machine, and test was carried out according to Japan Society of Mechanical Engineers (1981) test method. Finally, the number of cycles to failure, i.e., fatigue life, is noted from the display of the machine. After completion of the experiment, simulation was performed using ANSYS® software.

4 Fatigue Testing

4.1 Testing Method

There is a need for statistical S-N testing to predict fatigue life at various stress amplitudes. The S-N test method presented by the Japan Society of Mechanical Engineers (1981) is used in this project work. This method can be used as a guideline to determine an S-N curve with a reliability of 50% and a minimum sample size. This method requires 14 specimens. Eight specimens are used to determine the finite fatigue life region, and six specimens are used to find the fatigue limit. The recommended sample size for the number of test samples used to generate the S-N curve is 6–12 for preliminary and research tests. The number of specimens used in this project for testing is 7. The testing method is called staircase method. Number of cycles to failure is the fatigue life of the composite material (Fig. 2).

Fig. 2 S-N testing with a small sample size

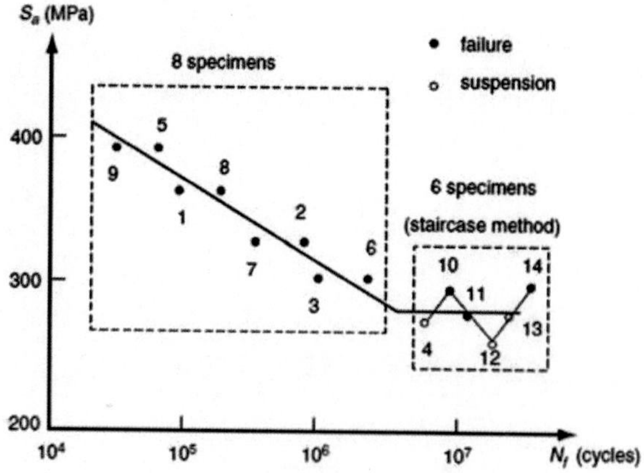

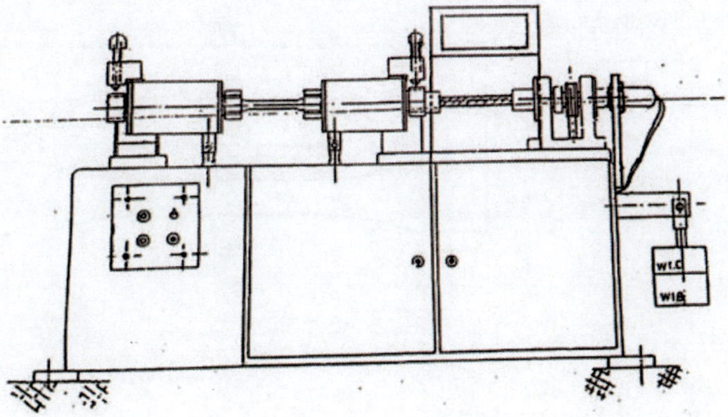

Fig. 3 Fatigue test machine

4.2 Rotating Bending Fatigue Test Machine

Fatigue testing of composite specimens was conducted on rotating bending fatigue test machine. The input parameter on the specimen is bending moment (M), and output parameter is number of cycles to failure (N). Bending stress is calculated using the formula

$$S = M/Z \tag{1}$$

$Z =$ Section modulus $= bt^2/6$
$b =$ Width of the specimen
$t =$ Thickness of the specimen

Bending moment taken for experimentation is from to 80 kg-cm. A line diagram of rotating bending [7] fatigue test machine used for experimentation is shown in Fig. 3.

4.3 Specimen Size

Fatigue test specimens used for experimentation were according to ISO 527B standard. The thickness of the specimen is 3 mm. The dog-bone-shaped specimens are used. Figure 4 shows the specimen prepared for the experiment.

Fig. 4 Fatigue test specimen

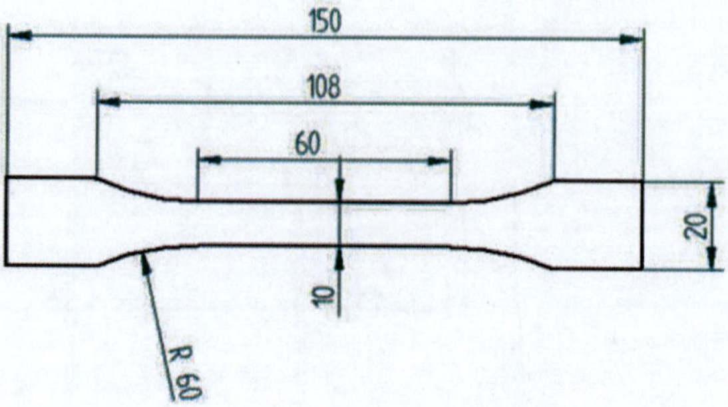

5 Results and Discussions

Both experimental work and simulation work were conducted on the composite specimen. The results of both the works have been compared.

5.1 Experimental Work

To conduct the experiment on a test specimen, a fixture attachment is designed to suit the machine. Fixture is a work-holding device used to locate and support the work. The designed fixture is of material mild steel. A rectangular slot at the center of the fixtures will hold the specimen and is tightened with the screw. Figures 5, 6, and 7 show the fixture attachments. Fixture was modeled in a modeling software CATIA®.

Table 1 shows the experimental results obtained on rotating bending fatigue test machine. The experiment was conducted till the failure of composite specimen is observed as a fiber breakage or delamination of matrix. It has been observed from the results that, as the bending stress increases, the number of cycles to failure decreases. For bending stress of 276.59 MPa, the number of cycles to failure is 7856. For bending stress of 377.325 MPa, the number of cycles to failure is 2922.

Fig. 5 Isometric view of fixtures

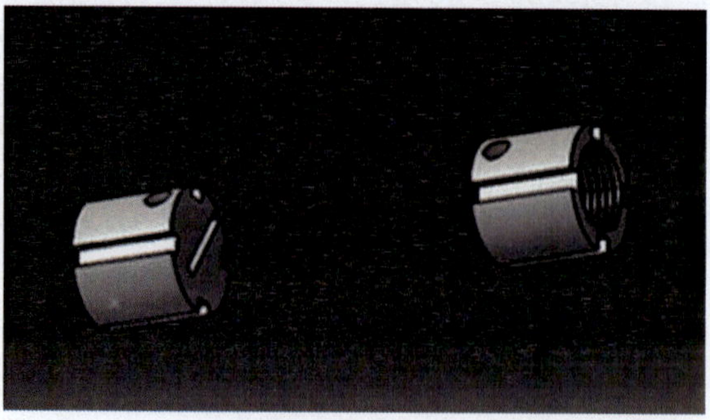

Fig. 6 Top view of fixture

Fig. 7 Front view of fixture

Table 1 Experimental results

Bending moment (kg-cm)	Bending stress (MPa)	Number of cycles to failure
50	251.55	9981
55	276.7	7856
60	301.86	6600
65	327.015	4500
70	352.17	3700
75	377.325	2922
80	402.48	2350

Fig. 8 Linear S-N graph from the experiment

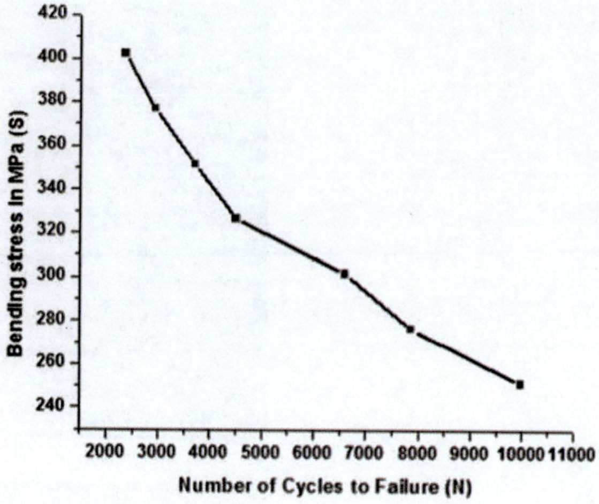

The S-N graph obtained from the experimental results is shown in Fig. 8.

5.2 *Simulation Results*

The simulation work was carried out using ANSYS® software. The dog-bone-shaped specimen was modeled in CATIA® software and imported into ANSYS workbench®. Figure 9 shows the model of the specimen. Static structural analysis was performed by taking composite material as E-glass and epoxy. The default properties of material were taken for further analysis.

Table 2 shows the results obtained by simulation. It has been observed from the results that, as the bending stress increases the number of cycles to failure decreases. For bending stress of 278 MPa, the number of cycles to failure is 7950.5. For bending stress of 381.41 MPa, the number of cycles to failure is 3437.8.

Results of von Mises stress or bending stress and number cycles to failure, i.e., fatigue life obtained by software are shown in Fig. 10 and Fig. 11, respectively, for bending moment 75 kg-cm.

Fig. 9 Solid model of the specimen

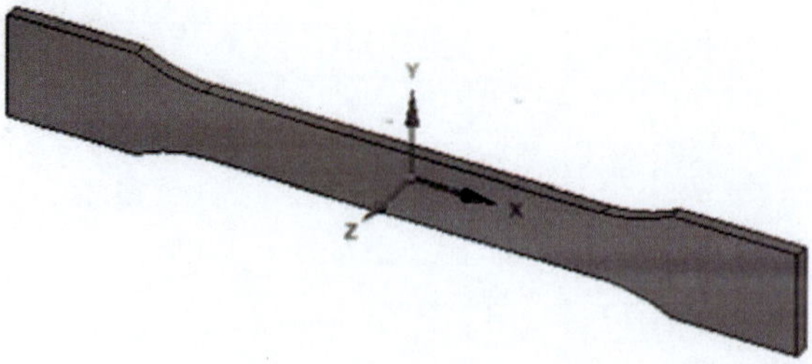

Table 2 Software Results

Bending moment (kg-cm)	Bending stress or alternating stress (MPa)	Number of cycles to failure
50	250	10300
55	278	7950.5
60	301	6900
65	328	5100
70	351	4150
75	381.41	3437.8
80	402	2100

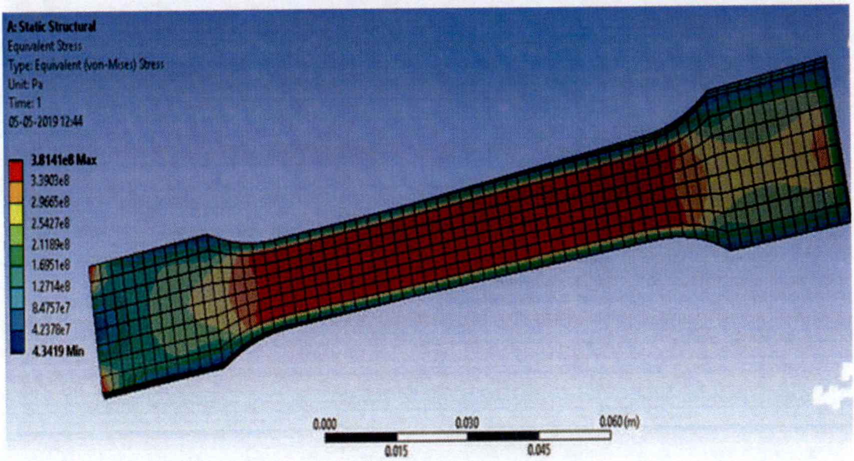

Fig. 10 Von Mises stress pattern plot

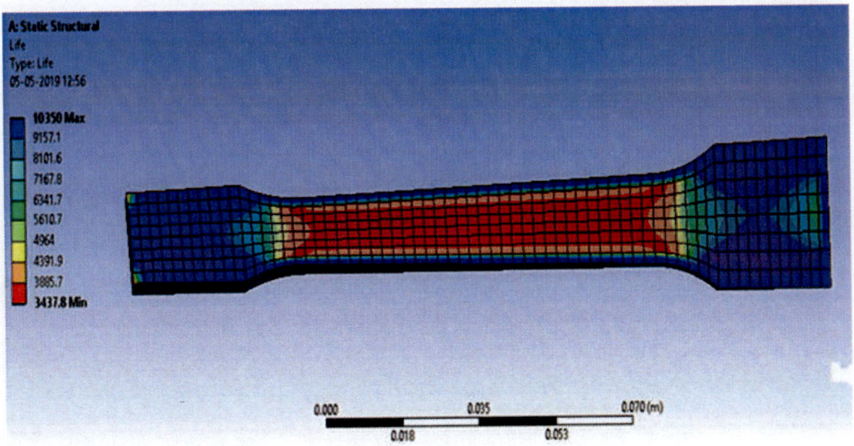

Fig. 11 Number of cycles to failure plot

For 55 kg-cm bending moment, the fatigue life obtained is 7950.5 cycles the plot of which is shown in Fig. 12.

The linear S-N graph obtained from simulation results is shown in Fig. 13.

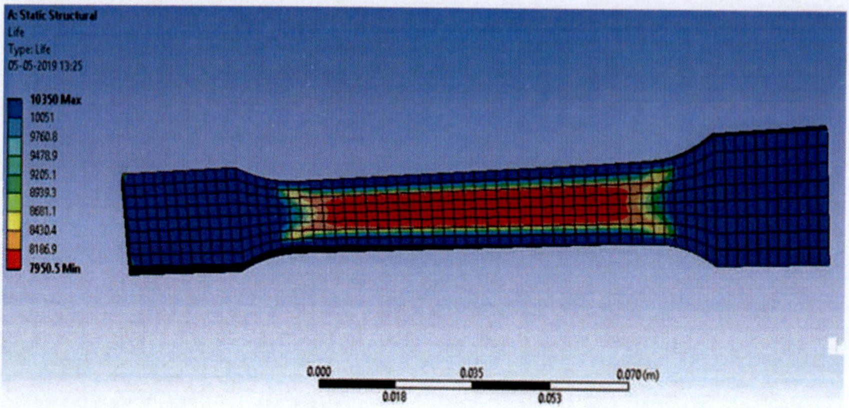

Fig. 12 Fatigue life plot

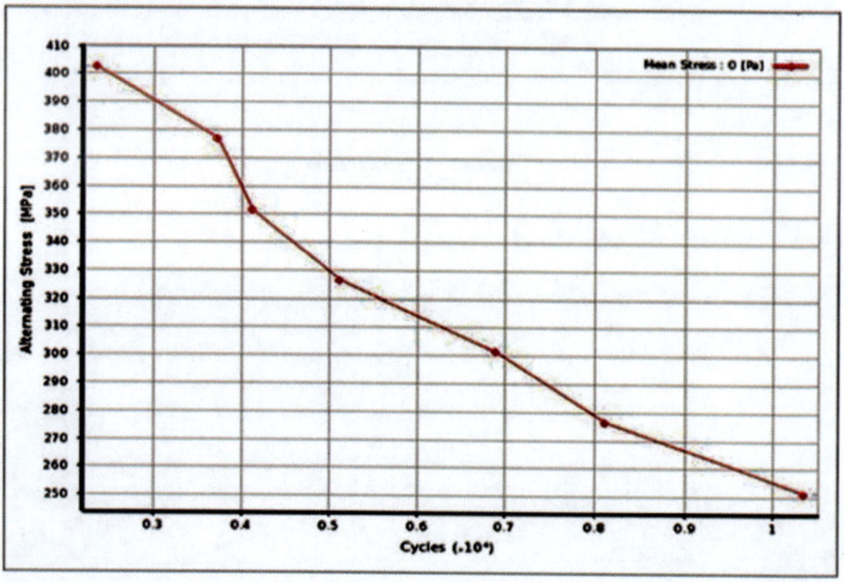

Fig. 13 Linear S-N graph from the simulation

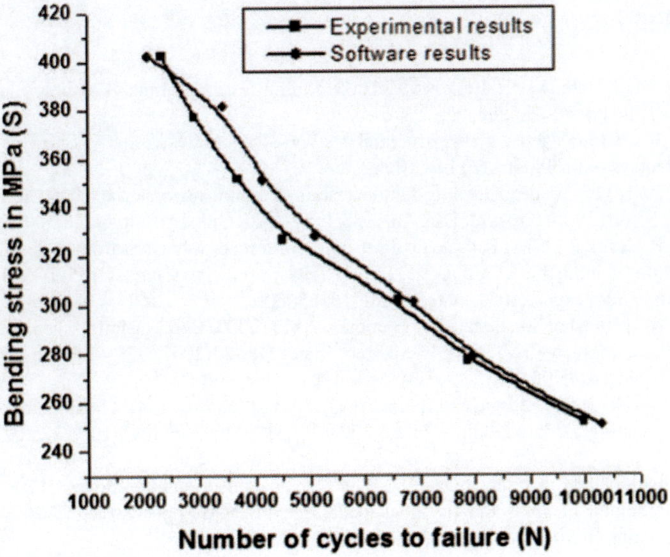

Fig. 14 Linear S-N graph

5.3 Comparison of Results

The results obtained by experiment and simulation are compared in Fig. 14. The results by both the methods are well matching without not much deviation.

6 Conclusions

It is found from the experiment that failure of material occurred in the form of matrix delamination. The maximum stress or bending stress obtained by manual calculation for 75 kg-cm bending moment is 377.1 MPa, and by simulation it is 381.41 MPa with an error of 1.13%. The number of cycles to failure for 55 kg-cm bending moment obtained by experiment is 7856 cycles, and by simulation it is 7950.5 cycles with an error of 1.19%. The fatigue life of the material falls under the category of low-cycle fatigue. As the bending stress increases, the fatigue life of material reduces. Increasing the amount of bending stress will damage more and leads to decreased fatigue life of the laminate.

Acknowledgements Authors thank Visvesvaraya Technological University, Belagavi for supporting this project work. Also thanks to S.D.M. college of engineering and technology, Dharwad for providing the infrastructure and laboratory equipment for conducting the experiment.

References

1. Lee YL, Pan J, Hathaway R, Barkey ME (2005) Fatigue testing and analysis (theory and practice). Elsevier Butterworth-Heinemann, USA
2. Sunder R (2017) Materials and structural testing—the knowhow and know why, Bangalore integrated system solutions (P) Ltd., Bangalore
3. Ueki Y (2018) High-speed bending-fatigue testing of composite materials. IOP Conf Ser: Mater Sci Eng 388:012008. https://doi.org/10.1088/1757-899X/388/1/012008
4. Wilson P, Heyes P (2018) Fatigue of short fibre reinforced polymers: from material process to fatigue life of industrial components. In: MATEC web of conferences 165, 08003 FATIGUE 2018, https://doi.org/10.1051/matecconf/20181650800
5. Khalid YA, Mutasher SA, Sahari BB, Hamouda AMS (2005) Bending fatigue behavior of hybrid aluminum/composite drive shafts. Sci Direct-Mater Des 28(2007):329–334. https://doi.org/10.1016/j.matdes.2005.05.021
6. Khashaba UA (2003) Fatigue and reliability analysis of unidirectional GFRP composites under rotating bending loads. J Compos Mater 2003(37):317. https://doi.org/10.1177/0021998303037004680
7. El-Assal Ahmed M, Khashaba UA (2007) Fatigue analysis of unidirectional GFRP composites under combined bending and torsional loads. Sci Direct-Compos Struct 79(2007):599–605. https://doi.org/10.1016/j.compstruct.2006.02.026

Damage Tolerance Approach for Estimation of the Remaining Life of 60 MW Turbine-Generator Shafts of Hydropower Plant—A Case Study

R. K. Kumar, M. Janardhana, K. Kaushik, and N. L. Santhosh

Abstract The structural integrity of the rotor generator and turbine shafts of a 60 MW hydro plant has been carried out through the application of advanced Phased Array Ultrasonic Technique (PAUT) and crack growth rate analysis. The size and orientation of the crack-like defects were evaluated by the PAUT technique. The multi-array ultrasonic sensor was scanned all over the accessible surfaces of the shaft through manual scanning. The scanning was performed in different diameter sections of the shaft. The complete volume of the shaft was evaluated through scanning in different patches made from the top to bottom portion of the shaft. The critical defects identified by this technique with their locations and orientation inside the rotor have been presented. The stress and fatigue life of the shaft under defect-free condition have been carried out by finite element analysis using the commercial code ANSYS Mechanical®. The calculation of the remaining life of the rotor shaft was carried out based on the principle of fracture mechanics using the code FRANC3D™. The residual life of the shaft was calculated based on the crack growth rate data of critical cracks close to the surface.

Keywords Hydroturbine generator · Phased array ultrasonic inspection · Stress analysis

1 Introduction

The refurbishment and uprating of hydro plants in India are being persused in a big way in units that have completed more than 40 years of service, and a number of case histories in this regard have been reported [1–3]. While the generator and

R. K. Kumar (✉) · M. Janardhana
Central Power Research Institute, 560080 Bangalore, India
e-mail: rkkumar@cpri.in

M. Janardhana
e-mail: janardhana@cpri.in

K. Kaushik · N. L. Santhosh
Dhio Research and Engineering (P) Ltd, Bangalore, India
e-mail: santhosh@dhioresearch.com

S. Seetharamu et al. (eds.), *Fatigue, Durability, and Fracture Mechanics*, Lecture Notes in Mechanical Engineering, https://doi.org/10.1007/978-981-15-4779-9_27

turbine components undergo a major change in the design during uprating in terms of class of insulation, rotor poles, turbine runner, etc., the shaft components are retained keeping in view the available margin on the factor of safety. The rotor shaft of hydro plants is generally very large in diameter and length. In view of the criticality of loading and high magnitude of torsional stresses during service, they are generally designed as a hollow section with a high factor of safety. As these rotors are manufactured through casting and forging route, they invariably considered to have process-induced defects like slag, blowholes, etc. [4, 5]. Owing to the non-availability of proven non-destructive evaluation techniques during late 70 s, the rotor components made with controlled process parameters have been put into use. The structural integrity assessment of turbine-generator rotor shafts of hydro plants with particular reference to remaining life prediction is considered critical from the viewpoint of achieving safe and reliable operation under the uprated conditions. Hence, the condition assessment of these rotors is being pursued on a regular basis by the utilities and forms an important exercise during the Renovation and Modernization program.

With the advent of advanced ultrasonic techniques in the late ’90 s, the improved quality control measures have been put into practice. While the conventional ultrasonic technique has been routinely applied for the identification of various defects such as crack, discontinuity, etc. along with depth inside the component, the evaluation of the orientation of the defect is not possible. The orientation of crack geometry with reference to the axis of rotation affects its growth during service. Any existing crack in the rotor is in phase with the direction of shear stress, grows at a rapid rate under torsional loading conditions. The rate of crack growth during service is affected by the material properties as well as the crack geometry. Any crack reaching the surface is considered detrimental as it leads to catastrophic failure of the rotor.

In this paper, the FEA based damage tolerance studies of generator shaft considering critical cracks identified by PAUT inspection of 60 MW turbine-generator shaft are presented.

2 Details of the Turbine-Generator Rotor Studied

In the present study, both the turbine and generator shafts have been considered as an integral part of the rotating system and thus considered as a single body for the purpose of simulations. The view of the 60 MW turbine-generator rotor shaft studied is shown in Fig. 1. The radius of the generator rotating mass component/disk was 3158 mm which depicts the rotor poles attached to the shaft through the spider web assembly. The details of the integrated shaft are as follows.

- Mass of generator rotor component: 110 tons
- Mass of turbine shaft: 18 tons
- Mass of turbine runner: 2.316 tons
- RPM of the generator: 500 rpm
- Turbine type: Pelton.

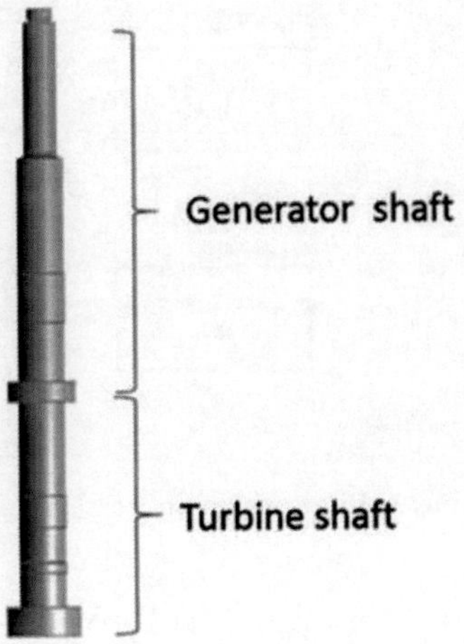

Fig. 1 3D geometrical model of Turbine-Generator shaft

The mass of the poles and other masses are modeled as a solid disk and attached to the shaft. The hydrodynamic bearing supports the top region of the generator shaft.

2.1 *Phased Array Ultrasonic Inspection of the Rotor Shaft*

The application of the Phased Array Ultrasonic Technique (PAUT) which has been developed primarily for the complex geometry of turbine components has been adopted in the present study. The 16-element ultrasonic array probe was used in an OLYMPUS make OMNISCAN SX equipment. Each of the elements in the multi-array probe is excited in a controlled manner by an independent pulser so as to achieve a focused beam of cylindrical wavefront which enters into the shaft material, and the reflections from the crack and other defects are analyzed. The multi-array elements are pulsed in such a way as to cause multiple-beam components to combine each other and form a single wavefront traveling in the desired direction. Similarly, the receiver function combines the input from multiple elements into a single presentation. The typical phased array pulse receiver system used during the ultrasonic inspection is shown in Fig. 2.

Multiplexing across many elements allows motionless high-speed scans from a single transducer position. More than one scan may be performed from a single location with various inspection angles.

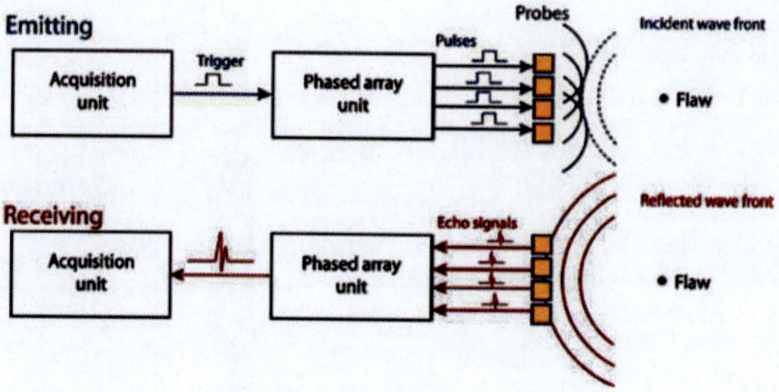

Fig. 2 Basic principle of phased array ultrasonic technique

3 Comparison Between Conventional UT and PAUT Imaging

The inherent feature of beam steering, focusing feature available in the PAUT technique enables scanning of complex geometries and identifies the defects that are present at different orientations. Also, the steering of the beam enables the capture of defects which are present close to each other. The conventional ultrasonic technique utilizes a single probe with a fixed angle and cannot be used to identify closely spaced defects. The scanning methodology as followed in PAUT enables the volume mapping of defects in a given component. The ability to sweep the beam from a fixed location of the probe enables the mapping of complex geometries like turbine blade steeples. The presentation of PAUT system output signals in the form of sectional view image (B-Scan) plan view (C-Scan) images in addition to the conventional A-Scan output is shown in Fig. 3.

4 Methodology of PAUT Inspection

In order to cover the entire volume of the shaft by PAUT, the shaft regions were divided into a number of patches of 50 mm length matching with that of the phased array probe length. Further, the complete circumference of the shaft was divided into four different quadrants in the clockwise direction. A total of 30 patches were made to cover the entire volume of the shaft (Fig. 4). The PAUT probe was scanned in the circumferential direction along with different patches, and data acquisition was carried out on four segments in a single patch. All the marked patches were scanned, and UT spectrum response data has been acquired.

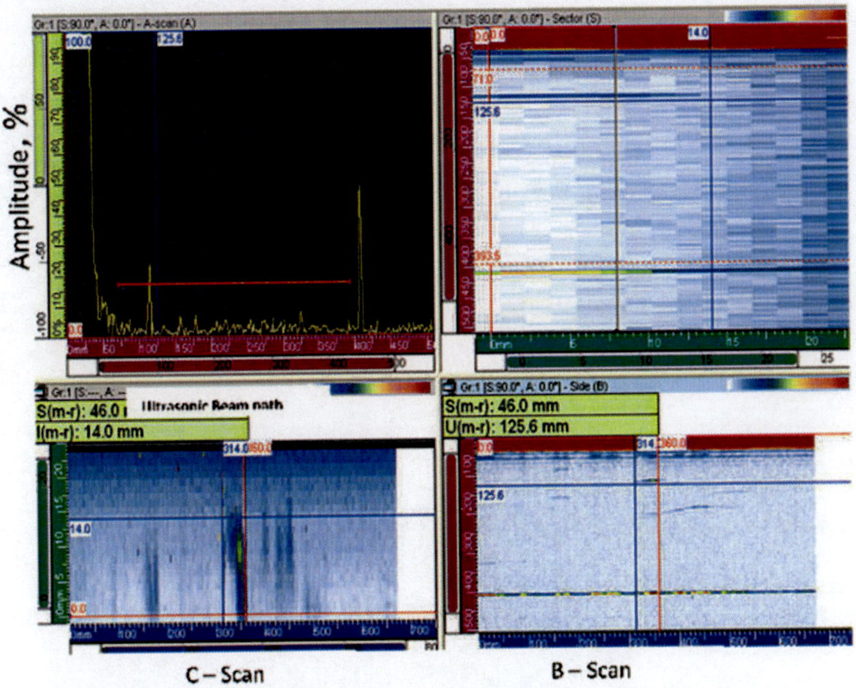

Fig. 3 Typical PAUT output results

Fig. 4 Patching made on turbine and generator shaft region for paut

5 Stress Analysis of the Shaft by Finite Element Method

The generator shaft supports the rotor part of the generator which houses the electric poles, the turbine shaft, and the runner. The top region of the generator shaft is supported by a hydrodynamic bearing. The self-weight of the shaft also acts as the load in the vertical gravity direction.

6 Results and Discussion

6.1 PAUT Crack Location Results

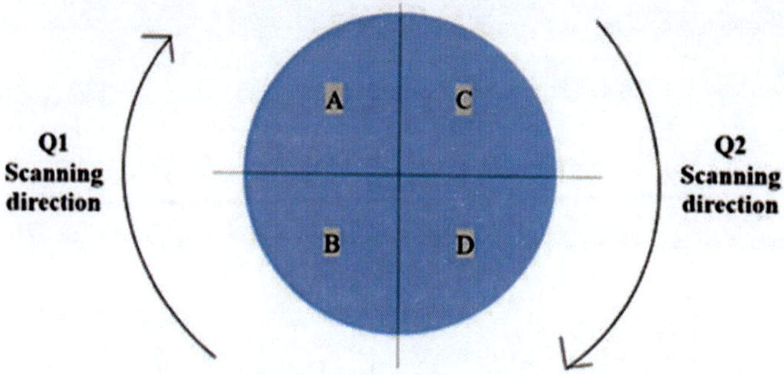

Q1& Q2 are divided into A, B and C, D segments

The B-Scan mapping clearly identified multiple cracked regions very close to the bottom of the turbine shaft. The multiple isolated minor defect indications were observed in the central region of the shaft which are presumably considered from the forging defects such as segregation of inclusion. The defect indications recorded with radial lengths in the near-surface regions of the shaft were considered critical as the operating stress levels are maximum on the surface and thus the increased rate of crack growth under cyclic loading. The details of various defect indications observed while scanning in different quadrant patch regions are shown in Fig. 5.

The near-surface defect indications could be easily identified by the application of miniature shear wave probe, and the scanning was done both in the axial direction and in the circumferential direction. Based on the criticality of defects observed, the crack growth rate analysis was carried out, and the summarized results on different patches scanned are given in Table 1. The indication of the green color code is indicative of defect-free/point defect regions and the red color indicative of critical-sized defects in the respective patches.

The indication of the green color code is indicative of defect-free or pointed defect regions and the red color indicative of critical-sized defects in the respective patches.

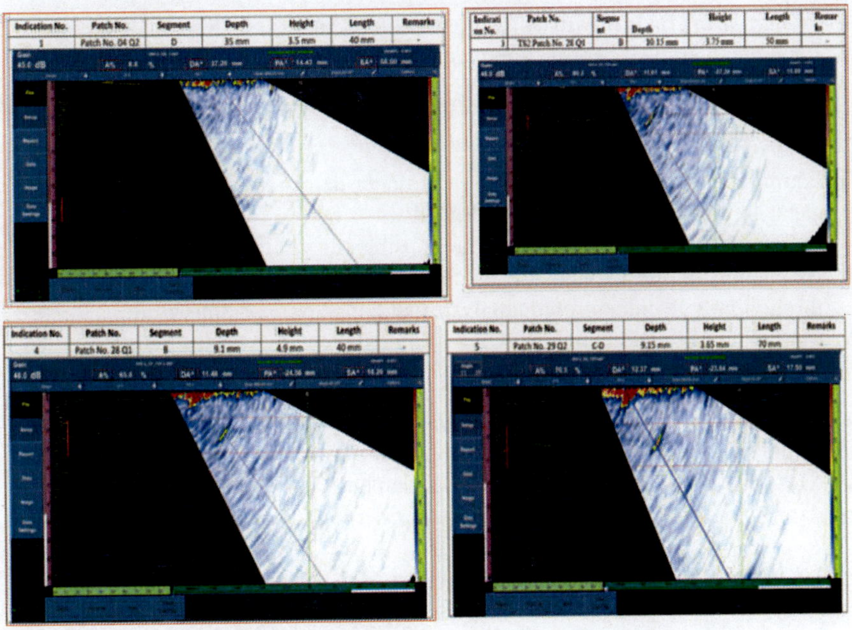

Fig. 5 PAUT output signal spectrum showing defect location during beam steering

Table 1 Criticallity of defects observed in different patches

Generator shaft

1	2	3	4	5	6	7	8	9	10

Turbine shaft

1	2	3	4	5	6	7	8	9	10
11	12	13	14	15	16	17	18	19	20
21	22	23	24	25	26	27	28	29	30

6.2 *Finite Element Modeling and Analysis of the Shaft*

The methodology followed in determination of the remaining life of the cracked shaft is summarized below.

- Identification of crack locations inside the rotor through PAUT.
- Detection of surface cracks through magnetic and dye penetrant methods.
- Geometric modeling of the shaft—CAD modeling of the rotor shaft based on the design drawing.
- The meshing of the shaft—Discretization into small elements for analysis.
- Incorporation of critical crack geometries observed by the PAUT.
- Stress analysis of rotor for the prediction of fatigue life.

- Prediction of crack growth using programs like Ansys and FRANC3D ™ modules for arriving at the extended life under the uprated condition.

As a part of life estimation, the following analysis were carried out on the meshed model of the shaft.

- Estimation of stress levels of the shaft under defect-free conditions for uprated capacity conditions—to identify critical stress regions.
- Estimation of fatigue life of rotor shaft using Ansys ™ program.
- Estimation of the service life of cracked shaft under uprated load conditions through fatigue crack growth analysis using FRANC3D™ program.
- Prediction of the remaining life of the rotor shaft at uprated load conditions so as to have a minimum service life of 1 year.

The 3D CAD model of the hydroturbine rotor assembly was created in system, as shown in Fig. 6. The turbine-generator shaft was meshed with tetrahedral elements in ABAQUS™, with a total of 1312237 elements as shown in Fig. 7. During the operating condition, the turbine torque is considered consumed by the generator rotor.

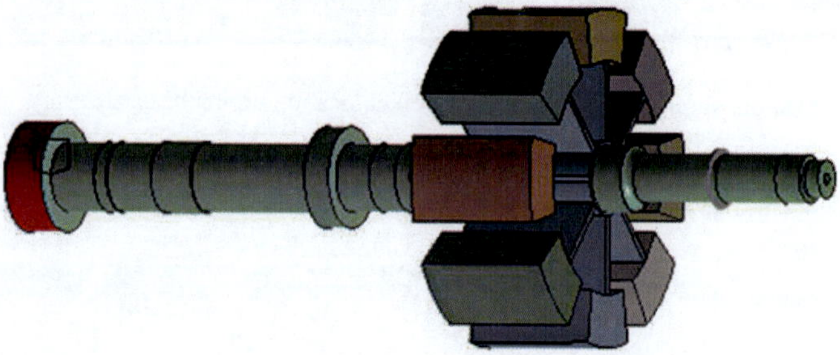

Fig. 6 View of generator shaft with rotating mass

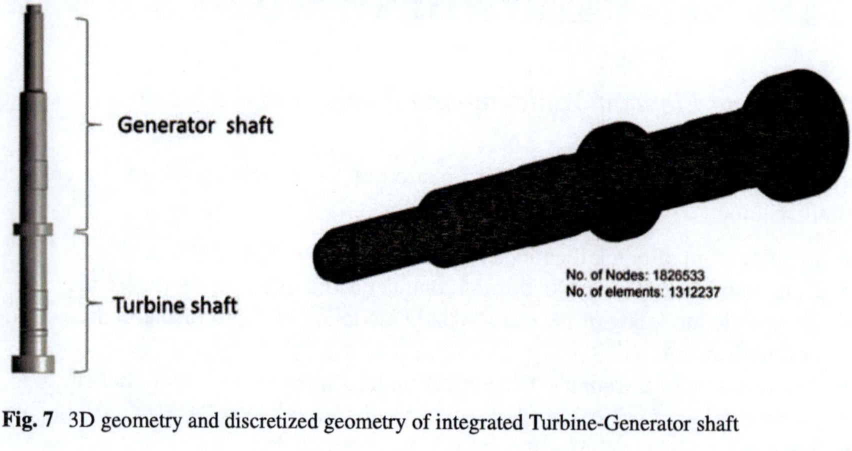

Fig. 7 3D geometry and discretized geometry of integrated Turbine-Generator shaft

The shaft is allowed to rotate in the Y-direction, and the translation displacements were constrained for the shaft.

FE analysis takes into account the geometric, material, and loading histories in the preprocessing stage. Stresses and strains are taken from the FE analysis to the fatigue calculations. The fatigue analysis was carried out by Ansys™ software tool adopting Brown–Miller algorithm. The calculation of fatigue life is due to each load block and compute the damage for each load block. Finally, the total damage for the defined load spectra and calculation of a number of cycles for failure is carried out.

6.3 Results of Fatigue Crack Growth Rate Studies

6.3.1 Stress Analysis of the Integrated Shaft

The von Mises and the shear stresses experienced during the running of the turbine were estimated through the static structural analysis considering all the parameters such as efficiency, misalignment, magnetic imbalance, etc. The assumptions made for the boundary conditions during the analysis are as follows.

- The calculated torque is generated at the turbine end. The generated torque will be consumed at the rotor end. Hence, the maximum torque was applied at the generator mounting end and at the rotor. The applied torques are opposite to each other in direction. Looking from the top view, the torque at the turbine end is clockwise, while it is anticlockwise at the rotor.
  - The centrifugal force at the shaft is also taken into account with a rotational velocity of 500 rpm.
  - As this is a Pelton wheel turbine, there is no upward thrust to balance the mass of the rotor was considered.
  - The shaft is suspended from the top but is free to rotate.

The turbine mass was considered at the bottom-most flange end as a lumped mass. The upward thrust is applicable only when the turbine is running. When the turbine is not running, the upward thrust does not come into the picture and only the dead weights cause the stresses. The yield stress of the turbine shaft material was calculated from the failed turbine shaft material. The calculated stresses at the smallest cross section of the shaft is about 20 MPa which is small compared to the yield stress of the shaft material, which is in excess of 700 MPa. The mass of the poles and other masses are modeled as a solid disk and attached to the shaft.

The finite element (FE) analysis of the generator shaft was carried out at a maximum rotational speed of 500 rpm. Global Y-direction is considered as the axis of rotation.

The average stress in a rotating component in all three directions of x, y, and z is expressed in terms of von Mises stress. The maximum von Misses stress observed from the static analysis with a rotational body force is observed to be 70.8 MPa

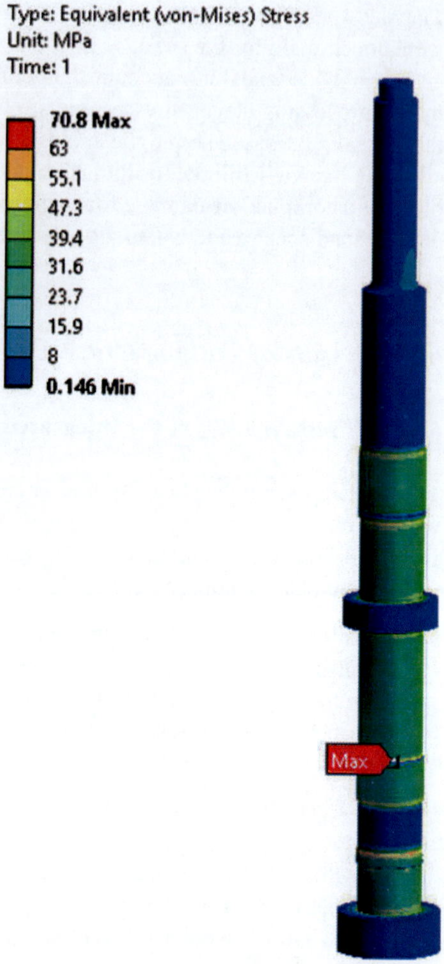

Fig. 8 Von misses stress contour on different sections of the shaft

against the allowable maximum tensile stress of 700 MPa, calculated based on the available shaft material, and the maximum stress location is observed in the region just above the turbine rotor bolting flange as shown by the green color contour region (Figs. 8 and 9). The top end of the generator shaft above the rotating mass mounting flange region has been observed to experience comparatively very low-stress levels.

6.4 Fatigue Analysis of Untracked Shaft

The Brown–Miller algorithm which is based on a strain-based approach, primarily used for the ductile material, is used for the determination of fatigue life [6]. The fatigue analysis of the generator shaft was carried out under defect-free situations

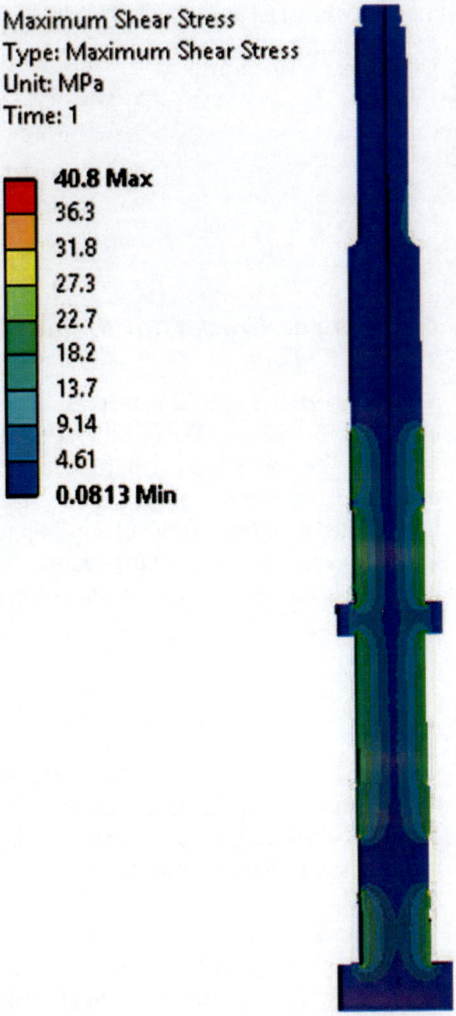

Fig. 9 Torsional shear stress profile

at the uprated loading condition. The fatigue life of the shaft with no cracks was observed to be 1.0×10^{12} cycles. From the plant operation data, it is found that the generator shaft has running cycles of 5.913×10^{9} cycles. Thus, under defect-free condition, the fatigue life would have been much more than 55 years. However, the actual life is limited by the number of cracks, size, and orientation as observed by the PAUT results.

Table 2 Material properties and crack growth constants used for the fracture analysis

Young's modulus	200 GPa
Poisson's ratio	0.3
Yield strength	275 MPa
Tensile strength	590 MPa
Fracture toughness	45 MPa $\sqrt{m}$
Paris constant, C	3×10^{-9}
Paris exponent (m)	3.0877

6.5 *Fatigue Crack Growth Rate Analysis*

The observed four critical defects close to the OD surface were considered for the crack growth analysis. The growth rate of these cracks under the uprated load conditions was monitored through finite element simulation, taking into account the results of the dynamic stress analysis profiles. The fatigue crack growth rate in a structure is correlated with the severity of crack tip loading, characterized by the elastic stress intensity factor, 'K' [7, 8]. The fatigue crack growth rate in the steel shaft material has been predicted from the known stress intensity factor through the well-established Paris equation

$$\frac{da}{dN} = C(\Delta K)^n \tag{1}$$

where 'ΔK' is the change in 'K' during each fatigue cycle in MPa$\sqrt{m}$; da/dN is the increase in crack length in m/cycle, and 'C' and 'n' are experimentally derived constants. The constants used for the crack growth used for the analysis are based on the yield strength properties of the rotor steel specification, and the values are given in Table 2.

The crack growth analysis involves incremental growth of the crack in different steps. After each growth, the modified crack dimension was inserted into the shaft, and re-analysis of the stress magnitude ahead of the crack front was estimated. With the newly calculated stress condition around the crack region, the growth rate of the crack was monitored. A total of 25 steps were followed until the crack reached the outer surface of the shaft. The stress intensity factor which is a function of the crack growth rate at the crack front for each of the steps is calculated.

The sub-model approach involving the creation of local sub-model for the crack insertion region was followed. The crack growth analysis of the inserted cracks was monitored considering the load spectra obtained from the static analysis. Depending upon the local stress intensities experienced in each of the crack tips, the growth of crack takes place in a preferential direction. The incremental growth of crack length (da) is estimated corresponding to the number of fatigue cycles (dN).

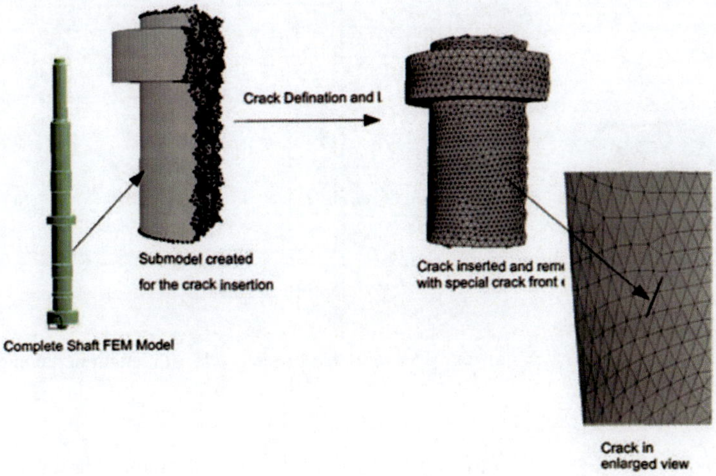

6.5.1 Material Properties and Boundary Conditions Used

The generator shaft is made up of UNS S41000 and the crack growth rate linked parameters such as Paris constant, and Paris exponent was selected matching with the strength properties of the rotor shaft material as given in Table 2.

Assuming the threshold stress intensity factor for the material to be 1 MPa$\sqrt{m}$, the calculated SIF is observed to be crossing the K_{th} value, and hence the crack growth was considered in multiple steps. After each step, the stress corresponding to the modified dimension of the crack was considered, and the growth of the crack was continued. This process was continued iteratively until the crack reaches the surface of the shaft, which is considered as the end of life. The crack was made to grow with the following material-specific constants such as Paris exponent and crack growth rate constant, obtained from the laboratory crack growth rate test conducted on the shaft material. The crack was allowed to grow until it reaches the surface of the shaft. For cracks that are deep inside the shaft, extrapolation of the crack growth rate data was considered for determination of the time required to reach the surface.

The shape of the crack and other fracture parameters was captured in various stages of crack growth. The increase in crack length toward the surface was monitored for various growth steps, and the typical crack growth cycle for the crack no. 1 is shown in Fig. 10.

Considering the continuous operation of the unit at 500 RPM in a day, the estimation of crack growth life was made based on the number of cycles required or the crack to reach the surface, which is estimated to be 993315700 cycles corresponding to the crack No. 4.

Number of Days = (993315700)/(500*60*24) = 1379 days (3.78 years).

Similarly, for all other cracks, its growth rate was analyzed for a typical of 30 growth steps, and the results were extrapolated, especially for the cracks deep inside the shaft.

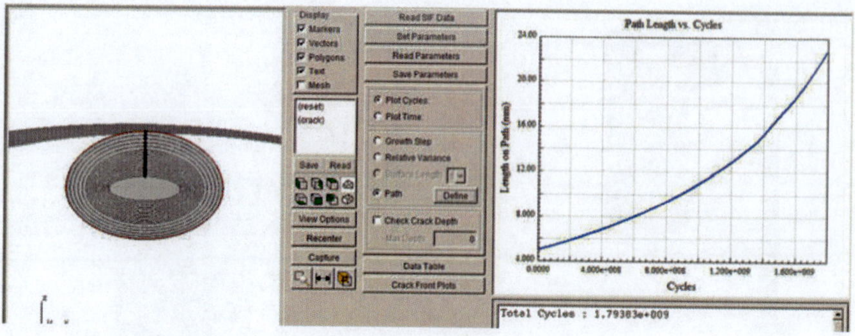

Fig. 10 Typical incremental crack growth cycles with that of the crack path length pertaining to crack no. 1

The results of crack growth rate data of all the four critical cracks observed are summarized in Table 3.

The rate of growth of each of these cracks has been observed on a continuous basis up to 25×10^6 cycles equivalent of 2500 h of operation. Extrapolation of the crack growth data is for cracks deep inside the shaft. The growth rate of crack No. 3 and 4 which are closer to the surface is much higher compared to other cracks in view of the increased stress intensity.

7 Conclusions

The systematic study on the cracked rotor shaft through a combination of advanced PAUT inspection and crack growth rate analysis under the uprated loading conditions indicated the following.

- The PAUT technique has identified all the crack-like defects present in the rotor shaft. Most of the cracks were observed to be close to the OD side of the shaft, and their orientation is observed to be slightly angular with respect to the shaft axis (less than 20°).
- The stress analysis of the shaft, the area-averaged maximum shear stress, was observed to be 40.8 MPa in regions close to generator-turbine bolting flange.
- The maximum von Mises stress under uprating conditions is observed to be 70.8 MPa against the estimated yield strength of 700 MPa, based on the sample collected from the failed shaft.
- For a given rating condition, the crack growth rate in turbine-generator shaft depends primarily on the magnitude of the crack, its location, and the stress intensity factor at the crack tip, etc.

Table 3 Crack growth rate data of four critical cracks and predicted remaining life

Critical cracks	Depth	Crack length (2a)	b	Angle of crack	No. of growth steps	Increase in crack length, mm	Total growth cycles	Growth in years	Remaining length to grow, mm	Predicted remaining life (Years)
1	18	30	5	30	19	18	1.7983E + 09	6.84	0	~9.6 years
2	35	40	5	30	24	20	1.3590E + 09	5.17	15	>20 years
3	10	40	5	30	19	10	1.2190E + 09	4.64	0	4.64 years
4	10	70	3	30	24	10	9.9332E + 08	3.78	0	3.78 years

- Based on the crack growth rate studies carried out on the critical crack considered, a life of 3.78 years was observed (crack reaching the surface of the shaft) under the conditions of fully reversed loading with assumed threshold K_I value of 1 MPa$\sqrt{m}$.

References

1. Rahi OP, Chandel AK (2015) Refurbishment and uprating of hydropower plants—a literature review. Renew Sustain Energy Rev 48:726–737
2. Oberoi BR, Naidu BSK (1988) Uprating and refurbishing hydro plants in India. Int Water Power Dam Constr 40(10):24–26
3. Naidu BSK (1989) Uprating hydro turbines: hydraulic and mechanical factors. Water Power 41(10):19–22
4. Citarella R, Lepore M, Shlyannikov M, Yarullin R (2015) Fatigue crack growth by FEM-DBEM approach in a steam turbine blade. Ind Eng Manag 4(160):2169–2173
5. Hou J, Wescott R, Attia M (2014) Prediction of fatigue crack propagation lives of turbine discs with forging-induced initial cracks. Eng Fract Mech 131, 406–418, 30 Nov 2014
6. Brown MW, Miller KJ (1979) Initiation, and growth of cracks in biaxial fatigue. Fatigue Fract Eng Mater Struct 1(2):231–246
7. Paris PC (1998) Fracture mechanics and fatigue: a historical perspective. Fatigue Fract Eng Mater Struct 21(5):535–540
8. Vasudevan AK, Sadananda K (1995) Classification of fatigue crack growth behavior. Metall Mater Trans A 26(5):1221–1234

Functionally Graded Material for Aircraft Turbine Disc on Fatigue Failure—An Overview

Mohammed Asif Kattimani, P. R. Venkatesh, and L. J. Kirthan

Abstract In aircraft turbine disc, cracks are originated at the trailing edge of the blades during operations. The literature review on microstructural studies revealed that the fatigue cracks are initiated due to high operating temperatures of turbine which led to failure of the disc and blades by stress rapture followed by fatigue crack propagation at the hub of the turbine. It is recommended that incorporation of functionally graded material (FGM) in the hub section of the turbine will impede the initiation and growth of crack at the hub of the turbine blades. A literature review on characterizing the effect of induction and laser hardening processes for fatigue life of steel alloy is systematically presented. The suggestion toward the improvement of fatigue life of an aircraft turbine disc and blades is addressed.

Keyword Trailing edge · Microstructural · Fatigue cracks · FGMs. turbine blades · Turbine blades

1 Introduction

The limitations of isotropic and homogeneous materials were overcome with the advent of composite material. However, in recent years, scientists and engineers have realized the limitations with composite materials and hence are moving toward materials with functionally graded properties. Functionally graded materials (FGM) are a new class of materials, whose properties are tailored to meet the property requirements of a particular application. For example, the high-pressure turbine section

M. A. Kattimani (✉)
Research Scholar, RVCE-VTU-RC, Bangalore, India
e-mail: mohammedasifkattimani@gmail.com

P. R. Venkatesh · L. J. Kirthan
Department of ME, RVCE, Associate Professors, Bangalore, India
e-mail: venkateshpr@rvce.edu.in

L. J. Kirthan
e-mail: kirthanlj@rvce.edu.in

S. Seetharamu et al. (eds.), *Fatigue, Durability, and Fracture Mechanics*, Lecture Notes in Mechanical Engineering, https://doi.org/10.1007/978-981-15-4779-9_28

reaches the highest temperature and is one of the highest stress parts of the engine, requiring very specialized superalloy materials [1].

There are different types of functionally graded materials (FGMs). These include the material gradient in functionally graded material, the microstructural gradient in functionally graded material, and the porosity gradient in functionally graded material [2].

From the previous century, numerous methods to enhance the fatigue life of steels were suggested, like heat and mechanical treatments, surface alloying, overlying coatings, etc. Amid them, surface treatments on steel pieces have various applications in many technological arenas. Usually, these are used to gain an enhancement of hardness, wear resistance, and corrosion properties. These processes allow steel pieces to achieve the requirement of high performance in the industrial field.

Most of the reviewed data in this paper are based on the classification of different heat treatment methods to obtain a high compressive residual stress to improve fatigue characteristics.

Following the above brief introduction, Sect. 2 describes an overview of gas turbine materials and fatigue failure of disc and blades, and Sect. 3 presents an overview of fatigue life evaluation using induction hardening process for various steels; Sect. 4 is concerned with fatigue behaviors of various steels on laser hardening, and Sect. 5 concludes with factors that ultimately control properties of graded materials.

2 Fatigue Failure of Gas Turbine Disc and Blades

Generally, gas turbines are used for the extraction of energy from warm gas flows in order to derive compressor and other accessories like gearbox. The disc of gas turbine is subjected to elevated temperature gradient and rotational velocity. High temperature leads to decrease in disc material strength, and high centrifugal forces arise due to high speed.

The life of important aerospace parts is controlled by means of degradation and failure known as fracture, wear, yielding, fatigue, corrosion, creep, erosion, etc.; the most important engine parts are considered to be gas turbine discs, and they have to bear substantial thermal and mechanical load. If anything happens in the section of turbine, then the complete engine function will be affected. Due to over speeding of the disc, turbine wheel would cause catastrophic failure and puncture of the engine casing by the large fragments of the disc. Under heavy operational conditions, aircraft is permitted for high rotational turbine speed.

The main focus is on the three critical sections of an aeroengine turbine disc: the assembly holes and the hub zone and the dovetail-rim area (fir-tree slots, serration fitting) as indicated in Fig. 1. The contact between the turbine blade and disc is subjected to centrifugal forces and thermal stresses which indicate the most prominent area from the point of view of the static and fatigue techniques [3]. The

Fig. 1 The dovetail-rim area (fir-tree slots, serration fitting) of turbine disc. Reference [3]

loads involved in these components are generally the centrifugal forces and thermal stresses.

There has been a lot of research toward the stress and failure of the jet engine turbine. Lucjan witek [3] explained the turbine disc critical area in case of both over speed and operational condition involved in the damage mechanisms analysis. Hou [4] presented the numerical and experimental investigation of fatigue fracture of turbine blades; Fig. 2a indicates the primary failure of blade at the top fir tree. Figure 2b and c depicts a typical crack in a used blade and fatigue progression marks. Park et al. [5] also focused on roblems of turbine parts connected to fatigue fracture. Bhaumik [6] described the orientation of columnar grains along the centerline of the blade through hub which are like a fountain as shown in Fig. 3. The result of that type of failure is very dangerous and leads to engine destruction and ultimately the loss of precious life.

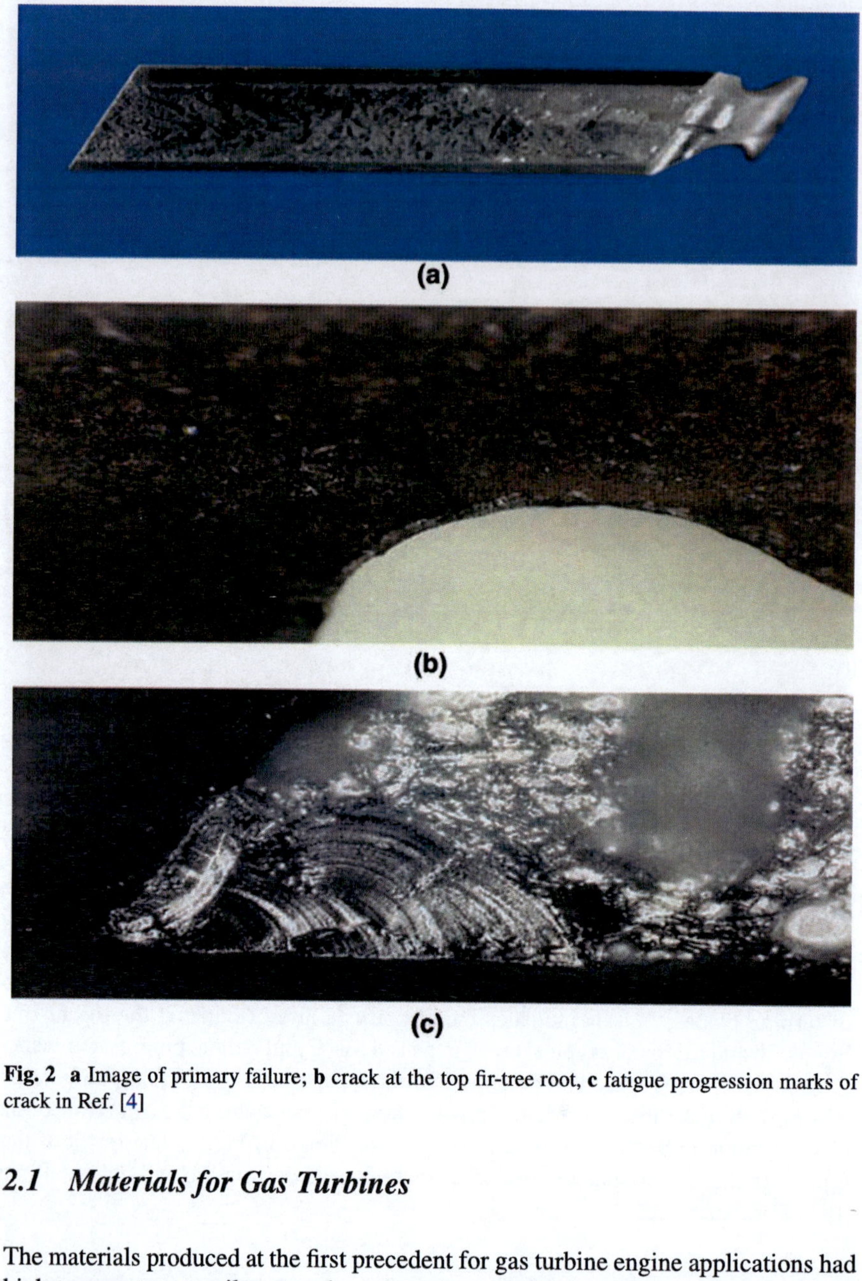

Fig. 2 **a** Image of primary failure; **b** crack at the top fir-tree root, **c** fatigue progression marks of crack in Ref. [4]

2.1 Materials for Gas Turbines

The materials produced at the first precedent for gas turbine engine applications had high-temperature tensile strength as the prime conditioned. This necessity quickly changed as operating temperatures rose. Stress rupture life and then creep properties became significant. In the subsequent years of development, low-cycle fatigue (LCF) life turns into one more significant parameter. Many parts in the aeroengines lead to

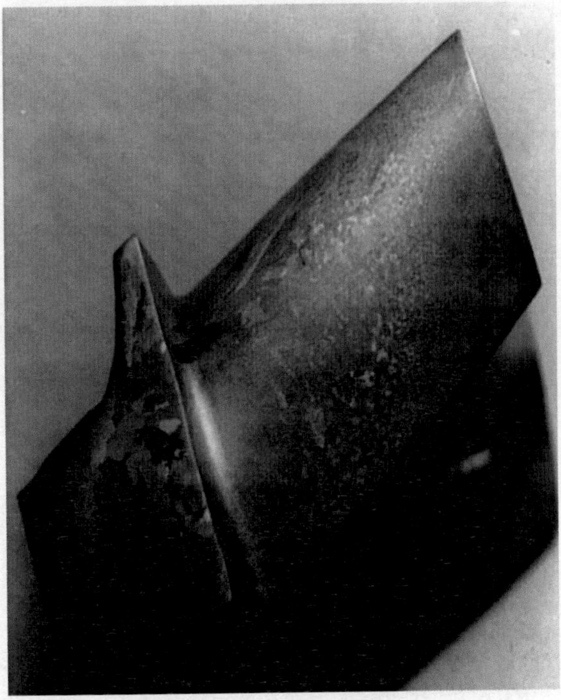

Fig. 3 Columnar grains at the blade root region. Reference [6]

fatigue and/or creep loading, and the selection of material is then based on the ability of the material to withstand such loads [7].

Austenitic Fe-base alloy has been used in Aeron engine turbine disc of A286. Disc of aircraft engine has been manufactured from superalloy 718 for more than 25 years [8]. Conventional ingot metallurgy route has been used to produce both of these alloys.

The significance of alloys such as 718 and 706 can be noticed from the fact that many authors paid heed on the importance and development of these alloys [9–15]. Udimet 720 is also considered as an advance wrought alloy which is useful in land-based gas turbines. In order to develop 720LI alloy, Cr content is being reduced to hamper sigma phase formation and also boron and carbon level to minimize stringers and clusters of borides, carbides, or carbonitrides. These alloys plug a vital role in land-based gas turbines and also have the applications in gas turbines of aircraft (1). Table 1 gives the details of superalloys and special steels used in the manufacturing of gas turbine disc.

Table 1 List of superalloys and special steels used in the manufacturing of gas turbine disc [7]

Grade name	Chemical composition—percentage												
	Ni	Cr	Co	Mo	W	Ti	Al	C	B	Zr	Fe	Cb	V
Wrought alloy-720	55	18	14.8	3	1.25	5	2.5	0.035	0.033	0.03	–	–	–
Ni-Fe alloy-718	52.9	19	–	3	–	0.9	0.5	0.03	–	–	18.5	5.1	–
Wrought alloy -720LI	57	16	15	3	1.25	5	2.5	0.025	0.018	0.03	–	–	–
Ni-Fe alloy-706	42.02	16	–	–	–	1.8	–	0.03	–	–	37	2.9	0.25
Austenitic Fe alloy	25	15	–	1.2	–	2	0.3	0.08	0.006	–	–	–	–
Medium-carbon low-alloy steel-CrMoV	0.5	1	–	1.25	–	–	–	0.30	–	–	96.7	–	0.25

3 Effect of Induction Heating on Steel Alloy for Fatigue Behavior

Induction hardening is a well-known heat treatment process which hardens only a material surface and results in a high degree of compressive residual stress over the surface. It has observed that an increase in fatigue life of various types of machine components subjected to surface hardening process [16–19].

Fatigue failure is normally examined as the main cause of affecting any specimen under dynamic loading condition. Approximately, 90% of failure situations in mechanical specimens are due to fatigue. According to the significance of this type of failure, many researcher scholars had investigated; few of important studies are reported in the following section.

The super-rapid induction hardening process was used to quench the specimens at five different thermal cycles as shown in Fig. 4 to obtain shallow hardened layer on ordinary structural steel with 45% content of carbon. Primarily, the specimens were subjected to annealed treatment at 1100 °C for 3 h and these quenched to 900 °C, and finally tempered at the temperature of 600 °C. The hardness and residual stress distribution in samples with five different thermal cycles exhibit a high compressive residual stress of almost 1000 MPa close to the surface as shown in Fig. 5. Material with a distinctly shallow hardened layer indicates only marginally higher fatigue strength as compared to the untreated specimen because non-hardened area lying beneath the hardened layer has a fatigue crack described in Fig. 6 [20]

A novel experimental technique was used to review the possible impact of residual stress on fatigue strength. The method of induction heating is appeared to be an effective way in order to introduce residual stress (RS) in thick-walled hollow fatigue AISI 304L stainless steel specimen without altering their geometry, surface roughness, and microstructure [21].

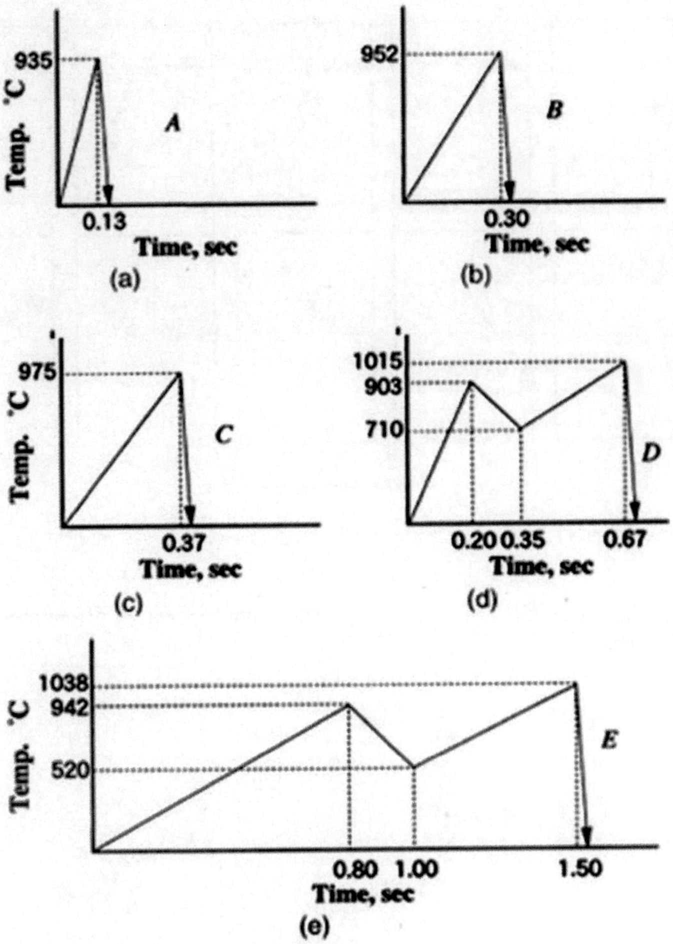

Fig. 4 Heat cycle for SRIH and quenching. Reference [20]

SAE 52100 specimens were subjected to heating and quenching one to three times so that martensitic microstructure can be refined. After doing rotating bending fatigue test under low stresses, the fisheye areas (internal crack propagation) by laser confocal and scanning electron microscopes were noticed [22].

An AISI E52100 shaft was repeatedly subjected to induction heating and quenching, and then the laser confocal microscopy was used to observe the microstructure of the material; the properties such as hardness, residual stresses, fatigue life, and content of retained austenite were measured using SEM and orientation image mapping. The structure of the repeated quenched specimens represents the reduction of prior austenite grain size and refined martensitic microstructure. The fatigue life of the repeatedly quenched specimen was three times increased due to refined martensitic structure and large content of retained austenite [23].

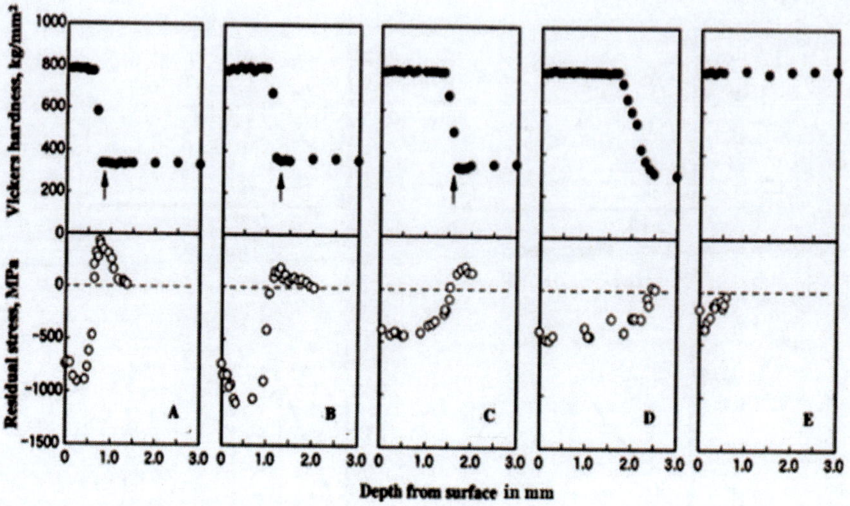

Fig. 5 Distribution of hardness and residual stress. Reference [20]

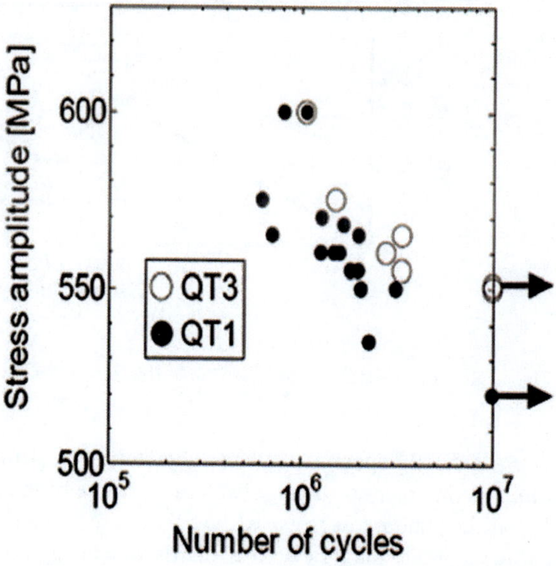

Fig. 6 Stress-life curve of induction heat treated specimen. Reference [24]

EN8 steel specimens were heat treated by induction hardening, nitriding, and combined nitriding-induction hardening. Fatigue testing was performed in cantilever-type fatigue testing machine. The results show improved strength (fatigue) and microhardness when combined heat treatment process of nitriding was followed by induction hardening [24] (Figs. 7 and 8).

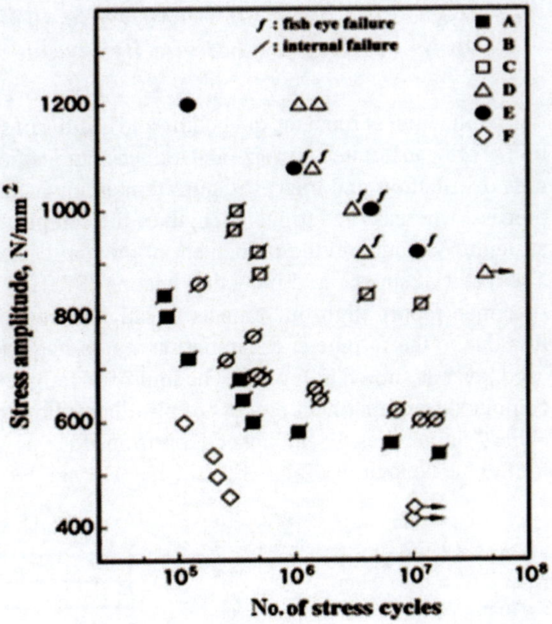

Fig. 7 Results of fatigue test. Reference [22]

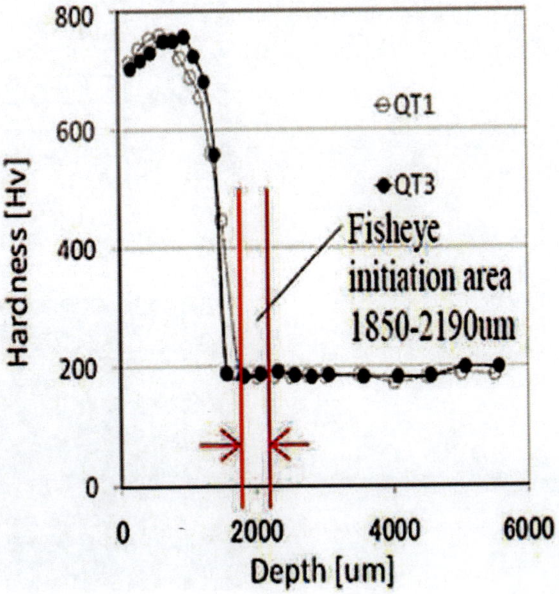

Fig. 8 Vickers' hardness distribution. Reference [22]

3.1 Highlights Summary of Effect of Induction Heating on Steel Alloy for Fatigue Behavior

The application of repeated quenching and heating of specimens represents the reduction of prior austenite grain size and refined martensitic microstructure [23]. The hardness distribution and microstructure remain unchanged by the repeated quenching process, whereas the fatigue strength of the samples increases with higher retained austenite content and the refinement of the martensite structure as shown in Fig. 9. The result of super-rapid induction heating (SRIHQ) to an ordinary structural steel specimen report slight increments in fatigue strength over untreated sample [20]. It is due to the fatigue crack initiation at non-hardened area just beneath the hardened layer as shown in Fig. 10. The improved fatigue strength was accessible by the combined heat treatment process of nitriding followed by induction hardening [24]. Table 2 ummarizes the published reports on effect of induction heating on steel alloy for fatigue behaviour (Table 3).

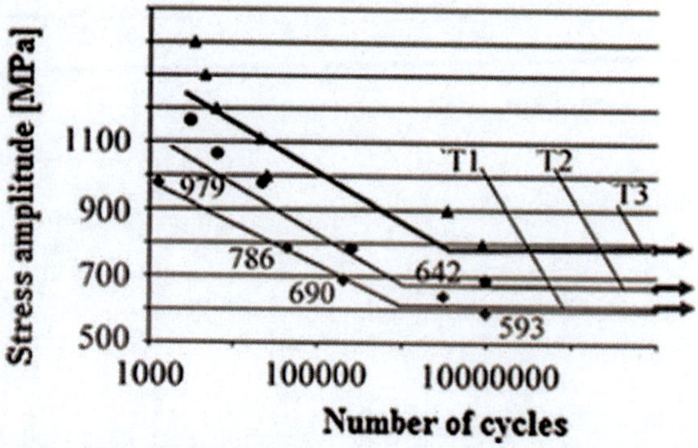

Fig. 9 Stress-life plot of samples after quenching. Reference [23]

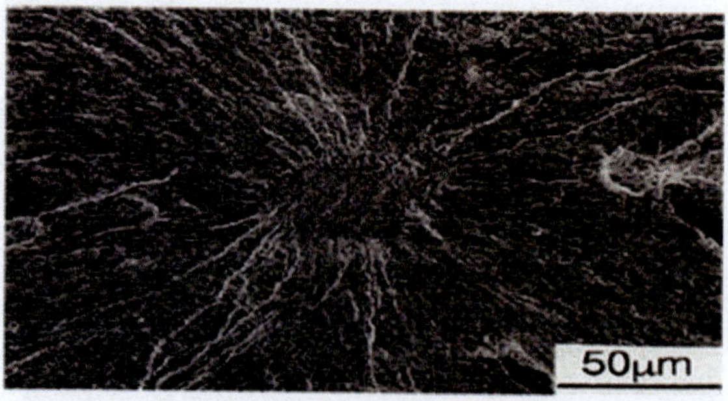

Fig. 10 Crack originate location for sample with hardened layer of 1.2 mm. Reference [20]

Table 2 Summarizes the published reports on effect of induction heating on steel alloy for fatigue behavior

S. No.	Reference No.	Materials	Heating method	Temperature	Type of mechanical testing	Effect on mechanical properties		
						Hardness (HV)	Fatigue loading	
							Stress (MPa)	Life (N)
1	[20]	structural steel 45% C	Rotation bending fatigue tests (Induction heating and quenching (IHQ))	Annealed = 1100 °C Quenched = 900 °C Tempered = 600 °C	Super-Rapid Induction Heating (SRIH) system	–	800 1000 1200	5.1E + 05 4.61E + 06 1.1E + 06
2	[21]	AISI 304L Stain less steel	High frequency induction heating	Below 1000 °C, i.e., between 20–1000 °C	High-cycle fatigue test [hydraulic instron machine]	–	210 230 255 ($S_a = 0$ & $R = -1$)	$>10^7$ 23052 9484
3	[22]	SAE 52100 bearing steel	Repeated induction heating process and quenching	–	Rotating bending fatigue test	757HV	510 To 734	Increased by 30 no's i.e. from 520–550
4	[24]	EN8 steel	1 Nitriding 2 Induction Hardening 3 Nitriding + IH	560 °C	Hardness Tensile Fatigue	550 525 620	362.08 336.01 387.93	1.05E + 07 9.56E + 06 1.30E + 07

Table 3 Summarizes the published reports on effect of laser hardening on steel alloy for fatigue behavior

S. No.	Reference No.	Materials	Heating method	Temperature/Power	Type of mechanical Testing	Effect on mechanical properties		
						Hardness (HV)	Fatigue loading	
							Stress (MPa)	Life (N)
1	[43]	AISI 4140 steel	Laser radiation + Vibration peening	600 °C 700 °C 800 °C	ASTM E468-08 Fatigue Test		$U_n = 230$ 330 335 350	1E + 08 1E + 09 1E + 09 1E + 07
2	[45]	AISI 1040	High power Diode Laser (200 W, 20 mm/s)	20 °C	Four point rotating bending	–	460	1.375E + 05
3	[47]	SAE9254 carbon low-alloy spring steel	Unpeened Laser Shock peening WLSP	25 °C 250 ± 15 °C	ASTM E468-08 Fatigue Test	196.14 601.38 673.49	–	1.83E + 05 14.11E + 05 28.84 E + 05
4	[46]	Cr12MoV steel	Laser quenching	–	Tension–Tension fatigue test R = 0.1	–	368 388 422	–
5	[44]	AISI 15B21H	2.5 kW CO_2 Laser hardening	1500 W	Rotating bending	950	405	2E + 07

4 Effect of Laser Hardening on Steel Alloy for Fatigue Behavior

The laser hardening is an industrial technology which utilizes high-intensity laser diode for enhancing the surface properties of engineering materials. The main benefits of using a laser are attributed to (1) its small size and (ii) its easy manipulation. The small diameter of laser beam allows a selective heat treatment without affecting unnecessary area, and the easy manipulation of its power and spot size allows an ability to tailor the thickness of heat-affected zone (HAZ) [25–30]. Many researchers [31–42] indicate improved fatigue properties of cast iron and steel using laser transformation hardening. Generally, high surface hardness and residual compressive stress values were found in those studies. Few of important studies are reported in the following section.

M.C. OH et al. study shows the improved fatigue behavior of AISI 4140 steel over their surface treatment using laser diode of 1kw power under fine control and constant temperatures at 600, 700, and 800 °C, respectively. Later, the laser hardened fatigue samples were subjected to vibration peening by magnetostrictive transducer, as presented in Figs. 11, 12 and 13 [43].

Pedro De la et al. reported the effect of laser heating on plane bending fatigue steel. The smooth and notch test samples were subjected to laser heating, then quenching and tempering result in 18 and 56% enhancement of fatigue limit. The initiation of crack at surface or subsurface due to impudently of load level and fatigue life results in decrease of fatigue strength of the material [44].

Stefano Guarino et al. also indicated the enhancement in fatigue behavior of AISI 1040 steel over their surface treatment using high power laser diode. The laser power and scanning speed were altered to study the significance of each operational parameter. Figure 14 shows the laser heat source treatment of the specimen on a CNC turning machine, and further the specimen was used to conduct the fatigue test on rotating bending machine. The Gaussian energy distribution curve generated for

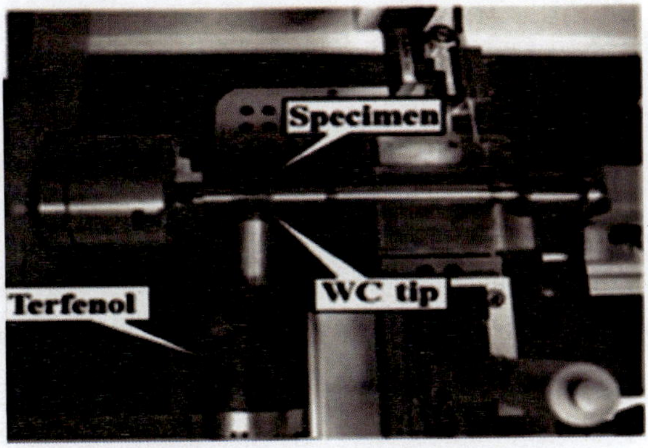

Fig. 11 Vibration peening system. Reference [43]

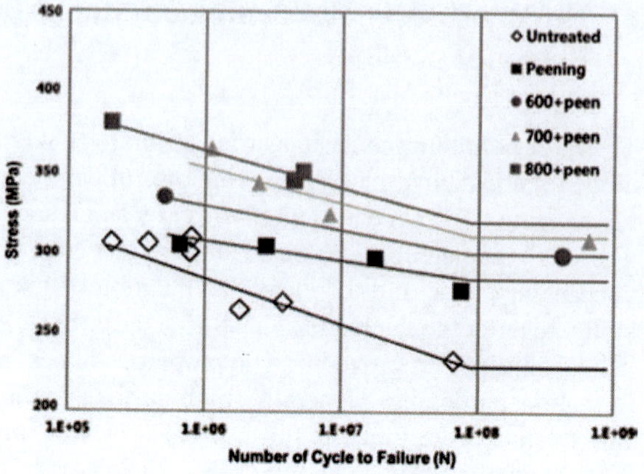

Fig. 12 Stress-life plot of AISI 4140 steel for various surface treatment conditions. Reference [43]

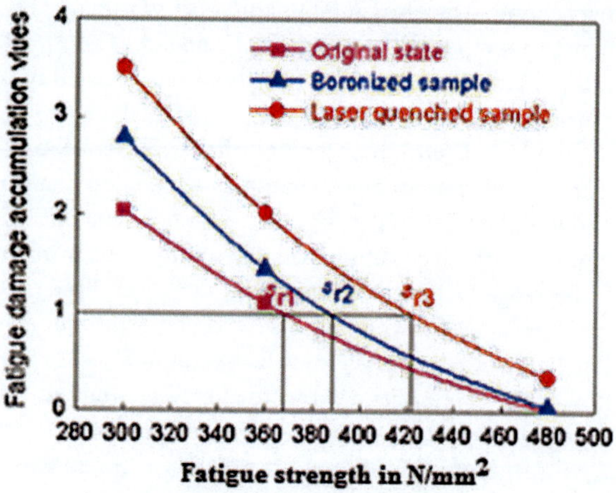

Fig. 13 Fatigue strength test result. Reference [46]

three-dimensional finite element models for transient analysis is shown in Fig. 15 [45].

5 Highlights Summary of Effect of Laser Hardening on Steel Alloy for Fatigue Behavior

The process of peening and originating compressive residual stress over surface of the specimen forms the refined grain structure and impedes the imitation and growth of crack which leads to the improvement of fatigue strength of the material [43]. Figure 16 shows the microphotograph of fatigue fracture specimen of untreated, peening, and laser hardening at 800 °C later subjected to peening process [43]. The

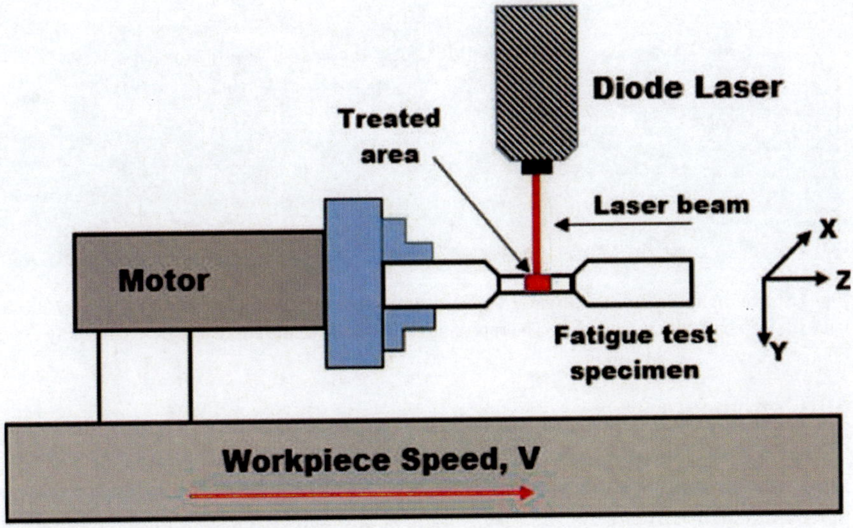

Fig. 14 High power laser diode system. Reference [44]

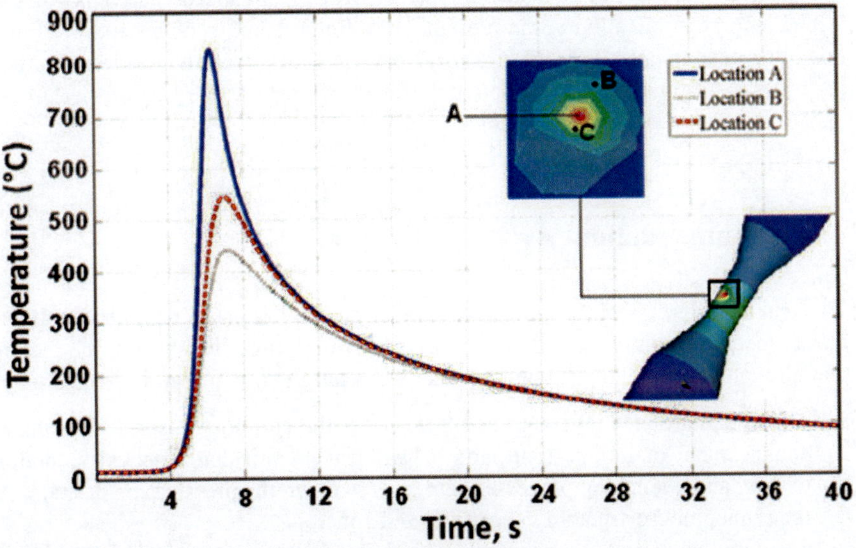

Fig. 15 Gaussian energy distribution curve. Reference [44]

results of laser heating of smooth and notch B-Mn steel specimen show 18% and 56% enhancement of the fatigue strength [44].

Table 2 summarizes the published reports on effect of laser hardening on steel alloy for fatigue behavior.

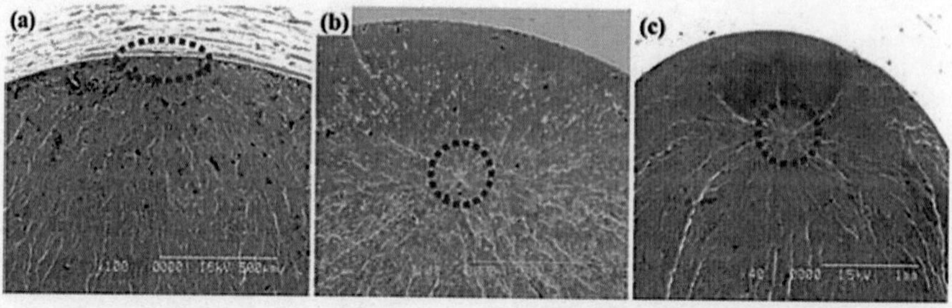

Fig. 16 Microphotograph of fatigue fracture **a** untreated, **b** peening, **c** laser hardened at 800 °C later subjected to peening process. The representation of crack initiation at sites is shown by circle

6 Conclusion

The breakdown of the turbine rotor disc of a gas turbine engine is characterized by an inaccurate microstructure at the blade airfoil section and high operating temperature, which led to crack initiation at the trailing edge by stress rupture. The subsequent crack growth is by fatigue under the cyclic loading conditions.

Research and development toward the growth of advanced materials for gas turbine engine application is the present study for research scholar, and the functionally graded materials (FGM) are also a new class of materials, whose properties are tailored to impede the initiation and growth of crack at the hub of the turbine blades and hence to increase the engine efficiency.

7 Recommendation

This extensive review of the research in graded steel alloy based on their heat treatment processes mainly induction, and laser provides some insight into the factors that ultimately control their properties mainly as fatigue life and grain structure.

1. The thick hardened layers up to 1.8 mm over the component surface results in enhancement of fatigue limit, and it has been proven that repeated induction heating and quenching processes help to slow down the propagation of crack by the refined microstructure in heat-affected zone.
2. The combined heat treatment process shows better mechanical properties in comparison to specimens without treatment.
3. The enhancement of fatigue limit by 40% on 800 °C laser heating and subsequently peened compared with untreated condition. The main mechanism for improving fatigue property is due to high compressive residual stress and contribution of the martensitic transformation on the laser hardened surface.
4. Fatigue enhancement could be guaranteed by taking supervision to optimize the laser parameters in order to avoid the damaged zone.

5. The operating temperature of the turbine blades is required to be maintained at the prescribed range of blade alloy temperature.

References

1. David Furrer and Hans Fetch (1999) Ni-based superalloys for turbines discs. JOM 51(1):14–17
2. Mahamood RM, Titilayo E (2017) Introduction to functionally graded materials, Springer international publisher
3. Witek L (2005) Failure of turbine discs of an aero engine
4. Hou J, Wicks BJ, Antoniou RA (2002) An investigation of fatigue failure of turbine blades in gas turbine engine by mechanical analysis. Eng Failure Anal 9:201–211
5. Park M, Hwang Y, Choi Y, Kim T (2002) Analysis of aJ69-T-25 engine turbine blades fracture. Eng Failure Anal 9:593–601
6. Bhaumik SK (2002) Failure of turbine rotor blisk of an aircraft engine. Eng Failure Anal 9:287–301
7. Mukeinutalatati NR (2011) Materials for gas turbines-an overview
8. Schilke PW (2004) Advanced gas turbine materials and coatings
9. Loria ED (1989) Proceeding of conference on superalloy 718-metallurgy and applications, TMS, ISBN 0-87339-097-0, Warrendale, PA
10. Loria ED (1991) Proceeding of conference on superalloy 718, 625 and various derivatives, TMS, ISBN 0-87339-173X, Warrendale, PA
11. Loria ED (1994) Superalloy 718,625 and various derivatives, the minerals, metals & materials society, ISBN 0-87339-235-3, Warrendale, PA
12. Loria ED (1997) Superalloy 718,625,706 and various derivatives, the minerals, metals & materials society, ISBN 0-87339-376-7, Warrendale, PA
13. Loria ED (2001) Proceeding of 5th international conference on superalloy 718, 625,706 and various derivatives, TMS, ISBN 0-87339-501-7, Warrendale, PA
14. Loria ED (2005) Proceeding of 6th international symposium on superalloy 718, 625,706 and various derivatives, TMS, ISBN 0-87339-602-8,Warrendale, PA
15. Poulose Neeta (2018) Proceeding of fatigue. Durability, Fracture Mechanics
16. Kurath P Jiang Y (1995) SAE technical paper series, 950707
17. Bertini L, Fontanari V, Bellin F, Fuganti A, Plano S (1998) Fatigue Des 1998(2):593
18. Jiang Y (1998) Metal mater 1998(4–3):520
19. Zhang HY, Stephen RI, Glinka G (2000) ASTM 2000(STP1360):240
20. Komotori J, Shimizu M, Misaka Y, Kawasaaki K (2001) Fatigue strength and fracture mechanism of steel modified by super-rapid induction heating and quenching. Int J Fatigue 23:225–230
21. Paquet D, Lanteigne J, Bernard M, Baillargeon C (2014) Characterizing the effect of residual stresses on high cycle fatigue (HCF) with induction heating treated stainless steel specimens. Int J Fatigue 59, 90–101
22. Koike H, Santos EC, Kida K, Honda T, Rozwadowska J (2011) Effect of repeated induction heating on fatigue crack propagation in SAE 52100 bearing steel. Adv Mater Res 217–218, 1266–1271
23. Santos EC, Kida K, Honda T, Koike H, Rozwadowska J (2011) Fatigue strength improvement of AISI E52100 bearing steel by induction heating and repeated quenching. Phys Chemical Mech Mater 5
24. Kumar MS, Ragunathan S, Srinivasan V (2016) Effect of heat treatment on the fatigue behavior of EN8 steel ARPN. J Eng Appl Sci11
25. Luo KY, Lu JZ, Zhang YK, Zhou JZ, Zhang LF, Dai FZ, Zhang L, Zhong JW, Cui CY (2011) Mater Sci Eng, A 528:4783–4788

26. Avil´es R, Albizuri J, Lamikiz A, Ukar E, Avil´es A (2011) Int J Fatigue 33, 1477–1489
27. Heitkemper M, Fischer A (2002) Proceedings of the 6th international tooling conference, vol 2, pp 935–945
28. Singh HB, Copley SM, Bass M (1981) Metall Trans A 12:138
29. Gariepy A, Larose S, Perron C, Bocher P, Levesque M (2013) Finite Elem Anal Des 69:48–61
30. Fe P, Jiang C (2013) Mater Des 56:1034–1038
31. Gnamuthu DS (1979) Laser surface treatment. In: Metzbower EA (ed) Applications of lasers in material processing. American Society for Metals, Metals Park, OH, pp 177–209
32. Singh HB, Copley SM, Bass M (1981) Fatigue resistance of laser heat-treated 1045 carbon steel. Metall Trans A 12A:138–140
33. Shur EA, Voinov SS, Kleshcheva II (1982) Increasing the design strength of steels by laser hardening. Metal Sci Heat Treat 24(5–6):341–344
34. Mitin VY, Tesker E, Gur'ev VA (1989) Influence of the surface relief of laser hardening on the cyclic strength of 45 steel. Metal Sci Heat Treat 30(9–10):769–772
35. Bello´ JM, Ferna´ndez BJ, Lopez V, Ruiz J (1994) Fatigue performance and residual stresses in laser treated 50CrV4 steel. J Mater Sci 29, 5213–5218
36. Yijun H, Deheng S, Hongying Y (1994) Effect of surface treatment with laser on the fatigue life of class 1 blade of WP-6A engine compressor. In: Da-Heng W (ed) Second international conference on optoelectronic science and engineering '94, Beijing, pp 600–602
37. Kato A (1985) Prevention of fracture of cracked steel bars using laser: part I. J Eng Mater Technol 107:195–199
38. Merrien P, Lieurade HP, Theobalt M, Baudry G, Puig T, Leroy F (1992) Fatigue strength of laser beam surface treated structural steels. Surf Eng 8(1):61–67
39. Lin R (1992) On residual stresses and fatigue of laser hardened steels. Doctoral thesis, Mechanical engineering, Linko¨ping University
40. Ericsson T, De la Cruz P (1987) Residual stresses and fatigue of a surface treated B-Mn steel. In: Lieurade HP (ed) Fatigue and stress of engineering materials. IITT Int, Paris, pp 113–129
41. Sun Y-T, Li J-Q, Lu J (1988) The effect of laser treatment of 60 Si2MnA steel on the distribution of residual stress and fatigue properties. In: Metzbower EA, Hauser D (eds) International power beam conference: power beam processing, ASM, Metals Park, OH, pp. 191–197
42. Doong J-L, Chen T-J, Tan Y-H (1989) Effect of laser surface hardening on fatigue crack gro growth rate in AISI-4130 steel. Eng Fract Mech 33(3):483–491
43. Oh MC, Yeom H, Jeon Y, Ahn B (2015) Microstructural characterization of laser heat treated AISI 4140 steel with improved fatigue behavior. Arch Metall Mater 60
44. De la Cruz Pedro, Oden Magnus, Ericsson Torsten (1998) Effect of laser hardening on the fatigue strength and fracture of a B-Mn steel. Int J Fatigue 5:389–398
45. Stefan Guarino and Gennaro Salvatore Ponticelli (2017) High power diode laser (HPDL) for fatigue life improvement of steel: numerical modelling. Metals 7:1–12
46. Kong DJ, Xie CY (2015) Effect of laser quenching on fatigue properties and fracture morphologies of boronized layer on Cr12MoV steel. Int J Fatigue 80
47. Warm laser shock peening without coating induced phase transformations and pinning effect on fatigue life of low-alloy steel

Literature Review of Fatigue Failure of Elastomeric Materials Under Complex Loading

Durgacharan Mishra, Vidit Bansal, Sharad Goyal, Nikhil kumar Singha, Vipul John, and Sujith Nair

Abstract In today's world, the elastomer is making an immense presence in many engineering applications like tires, etc. During service, the elastomeric component experiences complex loading cycles with varying amplitudes. Therefore, fatigue failure is a prime concern in their end-use application. The available effective techniques to predict fatigue life under complex loading is very crucial for designing the elastomeric component. The failure behavior of the elastomer depends on the various phenomena (Crack precursor size, crack growth rate, strain induce crystallization, fatigue threshold, load path) that governs the durability of the elastomeric component. In order to characterize the durability of the elastomer, several fatigue equivalence parameter-based studies have been introduced for multiaxial loading conditions. These fatigue life evaluation parameters include maximum principal strain, strain energy density (scalar parameter, independent of plane orientation), and crack energy density (plane-specific or critical plane parameter). This paper cites a comprehensive review of literature on fatigue behavior with prime focus on available methodology and models for crack initiation and growth prediction on the elastomer-based components. Different effective methods for fatigue failure prediction are reviewed and discussed.

Keywords Elastomer · Fatigue · Crack · Durability

1 Introduction

Elastomers have been widely utilized in many engineering applications due to their unique mechanical properties. Some of its applications include tire, conveyor belts, dampers, O rings, gaskets, dampers, etc. During its end-use application, elastomeric components are subjected to varying amplitude load cycles. This variable load cycle

D. Mishra · N. Singha · V. John · S. Nair
Rubber Technology Center, IIT Kharagpur, Kharagpur 721302, West Bengal, India

V. Bansal (✉) · S. Goyal
Advanced Engineering CEAT LTD (R&D), 389350 Halol, Vadodara, India
e-mail: vidit.bansal@ceat.com

S. Seetharamu et al. (eds.), *Fatigue, Durability, and Fracture Mechanics*, Lecture Notes in Mechanical Engineering, https://doi.org/10.1007/978-981-15-4779-9_29

leads to fatigue failure of the components. This reason makes fatigue in rubber as a prime concern. Generally, fatigue failure in elastomer starts from the initiation of cracks to the ultimate failure of components when exposed to multiaxial load paths.

There are two different approaches for the durability analysis of elastomer, crack nucleation and crack propagation approaches [1]. In crack nucleation approach, the fatigue life is determined by the appearance of small cracks, decrease in mechanical stiffness, or ultimate fracture [2–5], whereas in crack propagation approach, it is utilized when crack dimension and its position are well known. This method is applied when the component persists the crack growth before its ultimate failure [6, 7].

To increase the fatigue life cycles, there is a need for analyzing many phenomena that affect its fatigue failure behavior of elastomer. In the past few years, there had been significant work done in characterizing fatigue failure of elastomer. In this study, it has been pointed out that understanding the crack growth is a crucial tool for determining failure behavior.

Throughout this survey, various issues related to failure behavior have been addressed like material behavior during fatigue cycles, cracks appearance, crack dimension and its orientation, effect of load types and its path on crack growth, fatigue life evaluation parameter, the combined impact of these events on the durability of elastomer, and many more. To achieve life prediction target, it is very much necessary to address the abovementioned issues and many others.

This paper cites a comprehensive review of literature on elastomer fatigue behavior with emphasizing methodologies and models implemented in the life prediction of elastomer subjected to multiaxial loading condition.

2 Tearing Energy

Rivlin and Thomas [8, 9] extended the concept of tearing energy or energy release rate term stated in Griffith's theory. They stated it as a total amount of energy released due to the formation of new surfaces. This parameter depends upon the viscoelastic property and strain of material at crack tip irrespective of its specimen geometry. At crack tip, mechanism of energy release rate is localized and is defined as

$$T = -\frac{\mathrm{d}(U - V)}{\mathrm{d}A} \tag{1}$$

whereas U is the potential energy linked with different cracked configurations in the material, V is the work done by external forces as crack grows, and A is the surface area of crack.

Rivlin and Thomas [10, 11] confirmed the validity of energy-based criteria by several different specimens. For pure shear crack growth specimen, tearing energy is stated as

$$T = WL_0 \tag{2}$$

where W stands for strain energy density and L_0 is the original height of specimen. Various different specimens have been developed and used in finding the fatigue crack growth behavior in elastomer [12]. These include test specimens like simple tension, pure shear, trouser, and angle test specimen [9, 13].

Rivlin and Thomas [9] stated well-known relation of tearing energy

$$T = 2k(\lambda)Wa \tag{3}$$

where k is the proportionality factor that depends on strain and a is the crack length.

Lindley [14] gives the dependence of k on λ by the equation stated below:

$$k = \frac{2.95 - 0.08(\lambda - 1)}{\sqrt{\lambda}} \tag{4}$$

With the advancement of finite element analysis, the evaluation of tearing energy for cracks in components has been enabled [14]. Calculation of tearing energy has been implemented by various methods, and they are now immensely applied in many engineering applications.

3 Intrinsic Strength

Elastomer strength depends not only on its chemical structure but also on the viscoelastic behavior occurring near crack tip fields [15]. Due to this viscoelastic energy dissipation, the total energy required to propagate the crack is usually more significant than the intrinsic strength of the elastomer. Robertson et al. [16] have defined intrinsic strength as the residual strength remaining in the material after the strength-enhancing effects of energy dissipation in crack tip fields is removed. The practical reason for quantifying the intrinsic strength is that it makes a lower limit of fatigue crack growth rate. So at tearing energy below the intrinsic strength, the crack cannot propagate since there is no enough energy available to break the polymeric chain at the crack tip [17]. Fatigue crack growth and tear strength properties involve extra energy contribution due to viscoelastic energy losses, thus preventing the adequate characterization of the intrinsic strength of the material. Robertson et al. [16] in their work have used Coesfeld intrinsic strength analyzer that is operated with a testing methodology developed by Endurica LLC, which is based on the approach of Lake and Yeoh [18] to find the intrinsic strength of elastomer. This experiment procedure on the intrinsic strength analyzer executes in a few hours compared with fatigue crack growth testing near endurance limit, and it provides insight into molecular-level parameters thought to govern durability of rubber during end-use applications [16].

4 Material Characteristics

For the detailed evaluation of the fatigue life of elastomer, knowledge is required for the fatigue failure characteristics of elastomer [7]. The fatigue life depends not only upon the maximum energy release rate experienced during a load cycle, but it also depends upon the minimum value of the parameter that is used to define the loading cycles, or we can say on R ratio [19]. The R ratio can be defined as the ratio of the minimum load obtained during a cycle to the maximum load.

$$R = \frac{P_{\min}}{P_{\max}} \tag{5}$$

When specifying the R ratio, it is essential to indicate the type of loading parameter P on which ratio is based, such as stress, strain, energy release rate, etc. [19]. A periodic load history with cyclic limits is shown in Fig. 1.

The effect of change in R ratio is more profoundly shown by rubber, which undergoes strain-induced crystallization. In that rubber, the material near the crack tip crystallizes during straining, and also there will be crack tip blunting during the extensive amplitude loading [19].

When R = 0, i.e., thoroughly relaxing condition, material near the crack tip will be crystallizing and melting with each load cycle. During R > 0, i.e., non-relaxing condition, there is fluctuation of the crystallized region. However, the crystalized zone will always be present that will retard the crack growth propagation [20]. This type of scheme has a tremendous physical significance as the rubber component during its operation often experienced a large static load upon which smaller dynamic loads are superimposed [19]. The R ratio scheme can be used for the determination of fatigue life under loading condition during their application.

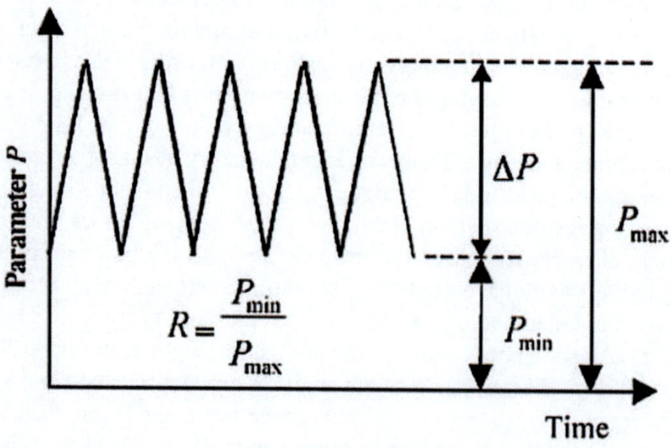

Fig. 1 Reference [19] A periodic load history P(t) with cyclic limits P_{max} and P_{min}

4.1 Fully Relaxing Condition

Mars et al. [7] proposed a model to estimate the behavior of crack growth under fully relaxing condition. When the maximum tearing energy $T_{\max}$ is below the threshold value T_O, there will be no crack growth due to mechanical fatigue [21]. But when $T_o < T\text{max} < T_t$, here T_t is transition point to power-law behavior. The crack growth $\mathrm{d}C/\mathrm{d}N$ will linearly increase with T_{max}, which can be related to the below equation:

$$\frac{\mathrm{d}C}{\mathrm{d}N} = A(T\text{max} - T_O) \tag{6}$$

where A is proportionality constant.

When $T_t < T\text{max} < T_c$, where T_c is the critical tearing energy, in this crack growth rate, tearing energy is related by a power-law relation having slope F [7, 8].

$$\frac{\mathrm{d}C}{\mathrm{d}N} = r_c\left(\frac{T\text{max}}{T_C}\right)^F \tag{7}$$

When $T\text{max} > T_C$, the crack growth rate increases rapidly, and r_c will have maximum fatigue crack growth rate corresponding to T_c [19].

4.2 Non-relaxing Condition

In this loading condition, load on the material is not fully relaxed during repeated loading cycles, i.e., in this condition, R > 0. The fatigue behavior of elastomer is strongly depended upon this non-relieving load cycle [22]. The R ratio effect depends primarily upon the degree of strain-induced crystallization exhibited by the elastomer. The material exhibiting these phenomena of strain crystallization than the crack growth rate will decrease during non-relaxing condition [22].

For strain crystallization elastomer, Mars and Fatemi [19] give a phenomenological model to describe the effect of R ratio valid for the range $10^{-8} < \mathrm{d}a/\mathrm{d}N < 10^{-3}$mm/cycles and $10^3 < N < 10^6$cycles. They observed that when the crack growth rate is plotted against $T\text{max}$ at a given level of R ratio, the plot exhibits a power-law behavior. Also, they point that power-law curves tend to converge approximately at a single point, and this point is estimated as an intersection of asymptote associated with static load (R = 1) and the power-law curve associated with the fully relaxing condition (R = 0). Observation was embodied in the following models [7, 19]:

$$\frac{\mathrm{d}a}{\mathrm{d}N} = r_c\left(\frac{T\text{max}}{T_c}\right)^{F(R)} \tag{8}$$

The power-law exponent has been expressed in two forms. The first form is a third-order polynomial given below:

$$F(R) = F_0 + F_1 R + F_2 R^2 + F_3 R^3 \quad (9)$$

In this above-stated equation, $F_0,\ F_1,\ F_2,\ and\ F_3$ are the constants which characterize the elastomer dependence on R.

The second version of F(R) include only single constant F_4 along with F_0.

$$F(R) = F_0 e^{F_4 R} \quad (10)$$

It was noted that second version sacrifices accuracy relative to first particularly for the small nonzero value of R ratio. The second version is preferred when limited data are available when a large range of data is available; the first form may provide better accuracy [19].

Harbour et al. [23] also reported the effect of R ratio on strain crystallizing rubber in which it was reported that the rate of crack growth was decreased by a certain degree of a factor than thoroughly relaxing condition. But in the case of non-crystallizing rubber, there is no much significant change in crack growth observed.

5 Dwell Effect Phenomenon During Cyclic Loading

The crack growth rate in the elastomer is profoundly affected by the dwell time delved during the constant amplitude cyclic loading. Harbour et al. [24] done a detailed study related to the dwell effect in filled NR and SBR. In their work, they have demonstrated that dwell effect has a more profound impact in SBR than in NR, as shown in Fig. 2. They have reported that the crack growth rate in SBR was almost ten times higher (Ten seconds of dwell period with five cycles between dwell periods) compared to constant amplitude cyclic loading condition. In their study, they have done experiments to see the effect of dwell period parameters (i.e., dwell time, dwell level strain, and number of cycles between the dwell period and cyclic frequency). In that experiment, one parameter was varied and other parameters remain constant throughout the test. Conclusion of this leads that dwell time(t_d) and the number of cycles(N_d) between the dwell period have a more profound impact on crack growth rate.

Using the obtained experimental results and applying the power-law fit into data, they form a model [24].

$$r_{\text{dwell}} = r_{\text{normal}}\left(1 + c_d \times t_d^{C1} \times N_d^{C2}\right) \quad (11)$$

where r_{dwell} is the crack growth rate for signal including the dwell period, and r_{normal} is the crack growth rate at constant amplitude cyclic loading. This model predicts the crack growth rate for specified dwell time, but it slightly overpredicts for small

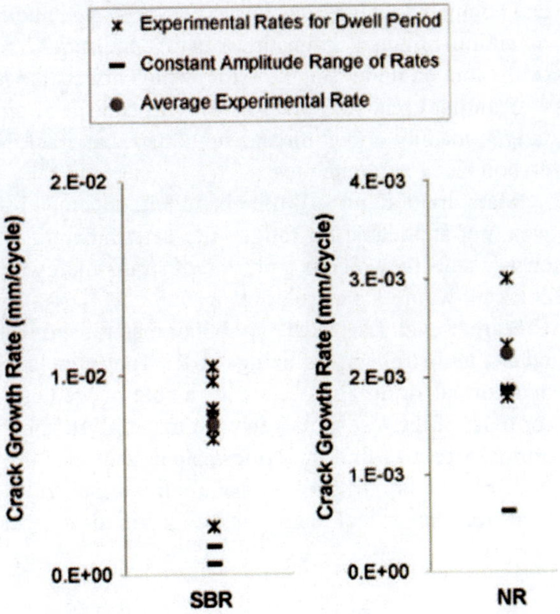

Fig. 2 Reference [24] Comparison of the experimental crack growth rates for a dwell period test signal for peak strains of 27.5% in filled SBR and 32.5% in filled NR

t_{dwell} and also the prediction of infinitely large crack at very long t_{dwell} is not very likely in a real experiment. This model is not limited to a particular elastomer as its constants are dependent on the material.

Dwell effect is assumed due to the time-dependent recovery in rubber microstructure at the crack tip, producing a localized and temporary elevated stress state [24]. This theory needs a deeper approach by studying the dwell effect at micro-scale near the crack tip region.

6 Multiaxial Fatigue Behavior

During a service application, a rubber component often encountered with time-varying load condition applied simultaneously from more than one direction (i.e., Multiaxial condition). Along with this in multiaxial fatigue, loading condition also plays an important role, i.e., in phase or out of phase.

Mars and Fatemi [25] developed a novel specimen for analyzing the fatigue behavior of elastomer under multiaxial loading condition. Using this specimen, they [5] found out that maximum principal strain is fatigue equivalence parameter while the strain energy density gave the worst correlation with the experimental result. In their another study [26], they suggested that crack nucleation in filled rubber starts from existing inherent flaws such as voids, surface cavities, and the interface between rubber and non-rubber components. They also reported that fatigue crack initiation and growth were observed to occur on the preferred failure plane. For simple and complex loading, they observed a preferred nucleation plane, but the relation to maximum principal strain direction was different. Crack closure was observed during cyclic torsion under static torsion, which affects the nucleation plane.

Saintlier et al. [27] invested the mechanism of crack initiation under multiaxial fatigue loading condition and suggested that crack initiate from inclusion or large carbon black agglomerates.

Mars and Fatemi [4] in their finding concluded that scalar equivalence criteria were not applicable for fatigue life determination. They suggested using cracking energy density, which is a portion of strain energy density that is available to cause crack growth in a particular plane.

Harbor et al. [28] used two rubber materials, one is strain crystallizing (NR) and other a non-strain crystallizing (SBR). They concluded that cracking energy density and normal strain approaches were able to predict the dominant crack orientation for most of the test signals in each material. In another study, they used maximum normal strain to find the critical plane and cracking energy density on that plane to determine fatigue life. As maximum normal strain is independent of constitutive behavior, the critical plane can be identified more easily than for cracking energy density.

7 Scalar Fatigue Life Equivalence Parameter

In elastomer maximum principal strain, octahedral shear strain, maximum shear strain, strain energy density, and maximum principal stress have all been used as parameters for correlating fatigue life of the component [7]. All of these stated parameters are scalar, having a single component value at a given material point.

Scalar fatigue life equivalence parameters are not specific to material plane [29]. Like in case of strain energy density as a fatigue life parameter states that the same amount of energy would be released for a given increment of crack growth, irrespective of crack orientation. No preferred cracking plane can therefore be predicted from strain energy density [1]. In case of maximum principal stress or strain, there is a normal plane associated with the loading condition but not with the material. This condition of normal plane can be considered in loading history when the maximum principal stress remains constant but direction associated with material plane changes. In some cases, like in tension test, the maximum principal stress direction coincides with the normal to the material cracking plane, but this is an exception case. Other limitation of scalar theories includes that in some multiaxial loading cycle, strain energy density can be maintained constant, while simultaneously cycling the individual strain components. So in this case of multiaxial loading cycle, strain energy density in tension, compression, torsion, and for other types loading are all equal [7]. And also the scalar parameter like strain energy density does not able to account for the effects of compression on fatigue. During compression, large amount of strain energy is stored without a damage effect due to crack closure [30]. Due to above all stated limitations, plane-specific approach is preferred over the scalar theories in the multiaxial loading condition.

8 Plane-Specific Fatigue Life Equivalence Parameter

In experimental studies carried out related to different loading conditions involved like [30] tension–compression, tension–torsion, torsion–compression, and other loading, histories show that crack precursor tends to occur on a specific plane that is related with applied loading history [26, 30, 31]. The crack growth rate or fatigue life depends on orientation and loading experience of such specific planes under the influence of mechanical load cycles [30–32].

The cracking energy density term proposed by Mars [33] represents the portion of strain energy density that may be said to be available for release by crack growth on a specific material plane. In general, only a part of strain energy density is available, which depends on loading states and orientation of crack.

The plane-specific approach entails the evaluation of the number of cycles required for crack precursor development on each possible failure plane and then identifying the plane for which the life is expected to be shortest. To find the life of critical plane, the loading history local to the plane must be estimated. This involves the calculation of cracking energy density W_c, by numerical integration of given below equation [30, 31]:

$$\mathrm{d}W_C = \overrightarrow{S}.d\vec{e} \tag{12}$$

where $\overrightarrow{S}$ is the traction vector associated with the specified material plane and $\mathrm{d}\vec{e}$ is the strain increment vector on the specified material plane.

This method is different from the prior fatigue life estimation method (strain energy density, maximum principal stress, maximum principal strain) as it tracks the damage on a material plane basis. Previous fatigue analysis methods fail to distinguish between material planes and spatial planes or even to predict the failure planes.

Also, to provide proper accounting of the effect of multiaxial loading condition, the plane-specific approach provides a way to consider the effect of crack closure due to compression as during compression, a significant portion of strain energy to remain unavailable for driving crack growth. So during the calculation of cracking energy density, only the tensile and shear part of $\overrightarrow{S}$ is used. Any part of strain work associated with the compressive component of $\overrightarrow{S}$ is excluded [31].

9 Variable Amplitude Loading Condition

The rubber components during service condition often experienced cyclic loads which are very much complicated than constant amplitude loading condition. Not only this, these variable amplitude test signals are often affected by several typical factors that frequently occur during service load condition. These factors include load level, load sequencing, R Ratio, and dwell period.

Harbour et al. [34], in his work model, proposed the effect of variable amplitude loading cycles. In this, the fatigue crack growth behavior at variable amplitude load cycles was evaluated by linear crack growth model equivalent to Miner's linear damage rule.

The linear crack growth model equates the crack growth rate r to be equal to the sum of the crack growth rates r_i of each individual component of the block based on the data from the constant amplitude experiment as per Eq. 13.

$$r = N_1 r_1 + N_2 r_2 + N_3 r_3 + N_4 r_4 + \cdots + N_i r_i \tag{13}$$

where N_i is the number of applied cycles for each component per sequence. Dividing the total crack growth rate r for the test sequence by the total number of cycles in sequence yields the average crack growth rate per cycles. That is, if the duty cycle is composed of n cycles, then the rate of crack growth per application of the entire duty cycle is equal to the sum of the rates of N individual cycles [34].

The rate of crack growth obtained from this calculation has units of crack growth per block application so that when this is used with fatigue life estimation equation (Eq. 14), the result of the integration is a repeat of the overall block required to grow the crack precursor from initial size (c_0) to the final failure size (c_f) [31].

$$N = \int_{c_0}^{c_f} \frac{\mathrm{d}C}{r(T)} \tag{14}$$

This crack growth rate prediction model was found to be an effective method of predicting variable amplitude loading cycles in their research. The drawback of this model was that this model was not able to predict the increased crack growth rate caused by the inclusion of a dwell period into the test signal [23].

10 Fatigue Crack Growth Measurements Procedures

For characterizing crack growth behavior during fatigue cycles, several different challenges faced by an individual like a selection of strain levels, cost budget, time constraint, and many other [35]. Goosens et al. [35] in his work compared two methods for fatigue crack growth measurements, i.e., Step method and ramp method. In step method [7, 35], test piece is imposed an increasing strain in a step-by-step way. In this, strain is held constant during each step, and amplitude is increased with each successive step as shown in Fig. 3.

Another advanced method for characterizing crack growth behavior is ramp method [35]. In this test procedure, the strain amplitude is increased continuously from a minimum strain to the maximum strain, so in this method crack driving force is enhanced with each successive cycle as shown in Fig. 4.

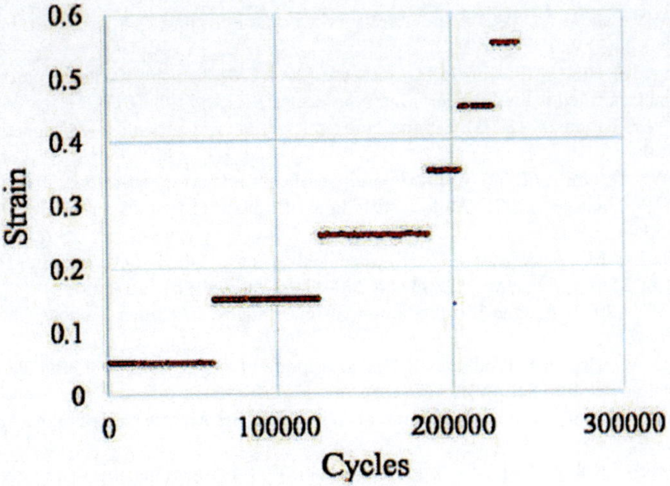

Fig. 3 Reference [35] strain history during step method

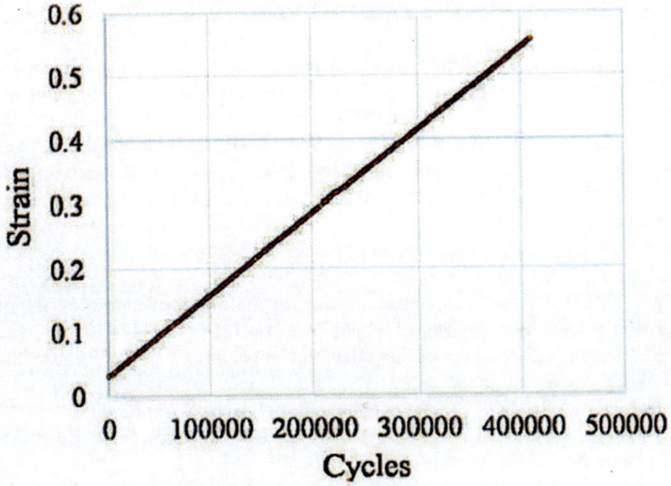

Fig. 4 Reference [35] Strain history in ramp method

Goosens et al. [35] concluded in their work that ramp method enables us to observe all crack growth rates during a test in a prespecified period, low operating cost, giving lower uncertainty compared to the step method.

References

1. Mars WV, Fatemi A (2002) A literature survey on fatigue analysis approaches for rubber. Int J Fatigue 24(9):949–961
2. Le Cam JB, Huneau B, Verron E, Gornet L (2004) Mechanism of fatigue crack growth in carbon black filled natural rubber. Macromolecules 37(13):5011–5017
3. Mars WV, Fatemi A (2003) Fatigue crack nucleation and growth in filled natural rubber, 779–789
4. Mars WV, Fatemi A (2005) Multiaxial fatigue of rubber: Part I: equivalence criteria, 515–522
5. Mars WV, Fatemi A (2005) Multiaxial fatigue of rubber: Part II: experimental observations, 523–538
6. Tee YL, Loo MS, Andriyana A (2018) Recent advances on fatigue of rubber after the literature survey by Mars and Fatemi in 2002 and 2004. Int J Fatigue 110:115–129
7. Mars WV (2007) Fatigue life prediction for elastomeric structures. Rubber Chem Technol 80:419–424
8. Thomas AG, British T, Producers R (1955) Rupture of rubber. II. The strain concentration at an incision. J Polym Sci XVIII:177–188
9. Rivlin RS, Thomas AG (1953) Rupture of rubber. I. Characteristic energy for tearing. J Polym Sci 10:291–318
10. Thomas AG, Rubber B (1958) Rupture of rubber. V. Cut growth in natural rubber vulcanizates XXXI:467–480
11. Thomas AG The development of fracture mechanics for elastomer. Rubber Chem Technol 67:G50–G60
12. Gent AN (2001) Engineering with rubber, 2nd edn. Hanser Publishers, Munich
13. Greensmith HW, Mullins L, Thomas AG (1960) Rupture of rubber. Trans Soc Rheol 4(1):179–189
14. Lindley PB (1972) Energy for crack growth in model rubber components. J Strain Anal 7(2):132
15. Lake GJ, Yeoh OH (1987) Effect of crack tip sharpness on the strength of vulcanized rubbers. J Polym Sci Part B Polym Phys 25(6):1157–1190
16. Robertson CG, Stoček R, Kipscholl C, Mars WV (2019)Characterizing the intrinsic strength (Fatigue Threshold) of natural rubber/butadiene rubber blends. Tire Sci Technol 47(4):292–307
17. Andrews EH (1963) Rupture propagation in hysteresial materials: stress at a notch. J Mech Phys Solids 11(4):231–242
18. Lake GJ, Yeoh OH (1978) Measurement of rubber cutting resistance in the absence of friction. Int J Fract 14(5):509–526
19. Mars WV (2003) A phenomenological model for the effect of r ratio on fatigue of strain crystallizing rubbers. Rubber Chem Technol 76(1241–1258):419–424
20. Mars WV, Fatemi A (2002) Factors that affect the fatigue life of rubber. Rubber Chem Technol 24:391–412
21. Lake GJ, Lindley PB (1966) Fatigue of rubber at low strans. J Appl Polym Sci 10(2):343–351
22. Lindley PB (1973) Relation between hysteresis and the dynamic crack growth resistance of natural rubber. Int J Fract 9(4):449–462
23. Harbour RJ, Fatemi A, Mars WV (2007) Fatigue crack growth of filled rubber under constant and variable amplitude loading conditions. Fatigue Fract Eng Mater Struct 30(7):640–652
24. Harbour RJ, Fatemi A, Mars WV (2007) The effect of a dwell period on fatigue crack growth rates in filled SBR and NR. Rubber Chem Technol 80:838–853

25. Mars WV, Fatemi A (2004) A novel specimen for investigating the mechanical behavior of elastomers under multiaxial loading conditions. Exp Mech 44(2):136–146
26. Mars WV, Fatemi A (2006) Nucleation and growth of small fatigue cracks in filled natural rubber under multiaxial loading, 7324–7332
27. Saintier N, Cailletaud G, Piques R (2006) Crack initiation and propagation under multiaxial fatigue in a natural rubber 28:61–72
28. Harbour RJ, Fatemi A, Mars WV (2008) Fatigue crack orientation in NR and SBR under variable amplitude and multiaxial loading conditions, 1783–1794
29. Mars WV, Fatemi A (2007) The correlation of fatigue crack growth rates in rubber subjected to multiaxial loading using continuum mechanical parameters. Rubber Chem Technol 80(1):419–424
30. Mars WV, Wei Y, Hao W, Bauman MA (2019) Computing tire component durability via critical plane analysis. Tire Sci Technol 47(1):31–54
31. Barbash KP, Mars WV (2016) Critical plane analysis of rubber bushing durability under road loads. In: SAE technical papers
32. Mars WV (2001) Multiaxial fatigue crack initiation in rubber. Tire Sci Technol 29(3):171–185
33. Mars WV (2002) Cracking energy density as a predictor of fatigue life under multiaxial conditions. Rubber Chem Technol 75:1–18
34. Harbour RJ, Fatemi A, Mars WV (2008) Fatigue life analysis and predictions for NR and SBR under variable amplitude and multiaxial loading conditions. Int J Fatigue 30:1231–1247
35. Goossens JR, Mars WV (2018) Finitely scoped, high reliability fatigue crack growth measurements. Rubber Chem Technol 91(4):644–650

Development of Fatigue Life Equation for NBR Material and Estimating Fatigue Life Using Fea

Rohit Rajendra Shitole, Onkar Pradip Haval, and Litesh Govind Pandharpurkar

Abstract Rubber materials are used in many pneumatic applications. Rubber has excellent damping characteristics and it experiences large reversible elastic deformation. For the safety and reliability of rubber components, it is important to predict the behaviour of rubber. In this research, fatigue life of rubber components is predicted by finite element analysis and developed fatigue life equation. Mechanical tests such as uniaxial tension, uniaxial compression and pure shear test were performed on NBR material for the material characterisation in ANSYS. A hyperelastic model was selected by performing the curve fitting in ANSYS. Fatigue tests on a standard dumbbell specimen of NBR material were performed at different strains, and fatigue life equation was derived with least square fit method and fatigue test data. Then the maximum principal elastic strain obtained through FEA was used as the fatigue damage parameter as mentioned in the literature. Maximum principal elastic strain was substituted in the obtained fatigue life equation and fatigue life of rubber spring was predicted. The results were validated by performing the fatigue test on rubber spring.

Keywords Fatigue life equation · Finite element analysis (FEA) · Maximum principal elastic strain · Nitrile butadiene rubber (NBR) · ε-N curve

1 Introduction

Rubbers are used in many industrial as well as automobile applications because of their excellent sealing characteristics [1]. They are used in various applications such as tyres, O-rings, gaskets, boots and seals. Many types of rubber such as natural

R. R. Shitole (✉) · O. P. Haval · L. G. Pandharpurkar
Knorr-Bremse Technology Center India Pvt. Ltd, Pune, India
e-mail: rohit.shitole2@knorr-bremse.com

O. P. Haval
e-mail: Onkar.Haval@knorr-bremse.com

L. G. Pandharpurkar
e-mail: Litesh.Pandharpurkar@knorr-bremse.com

S. Seetharamu et al. (eds.), *Fatigue, Durability, and Fracture Mechanics*, Lecture Notes in Mechanical Engineering, https://doi.org/10.1007/978-981-15-4779-9_30

rubber (NR), Nitrile Butadiene rubber (NBR), Neoprene (CR) and Styrene-Butadiene rubber (SBR) are used in industrial applications [1, 2]. One important characteristic of rubber is that it does not experience permanent deformation even at large elongations. Rubbers also provide better damping for vibrations and good resistance to abrasion [3]. Rubber experiences various failures due to incorrect design and complex behaviour. Rubber being a critical product in automobile industry, it is important to design, analyse and evaluate the strength to ensure the safety and quality of the rubber products [1, 4].

Fatigue life of rubber components can be estimated by various methods such as principal strain method, Green-Lagrange strain method, stress method and fracture mechanics method. Kim [1] determined the fatigue life equation for the natural rubber by performing the fatigue test on dumbbell specimens. He estimated the fatigue life of engine rubber mount by using the maximum Green-Lagrange strain and the fatigue life equation of the rubber. Suryatal [4] has explained the evaluation method for fatigue by using the principal strain as the damage parameter. In this paper, the hyperelastic model was selected by using curve fitting on uniaxial tension test data of chloroprene rubber. The maximum principal elastic strain value was obtained through FEA was substituted in the obtained equation to predict the life of rubber mounts. Li [5] explains a similar method in his research paper. The equation of fatigue life can be obtained through the fatigue life tests and this equation can be used to predict the life of rubber components through FEA obtained principal strain. In my previous work [6], various hyperelastic models and the methodology to carry out the finite element analysis of rubber components have been explained.

In this paper, equation of fatigue life of Nitrile Butadiene Rubber (NBR) was obtained by performing the fatigue tests. Finite element analysis of rubber spring was carried out. Through the obtained fatigue life equation and FEA, fatigue life of rubber spring was derived.

The rubber spring considered in this study is used in the commercial vehicles. Figure 1 shows the rubber spring, which was manufactured by vulcanizing NBR with Shore A Hardness 70.

Fig. 1 Rubber spring

2 Material Definition

2.1 Selection of Hyperelastic Model

The rubber material used for the rubber spring is made from nitrile butadiene rubber of Shore A hardness 70. As rubber behaviour is complex, there are various challenges faced while designing rubber components. Rubber shows viscoelastic behaviour, but rubber is modelled as a hyperelastic material, with the help of hyperelastic models.

There are various hyperelastic models available in FE software. To define the appropriate hyperelastic model, generally the following test data is needed:

1. Uniaxial Tension Test Data
2. Biaxial Tension Test Data
3. Pure Shear Test Data
4. Volumetric Test Data

In this paper, NBR was considered as an incompressible material. Hence, a volumetric test was not performed in this research.

Uniaxial Tension Test. Uniaxial tension test was performed on the standard dumbbell-shaped specimen until the complete failure of the specimen as shown in Fig. 3a. This test was performed according to ASTM D412 standard. Type C specimen was used for this test as shown in Fig. 2 [2]. The test was carried out at room temperature of 23 °C with test speed of 500 mm/min and stress–strain data was obtained. The following results were obtained as shown in Fig. 3.

Uniaxial Compression Test. Uniaxial compression test was performed on the standard button-shaped specimen according to ASTM D595 standard as shown in Fig. 4a. Stress–strain data is obtained from this test. Results obtained from the test are shown in Fig. 4b.

Due to the unavailability of biaxial test setup, the test was not performed. In biaxial tension, the material is elongated in two directions while it is compressed in

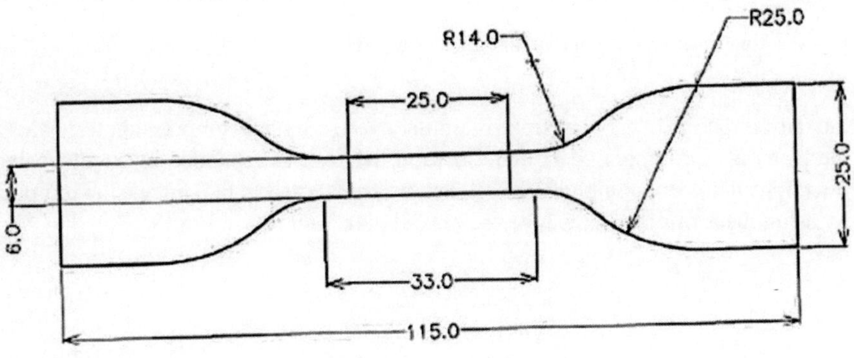

Fig. 2 ASTM D412 Type C specimen [2]

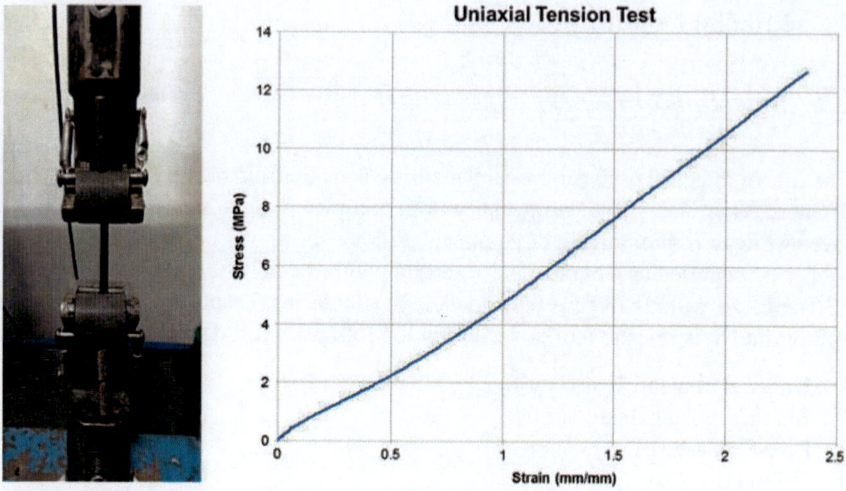

Fig. 3 **a** Uniaxial tension test setup. **b** Uniaxial tension test result

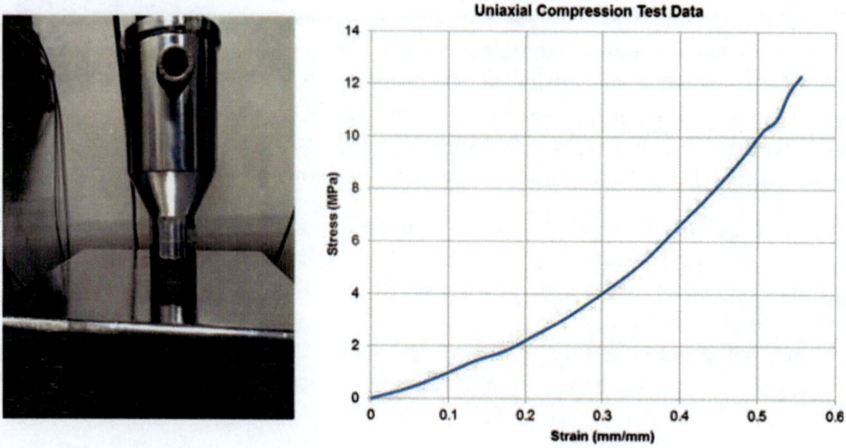

Fig. 4 **a** Uniaxial compression test. **b** Uniaxial compression test result

the third direction. This behaviour is similar to the uniaxial compression test, where the material is compressed in one direction, while it is expanded in the other two directions. Uniaxial compression test data was converted to biaxial tension test data by using the formulae [7]. Converted biaxial data is shown in Fig. 5.

$$\varepsilon b = \sqrt{\frac{1}{-\varepsilon c + 1}} - 1$$

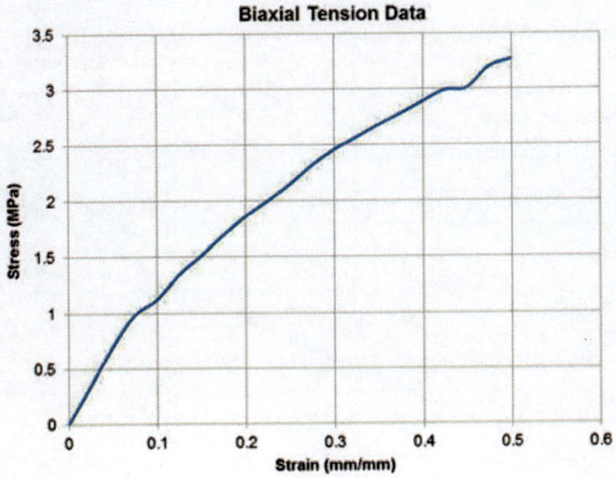

Fig. 5 Biaxial tension data

$$\sigma b = \frac{\sigma c}{(\varepsilon c + 1)^3}$$

Pure Shear Test. Pure shear test is carried out on planar specimen as shown in Fig. 6a. The width of the specimen was 10 times large than the length of the specimen [2, 8]. The test was performed till 50% elongation. The test was performed at 5 mm/min test speed and results were obtained as shown in Fig. 6b.

Curve Fitting. By using the test data and performing the hyperelastic curve fitting in ANSYS, Ogden 2nd order hyperelastic model was selected as shown in Fig. 7. In rubber spring application, maximum strain achieved is less than 100%. Hence, during the curve fitting, uniaxial test data up to 100% strain was used.

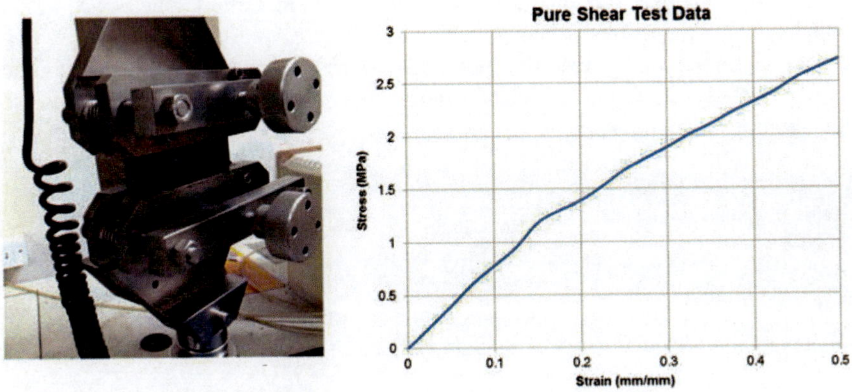

Fig. 6 **a** Pure shear test. **b** Pure shear test result

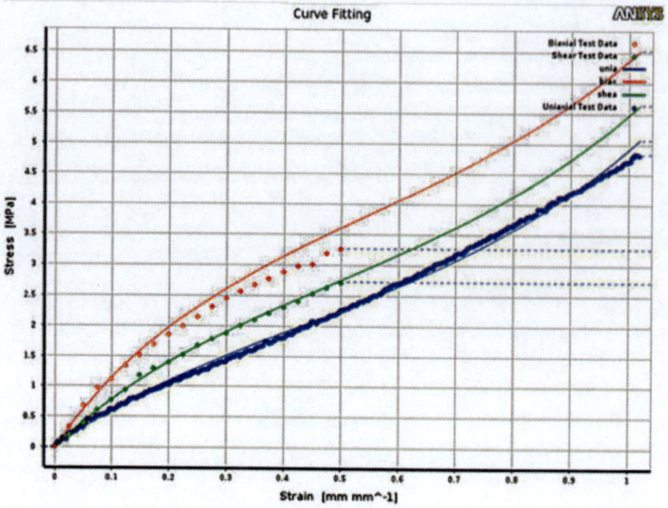

Fig. 7 Curve fitting using Ogden 2nd order model

2.2 *Fatigue Properties of NBR Material*

To calculate fatigue properties, two methods are available. One is the SN curve method and the other is the ε-N curve method. Due to the behaviour of rubber, ε-N curve method was selected for this research.

To calculate fatigue properties, fatigue tests were performed according to ISO 6943 till the complete failure of the specimen. This test was performed on dumbbell specimen which is similar to the uniaxial tension test specimen. These tests were performed at room temperature of 23 °C by using tensile fatigue testing machine. The fatigue tests were performed at five different strains (100, 80, 55, 45 and 30%). The results obtained from the fatigue tests are shown in Table 1. The ε-N curve was obtained from the results as shown in Fig. 8.

When the fatigue test results are plotted in double logarithmic co-ordinates, they can be treated as a straight line. Hence, fatigue life equation of NBR can be expressed in Basquin's equation form as

Table 1 Fatigue test results of NBR

Sr. No.	Strain value (%)	Fatigue life in number of cycles (N)
1	100	4880
2	80	6380
3	55	29978
4	45	93125
5	30	570000

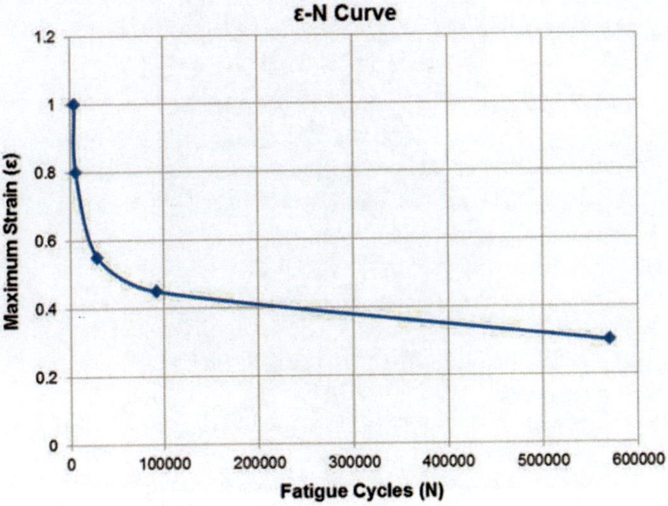

Fig. 8 ε-N curve

$$\varepsilon^{m}\,N = C \quad (1)$$

where m and C are the material constants.

The above Eq. (1) can be transformed as

$$m\log\varepsilon + \log N = \log C \quad (2)$$

By solving Eq. (2) by least square method, values of m and C were determined as 4.1127 and 3391.

So the fatigue life equation of the NBR material can be expressed as

$$\varepsilon^{4.1127}N = 3391.381 \quad (3)$$

Equation (3) can be used to predict the life of rubber components.

3 FEA of Rubber Spring

The rubber spring was analysed in ANSYS Workbench 17.2 software. Geometry is symmetric, so only one-fourth model was modelled by applying the cyclic symmetry as shown in Fig. 11. Tetrahedral elements were used for the meshing of the model as shown in Fig. 9. Ogden 2nd order hyperelastic model was selected for the non-linear analysis. The upper plate above the rubber spring was given the displacement of 6.4 mm in the downward direction, which causes the deformation of rubber spring.

Fig. 9 Mesh

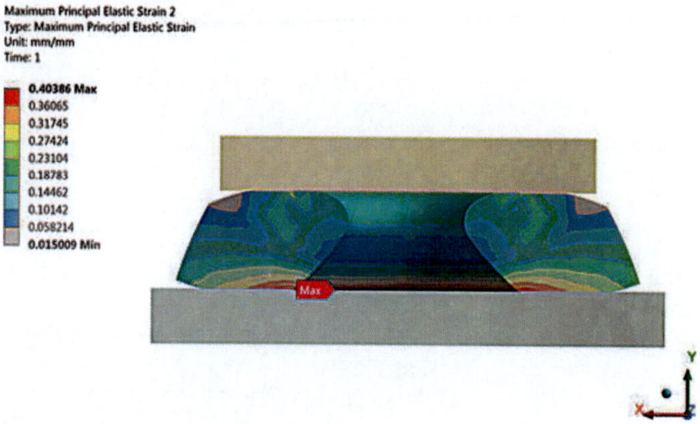

Fig. 10 FEA of rubber spring

The deformed shape and strain of the rubber spring is shown in Fig. 10. The regions showing the maximum strain are the critical regions.

4 Fatigue Life Prediction

Maximum principal strain shows the possible failure location of the rubber spring. The crack initiation will start at those locations predicted by FEA and the complete failure of the specimen occurs when the crack is propagated throughout the specimen.

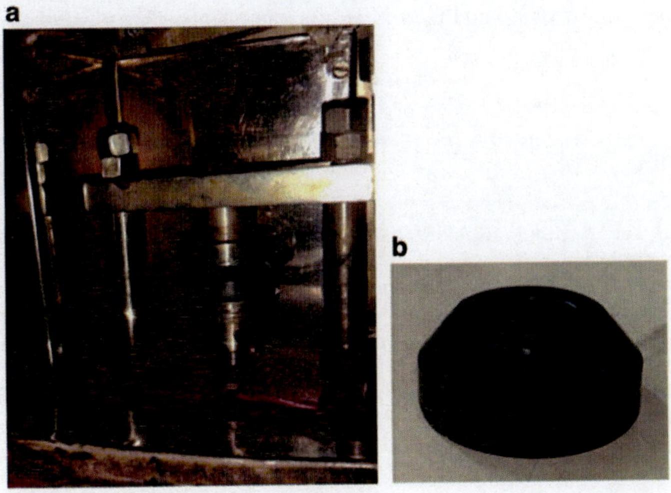

Fig. 11 **a** Fatigue test rig. **b** Failure of rubber spring

Hence, maximum principal elastic strain value is used to predict the fatigue life of NBR. The fatigue life of rubber spring was determined by substituting the FEA obtained maximum principal strain value from Fig. 10 into the obtained fatigue life Eq. (3) of NBR.

The fatigue life of rubber spring was predicted as 141204 cycles.

5 Experimental Validation

The rubber spring was subjected to the fatigue load on fatigue test rig as shown in Fig. 11a to validate the accuracy of the prediction method. The specimen was subjected to pulsating load till the failure of the specimen. The numbers of cycles till the failure of the specimen were measured, which were 117953 cycles. The failure of the specimen was observed in a similar way as predicted through FEA as shown in Fig. 11b.

6 Conclusions

The fatigue life of rubber spring was predicted by FEA and the fatigue life equation. The fatigue life predicted by the method is 141204 cycles while the fatigue life of rubber spring obtained from the test is 117953 cycles. Hence, result was validated experimentally with difference of 19.71% between experimentation and predicted results. The obtained fatigue life equation of NBR material from this research can be used to predict the fatigue life of NBR70 components. This satisfies the research of this work.

References

1. Kim WD, Lee HJ, Kim JY, Koh SK (2004) Fatigue life estimation of an engine rubber mount. Int J Fatigue 26:553–560
2. Bauman JT (2008) Fatigue stress and strain of rubber components—guide for design engineers. Hanser Publications
3. Ali A, Hosseini M, Sahari BB (2010) A review of constitutive models for rubber-like materials. Am J Eng Appl Sci 3, 232–239
4. Suryatal B (2015) Fatigue life estimation of an elastomeric pad by ε-N curve and FEA. J Surf Eng Mater Adv Technol 5:85–92
5. Li Q, Zhao J-C, Zhao B (2009) Fatigue life prediction of a rubber mount based on test of material properties and finite element analysis. Eng Fail Anal 16:2304–2310
6. Shitole RR, Chavan US, Dolse S, A review on methodology of material characterization and finite element modelling of rubber-like materials. IOSR J Eng (IOSRJEN), ISSN (e): 2250–3021, ISSN (p): 2278–8719
7. Lobdell M, Croop B, 5–7 Oct (2016) A mechanism for the validation of Hyperelastic materials in ANSYS. In: CADFEM ANSYS simulation conference
8. Moreira D, Nunes L (2013) Comparison of simple and pure shear for an incompressible isotropic hyperelastic material under large deformation. Polym Testing 32:240–248

Fatigue Properties of Adhesively Bonded Teakwood Filled Circular Hollow Steel Tube in Rotary Bending Application

Bharatesh Adappa Danawade and Ravindra R. Malagi

Abstract The objective of this paper is to present experimental work on fatigue life and fatigue strength of adhesively bonded teakwood filled circular hollow tube. Adhesively bonded teakwood filled circular hollow sectioned steel tube was subjected to 4-point rotating bending fatigue to determine the fatigue strength and fatigue life. S-N curve was drawn and relations have been built to predict the number of cycles for the applied stress or vice versa. The relations built in this paper can safely be applied for design of the fatigue life or fatigue strength of adhesively bonded teakwood filled circular hollow steel tubes. Experimental results were validated by static strength in bending by Euro code (EC4).

Keywords Fatigue life · Fatigue strength · Rotary bending · S-N curve · Static strength in bending

1 Introduction

Composite materials posses longer fatigue life compared to steel and give warning that it is losing its stiffness. Fatigue has a strong correlation with stiffness of composites. These characteristics of composites can be used favorably for structural applications. Fatigue is among the most common cause of failures in rotary shafts which are subject to various loading conditions. Fatigue failures start at the most vulnerable point, it may be mechanical or metallurgical or combination of the two [1]. Bhaumik et al. [2] presented a case study on fatigue failure of hollow power transmission shaft and presented prevention measures to avoid such failures. Miscow et al. [3] designed and built a fatigue simulator to test full-scale drill pipes under rotating cycling bending and constant tension loading. West system [4] on fatigue explains

B. A. Danawade (✉)
Department of Mechanical Engineering, Gharda Institute of Technology, University of Mumbai, Lavel 415708, India
e-mail: bharatesh23@gmail.com

R. R. Malagi
Department of PDM, Visvesvaraya Technological University, Belagavi 590014, India
e-mail: rrmalagi@rediffmail.com

S. Seetharamu et al. (eds.), *Fatigue, Durability, and Fracture Mechanics*, Lecture Notes in Mechanical Engineering, https://doi.org/10.1007/978-981-15-4779-9_31

that wood exhibits unusually good fatigue resistance. Chen et al. [5] conducted low cycle fatigue tests to predict the fatigue life of 304 stainless steel tubular geometry. The fatigue loading applied to the specimen was tension-compression followed by torsion. Dawood et al. [6] investigated the steel-concrete bridge beams strengthened with high modulus carbon fiber-reinforced polymer under the overloading and fatigue conditions. Kim and Heffernan [7] presented progress and achievement in the application of FRP on strengthening reinforced concrete beams subjected to fatigue. Zhenyu et al. [8] conducted field applications and laboratory research which shows the feasibility of concrete-filled fiber-reinforced polymer tubes in bridges. An analytical tool is developed to trace the response of CFFTs under fatigue loading. A detailed parametric study shows that fatigue response of CFFT beams can improve by either increasing the reinforcement index or the effective modulus of FRP tube in the longitudinal direction. Malagi et al. [9] observed that wood filling of circular hollow tube improves the fatigue behavior as compared to hollow circular sections.

Wood has the ability to withstand high cyclic loads year after year. This ability of wood is developed by nature for dynamic and adverse environmental conditions. Hence, specific strength and stiffness of woods are greater than metals. When wood is used in building composite system for fatigue loads, naturally life and strength of composite give better results than individual metals. Very little or no information is available on the fatigue behavior and properties of adhesively bonded wood filled hollow sections. A 4-point rotating bending fatigue testing machine was used to determine the fatigue strength and fatigue life of adhesively bonded wood filled circular hollow sections. The specimens were subjected to completely reversed stress cycles with constant amplitude. A statistical approach for predicting the fatigue bending strength and fatigue life is discussed in this paper. An attempt has been made to use Euro code EC4 for in-filled tubes to determine the properties of wood filled tubes using static bending tests for bending conditions.

2 Materials

The materials selected for compositing were teakwood, steel tube, and Fevitite superfast adhesive—commercially available teakwood, circular hollow sectioned carbon steel tube, and Fevitite superfast adhesive procured from local market. The air-dried and seasoned teakwood with moisture content less than 12% was brought and stored. The general structural purpose rectangular hollow sectioned carbon steel tube categorized under IS2062/2006 (Grades A, B, and C), ASTM-A36M M, DIN 17100 was considered. To manufacture adhesively bonded wood filled steel tubes a Fevitite superfast structural grade adhesive was procured and used based on suitability for application.

The average longitudinal modulus of elasticity of teakwood material was 12328.90 MPa (with standard deviation of 541.48), The average value of modulus of rupture (MOR) was 80.65 MPa (with standard deviation of 7.80). The average maximum crushing strength was 48.04 MPa (with standard deviation of 0.28), the

Table 1 Properties of adhesive Fevitite super fast

Test property	Value
Overlap shear strength (MS-MS), Bond cured for 24 h/RT	>160 kg/cm^2
Cleavage strength (MS-MS), Bond cured for 24 h/RT	>60 kg/cm^2
Hardener density	0.9 g/cm^3
Resin density	1.15 g/cm^3

Poissons ratio of 0.52. These properties of teakwood material were determined by preparing specimens and testing them as per the American Society for Testing and Materials standards ASTM 143 [10].

Average yield stress, ultimate stress, and Young's modulus that were determined by experimentation are 367.74 MPa (with Std. Dev. of 14.19), 477.76 MPa (with Std. Dev. of 34.85), and 209.02 GPa (with Std. Dev. of 9.55). Poisson's ratio was taken as 0.29. These steel properties were determined by tensile testing of sheets cut from the tube faces as per the American Society for Testing and Materials standards ASTM E8 [11].

Fevitite superfast adhesive is a faster setting two-component multipurpose epoxy adhesive system. Resin and hardener are mixed thoroughly in the ratio 1:1 by volume. Faster setting makes it ideal for those jobs which have to be finished in 3 h. It can set within 2–3 h at 30 °C. This system does not contain volatile matter; hence there is negligible shrinkage on curing. Fevitite superfast which is internally flexible provides strong joints with excellent low creep properties hence makes it ideal for structural bonding applications [12]. The properties of Fevitite superfast adhesive are taken from manufacturer manual and are shown in Table 1.

3 Preparation of Adhesively Bonded Teakwood Filled Steel Tube

Thin coat of approximately 0.05–0.10 mm of Fevitite superfast adhesive was applied to teakwood surface using a spatula. Teakwood was fitted on lathe as shown in Fig. 1. The assembly is allowed to cure for 12 h, and the specimen ready for test is shown in Fig. 2.

Figure 3 shows the adhesively bonded wood filled steel tube specimen diagram, where D is external diameter of steel tube for loading, d_o is the outside diameter, d_i is internal diameter of steel tube, L is the length of steel tube, Lp is parallel length, and R is radius.

The amount of steel, teakwood, and adhesive used in manufacturing the adhesively bonded wood-filled steel tubes (AWFT) specimens is shown in Table 2. Where m_{ST}, m_{TW}, m_{AD}, m_{AWFT}, are mass of the steel, teakwood, adhesive, and teakwood filled steel tube, respectively.

Fig. 1 Press fitting of teakwood into steel tube on lathe

Fig. 2 Adhesively bonded wood filled steel tube

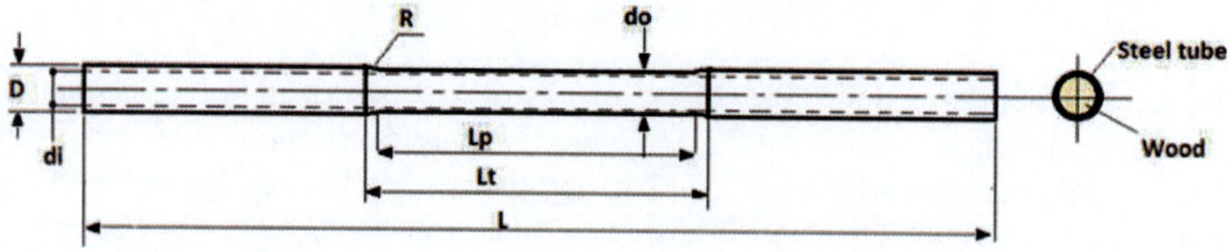

Fig. 3 Adhesively bonded wood filled steel tube specimen

Table 2 Proportion of teakwood and steel in AWFT

Specimen code	m_{ST} (kg)	m_{TW} (kg)	m_{AD} (kg)	m_{AWFT} (kg)	%ST	%TW	%AD	%AWFT
AWFT_200_1	0.079	0.007	0.002	0.088	89.77	7.95	2.27	100
AWFT_200_2	0.076	0.009	0.002	0.087	87.36	10.34	2.3	100
AWFT_200_3	0.08	0.006	0.003	0.089	89.89	6.74	3.37	100
AWFT_175_1	0.077	0.007	0.003	0.087	88.51	8.05	3.45	100
AWFT_175_2	0.077	0.006	0.002	0.085	90.59	7.06	2.35	100
AWFT_175_3	0.078	0.008	0.003	0.089	87.64	8.99	3.37	100
AWFT_150_1	0.078	0.007	0.002	0.087	89.66	8.05	2.3	100
AWFT_150_2	0.08	0.007	0.002	0.089	89.59	7.84	2.58	100
AWFT_150_3	0.078	0.006	0.002	0.086	90.7	6.98	2.33	100
AVG	**0.078**	**0.007**	**0.002**	**0.087**	**89.299**	**8**	**2.7**	**100**
SD	**0.0014**	**0.001**	**0.0005**	**0.0015**	**1.2017**	**1.12**	**0.53**	**0**

Table 3 Physical and geometrical properties of AWFT

Specimen code	D (mm)	d_o (mm)	d_i (mm)	L (mm)	Lt (mm)	Lp (mm)	ρ (kg/m^3)	Ii (mm^4)
AWFT_200_1	12.03	10.94	8.75	226	96	72	3738.363	703.14
AWFT_200_2	11.99	10.92	8.64	226	96	72	3712.096	698.01
AWFT_200_3	12.01	10.93	8.69	226	96	72	3767.039	700.57
AWFT_175_1	12.1	10.96	8.55	226	96	72	3508.399	708.29
AWFT_175_2	12.02	10.93	8.6	226	96	72	3431.214	700.57
AWFT_175_3	12.06	10.94	8.57	226	96	72	3480.066	703.14
AWFT_150_1	12.06	10.99	9.06	226	96	72	3454.135	716.08
AWFT_150_2	12	10.95	8.86	226	96	72	3621.943	705.71
AWFT_150_3	12.03	10.97	8.96	226	96	72	3429.997	710.88

Measured values of Physical and Geometrical properties of adhesively bonded teakwood filled steel tube (AWFT) specimens are shown in Table 3. Where ρ is density and Ii is moment of inertia of specimen.

4 Experimental Evaluation of Fatigue Strength Adhesively Bonded Teakwood Filled Steel Tube

In order to find out fatigue strength and fatigue life of adhesively bonded teakwood filled steel tubes (AWFT), 4-point rotating bending fatigue testing machine used is shown in Fig. 4. The specimens were subjected to completely reversed stress cycles with constant amplitude. During each cycle, the rotating specimen was

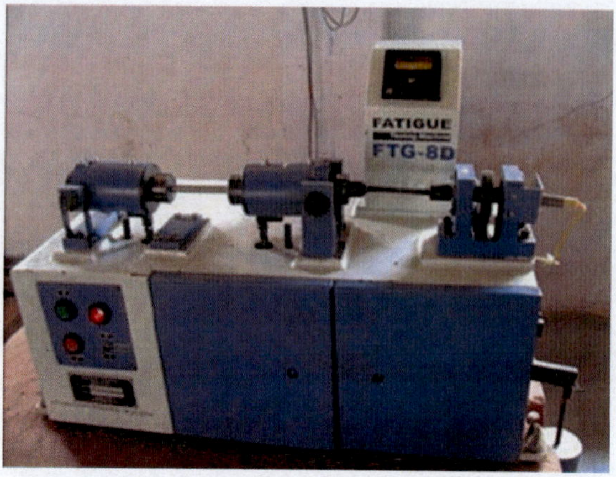

Fig. 4 Fatigue testing machine FTG-8D [13]

subjected to both tensile and compressive stresses alternatively. The specimen loading arrangement is shown in Fig. 5.

A single test consists of applying a known, constant bending load on a rotating specimen. During one cycle of rotation, the specimen was subjected to the two types of bending stresses (tensile and compressive) having the same magnitude but opposite signs and due to these alternative stresses the specimen fails after certain number of cycles. The number of cycles sustained by the specimen up to the failure is recorded by the counter. The above procedure is repeated for all the specimens by applying different bending moments at constant spindle rpm. The number of cycles sustained by each specimen up to the failure is recorded.

The specimens were first subjected to maximum bending moment. The bending moment was reduced in each step and the number of cycles to failure was noted. The results obtained were plotted to generate S-N. The bending strength was calculated using the following Eqs. 1 and 2.

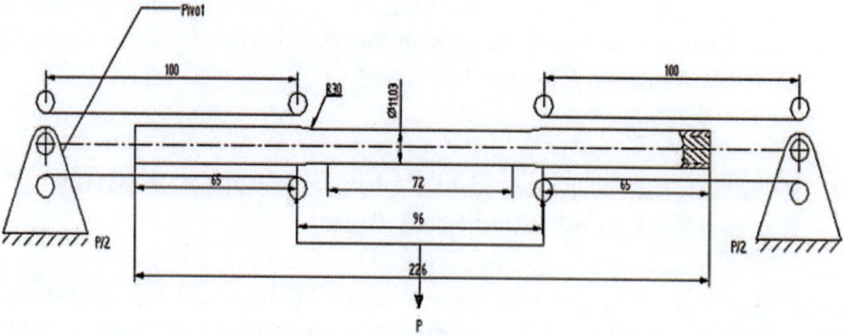

Fig. 5 Specimen loading arrangement [13]

$$Ii = \pi/64\left(do^4 - di^4\right) \tag{1}$$

$$\frac{M}{Ii} = \frac{\sigma b}{y} = \frac{E}{R} \tag{2}$$

The specimens were first subjected to maximum bending moment of 19,620 N mm. The bending moment was reduced by in each step and the number of cycles to failure was noted. The results of fatigue test results of adhesively bonded teakwood filled steel tube (AWFT) are summarized in Table 4. The nature of failure of specimen is shown in Figs. 6 and 7. The fatigue bending strength versus number of cycles before failure is plotted on the logarithmic graph and is shown in Fig. 8. This graph is also called as S-N curve. To determine the fatigue life or fatigue strength following Eq. 3 is established with coefficient of correlation of 0.998

$$\sigma b = -8.28\ln(Nf) + 252.1 \tag{3}$$

5 Comparison of Experimental Results with Euro Code

S-N curves obtained under torsion or bending load-control test conditions often do not have data at the shorter fatigue lives (say 10^3 or 10^4 cycles and less) due to significant plastic deformation. The life of specimen is less when the service load is maximum and life of specimen is large toward minimum service loads; sometimes infinite or too long life in specific materials. When the material is subjected to maximum load in bending, tensile, compression, or torsion, the material fails purely at static load of failure, i.e., ultimate strength of material as shown in Fig. 9 [14]. It means that fatigue strength approaches the static strength when components are subjected to higher cyclic loads. And when the service load conditions are far lower than the load the curve becomes parallel to *x*-axis resulting in large fatigue life, and load approaches the endurance limit or fatigue limit.

Euro code (EC4) [15] code determines moment resistance of the concrete-filled tube (CFT) beams based on plastic stress distribution and full strain compatibility of the cross section of both steel and concrete. Strength in tension was omitted in the calculation and this code becomes applicable to the design of CFT beams with concrete cylinder strength and yield strength of not higher than 50 and 355 MPa. The ultimate moment capacity as per EC4 is calculated using the following Eqs. 4–6.

$$Mpl.Rd = Mmax.Rd - Mn.Rd \tag{4}$$

$$Mmax.Rd = W_{pa} * f_{yd} + W_{ps} * f_{sd} + W_{pc} * f_{cd}/2 \tag{5}$$

Table 4 Results of fatigue tests of AWFT

Specimen	D (mm)	d_o (mm)	Lt (mm)	Lp (mm)	ρ (kg/m^3)	Ii (mm^4)	M (N mm)	Nf	σb (MPa)
AWFT_200_1	12.03	10.94	96	72	3653.4	703.14	19620	169,177	152.58
AWFT_200_2	11.99	10.92	96	72	3626.76	698.01	19620	167,559	153.42
AWFT_200_3	12.01	10.93	96	72	3640.06	700.57	19620	155,941	153
AWFT_175_1	12.1	10.96	96	72	3387.42	708.29	17167.5	1,636,066	132.78
AWFT_175_2	12.02	10.93	96	72	3350.48	700.57	17167.5	1,641,444	133.87
AWFT_175_3	12.06	10.94	96	72	3362.76	703.14	17167.5	1,689,955	133.51
AWFT_150_1	12.06	10.99	96	72	3374.73	716.08	14715	19,020,237	112.88
AWFT_150_2	12	10.95	96	72	3325.86	705.71	14715	18,099,502	114.12
AWFT_150_3	12.03	10.97	96	72	3350.23	710.88	14715	18,936,972	113.5

Fig. 6 Nature of failures in adhesively bonded wood filled tubes (AWFT)

Fig. 7 Brittle failure of adhesively bonded wood-filled tube

$$Mn.Rd = W_{pan} * f_{yd} + W_{psn} * f_{sd} + W_{pcn} * f_{cd}/2 \tag{6}$$

where, $W_{pcn} = (d - 2t) * hn^2$, $W_{pan} = b * hn^2 - W_{pcn} - W_{psn}$, $hn = \frac{Npm.Rd - A_{sn}(2f_{sd} - f_{cd})}{(2b * f_{cd} + 4t(2f_{yd} - f_{cd}))}$, $Npm.Rd = A_c * f_{cd}$, $f_{cd} = 1.0 * f_{ck}$

where W_{pan} is the plastic section modulus of the steel tube, W_{psn} is the plastic section modulus of steel reinforcement, W_{pcn} is the plastic section modulus of the concrete core, f_{yd} is the yield strength of the steel, f_{sd} is the yield strength of the reinforcement, f_{cd} is the yield strength of the concrete, f_{ck} is the concrete cylinder strength, A_{sn} is

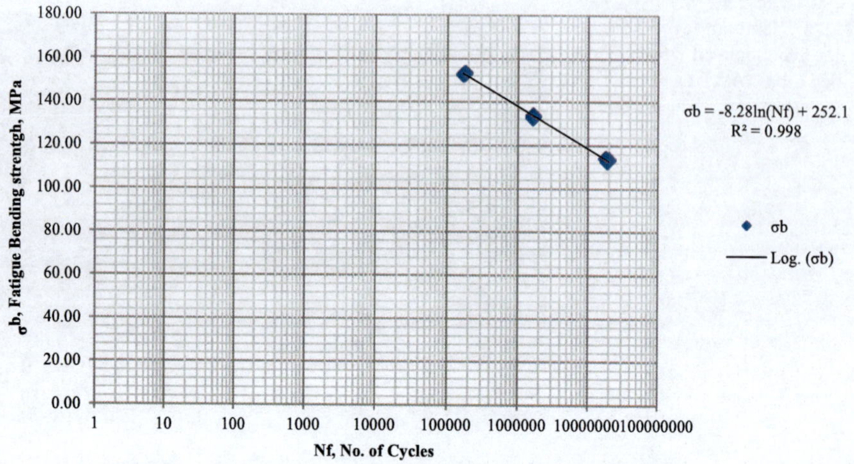

Fig. 8 S-N curve for adhesively bonded teakwood filled steel tube

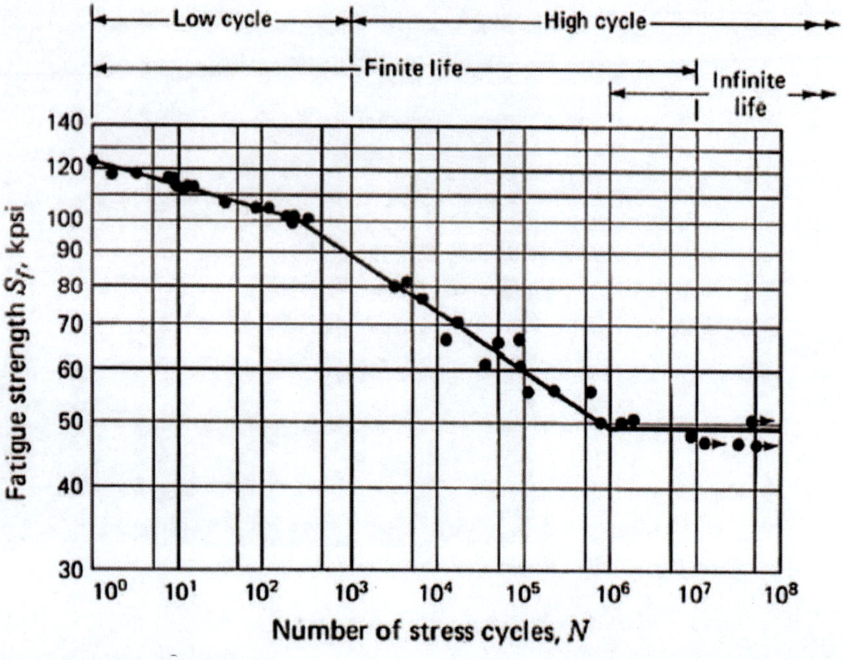

Fig. 9 An S-N diagram plotted from the results of completely reversed axial fatigue tests. Material: UNS G41300 steel normalized; $Sut = 116$ kpsi; maximum $Sut = 125$ kpsi (Data from NACA Technical Note 3866, December 1966) [14]

Table 5 Bending moment and bending strength by Euro code (EC4)

Specimen code	Mec4 (N mm)	σbec4 (MPa)
M/AWFT_200_1	83460.29	237.39
M/AWFT_200_2	84781.38	242.92
M/AWFT_200_3	84233.76	240.47
M/AWFT_175_1	88461.6	249.79
M/AWFT_175_2	86054.33	245.67
M/AWFT_175_3	87120.74	247.8
M/AWFT_150_1	78948.53	220.5
M/AWFT_150_2	81582.58	231.21
M/AWFT_150_3	80284.04	225.87
AVG	**83880.81**	**237.96**
SD	**3159.58**	**10.15**

Table 6 Fatigue bending strength for cycles before failure

Nf (cycles)	10^0	10^1	10^2	10^3	10^4	10^5	10^6	10^7	10^8	10^9	10^{10}
σbf (MPa)	252.1	233.03	213.97	194.9	175.84	156.77	137.71	118.64	99.58	80.51	61.45

the area of reinforcement, A_c is the area of the concrete, t is the thickness of the steel tube and b is the diameter of the steel tube. The notations used are from original code and the values for the same are substituted from the steel and teakwood used in this work.

Considering beam under bending load and applying the Euro code (EC4) static strength in bending is designed and is shown in Table 5 for all specimens.

The fatigue bending strength and cycles before failure are shown in Table 6 using Eq. 7

$$\sigma bf = -8.28\ln(Nf) + 252.1 \tag{7}$$

Fatigue bending strength is 252.10 and 233.03 for 10^1 and 10^2 Cycles before failure from Eq. 7. Hence, considering minimum life of 10^1 the fatigue strength values of both Euro code (EC4) and experimental values can be compared. The values of static bending strength obtained using Euro Code (EC4) are in good agreement with these experimental values of fatigue bending strength as shown in Table 7.

Table 7 Comparison of fatigue bending strength of EC4 and experimental method

Specimen code	*Nf* (cycle)	*σbf* (MPa)	σec4 (MPa)	σec4/*σbf*
AWFT_200_1	10^1	233.03	237.39	1.02
AWFT_200_2	10^1	233.03	242.92	1.04
AWFT_200_3	10^1	233.03	240.47	1.03
AWFT_175_1	10^1	233.03	249.79	1.07
AWFT_175_2	10^1	233.03	245.67	1.05
AWFT_175_3	10^1	233.03	247.8	1.06
AWFT_150_1	10^1	233.03	220.5	0.95
AWFT_150_2	10^1	233.03	231.21	0.99
AWFT_150_3	10^1	233.03	225.87	0.97
AVG	**10^1**	**233.03**	**237.96**	**1.02**
SD	**0**	**0**	**10.15**	**0.04**

6 Conclusion

The Fevitite superfast adhesive used for bonding provides sufficient confinement between teakwood and steel. It was observed that the number cycles before fatigue failure was minimum for maximum bending moment applied. The number of cycles before failure increase exponentially as the applied bending moment is reduced. The specimen subjected to fatigue test fail in a brittle manner. All the specimens failed due to crack initiation on the surface of the specimen and propagation of the crack ending in failure. A relation between fatigue life and fatigue strength was established with coefficient of correlation of 0.998 and can be used safely to predict fatigue life or fatigue strength of adhesively bonded teakwood filled carbon steel tubes. The static bending strength designed using Euro code 4 for adhesively bonded teakwood filled carbon steel tube is in good agreement with the experimental fatigue strength determined for fatigue life 10^1 cycles.

References

1. American Society of Metals (1975) Failures of shafts, failure analysis and prevention. Metals Handbook 10:373–397
2. Bhaumik SK, Rangaraju R et al (2002) Fatigue failure of a hollow power transmission shaft. Eng Fail Anal 9:457–467
3. Miscow GF, De Miranda PEV et al (2004) Techniques to characterize behavior of full size drill pipes and small scale samples. Int J Fatigue 26:575–584
4. West System Inc. (2005) Fatigue, technical information, Catalog Number, 000-545
5. Chen X, Dan J et al (2006) Fatigue life prediction of type 304 stainless steel under sequential biaxial loading. Int J Fatigue 28:289–299

6. Dawood M, Rizikalla S et al (2006) Fatigue and overloading behavior of steel-concrete composite flexural members strengthened with high modulus CFRP materials. J Compos Constr 11(6):659–669
7. Kim Y, Heffernan P (2008) Fatigue behavior of externally strengthened concrete beams with fiber reinforced polymers: state of the art. J Compos Constr 12(3):246–256
8. Zhenyu Z, Iftekhar A et al (2009) Fatigue modeling of concrete filled fiber reinforced polymer tubes. J Compos Constr 13:582–590
9. Malagi RR, Danawade BA (2015) Fatigue behaviour of circular hollow tube and wood filled circular hollow steel tube. Steel Compos Struct 19(3):585–599
10. ASTM D143 (1994) Standard method af testing small clear specimens of timber. American Society for Testing of Materials, West Conshohocken
11. ASTM E8 (2011) Standard test methods for tension testing of metallic materials. American Society for Testing of Materials, West Conshohocken
12. Pidilite Industries Limited (2009) Fevitite superfast product catalogue, Cc-Rf-Superfast-Eng-09/08
13. Fine Testing Machines (2008) Product catalog, Model Ftg-8(D), 2008/163
14. Shigley JE, Mischke CR et al (2004) Mechanical engineering design, Tutorial 4–17. The Mcgraw Hill, pp 1–11
15. British Standards Institute, London (1994) Euro-Code 4, Design of composite steel and concrete structures

Effect of Steel Fibers on the Sorptivity of Concrete

Yuvraj Singh, Sukhwant Singh, and Harvinder Singh

Abstract Due to an increasing need for sustainability, durability of concrete structures has become a matter of prime importance. And as far as durability of concrete is concerned, it is inversely proportional to its tendency to allow the ingress of water. It is thus quite instrumental in relating the measure of concrete's resistance to exposure in aggressive conditions. This paper presents the effect of adding steel fibers in concrete on the surface water absorption using sorptivity test. The variation in the sorptivity with varying percentage of steel fibers in concrete is determined by measuring the increase in the mass of specimens resulting from absorption of water as a function of time when only one surface of the specimen is exposed to water as per ASTM C1585. Test results indicate that the presence of steel fibers in concrete reduces the surface water absorption significantly as compared to the plain concrete indicating increased resistance to the uptake of water and hence increased durability.

Keywords Concrete · Corrosion · Durability · Steel fibers · Sorptivity

1 Introduction

Due to the increasing population and hence the rapidly increasing demand for infrastructure, there is undoubtedly an indispensible need for sustainable and innovative construction techniques. Concrete, a widely used construction material is being researched widely to achieve the desired quality in construction. And to have a sustainable concrete structure, it becomes important to take its durability into consideration. A durable concrete can be referred as one which resists weathering action or any sort of chemical exposure and the one which performs as expected under

Y. Singh (✉) · S. Singh · H. Singh
Guru Nanak Dev Engineering College, Ludhiana, India
e-mail: uvraj_23@yahoo.co.in

H. Singh
e-mail: hvs1221@gmail.com

Y. Singh
I K Gujral Punjab Technical University, Kapurthala, India

S. Seetharamu et al. (eds.), *Fatigue, Durability, and Fracture Mechanics*, Lecture Notes in Mechanical Engineering, https://doi.org/10.1007/978-981-15-4779-9_32

the prevalent conditions. Moreover, any kind of composite that can hold its original arrangement and serviceability in spite of any environmental conditions can be referred as durable. Selection of concrete's materials for a specific application is a crucial step in avoiding durability problems. And as far as durability of concrete is concerned, it is inversely proportional to its tendency to allow the ingress of water. Especially, the concrete structures that are exposed to marine water, de-icing salts, corrosive environments, etc. need low permeability of water for enhanced durability. Apart from providing safety and performing well throughout the service life of a structure, the enhanced durability of a structure also eliminates the expenses or troubles that may arise due to recurring need for repairs and maintenance. Thus, a durable structure prevents the undue wastage or exploitation of the resources.

Concrete is well known to be Quasi-brittle material. And concrete alone cannot solve the key requirement of being a usable structural material. Reinforcement within concrete creates a composite material, with the concrete providing strength against the compressive stresses while the reinforcement provides the strength against the tensile stresses. Thus, concrete reinforced with steel is the foundation of our modern society. On the contrary, while steel reinforcement solves one of the concrete's greatest limitations, it creates an entirely new problem. Corrosion of steel rebar is the most common form of concrete deterioration. And it is accelerated with the ingression of water, salts, and contaminants into the concrete. Several monitoring techniques and preventive measures have been reported in literature [12]. The structure subjected to worse exposure conditions is able to maintain its durability through two main performance characteristics. The one is low permeability and the other is shrinkage. Lower permeability can be achieved by better microstructure and by preventing shrinkage cracks that may appear due to heat of hydration. Fibers in concrete are known for their crack arresting characteristic. On the other hand, if not skillfully incorporated, fibers may lead to porous concrete due to decreased workability. Transport properties of materials like permeability might disturb the durability and structure of concrete composite. Therefore, it becomes important to investigate the role that fibers can play in regard to porosity or permeability of concrete.

In order to investigate the transport properties of concrete, researchers have been using enormous techniques, namely, water absorption test, permeability test, Rapid chloride penetration test, etc. [1, 3, 5–11]. Sabir et al. presented a test to measure the mortar and concrete's water absorption in one direction of the specimens by detecting the changes between the sorptivities of specimens [10]. Lockington et al. also presented the study of unsaturated hardened concrete by determining the amount of water soaked by solid concrete at different intervals of time to check the sorptivity index of concrete [8]. It was stated that this is the vital parameter which is instrumental to evaluate the concrete's resistance to ingress of water or contaminants [1, 3, 4, 8]. Though abundant investigation has been completed to recognize, examine, and investigate the mechanical properties of fiber strengthened concrete, slight exploration has focused on the transport qualities of this material. Concrete's transport properties, particularly permeability, may disturb the durability and internal structure of the composite. The rise in concrete permeability, due to the onset and advancement of cracks, delivers entrance of water, chlorides, and other destructive deteriorating agents. Thus, the present investigation aims to examine the effect of adding steel fibers into concrete on the rate of absorption of water by performing sorptivity test.

2 Experiment Design

2.1 Preparation of Samples

In order to investigate the effect of adding steel fibers in concrete on the rate of absorption of water, sorptivity test has been performed on different sets of specimens as per ASTM C1589-04. According to it, this test strategy is utilized to decide the rate of assimilation (sorptivity) of water by the pressure driven in cement concrete by estimating the expansion in the mass of a sample coming about because of ingestion of water as an element of time when just one surface of the sample is presented to water [2, 9]. In the present work, the fibers are added in concrete mix of M25 grade, represented with percentage of the entire volume of composite known as volume fraction (V_f). Steel fibers of hooked end type of dia 1 mm and length 50 mm are used.

In order to examine the effect of steel fibers on the sorptivity of concrete, the experimentation has been carried out on specimens with varying percentage of steel fibers (0, 0.5, and 1%). The standard test sample is a 100 ± 6 mm diameter disc, with a length of 50 ± 3 mm [2]. The absorption can be assessed at various separations from the uncovered surface. For obtaining the standard control samples, 150 mm × 150 mm × 150 mm cubes of grade M25 containing 0, 0.5, and 1% hooked end steel fibers were cast and cured for a period of 28 days. A total of 12 test specimens consisting of a set of three samples for each case (0, 0.5, and 1% steel fibers) were obtained from the drilled cores of the cast cubes as shown in Fig. 1.

Fig. 1 Core cutting technique for standard test samples

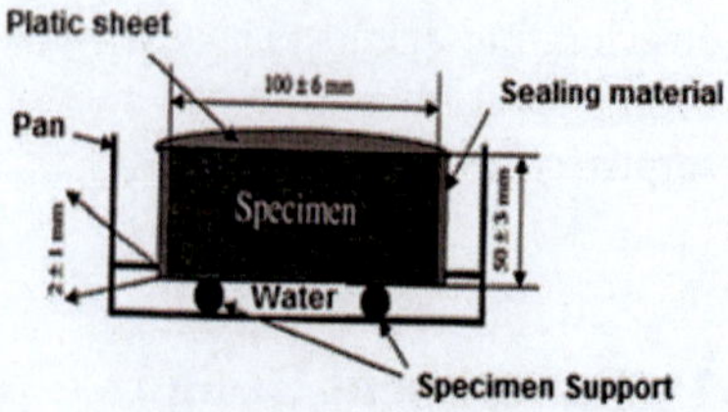

Fig. 2 Representation of sorptivity test process as illustrated in ASTM C1585 [2]

2.2 *Testing Procedure*

Before the test is performed, the obtained test specimens are pre-conditioned as per ASTM C1585. Followed by which, diameter and mass of the conditioned specimens are recorded. The side surface of each specimen is sealed with a suitable sealing material. In the present study, the side surfaces of the specimens were sealed with epoxy coating. The end of the specimen that will not be exposed to water is sealed using a loosely attached plastic sheet. Following which the mass of the sealed specimen is measured and recorded as the initial mass for water absorption calculation. The support device is placed at the bottom of the pan and the pan is filled with tap water such that the water level is 1–3 mm above the top of the support device as illustrated in Fig. 2. The water level is maintained throughout the duration of the test. After the timing device is started, the specimens are placed immediately on the support device and the time of initial contact with water is recorded [2]. The recording post the start of the test is taken at different intervals as recommended in ASTM C1585 for a period of 6 h. Figures 3 and 4 show the prepared sorptivity test setup.

Fig. 3 Epoxy applied sample and air tightened weighing scale

Fig. 4 Sorptivity test setup

Calculations

The absorption, I, is calculated in 'mm' as

$$I = \frac{m_t}{a \times d}.$$

where

I the water absorption,
m_t the change in specimen mass, at the time t,
a the uncovered zone of the sample, in mm^2, and
d the water density in g/mm^3.

After determination of the initial absorption (I) in terms of 'mm', the initial rate of water absorption (in $mm/s^{1/2}$) is worked out from the slope of the line that is the best fit to I plotted against the square root of time ($s^{1/2}$). This slope is obtained using linear regression analysis by using points from 60 s to 6 h [2, 9, 12].

3 Results and Discussion

As per the procedure described in the previous section, the standard test samples as extracted from the cast cubes are 97 mm wide with a length of 50 mm. Sorptivity test has been performed on these samples as per the procedure described in the previous section and as prescribed in *ASTM C1585* for a period of 6 h. After obtaining the increment in mass at different time intervals, the absorption has been measured in 'mm' as the ratio of increment in mass to that of the product of exposed area and the density of water [2]. The observations and the results so obtained for three different cases, i.e., PCC (0% steel fibers), 0.5% steel fibers, and 1% steel fibers are listed in Tables 1, 2 and 3.

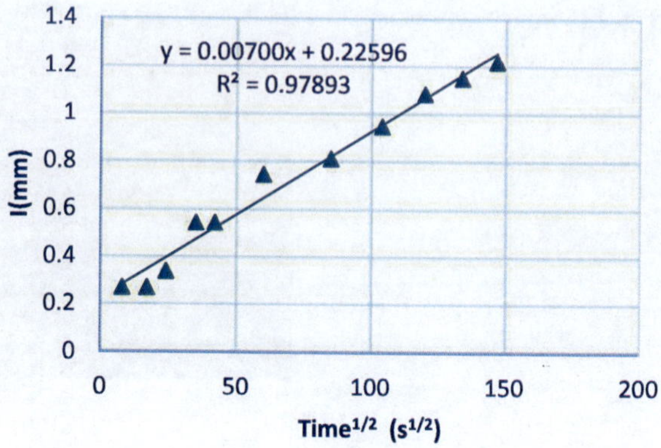

Fig. 5 Absorption versus $\sqrt{time}$ plot of PCC (0% SF)

Figure 5 shows the comparison of I (mm) versus time ($s^{1/2}$) plot obtained from the observations listed in Table 1 for plain cement concrete specimens. From the figure, it can be seen that the average absorption of water after a period of 6 h is 1.219 mm. From the linear regression analysis, the value of initial rate of absorption can be obtained as coefficient of correlation that is more than 0.98, i.e., $R = \sqrt{0.97893} \approx 0.989$.

As, Initial absorption is given by $I = S_i\sqrt{t} + b$, therefore from the linear regression analysis, the value of initial rate of absorption in case of PCC is obtained as 7×10^{-3} mm/$s^{1/2}$.

Figure 6 shows the comparison of I (mm) versus time ($s^{1/2}$) plot obtained from the observations listed in Table 2 for 0.5% SFRC specimens. From the figure, it can be seen that the average absorption of water after a period of 6 h is 0.947 mm which is comparatively lesser than plain cement concrete. From the linear regression analysis, the value of initial rate of absorption can be obtained as coefficient of correlation that is more than 0.98, i.e., $R = \sqrt{0.98644} \approx 0.993$.

As initial absorption is given by $I = S_i\sqrt{t} + b$, therefore from the linear regression analysis, the value of initial rate of absorption in case of 0.5% SFRC is obtained as 5.45×10^{-3} mm/$s^{1/2}$, which is much lower than in case of PCC.

Figure 7 shows the comparison of I (mm) versus time ($s^{1/2}$) plot obtained from the observations listed in Table 3 for 1% SFRC specimens. From the figure, it can be seen that the average absorption of water after a period of 6 h is 0.880 mm which is comparatively lesser than plain cement concrete and 0.5% SFRC specimens. From the linear regression analysis, the value of initial rate of absorption can be obtained as coefficient of correlation that is more than 0.98, i.e., $R = \sqrt{0.97630} \approx 0.988$.

As, Initial absorption is given by $I = S_i\sqrt{t} + b$, therefore from the linear regression analysis, the value of initial rate of absorption in case of 1% SFRC is obtained as 5.13×10^{-3} mm/$s^{1/2}$.

Table 1 Results of 6 h water sorptivity test of PCC (0% steel fibers)

Sr. No.	Time (s)	$s^{1/2}$	Mass of specimen, *m* (g)				Increment in mass (g)				Avg Δm	*I* (mm) = Δm/area/water density
			PCC1	PCC2	PCC3	PCC4	PCC1	PCC2	PCC3	PCC4		
1.	0	0	740	744	730	740	0	0	0	0	0	0
2.	60	8	742	746	732	742	2	2	2	2	2.0	0.270
3.	300	17	742	746	732	742	2	2	2	2	2.0	0.270
4.	600	24	742	746	732	744	2	2	2	4	2.5	0.338
5.	1200	35	744	748	734	744	4	4	4	4	4.0	0.542
6.	1800	42	744	748	734	744	4	4	4	4	4.0	0.542
7.	3600	60	746	748	736	746	6	4	6	6	5.5	0.745
8.	7200	85	746	750	736	746	6	6	6	6	6.0	0.812
9.	10,800	104	748	750	738	746	8	6	8	6	7.0	0.947
10.	14,400	120	748	752	738	748	8	8	8	8	8.0	1.083
11.	18,000	134	748	752	740	748	8	8	10	8	8.5	1.150
12.	21,600	147	750	752	740	748	10	8	10	8	9.0	1.219

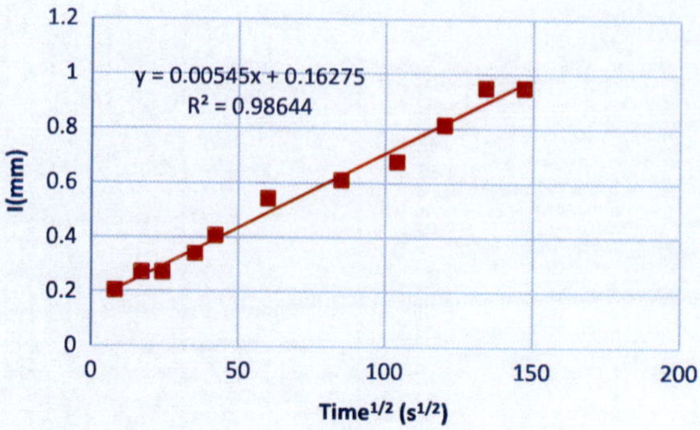

Fig. 6 Absorption versus $\sqrt{time}$ plot of 0.5% SFRC

Figure 8 shows the comparison of I (mm) versus time ($s^{1/2}$) plot of all sets of specimens. In the present work, the test has been carried out for a period of 6 h to get the initial rate of absorption in each case. The slope which represents early-age or initial absorption is steeper and thus indicates the greater sorptivity or rate of absorption during the early periods of exposure. From the figure, it can be seen that the initial rate of absorption from 60 s to 6 h is much higher in case of plain cement concrete as compared to steel fiber reinforced concrete. Table 4 summarizes the sorptivity (or rate of absorption) of each set of specimen. It has been observed that the infiltration of water reduces on addition of 0.5% and 1% steel fibers in plain cement concrete to amount of 22.31% and 28.21%, respectively, as compared to that of plain cement concrete with 0% steel fibers specimen. From the linear regression analysis, it has been observed that the initial rate of absorption is evidently reduced on addition of 0.5% and 1% steel fibers by 22.14% and 26.71%, respectively, as compared to plain cement concrete. This decrease in the rate of initial absorption of water is accounted due to the presence of steel fibers which provide a time lag for the water to infiltrate by acting as a barrier to the flow of water, and steel fibers also tend to eliminate the micro-cracks that may arise due to heat of hydration. There is an appreciable difference in the sorptivity of PCC and SFRC, while the difference in the rate of absorption in case of 0.5 and 1% SFRC is minute. Furthermore, from visual inspection it has been observed that the surface water percentage value is higher in plain concrete as compared to SFRC due to the reduced infiltration and rate of absorption. This reduced infiltration on addition of steel fibers to concrete indicates the improved microstructure and hence the improved durability of the concrete.

Table 2 Results of 6 h water sorptivity test of 0.5% SFRC

Sr. No.	Time (s)	$s^{1/2}$	Mass of specimen, m (g)				Increment in mass (g)				Avg Δm	I (mm) = Δm/area/water density
			0.5% SFRC1	0.5% SFRC2	0.5% SFRC3	0.5% SFRC4	0.5% SFRC1	0.5% SFRC2	0.5% SFRC3	0.5% SFRC4		
1.	0	0	802	810	808	812	0	0	0	0	0	0
2.	60	8	804	812	808	814	2	2	0	2	1.5	0.203
3.	300	17	804	812	810	814	2	2	2	2	2.0	0.270
4.	600	24	804	812	810	814	2	2	2	2	2.0	0.270
5.	1200	35	804	814	810	814	2	4	2	2	2.5	0.338
6.	1800	42	804	814	810	816	2	4	2	4	3.0	0.406
7.	3600	60	806	814	812	816	4	4	4	4	4.0	0.541
8.	7200	85	806	816	812	816	4	6	4	4	4.5	0.609
9.	10,800	104	806	816	812	818	4	6	4	6	5.0	0.677
10.	14,400	120	808	816	814	818	6	6	6	6	6.0	0.812
11.	18,000	134	808	818	814	820	6	8	6	8	7.0	0.947
12.	21,600	147	808	818	814	820	6	8	6	8	7.0	0.947

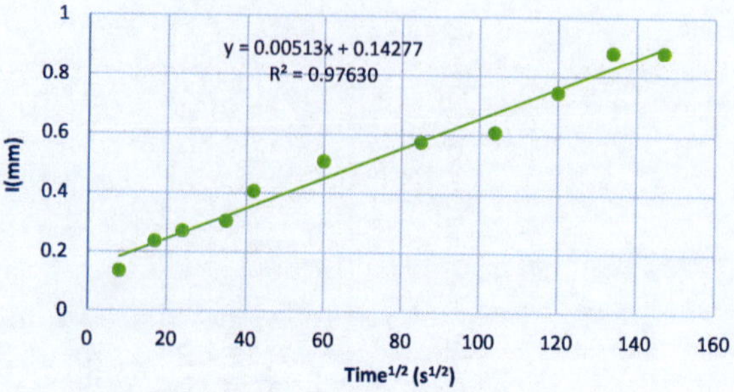

Fig. 7 Absorption versus $\sqrt{time}$ plot of 1% SFRC

4 Conclusions

Following are the conclusions that have been drawn from current study:

1. The infiltration of water in case of steel fiber reinforced concrete is less than plain cement concrete. The absorption reduces by 22.31% and 28.21% in case of 0.5% and 1% SFRC, respectively, as compared to the plain cement concrete. This is due to improved microstructure of concrete due to elimination of micro-cracks on addition of steel fibers and hence results in higher resistance to ingress of water leading to higher durability.
2. The rate of absorption or sorptivity in case of 0.5 and 1% SFRC is 22.14 and 26.71% less than that of plain cement concrete which indicates improved service life of SFRC.
3. The percentage decrease in the sorptivity is comparatively less when the dosage of steel fibers is increased from 0.5 to 1% as compared to that between 0 and 0.5%. This indicates the possibility of change of scenario on further increase of dosage of steel fibers. Thus, the improved microstructure leading to reduced sorptivity depends on the dosage of steel fibers and precise workmanship.

Table 3 Results of 6 h water sorptivity test of 1% SFRC

Sr. No.	Time (s)	$s^{1/2}$	Mass of specimen, *m* (g)				Increment in mass (g)				Avg Δm	*I* (mm) = Δm/area/water density
			1% SFRC1	1% SFRC2	1% SFRC3	1% SFRC4	1% SFRC1	1% SFRC2	1% SFRC3	1% SFRC4		
1.	0	0	862	866	860	870	0	0	0	0	0	0
2.	60	8	863	867	862	870	1	1	2	0	1.0	0.135
3.	300	17	864	868	862	871	2	2	2	1	1.75	0.237
4.	600	24	864	868	862	872	2	2	2	2	2.0	0.271
5.	1200	35	864	868	863	872	2	2	3	2	2.25	0.305
6.	1800	42	864	869	864	873	2	3	4	3	3.0	0.406
7.	3600	60	865	870	864	874	3	4	4	4	3.75	0.508
8.	7200	85	866	870	865	874	4	4	5	4	4.25	0.575
9.	10,800	104	866	870	866	874	4	4	6	4	4.5	0.609
10.	14,400	120	868	870	866	876	6	4	6	6	5.5	0.745
11.	18,000	134	868	872	868	876	6	6	8	6	6.5	0.880
12.	21,600	147	868	872	868	876	6	6	8	6	6.5	0.880

Table 4 Summary of sorptivity (or rate of absorption) of each set of specimen

Specimen type	0% SF	0.5% SF	1% SF
The rate of absorption ($\times 10^{-3}$ mm/s$^{1/2}$)	7.00	5.45	5.13

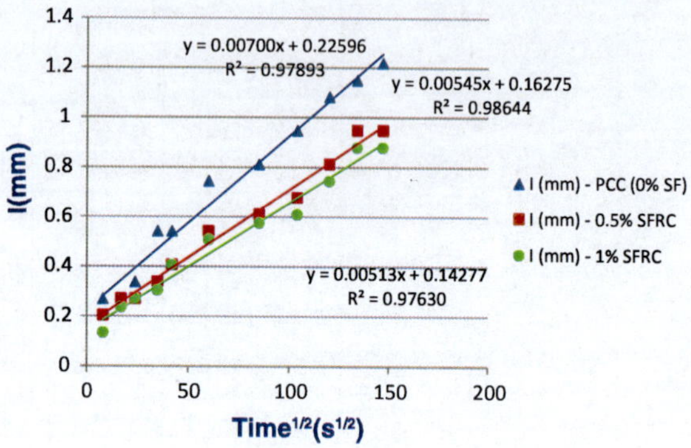

Fig. 8 Comparison of absorption versus $\sqrt{time}$ plot of all three cases

Acknowledgements The experimental work was carried out in the concrete technology laboratory of Guru Nanak Dev Engineering College Ludhiana. The authors gratefully acknowledge the assistance provided by the team of the college and I.K. Gujral Punjab Technical University, Kapurthala.

References

1. Afroughsabet V, Ozbakkaloglu T (2015) Mechanical and durability properties of high-strength concrete containing steel and polypropylene fibers. Constr Build Mater 94:73–82
2. ASTM C 1585:2004 Standard test method for measurement of rate of absorption of water by hydraulic-cement concretes. American Society for Testing and Materials, pp 1–6
3. Basheer L, Cleland DJ (2011) Durability and water absorption properties of surface treated concretes. Mater Struct/Materiaux et Constr 44(5):957–967
4. Depaa RAB, Felix Kala T (2015) Experimental investigation of self healing behavior of concrete using Silica fume and GGBFS as mineral admixtures. Indian J Sci Technol 8(36)
5. Ding Y, Li D, Zhang Y (2018) Quantitative analysis of macro steel fiber influence on crack geometry and water permeability of concrete. Compos Struct 187:325–335
6. Durán-Herrera A, Mendoza-Rangel JM, De-Los-Santos EU, Vázquez F, Valdez P, Bentz DP (2015) Accelerated and natural carbonation of concretes with internal curing and shrinkage/viscosity modifiers. Mater Struct/Materiaux et Constr 48(4):1207–1214
7. Leung HY, Kim J, Nadeem A, Jaganathan J, Anwar MP (2016) Sorptivity of self-compacting concrete containing fly ash and silica fume. Constr Build Mater 113:369–375
8. Lockington D, Parlange J-Y, Dux P (1999) Sorptivity and the estimation of water penetration into unsaturated concrete. Mater Struct 32(5):342–347
9. Patel VN (2009) Sorptivity testing to access durability of concrete against freeze-thaw cycling. M.E. thesis, McGill University
10. Sabir BB, Wild S, O'Farrell M (1998) A water sorptivity test for mortar and concrete. Mater Struct 31:568. https://doi.org/10.1007/BF02481540

11. Şahmaran M, Li VC (2009) Durability properties of micro-cracked ECC containing high volumes fly ash. Cem Concr Res 39(11):1033–1043
12. Singh Y, Kaur I, Gill AS (2014) Corrosion monitoring and service life prediction of rebars in structural concrete: state-of-review. In: Lecture notes in mechanical engineering. Springer, pp 133–138

Estimation of Residual Life and Failure Mechanism of Cracked Aircraft Wing Skin

Mahantesh Hagaragi, M. Mohan Kumar, and Ramesh S. Sharma

Abstract Many failures in aircraft structures are due to fatigue cracks initiating and developing from fastener holes at which there are large stress concentrations. In a typical wing skin, in the zone of riveted joint of rib/skin, the combination of high stress concentration could potentially lead to the appearance of the crack initiation and then crack growth under cyclic loading. Stress Intensity Factor (SIF) solutions are required for the assessment of fracture strength and residual fatigue life for defects in structures. In this context, many research works focused on evaluating the residual life of various cracked aircraft structures but only a few works have been done on light transport aircraft wing skin. The material used for wing skin is AL 2024-T351. A computational model for estimating the residual fatigue life of cracked wing skin is proposed. The complete computation procedure for the crack propagation analysis using low-cycle fatigue material properties is illustrated with the damaged wing skin. Initially, stress concentration effects at the cracked wing skin rivet holes are determined by applying analytical and numerical methods. Further, residual life and the failure mechanism in the cracked rivet holes of the wing skin are estimated. The wing skin with two cracked rivet holes for a pitch of 26 mm was analyzed using MSC NASTRAN/PATRAN for different crack lengths using MVCCI (Modified Virtual Crack Closure Integral) method by which strain energy release rate as well as stress intensity factors are calculated for different crack lengths, and fatigue crack growth life for progressive cracks for different *R* ratios is computed using a MATLAB program. Comparisons of the stress intensity factors estimated by FE analysis were in good agreement with the analytical solutions. Further, using SIF solutions, the residual life was estimated for the cracks emanating from the two rivet holes using crack growth models. The work also investigates the first failure

M. Hagaragi (✉) · R. S. Sharma
Department of Mechanical Engineering, RV College of Engineering, Bengaluru, India
e-mail: mahanteshh.mmd17@rvce.edu.in

R. S. Sharma
e-mail: rameshssharma@rvce.edu.in

M. Mohan Kumar
FSIG-STTD, CSIR-National Aerospace Laboratories, Bengaluru, India
e-mail: mmk@nal.res.in

S. Seetharamu et al. (eds.), *Fatigue, Durability, and Fracture Mechanics*, Lecture Notes in Mechanical Engineering, https://doi.org/10.1007/978-981-15-4779-9_33

mechanism out of two competing mechanisms of failure; Failure due to fracture or Failure due to plastic collapse at the net section between two advancing crack tips of the rivet holes of the wing skin. It was observed that the wing skin with crack rivet holes would fail by plastic collapse due to net section yielding. Further, the study can be extended to a multi-axial stress condition.

Keywords Wing skin · Stress concentration · Residual life · Net section failure · Strain energy release rate · Stress intensity factor

1 Introduction

For better performance and stability of aircraft, the manufacturing of aircraft needs to be carried out in such a way that it should have lower weight so that the thrust force required is minimum and structures should have a longer life. Aircraft industries prefer riveted joints for assembling various structures, these joints act as stress concentrator which leads to crack initiation that further leads to multi-site damage, i.e., crack formation at a number of rivet holes because of uniform stress distribution; thus, it requires a lot of efforts to predict the fatigue life of structure when it's multi-site damage.

Fatigue crack initiation, crack propagation, and failure are three periods of fatigue life; refer Fig. 1. Crack initiates at rivet hole locations or at the location where stress is concentrated. In the crack initiation period, the crack length is very small and never exceeds more than two to five grain size. Crack initiation stage is bigger than crack propagation and failure stages; over the large number of repetitive load application shear stresses are generated in the structure which causes irreversible changes in the metal structure, known as "cumulative fatigue damage". During the crack propagation period, over the number of repeated load application macro crack growth take place in the direction perpendicular to load application. As the crack propagates, the cross-sectional area of specimen reduces proportionally reducing the strength. At a certain stage, due to the weak cross-sectional area of specimen, complete failure occurs [1].

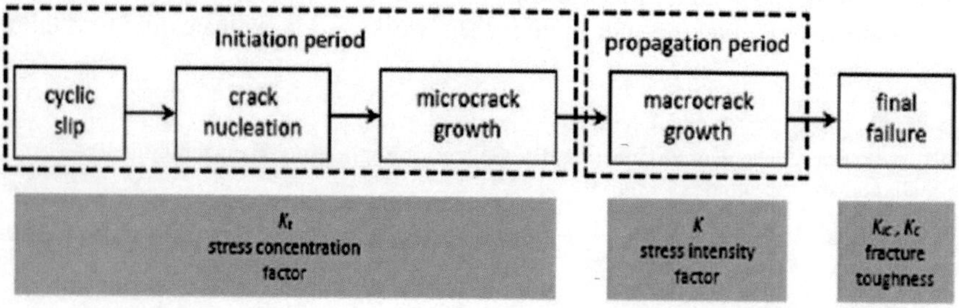

Fig. 1 Fatigue life stages [17]

The fatigue life of materials and structures relies upon a few parameters. In crack initiation, fatigue life is connected to metallurgical, geometrical, and loading parameters and residual stress. During fatigue crack growth, plasticity induced crack closure is watched. Fatigue crack closure is a phenomenon that comprises of the contact between crack surfaces during a portion of the load cycle. This contact influences the neighborhood stress and plastic distortion fields close to the crack tip, and subsequently, the micro-mechanisms responsible for fatigue propagation.

2 Proposed Model

2.1 *Modified Virtual Crack Closure Integral Method for Stress Intensity Factor Calculation*

Modified Virtual Crack Closure Integral (MVCCI) method is used to determine the Stress Intensity Factor (SIF) for different crack lengths in the wing skin. MVCCI method is based on the energy balance [2, 3]. In this technique, SIF is obtained for fracture mode from Eq. 1.

$$G = \frac{K_I^2}{E}\beta \quad (1)$$

where G is the energy release rate for mode I, K_I is the SIF for mode I, E is the modulus of elasticity, and β is 1 for plane stress condition. Calculation of the energy release rate is based on the Irwin assumption that the energy released in the process of crack expansion is equal to work required to close the crack to its original state as the crack extends by a small amount Δa. Irwin computed this work as shown in Eq. 2.

$$W = \frac{1}{2}\int_0^{\Delta a} u(r)\sigma(r - \Delta a)\mathrm{d}r \quad (2)$$

where u is the relative displacement, r is the distance from the crack tip, and Δa is the change in virtual crack length. Therefore, the strain energy release rate is given by Eq. 3.

$$G = \ln \Delta a \to 0\frac{1}{2\Delta a}\int_0^{\Delta a} u(r)\sigma(r - \Delta a)\mathrm{d}r \quad (3)$$

After simplification, the modified strain energy release rate is given by Eq. 4.

$$G = \frac{F \times \Delta u}{2 \times \Delta a \times t}\mathrm{N/mm} \quad (4)$$

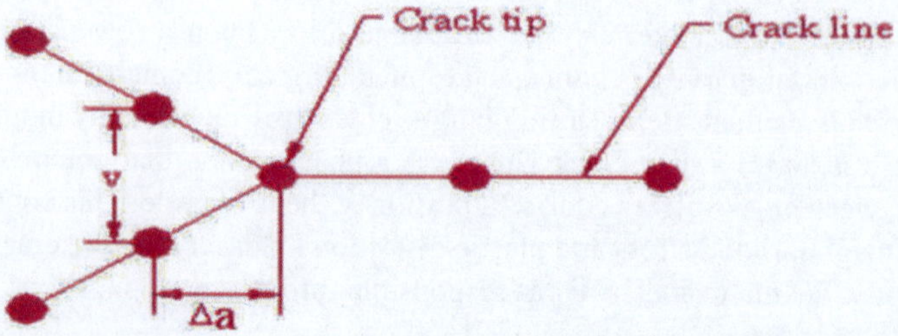

Fig. 2 MVCCI method illustration

where F is the forces at the crack tip, Δu is Crack Opening Displacement (COD), t is the thickness of the skin, and Δa is elemental edge length near the crack tip; refer Fig. 2 [4–7].

2.2 *Fatigue Life Estimation and Failure Mechanism in the Presence of Crack*

The fatigue life of cracked wing skin is obtained by the Paris law; this Eq. 5 relates crack growth rate with stress intensity factor range.

$$\frac{\mathrm{d}a}{\mathrm{d}N} = C\Delta K^m \tag{5}$$

where $\mathrm{d}a/\mathrm{d}N$ is fatigue crack growth rate, C and m are experimentally determined material constants, and ΔK is stress intensity factor range. ΔK is obtained by Eq. 6.

$$\Delta K = K_{\max} - K_{\min} \tag{6}$$

Failure mechanism in cracked wing skin is obtained by comparing SIF results with Fracture Toughness (K_{IC}) of material; if SIF is greater than the K_{IC} value, then wing skin fails due to fracture. The other mode of failure is plastic collapse at the net section between two advancing crack tips of the rivet holes of wing skin; if net section stress is greater than the yield strength of the material, then wing skin fails due to plastic collapse. Net section stress is calculated by Eq. 7.

$$\sigma_{\text{net section}} = \frac{\text{Forces at Crack Tip}}{\text{Net Section Area}} = \frac{F_{\text{crack tip}}}{(W - 2a_{\text{eff}}) \times t}\ \text{MPa} \tag{7}$$

where W is the pitch between two riveted holes, a_{eff} is the effective crack length, and t is the wing thickness [8–14].

3 Finite Element Analysis

Aircraft wing produces upward aerodynamic force; much of the outer surface of wings and fuselage is covered with skin commonly made up of aluminum and its alloys (include Copper, magnesium, and Zinc). In the service life of aircraft, multiple small cracks are formed in its structure, which grow slowly and join one another, and over the period of time, a bigger crack is formed. Due to discontinuities present in structure, there will be a generation of stress concentration in it, which significantly reduces the service life of aircraft.

In the present study, a cracked wing skin of aircraft is analyzed under tensile loads. Stress analysis was carried out on this panel to identify the maximum stress contour locations. Fatigue cracks will always get initiated from the location of maximum stress contour. Hence, the finite element tools were employed to the residual life estimation of wing skin, and the SIFs for the progressive cracks were estimated. Also, the fatigue life of the panel is evaluated along with the failure criteria.

3.1 *Wing Skin Geometry*

The wing skin is having a size of 60 $\times$ 120 $\times$ 1.5 mm thickness. The skin is joined to the airframe by rivets having a size of 4 mm diameter. The two riveted holes are separated by a center pitch of 26 mm. The wing skin is subjected to a tensile load of 7000 N. The details of wing skin geometry are shown in Fig. 3.

3.2 *Finite Element Modeling of Wing Skin*

For the analysis of wing skin, commercial FE softwares like MSC Patran and MSC Nastran is used. The modeling of wing skin is done with two riveted holes in it, and two-dimensional four-node quadrilateral elements are used; refer Fig. 4.

3.3 *Loading and Boundary Conditions*

During the flight, there is a change in atmospheric pressure which leads to pressurization of the wing box (internally) of aircraft and there will be air drag acting on the wing skin. Due to these loads, wing skin is subjected to tensile loading. In this study, we are considering air drag acting on wing skin which develops tensile stress in the structure.

In the present study, the air drag is considered as the load case. It is observed that wing skin will undergo some pressure value. The load is extracted from this stress

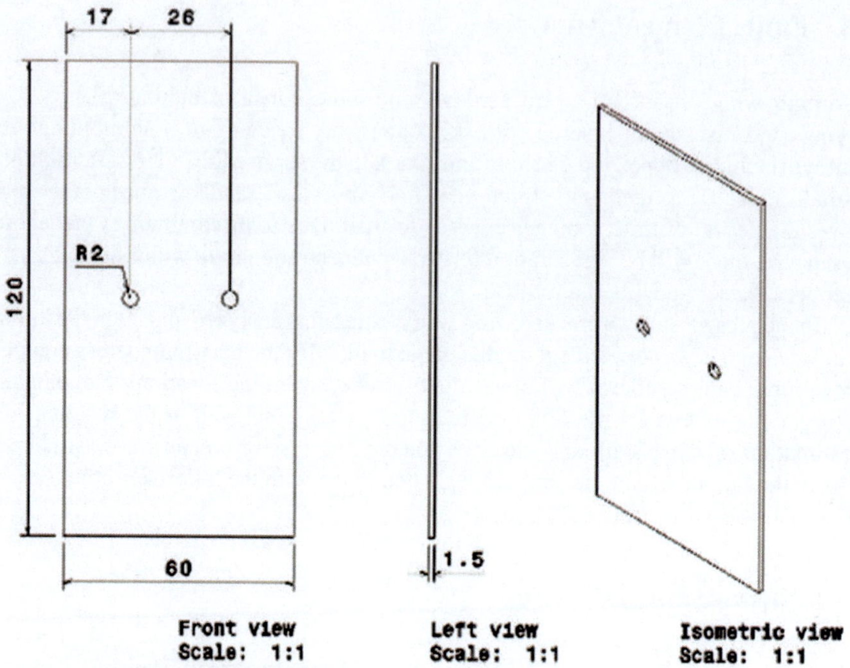

Fig. 3 Dimensional details of wing skin

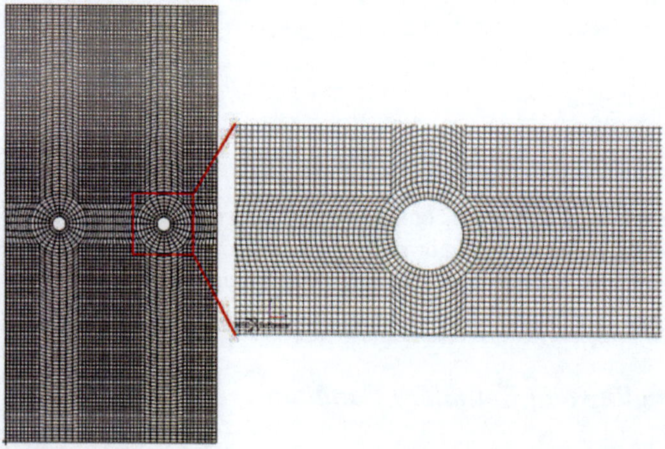

Fig. 4 FE model of wing skin

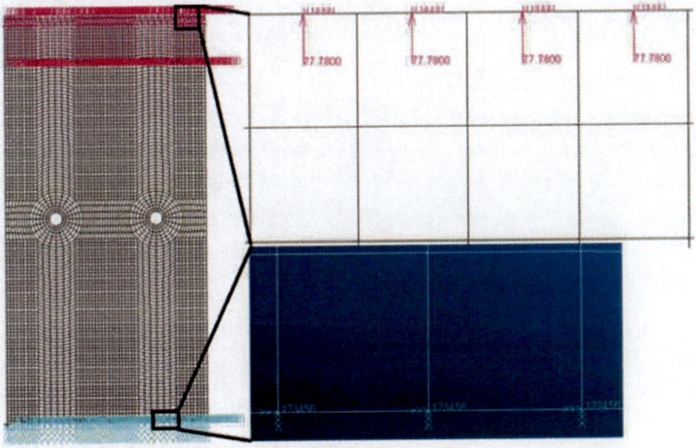

Fig. 5 Loads and boundary conditions of wing skin

and is applied on the panel. This load will generate 77.78 N/mm^2 of tensile stress on the wing skin. Tensile stress will be distributed uniformly along the cross section, and this stress is applied in the transverse axial direction as shown in Fig. 5.

3.4 Material Properties

Figure 4 represents the finite element model of the wing skin. It is made up of AL 2024-T351 material having a thickness of 1.5 mm with two riveted holes in it of diameter 4 mm each. The pitch between two rivet holes is 26 mm. Some of the material properties which are essential for this work are listed below; refer Table 1 [15, 16].

3.5 Stress Analysis of Wing Skin Rivet Holes Under Tensile Load (for Identification of Maximum Stress)

This section describes the identification of the maximum value of stress that acts at the wing skin rivet holes. The finite element model of wing skin with rivet holes is developed. Finite element analysis is performed using MSC Nastran–Patran finite element software. Two-dimensional four-node quadrilateral elements that were used to represent the wing skin are presented in detail in Sect. 3.2. The applied load was 77.78 MPa. Once the load distribution acting on wing skin is applied, the maximum stress value acting on wing skin was obtained. The maximum stress is 153 MPa at the notch root as presented in Fig. 6. And maximum displacement is found to be 8.82^{-02} mm as presented in Fig. 7.

Table 1 Material properties for AL 2024-T351

E	71,000 MPa	Modulus of elasticity
ν	0.33	Poisson's ratio
S_y	334 MPa	Material tensile yield strength
$\sigma_f{}'$	613 MPa	Fatigue strength coefficient
S_u	469	Ultimate tensile strength
$\varepsilon_f{}'$	0.35	Fatigue ductility coefficient
n'	0.121	Cyclic strain hardening exponent
k'	710 MPa	Cyclic strength coefficient
K_{IC}	91 MPa $\sqrt{\text{m}}$	Fracture toughness
$I_n{}'$	3.067	Parameter that is dependent on n
Ψ	0.95152	Non dimensional parameter
ΔK_{thO}	8 MPa $\sqrt{\text{m}}$	Range of threshold stress intensity factor
C	6E-9	The Paris equation constants
m	4.5849	The Paris equation constants

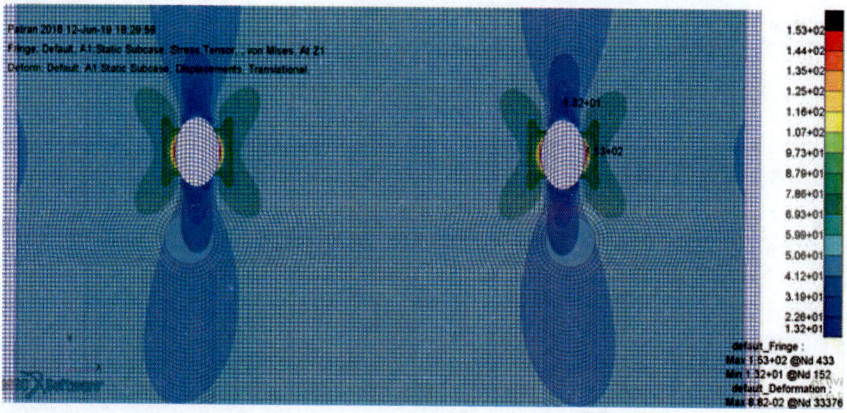

Fig. 6 Maximum stress distribution of wing skin at the root of rivet hole

3.5.1 Convergence Criteria

The term convergence is an important issue that has to be considered in the finite element analysis. Convergence is nothing but the smallest size of the element for which the results of finite element analysis are not affected by the change in element size.

Figure 8 shows the change in stress values with respect to the number of elements. It can be seen that as the number of elements is increasing, the values of stress are almost the same; so the smallest mesh size for which the stress values are same, even after changing the mess size, is taken for the analysis. In this particular problem, stress values are converging at a mesh size of 0.33 (Table 2).

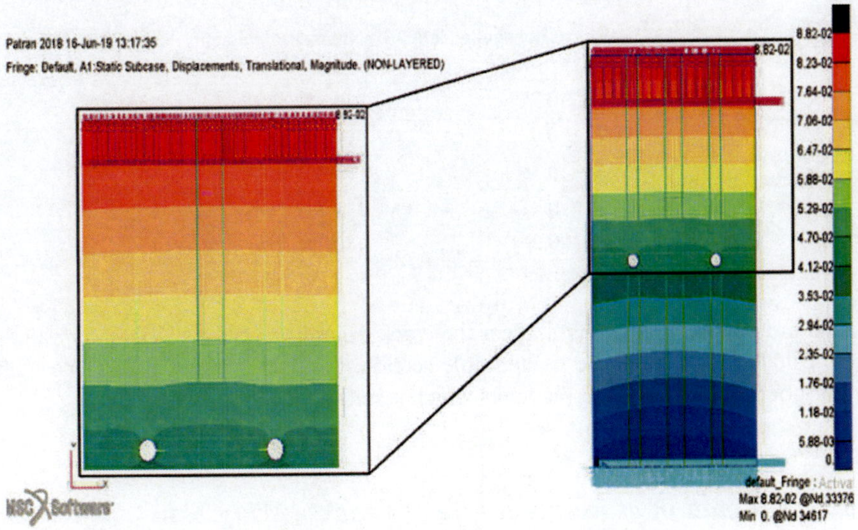

Fig. 7 Displacement plot of wing skin with rivet holes

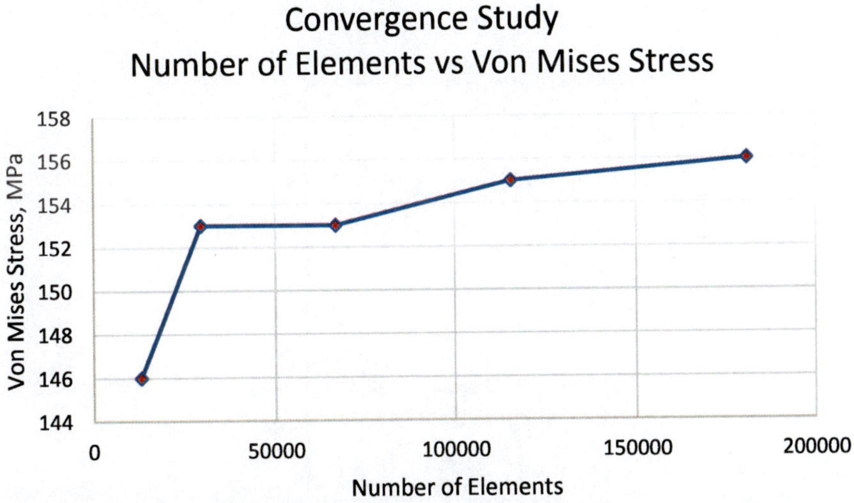

Fig. 8 Convergence criteria

3.6 *Estimation of Stress Intensity Factor for the Cracks in the Wing Skin Rivet Holes*

The progressive increment in crack length has been applied for the wing skin. By applying the same boundary conditions, the stress analysis has been carried out for

Table 2 Stress values for different mesh sizes

Global edge length	Number of elements	Von Mises stress (MPa)
0.75	12,983	146
0.5	29,408	153
0.33	66,804	153
0.25	115,799	155
0.2	180,894	156

each progressive crack length. Near the crack tip, the stress intensity factor values are calculated. Around the riveted hole section, fine meshing has been done with four-node quadrilateral 2D elements which results in accurate stress values.

3.7 *Estimation of Residual Life of Cracked Wing Skin*

After modeling the wing skin panel, crack is introduced in it as shown in Fig. 9. Stress distribution in the presence of crack is observed. Usually, maximum stress is

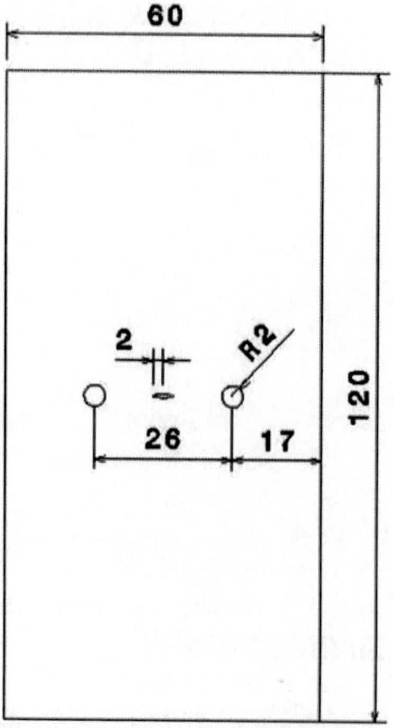

Fig. 9 Center crack in wing skin

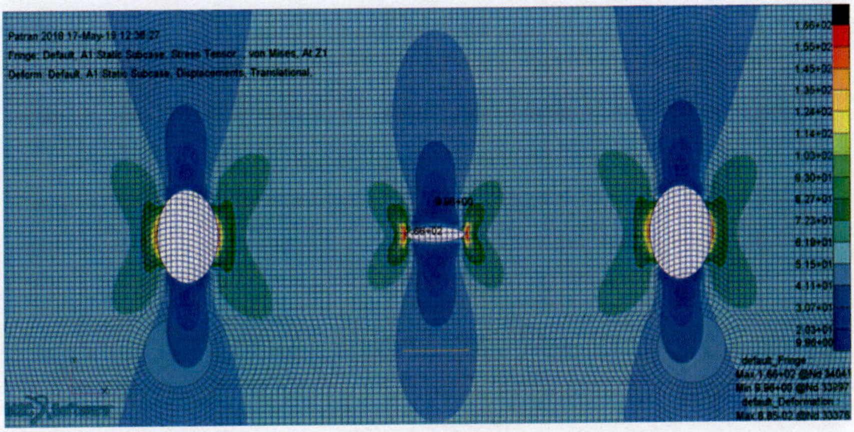

Fig. 10 Stress plot of the center crack between rivet holes of the wing skin

at the crack tip; refer Fig. 10. The procedure used to obtain incremental crack growth, stress intensity factor, and the number of cycles for crack propagation is as given below.

In order to calculate stress intensity factor for a crack, MVCCI method is used, and elemental displacements and elemental forces at the crack tip are measured. The SIF value at the crack tip is calculated by adding two elemental forces above crack tip and subtracting elemental displacement values at the crack tip.

From FE software (MSC Nastran–Patran)

For a crack length of 2 mm ($2a = 4$ mm, $a = 2$ mm)
Crack Opening Displacement (COD) = 0.001935 mm
Forces at the crack tip opening = 319.5 N
Elemental edge length at the crack tip = 0.33 mm
Thickness of wing skin panel = 1.5 mm

Then, Strain energy release rate

$$G = \frac{F \times \Delta u}{2 \times \Delta a \times t}\ \text{N/mm}$$

F = forces at the crack tip
Δu = Crack Opening Displacement (COD)
Δa = elemental edge length near the crack tip
t = thickness of the panel

$$G = \frac{319.5 \times 1.935 \times 10^{-3}}{2 \times 0.33 \times 1.5} = 0.624\,\text{N/mm}$$

Energy release rate for 2 mm crack is 0.624 N/mm.
Now, SIF is obtained for fracture mode from the below equation

$$G = \frac{k_I^2}{E}\beta$$

$G =$ energy release rate for mode 1
$k_I =$ stress intensity factor for mode 1
$E =$ modulus of elasticity
β is 1 for plane stress condition

$$0.624 = \frac{k_I^2}{71{,}000}$$

$$k_I = 6.65\,\text{MPa}\sqrt{\text{m}}$$

Stress intensity factor for 2 mm crack is 6.65 MPa$\sqrt{\text{m}}$.

A linear static stress analysis is performed for the wing skin for various crack lengths keeping the same loading condition. Crack propagates in the longitudinal direction due to loading in the transverse direction. The stress is maximum at the crack tip. The SIF is tabulated in step of 2 mm crack length up to 12 mm shown in Table 3.

Figure 11 represents the Analytical and FEA results, where X-axis represents the varying crack length and Y-axis represents the SIF, from Fig. 11. It is observed that analytical results are in good coordination with FEA results.

The fatigue life of cracked wing skin is obtained by the Paris law; this equation relates crack growth rate with stress intensity factor range. MATLAB is used to code a program for this; refer below program.

Table 3 Stress intensity factor of the wing skin

Crack length ‘a’ (mm)	Corrective function ‘Y’	Analytical SIF (MPa$\sqrt{\text{m}}$)	COD (mm)	Elemental forces at crack tip ‘F’ (N)	G (N/mm)	FEA SIF (MPa$\sqrt{\text{m}}$)	Difference (%)
2	1.079	6.65	0.001935	319.5	0.624	6.65	0
4	1.087	9.48	0.002792	449.02	1.266	9.48	0
6	1.116	11.9154	0.003496	558.32	1.972	11.83	0.7
8	1.153	14.2154	0.004181	666.87	2.816	14.14	0.5
10	1.266	17.4533	0.005004	802	4.054	16.965	2.797
12	1.701	25.6884	0.007002	1265.23	8.949	25.206	1.877

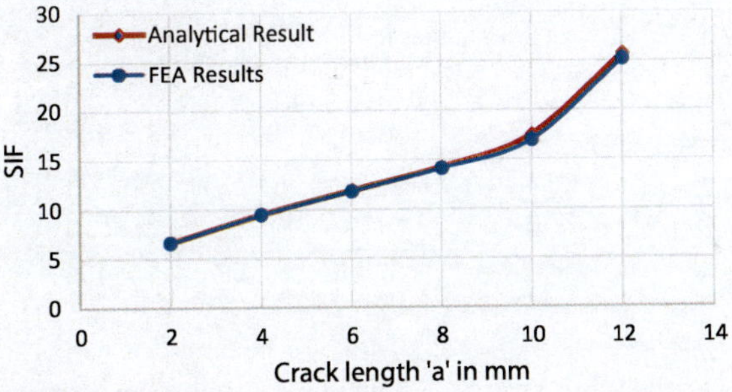

Fig. 11 Numerical and analytical SIF values comparison of cracked wing skin

MATLAB Code for Residual Life Estimation of Cracked Aircraft Wing Skin

```
%%---Wing Skin Under Constant Amplitude Tensile Fatigue Loading--- %%
clear all

c = 6E-9;
m = 4.5849;


k1c = 34.1;


R = 0.6;
Smax = 77.78;
deltaS = Smax*(1-R);

a = 2;
b = 12;
t = 1.5;
N = 0;
i = 1;

while(a<b)
y = 1.08899+0.04369*(a/b)-1.77302*(a/b)^2+9.21212*(a/b)^3-
15.8683*(a/b)^4+9.48718*(a/b)^5;
dk = deltaS*y*sqrt(3.142*0.001*a);

da = c*(dk)^m;
a = a+da;

dN = da/(c*(dk)^m);
N = N+dN;

NumberofCycle(i)= N;
CrackLength(i)= a;

i = i+1;
end

plot(NumberofCycle,CrackLength,'b')
xlabel('Number of Cycle,N')
ylabel('Fatigue Crack Length,a (mm)')
```

Wing skin panel is subjected to a constant amplitude loading of 77.78 MPa with a different R ratio, which is the ratio of minimum stress by maximum stress. Table 4 gives a brief description about the effect of stress ratio on the number of cycles for failure or residual life of the panel. As the R ratio is increased from 0 to 0.1, 0.2 etc.,

Table 4 Effect of stress ratio on residual life

Stress ratio, *R*	Number of cycles
0	3.6 × 104
0.1	5.8 × 104
0.2	9.9 × 104
0.4	3.75 × 105
0.6	2.4 × 106

the residual life of the wing skin is also increased, while the maximum stress acting on the wing skin panel is kept constant. The effect of stress ratio on residual life of wing skin is presented in the below graphs (refer Fig. 12).

Figure a. represents the fatigue crack length versus number of cycles for the failure of cracked wing skin. When $R = 0$, the minimum stress acting on wing skin is 0, so the stress amplitude is 77.78 MPa. The number of loading cycles that the wing skin can sustain in the presence of crack is 3.6 × 104 cycles. Figure b. represents the crack

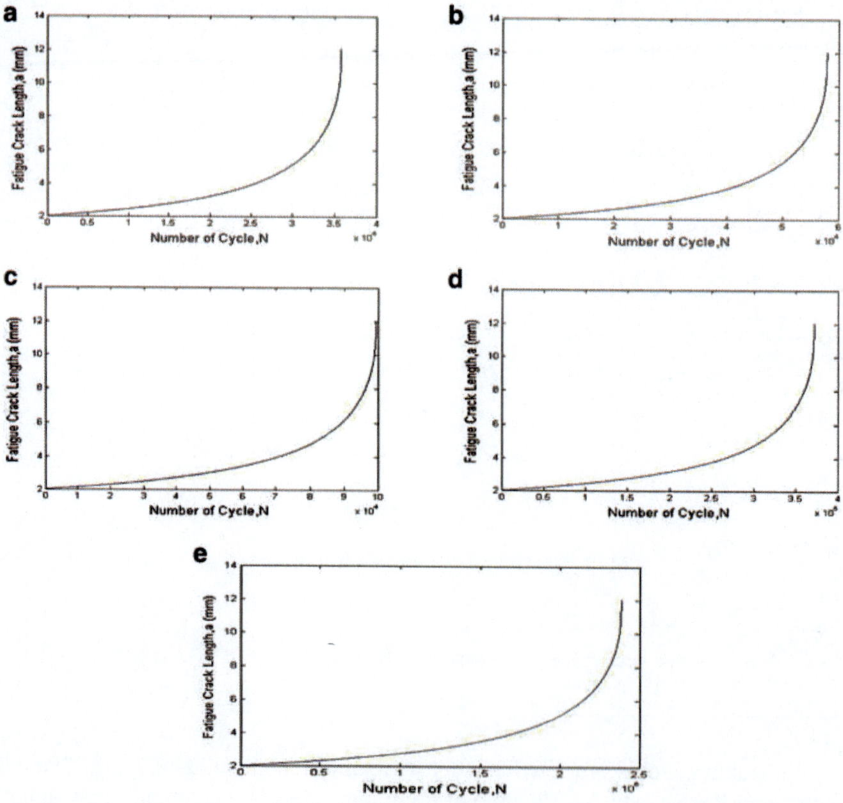

Fig. 12 a-N curve of the cracked wing skin for different stress ratios

length versus residual fatigue life. When $R = 0.1$, the minimum stress acting on wing skin is 7.778 MPa, so the stress amplitude is 35.001 MPa. The number of loading cycles that the wing skin can sustain in the presence of crack is 5.8 × 104 cycles. Figure c. represents the fatigue crack length versus residual fatigue life. When $R = 0.2$, the minimum stress acting on wing skin is 15.556 MPa, so the stress amplitude is 31.112 MPa. The number of loading cycles that the wing skin can sustain in the presence of crack is 9.9 × 104 cycles. Figure d. represents the fatigue crack length versus residual fatigue life. When $R = 0.4$, the minimum stress acting on wing skin is 31.112 MPa, so the stress amplitude is 23.334 MPa. The number of loading cycles that the wing skin can sustain in the presence of crack is 3.75 × 105 cycles. Figure e. represents the fatigue crack length versus residual fatigue life. When $R = 0.6$, the minimum stress acting on wing skin is 46.668 MPa, so the stress amplitude is 15.556 MPa. The number of loading cycles that the wing skin can sustain in the presence of crack is 2.4 × 106 cycles.

As observed from Table 4, as the stress ratio increases, there will be a proportionate increase in minimum stress acting on wing skin which in turn gradually reduces the stress amplitude, and the number of loading cycles that the wing skin can sustain in the presence of crack increases.

3.8 *Failure Mechanism of Cracked Wing Skin*

The net section means the cross-sectional area available between two rivet holes to sustain the load applied on the component. Stress concentration will be more around the rivet holes; this can be seen by the strength analysis of component, due to stress concentration, a crack will be generated at the rivet hole perpendicular to the direction of loading as shown in Fig. 13. As the aircraft flies due to varying conditions, the crack will propagate. As the crack propagates, the net section area available between two advancing crack tips simultaneously reduces and at certain crack length, if SIF at crack tip exceeds the fracture toughness value of the material, then the panel fails due to fracture. The other mode of failure is by plastic collapse; as the crack grows at its tip, stress concentration increases progressively and there will be some amount of plastic yielding around it. When the plastically yielded zones at two advancing crack tips consolidate, the catastrophic failure of the wing skin panel occurs.

After modeling the wing skin panel, strength analysis is carried on it, which results in maximum stress around the rivet holes. Then, the crack is introduced in it as shown in Fig. 13. Stress distribution in the presence of crack is observed. Now, the maximum stress is at the crack tip; refer Fig. 14. The procedure used to obtain incremental crack growth, SIF, and the number of cycles for crack propagation is as given below.

In order to calculate SIF for a crack, MVCCI method is used, and elemental displacements and elemental forces at crack tip are measured. The SIF value at the crack tip is calculated by adding two elemental forces above crack tip and subtracting elemental displacement values at the crack tip, and the net section stress is calculated

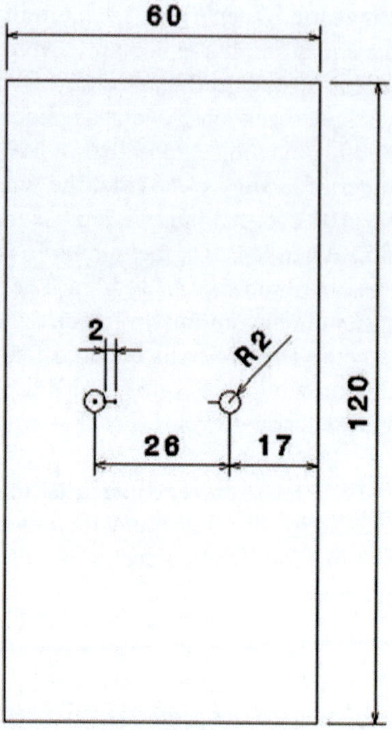

Fig. 13 Cracks around rivet holes

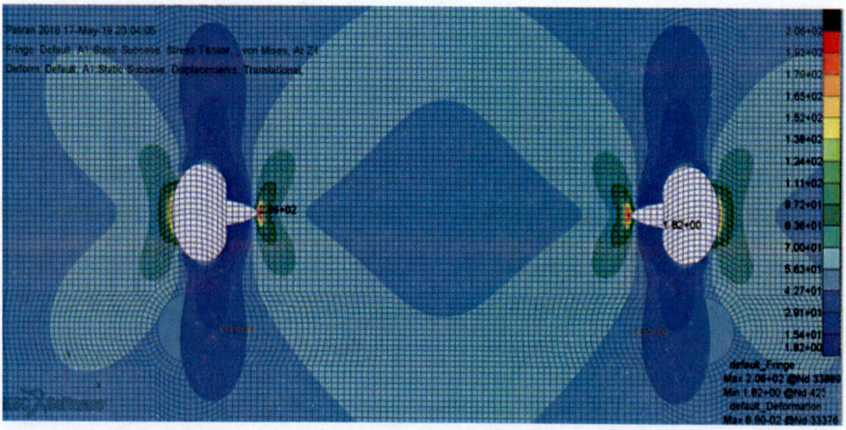

Fig. 14 Stress plot of cracks around rivet holes of wing skin

by

$$\sigma_{\text{net section}} = \frac{\text{Forces at Crack Tip}}{\text{Net Section Area}} = \frac{F_{\text{crack tip}}}{(W - 2a_{\text{eff}}) \times t}\ \text{MPa}$$

A linear static stress analysis is performed for the wing skin for various crack lengths keeping the same loading condition. Crack propagates in longitudinal direction due loading in transverse direction. The stress is maximum at crack tip. The SIF, effective crack length values are tabulated in step of 2 mm crack length till failure of wing skin panel.

From FE software (MSC Nastran–Patran)

For a crack length of 2 mm ($2a = 4$ mm, $a = 2$ mm)
Crack Opening Displacement (COD) = 0.00257 mm
Forces at the crack tip opening = 405.98 N
Elemental edge length at the crack tip = 0.33 mm
Thickness of wing skin panel = 1.5 mm

Then, Strain energy release rate

$$G = \frac{F \times \Delta u}{2 \times \Delta a \times t}\ \text{N/mm}$$

F = forces at the crack tip
Δu = Crack Opening Displacement (COD)
Δa = elemental edge length near the crack tip
t = thickness of the panel

$$G = \frac{405.98 \times 2.57 \times 10^{-3}}{2 * 0.33 * 1.5} = 1.0539\ \text{N/mm}$$

Energy release rate for 2 mm crack is 1.0539 N/mm
Now, SIF is obtained for fracture mode from the below equation

$$G = \frac{k_I^2}{E}\beta$$

G = energy release rate for mode 1
k_I = stress intensity factor for mode 1
E = modulus of elasticity
β is 1 for plane stress condition

$$1.0539 = \frac{k_I^2}{71{,}000}$$

$$k_I = 8.65\,\text{MPa}\sqrt{\text{m}}$$

Stress intensity factor for 2 mm crack is 8.65 $\text{MPa}\sqrt{\text{m}}$.
From the Irwin Plastic zone formula,

$$r_y = \frac{1}{2 \times 3.14} \times \left(\frac{K}{\sigma_y}\right)^2 = \frac{1}{2 \times 3.14} \times \left(\frac{273.546}{334}\right)^2 = 0.10681\,\text{mm}$$

Effective crack length is given by
$a_{\text{effective}} = a + 2r_y = 2 + (2 \times 0.10681) = 2.21362$ mm.
Net section stress is calculated by

$$\sigma_{\text{net section}} = \frac{\text{Forces at Crack Tip}}{\text{Net Section Area}} = \frac{F_{\text{crack tip}}}{(W - 2a_{\text{eff}}) \times t}\text{MPa}$$

$$\sigma_{\text{net section}} = \frac{328.55}{\{26 - (2 \times 2.21362)\} \times 1.5} = 10.15\,\text{N/mm}$$

A linear static stress analysis is performed for the wing skin for various crack lengths keeping the same loading condition. Crack propagates in the longitudinal direction due to loading in the transverse direction. The stress is maximum at the crack tip. The SIF and effective crack length values are tabulated in step of 2 mm crack length till the failure of wing skin panel as shown in Tables 5 and 6.

The above graph in Fig. 15 shows that stress intensity factor does not exceed the fracture toughness of material; so the material does not fracture. Similarly, the net section yielding calculations are done for the stress value between the two advancing crack tips, and these stresses are compared with the yield strength of the material. It is found that at the crack length of 12 mm, the net section stresses exceed the yield strength of material, refer Fig. 16, where it leads to yielding failure of the wing skin.

Table 5 Crack opening displacements and forces at crack tips

Crack length, *a* (mm)	Crack opening displacement (COD) (mm)	Forces at the crack tip opening (N)
2	2.57×10^{-3}	405.98
4	2.91×10^{-3}	462.84
6	3.29×10^{-3}	527.47
8	3.731×10^{-3}	604.25
10	4.346×10^{-3}	477.07
12	5.863×10^{-3}	534.03

Table 6 Net section average stress and effective crack length for plastic collapse of 26 mm pitch rivet hole

Crack length, a (mm)	K(FEA) (MPa$\sqrt{m}$)	Net section avg. stress (MPa)	Yield strength (MPa)	The Irwin formula for plastic zone length, r_y (mm)	Effective crack length, a_{eff} (mm)
2	8.65	10.15	334	0.10681	2.21362
4	9.8282	10.21062	334	0.13788	4.275755
6	11.156	15.3581	334	0.17765	6.3553
8	12.71547	26.059	334	0.2308	8.461575
10	12.194	47.144891	334	0.21225	10.4245
12	14.9849	376.48577	334	0.32052	12.64104

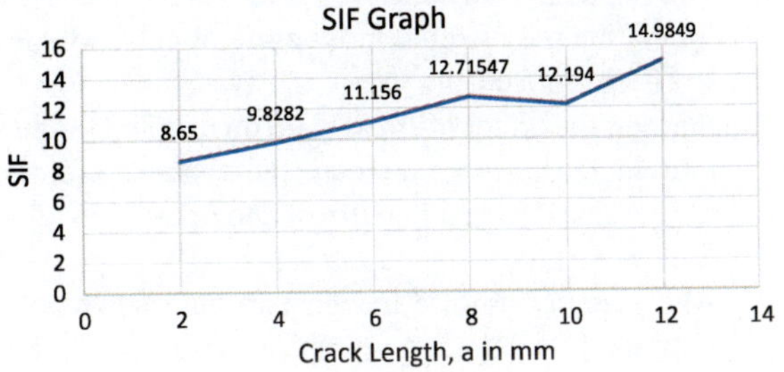

Fig. 15 SIF versus crack length for 26 mm pitch rivet hole

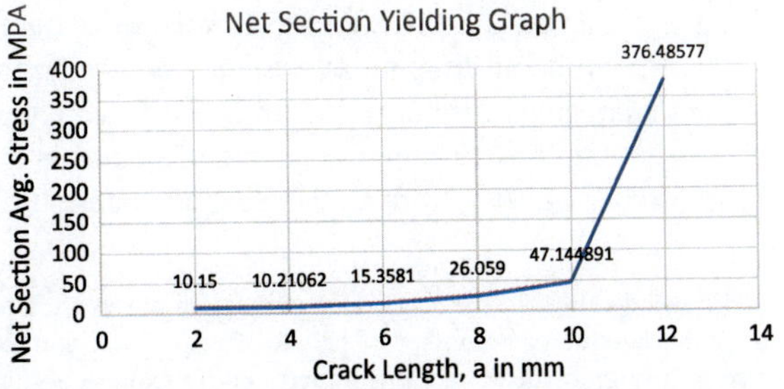

Fig. 16 Net section yielding for 26 mm pitch rivet hole

4 Conclusions

A wing skin with a center crack between the rivet holes for a pitch of 26 mm was analyzed using MSC NASTRAN/PATRAN for different crack lengths using MVCCI (Modified Virtual Crack Closure Integral) method by which strain energy release rate as well as stress intensity factors are calculated for different crack lengths, and fatigue crack growth life for progressive cracks for different R ratios is computed using a MATLAB program Also, the work investigates the first failure mechanism out of two competing mechanisms of failure; Failure due to fracture or Failure due to plastic collapse at the net section between two advancing crack tips of the rivet holes of wing skin, and the following conclusions can be drawn as stated below.

1. A linear static stress analysis is performed on wing skin with two riveted holes for tensile loading condition. The maximum stress of magnitude 153 MPa is found at the root of riveted holes.
2. The stress intensity factor is calculated for center cracked wing skin with rivet holes. Initially, the center crack of 2 mm is assumed and is emanating toward the rivet holes in step of 2 mm up to crack length of 12 mm, perpendicular to tensile loading condition of magnitude 77.78 MPa. It is found that SIF at 2 mm crack is 6.65 $\text{MPa}\sqrt{\text{m}}$, and the maximum value of SIF is 25.6884 $\text{MPa}\sqrt{\text{m}}$ at a crack length of 12 mm.
3. SIF values obtained by FEM are compared with analytical values. Good agreement between FEM and analytical result is obtained.
4. The residual fatigue life of wing skin is calculated using the Paris law. It was found that the residual fatigue life of cracked wing skin was estimated to be 3.6 $\times 10^4$ when $R = 0$.
5. Further, the effect of stress ratio on residual fatigue life was studied. As the stress ratio increases (which means there is an increment in minimum stress), there is an increment in residual life of center cracked wing skin. At $R = 0.6$, the residual life of cracked wing skin is 2.4×10^6.
6. Also, the first failure mechanism out of two competing mechanisms of failure is studied. In this case, initially, a crack of length 2 mm is taken at the root of rivet holes, and these cracks are emanating toward the center of the wing skin. Here, SIF values are compared with the fracture toughness of the material, and it is observed that SIF does not exceed the fracture toughness of material; so there will be no fracture of the material. Then, the average net section stresses are compared with the yield strength of the material, and it is found that at the crack length of 12 mm, the net section stresses exceed the yield strength of the material, where it leads to the plastic collapse of the wing skin.

Acknowledgements The successful completion of this work was made possible through the valuable contribution of a number of people. To say thank you to all of them is not even enough to express my gratitude. I take this opportunity to express my profound gratitude and deep regards to Mr. M. Mohan Kumar, Principal Scientist, for his exemplary guidance, patience, motivation, constant encouragement, and immense knowledge throughout the course of this work. I would also like to thank, Dr. Ramesh S Sharma, Associate Professor, for his wholehearted support and suggestions throughout this work and his immense help in the preparation of this work. I am thankful to Mr. Jitendra J Jadav, Director, CSIR-NAL, Bengaluru for providing me an opportunity to carry out this work in the laboratory. My sincere thanks to Dr. M. Krishna, Professor and Head, Department of Mechanical Engineering, for his support and encouragement. I express sincere gratitude to our beloved Principal Dr. K. N. Subramanya for his appreciation toward this work. Lastly, I take this opportunity to thank my parents, family members, and friends who provided all the backup support throughout the work.

References

1. TEC Eurolab. https://www.tec-eurolab.com/eu-en/company-profile.aspx
2. Venkatesha BK, Prashanth KP, Deepak Kumar T (2014) Investigation of fatigue crack growth rate in fuselage of large transport aircraft using FEA approach. Glob J Res Eng 14
3. Vashisht KV (2016) Damage tolerance evaluation of an aircraft skin structure by MVCCI technique (2d analysis) using MSC–Nastran/Patran. Int J Res Eng Technol 5:510–516
4. Rybicki EF, Kanninen MF (1977) A finite element calculation of stress intensity factors by a modified crack closure integral. Eng Fract Mech 9(4):931–938
5. Elisa P (2011) Virtual crack closure technique and finite element method for predicting the delamination growth initiation in composite structures. In: Advances in composite materials-analysis of natural and man-made materials. IntechOpen
6. Krueger R (2004) Virtual crack closure technique: history, approach, and applications. Appl Mech Rev 57(2):109–143
7. Dattaguru B (2013) Effect of non-linear behavior of joints on the damage tolerance analysis in aerospace structures. Proc Indian Natl Sci Acad (Spec Iss Part A) 79(4):553–562
8. Silva LF, Gonçalves JPM, Oliveira FMF, De Castro PMST (2000) Multiple-site damage in riveted lap-joints: experimental simulation and finite element prediction. Int J Fatigue 22(4):319–338
9. Partl O, Schijve J (1993) Multiple-site damage in 2024-T3 alloy sheet. Int J Fatigue 15(4):293–299
10. Schijve J (1995) Multiple-site damage in aircraft fuselage structures. Fatigue Fract Eng Mater Struct 18(3):329–344
11. Molent L, Jones R (1993) Crack growth and repair of multi-site damage of fuselage lap joints. Eng Fract Mech 44(4):627–637
12. Brown AM, Straznicky PV (2009) Simulating fretting contact in single lap splices. Int J Fatigue 31(2):375–384
13. Furukawa CH, Bucalem ML, Mazella IJG (2009) On the finite element modeling of fatigue crack growth in pressurized cylindrical shells. Int J Fatigue 31(4):629–635
14. Naveen Kumar HS, Suresh BS, Girish KE (2012) A study of net section failure between two equal cracks in an infinite plate. Int J Mod Eng Res 2655–2661
15. Maksimovic MS, Vasovic IV, Maksimovic KS, Trisovic N, Maksimovic SM (2018) Residual life estimation of cracked aircraft structural components. FME Trans 46(1):124–128
16. Khelil F, Aour B, Belhouari M, Benseddiq N (2013) Modeling of fatigue crack propagation in aluminum alloys using an energy based approach. Eng Technol Appl Sci Res 3(4):488–496
17. Ozturk F, Correia JAFO, Rebelo C, De Jesus AMP, da Silva LS (2016) Fatigue assessment of steel half-pipes bolted connections using local approaches. Procedia Struct Integr 1:118–125

Performance Comparison of Nitrided and Cryogenically Treated H13 Steel in Rotating Bending Fatigue

Tarang Shinde, Indrajit Patil, Omkar Walke, Gauri Wagh, and Pranali Pawar

Abstract The cryogenic treatments improve fatigue life of tool and die steels. This work involves heating the H13 steel specimens to 1020 °C, quenching in oil thereafter double tempering at 520 °C for 2 h. Thereafter these specimens are subjected to minus 185 °C in a cryobath for 16 h and 1 h soft tempering at 100 °C. The nitriding was performed at 550 °C in the atmosphere of atomic nitrogen for 2 h to get the effective case depth of 200 μm. These specimens were subjected to rotating bending fatigue at constant amplitude with room temperature conditions at a speed of 3000 rpm. The H13 specimens were characterised for hardness, surface roughness and analysis of fractured specimens to investigate the improved fatigue life for cryogenically treated specimens. It was established that improved hardness, reduced surface roughness and moderate distribution of alloy carbides were accountable for the increased fatigue life for 16 h cryo-treated steel.

Keywords Fatigue · Nitriding · Cryogenic treatment · Hardness · Carbides

1 Introduction

The fatigue is the major failure mechanism in hot forging die applications. AISI H13 hot work die steel is a widely suggested material for forging and extrusion processes which possesses high hardenability and tough structure on account of alloying constituents like Cr, Mo, V. The die steels are incessantly subjected to fatigue loading conditions specifically during hot working operations leading to the reduction in die life, demanding regular changes and adding to the cost of operation. Surface modification methods are used to make the surface harder since fatigue cracks initiate from the surface, mostly. Some authors have elaborated upon the influence of carbide

T. Shinde (✉)
Assistant Professor, Department of Mechanical Engineering, Vishwakarma University, Pune, India
e-mail: tarangshinde@gmail.com

I. Patil · O. Walke · G. Wagh · P. Pawar
Graduate Research Scholars, Sinhgad Academy of Engineering, Pune, India
e-mail: indrajitpatil2395@gmail.com

S. Seetharamu et al. (eds.), *Fatigue, Durability, and Fracture Mechanics*, Lecture Notes in Mechanical Engineering, https://doi.org/10.1007/978-981-15-4779-9_34

precipitation on the fatigue life of dies. Fukaura et al. [1] have discussed the fatigue crack initiation considering the morphology of carbides in cold work die steel. The authors explained about the refinement of primary carbides results in improvement of fatigue limit. It was also established that when the stress values are larger, the fractures initiate from the surface. In case of lower stress magnitudes, the cracks may initiate from the interior of the surface and propagate towards the surface. Ebara et al. [2] focussed upon the fatigue crack initiation and propagation of hot forging and cold forging die steels in different fatigue regimes such as low cycle, high cycle and giga cycle. It was proposed that the surface with the least value of surface roughness had indicated higher fatigue strength specifically in low cycle fatigue region. The vacuum heat treatment was useful for the improvement in fatigue strength and it was ascribed to smaller sized carbides. The fatigue cracks begin at the surface but the smaller carbides delay the crack initiation and it could originate at the subsurface area rather than at the surface. Mellolii et al. [3] had reported that the presence of chromium in H13 assisted in the improvement in oxidation resistance at room temperature as well as for fatigue loading conditions. The improved hardness readings were responsible for the decrease in crack growth rate on account of carbide precipitation as well as secondary hardening post-tempering. Baldissera et al. [4] had talked about the deep cryogenic treatment affecting the mechanical properties of different die steels. There had been improvement in the mechanical properties on account of complete transformation of retained austenite into martensite and fine dispersion of carbides as per authors' findings. The deep cryogenic treatment helped in increasing the carbide fraction which resulted in increased hardness. The distribution of fine carbides caused delayed crack nucleation and even some retained fraction of ductile austenite will arrest the crack in propagation phase. The authors have also reported that the fatigue limit shows improvement in high cycle fatigue regime on account of delay in crack propagation. Koneshlou et al. [5] had carried out deep cryogenic treatment on AISI H13 wherein, it was reported that, there was an improvement in mechanical properties of the steel on account of conversion of retained austenite into martensite as well as by precipitation of fine tertiary carbides which resulted in the improvement in wear characteristics. Based upon the discussion over the fatigue failure, it could be said that the carbide precipitation could play a significant role in increasing fatigue life of die steel for which deep cryogenic treatment may be considered as an add-on heat treatment for AISI H13 thereby, required to be elaborated in detail.

The present project work is aimed to compare the performance AISI H13 steel under various conditions like conventional treatment, cryogenic treatment and nitriding with respect to their response to rotating bending fatigue testing. The failure analysis of H13 specimen has been carried out to establish the best possible treatment for H13 steel when it is subjected to fatigue loading conditions.

2 Material and Heat Treatment

AISI H13 material was purchased to manufacture the fatigue specimens with the dimensions according to DIN50113 standards (Chemical composition of AISI H13: 0.39C–0.80Mn–5.42Cr–1.4Mo, –1.0 V percent by weight). One group of specimens were subjected to heating at 1020 °C in a stepwise way (heating to 500 °C at a rate of 7 °C/min, holding for 20 min and heating to 860 °C, holding for 20 min) to avoid development of thermal stresses. These specimens were soaked for 20 min at 1020 °C which was followed by oil quenching. The quenched specimens were double tempered at 520 °C for 2 h [6]. This was termed as conventional heat treatment for AISI H13 steel and the specimens were named as 'HT' specimens.

The other group of specimens was subjected to conventional heat treatment (as explained earlier) and cryogenic treatment at 185 °C for 16 h in a cryo-chamber supplied with liquid nitrogen (the cooling rate was 3 °C/min) [7]. After 16 h, these specimens were shifted to an insulated container till they acquired normal temperature after which these specimens were soft tempered for 1 h at 100 °C to relieve the cold stresses produced during the cryogenic treatment (Fig. 1). Sixteen hours of cryogenically treated specimens were identified as HTC16 specimens.

The third group of specimens was subjected to conventional treatment followed by pressurised gas nitriding process under the pressure of 5 atm at a temperature of 550 °C for 5 h. A calculated amount of ammonia gas was passed continuously

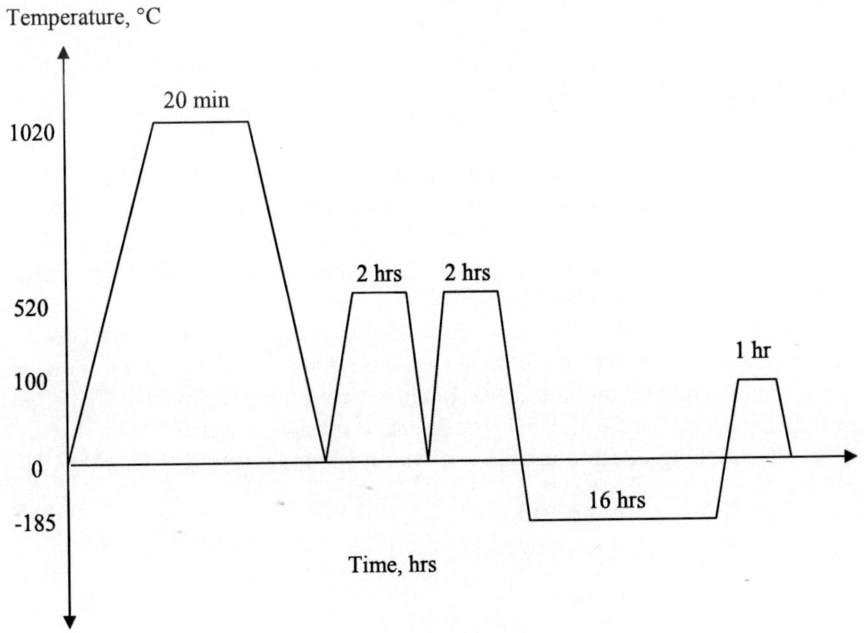

Fig. 1 Conventional treatment and cryogenic treatment for AISI H13

through the nitriding furnace and the conditions were maintained so as to achieve the case depth in the range of 140–160 μm and surface hardness in the range of 50–70 HRC. These specimens were specified as HT-N specimens.

2.1 *Characterization*

A Rockwell hardness measurement was carried for H13 specimens (C scale). The end faces of the specimen were polished on 1/0 to make them perfectly flat. First minor load was applied, followed by the application of major load of 150 kg for 12 s, and resistance offered by the material to indentation was recorded on the dial gauge. Five readings were captured to mark the average hardness for each specimen.

The fatigue limit is affected by the surface roughness as the fatigue failure starts at the superficial areas. The surface roughness of HT and HTC16 specimens was recorded by a contact type roughness tester (make: Mitutoyo: model: SJ-410) according to DIN EN ISO 4287:1998 with a total path length of the probe of 10 mm. Three readings were reported to register the average surface roughness.

The fractographic analysis of the specimens failed during the fatigue test was done by scanning electron microscope (make: ZEISS). The broken parts of the specimen were cleaned in an ultrasonic cleaner without disturbing the fractured surface to remove the debris and subsequently these specimens were subjected to SEM analysis to detect the crack initiation.

2.2 *Fatigue Test*

The HT and HTC16 specimens were tested on a rotating bending fatigue testing machine, (Make: Magnum, motor: 3000 rpm, extreme load: 50 kg) at 3000 rpm with a variation of stress starting from 946 MPa (corresponding to the load of 10 kg). The starting stress (and load) was selected at 75% of static tensile strength of the material under test [8] with a load ratio of -1 ($R = -1$). Three specimens were tested at each stress level to mark the number of cycles to failure. If no failure occurred after 10^7 cycles the test was terminated. If three specimens finished 10^7 cycles for the given stress level without failure, the corresponding stress value was noted as the fatigue or endurance limit for the material. The fractured specimens were observed in detail under the scanning electron microscope with a voltage range of 0–20 eV (Make: ZEISS).

3 Results and Discussion

3.1 Hardness and Surface Roughness Measurements

The conventional treatment was followed by cryogenic treatment. As the material is subjected to the cryogenic temperature by placing it in cryo-chamber, the residual stresses are induced in the material on account of the thermal shock and the phase change from austenite to martensite. At lower temperatures, the carbon diffusion rates are very insignificant. The alloying elements are separated but they do not combine with any other due to least degree of disorderness. The specimens were taken out from the cryo-chamber at the end of 16 h of cryogenic treatment and shifted to an insulated box till they reached room temperature conditions and thereafter these specimens had undergone soft tempering at 100 °C (Fig. 1). The soft tempering increases the thermal energy of the material causing to initiate the movement of carbon atoms. The carbon atoms combine with the alloying constituents (Cr, Mo, V) to produce their respective fine metal carbides. The driving force for this process is the residual stress which was generated during the cryogenic treatment. This is how the cryogenic treatment enhances the precipitation of fine alloy carbides in the tempered martensitic matrix [9].

Figure 2 indicates the plot for hardness measurements with respect to three treatments, namely, conventional treatment, cryogenic treatment and nitriding. It could be seen that the hardness values for HTC16 specimens are maximum amongst the three which are due to transformation of retained austenite into martensitic structure and precipitation of fine alloy carbides. It is a well-known fact that the retained austenite is accountable for the drop-in hardness in case of most of the tool steels and it is essential to eradicate the same. The cryo-treatment subjects the material to a temperature below martensite formation (Mf) temperature which leads to this

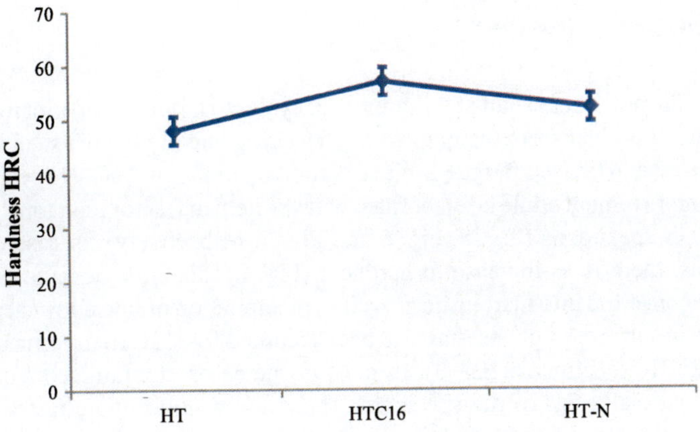

Fig. 2 Hardness values for HT, HTC16 and HT-N specimens

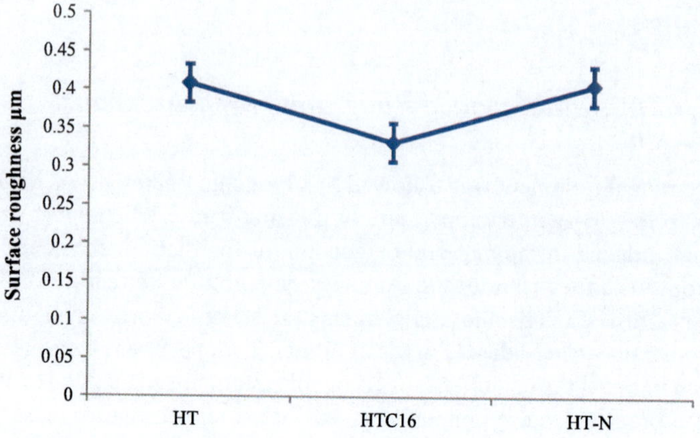

Fig. 3 Surface roughness plot for different specimens indicating minima for HTC16

retained austenite transforming into the martensite [10]. It benefits in improvement in the hardness of the part. The nitriding treatment is useful to improve the surface hardness with controlled parameters during the process.

The variation in surface roughness is clearly visible from Fig. 3. The 16 h cryogenically specimens (HTC16) indicate drop in surface roughness values as compared to conventional treatment (HT specimens) and nitride specimens (HT-N). This could be attributed to the fine dispersion of alloy carbides in the tempered martensitic matrix could have occupied the narrow valley regimes present on the surface resulting in the decrease in roughness values [11]. The fine carbide particles can accommodate themselves in these valleys thereby closing the micro-pits and improving the surface finish.

3.2 *Fatigue Behaviour*

Figure 4 indicates stress against the number of cycles to failure (S-N) plot for different specimens. It is clear from the graph that 16 h cryogenically treated steel specimens (HTC16) have increased fatigue limit as compared to HT and HT-N specimens.

This improvement could be elaborated with the help of fractograph for HT, HTC16 and HTC-N specimens (Fig. 5, Fig. 6 and Fig. 7, respectively). In case of HTC16 specimens, there is as increase in hardness (Fig. 2) this is on account of retained austenite converted into martensite as well as precipitation of fine alloy carbides. The reduction in surface roughness has also been recorded for HTC16 in comparison with HT and HT-N specimens (Fig. 3). In most of the cases, the fatigue cracks initiate at the superficial areas of the specimen. The crack would initiate faster if it finds favourable situations like low hardness or rough surface [12, 13]. In case of HT and

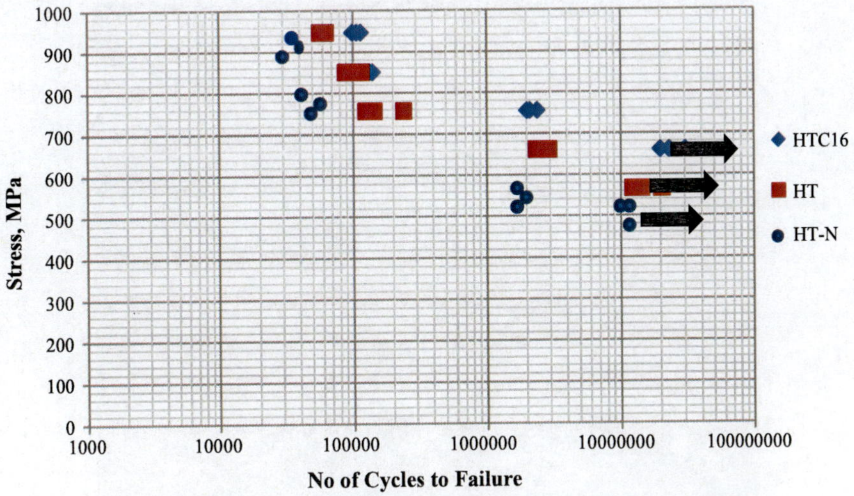

Fig. 4 S-N curve for different specimens indicating increased fatigue limit for HTC16 specimens

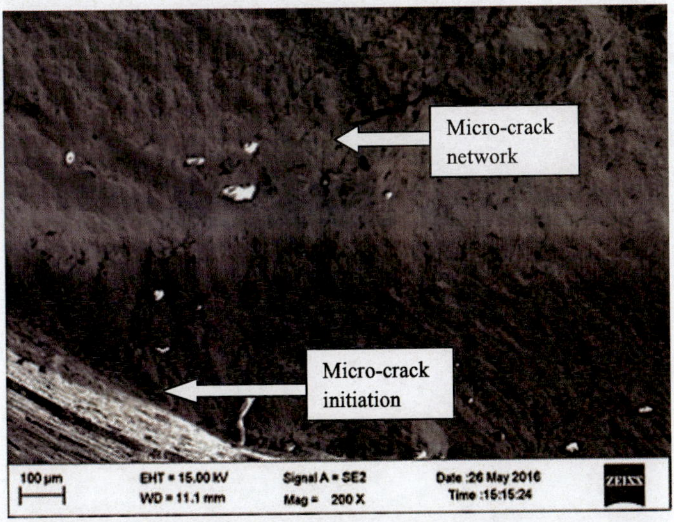

Fig. 5 Fractograph of conventionally treated (HT) specimen with microcrack initiated

HT-N specimens, the hardness values are low so as higher surface roughness. A rough surface would be a potential site for crack initiation.

The HT specimens indicate microcrack network developed just below the surface (Fig. 5) which could be due to a comparatively softer surface as well as irregular topography.

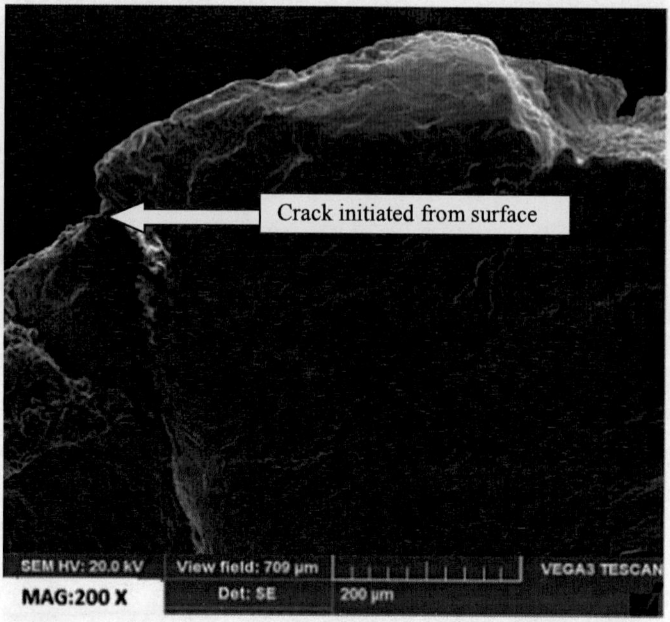

Fig. 6 Fractograph of nitrided (HT-N) H13 steel indicating crack initiation region

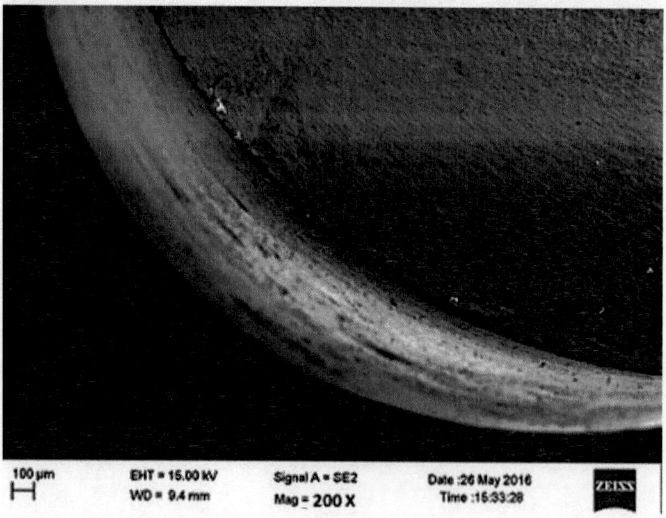

Fig. 7 Fractograph of HTC16 specimen indicating no surface crack initiation

In case of HT-N specimens, the fatigue crack could have started at the specimen surface on account of high roughness and low hardness or even at the junction between the nitrided layer and steel matrix. This crack could have increased in length during the running of specimen ultimately leading to failure (Fig. 6).

The fine alloy carbide precipitation in the tempered martensitic structure has prevented the early crack initiation in case of HTC16 specimens (Fig. 7). These fine carbides assist ion holding the matrix firmly and prevent the development of shear stresses necessary to initiate the failure. In addition to that, the fine carbide particles also act as crack arresters thereby delaying the propagation of crack once it is initiated. This is how it could be justified that the cryogenic treatment is helpful in not only delaying the crack initiation but also its propagation and subsequently prolonging the failure of the specimen [14].

4 Conclusion

The following conclusion points are proposed at the end of fatigue analysis of HT, HTC16 and HT-N specimens.

1. The cryogenic treatment improves the precipitation of fine alloy carbides as well as transformation of retained austenite into martensitic structure thereby improving the hardness, decreases surface roughness of H13 steel.
2. The fractographic analysis indicates that the crack propagation is hindered on account of fine alloy carbides which delays the failure of the part and subsequently improves the fatigue limit of H13 steel specimens with 16 h cryogenic treatment.
3. The specimens for conventional treatment and nitriding indicate the lower values for fatigue which could be attributed to shear-induced failure in case of conventional treatment and the interface between the nitride layer and the steel matrix might be a possible site for initiation of crack.

References

1. Fukaura K, Yokoyama Y, Yokoi D, Tsujii N, Ono K (2004) Fatigue of cold-work tool steels: effect of heat treatment and carbide morphology on fatigue crack formation, limit and fracture surface observations. Metall Mater Trans A 35A:1289–1300
2. Ebara R (2010) Fatigue crack initiation and propagation behaviour of forging die steels. Int J Fatigue 32:830–840
3. Mellouli D, Haddara N, Köster A, Ayedi HF (2014) Hardness effect on thermal fatigue damage of hot working tool steels. Eng Fail Anal 45:85–95
4. Baldissera P, Delprete C (2008) Deep cryogenic treatment: a bibliographic review. Open Mech Eng J 2:1–11
5. Koneshlou M, Asl KM, Komamizadeh F (2011) Effect of cryogenic treatment on microstructure, mechanical and wear behavior of AISI H13 hot work tool steel. Cryogenics 51:55–61
6. Mesquita RA, Consult RAF, Arbor A (2017) Heat treatment of tool steels, vol 2. Elsevier Ltd

7. Shinde T, Dhokey NB (2017) Influence of carbide density on surface roughness and quasi-stable wear behaviour of H13 die steel. Surf Eng 1–9
8. (2008) Descriptions of specimens and test procedures. ASM Int 5–8
9. Dhokey NB, Hake A, Kadu S, Bhoskar I, Dey GK (2013) Influence of cryoprocessing on mechanism of carbide development in cobalt-bearing high speed steel. Metal Mater Trans A ASM Int
10. Dhokey NB, Dandwate J, Gangurde H, Harle A (2012) Metallurgical investigation of cryogenically cracked M35 tool steel. Eng Fail Anal 21:52–58
11. Dhobe MM, Chopde IK, Gogte CL (2014) Optimization of wire electro discharge machining parameters for improving surface finish of cryo-treated tool steel using DOE. Mater Manuf Process 29(11–12):1381–1386
12. Suhr RW (1986) The effect of surface finish on high cycle fatigue of low carbon steel. Mech. Eng. Pub, London, pp 69–86
13. Ryu JH, Nam SW (1989) Effect of surface roughness on low cycle fatigue limit of Cr-Mo-V steel at 550 °C. Int J Fatigue 11(6):433–436
14. Shinde T, Dhokey NB (2017) Influence of tertiary carbides on improving fatigue limit of H13 die steels. J Metal Micro Ana Springer 6–5:398–406

Effect of Loading Angle on 3D Stress Intensity Factor and T-stress in a Compact Tension Shear (CTS) Fracture Specimen

C. M. Sharanaprabhu, Shashidhar K. Kudari, and Mujebur Rehaman

Abstract In elastic fracture analysis, stress intensity factors (K) and T-stress (T) are the two-important stress field characterizing parameters ahead at the crack-tip/front. In this paper, 3D finite element analyses were carried on Compact Tension Shear (CTS) mixed mode (I/II) fracture specimen with various thickness to width (B/W) ratios and loading angle (β) to evaluate the effect of loading angle on K and elastic T. For assort B/W ratios and β of the CTS mixed mode (I/II) specimen, the effective stress intensity factors (K_{eff}) and T results predominantly vary from center to surface along the crack-front. The K_{eff} and T results obtained in the present analyses were used to characterize the constraint effects in CTS mixed mode (I/II) fracture specimen.

Keywords Stress intensity factor · Elastic T-stress · Finite element method · Constraint issue

1 Introduction

The presence of mixed mode (I/II) fractures occur in many service structures such as pressure vessels, offshore structures, and aircraft components. These service structures for safe operation require a detailed understanding of stresses around the crack-tip. Earlier, the analytical or the numerical studies for the three-dimensional state of stresses near the crack-front problems were solved through either 2D or axisymmetric problems. Larsson and Carlsson [1] considered the second term of Williams [2] series

C. M. Sharanaprabhu (✉) · M. Rehaman
Department of Mechanical Engineering, Research Center, PES Institute of Technology & Management, Shivamogga, Karnataka, India
e-mail: cmsharanaprabhu@rediffmail.com

M. Rehaman
e-mail: mujeeb.312@gmail.com

S. K. Kudari (✉)
Department of Mechanical Engineering, Research Center, CVR College of Engineering, Hyderabad, India
e-mail: s.kudari@rediffmail.com

S. Seetharamu et al. (eds.), *Fatigue, Durability, and Fracture Mechanics*, Lecture Notes in Mechanical Engineering, https://doi.org/10.1007/978-981-15-4779-9_35

or T-stress for the analyzes of stresses and plastic zone around the crack-tip. Their results analyzed that the plastic zone size and shape around the crack-tip will depend on Williams [2] second term or T-stress. Later, Rice [3] and Bilby et al. [4] worked on the influence of crack-tip constraint depends on the sign and magnitude of T-stress. They confirmed that the high crack-tip constraint occurs due to the positive T-stress and the loss of crack-tip constraint takes place due to negative T-stress. For 2D or axisymmetric problems, researchers [5–8] have shown that an effective crack configuration under various steps of Mode-I loading for elastic or elastic-plastic crack-tip fields includes K or J-integral with the addition of the elastic T-stress. Accordingly, in fracture mechanics, the assessment process of constraint effect failure includes two parameters around the crack-tip/front for elastic and elastic-plastic materials. In addition, for mixed mode (I/II) Cotterell and Rice [9] illustrated that the path stability of slightly curved or kinked cracks will be affected by the T-stress.

Over the last two decades, the elastic T-stress was determined by different numerical methods such as a boundary layer method, the interaction I-integral method, displacement field method, and stress difference method which are found in the literature. Carlsson [1] and Leevers and Radon [10] determine T-stress along crack flanks utilizing a boundary layer method. They [1, 10] showed that the extraction of T-stress is obtained by considering the mean difference between σ_{xx} along the crack face. Cardew et al. [11] and Kfouri [12] based on a theorem of Eshelby [13], proposed interaction I-integral method to find T-stress for the 2D infinite plate and actual geometries around crack-tip under mode-I loading condition. For mixed mode I/II test specimen, Hallback and Jonsson [14] used an I-integral method to extract T-stress around the crack-tip. Al-Ani and Hancock [15], Ayatollahi et al. [16] used displacement field method for the extraction of T-stress for any mixed mode I/II loading.

The stress difference method at the crack-tip under mixed mode I/II loading was proposed by Yang and Ravi-Chandar [17] to calculate the elastic T-stress. They [17] examined that the extraction of the elastic T-stress by stress difference method gives accurate results, incorporating with the boundary layer method. The 3D stress field analyses near the crack-front play a vital role in the strength of materials [18]. The 2D or axisymmetric problems will give an inconvenient variation of fracture parameters such as stress intensity factors and T-stress along the crack-front (thickness). This leads to the development of three-dimensional problems involving singularities. Several researchers [19–22] in fracture mechanics analyzed the 3D crack-front fields to estimate the behavior of elastic materials under mode-I loading. Nakamura and Parks [23] extended 2D I-integral method to 3D crack problems for the extraction of T-stress. Kong et al. [24] have revealed that the fracture behaviors of elastic materials depend on the triaxial stress a head at the crack-tip/front under mixed mode I/II loading condition. Schutte and Abbasi [25] have addressed the complete set of the T-stress terms for 3D mixed mode internal cracks. Liu et al. [26] have carried out 3D finite element analyses on compact tension shear (CTS) specimen to study the stress contour. Consequently, the dependence of stress intensity factors and T-stress are important parameters to analyze stresses around the crack-front for elastic mixed mode (I/II) fracture. Therefore, the effect of loading angle on 3D stress intensity factor and T-stress in the CTS fracture specimen with various *B*/*W* ratios is presented in this paper to characterize the materials under elastic mixed mode I/II step loads using finite element analyses.

2 Finite Element Analysis

The CTS specimen size for various *B*/*W* ratios with $W = 90$ mm shown in Fig. 1 is used in the finite element analyses (FEA). The force or load applied at the six holes and boundary conditions utilized throughout the FEA for CTS mixed mode I/II fracture specimen were taken from the work of Borrego et al. [27]. The stress intensity factors and T-stress around, the crack-tip/front were calculated for the CTS specimens loaded at 6 loading angles (β) varying from 0° (mode-I) to the 90° (mode-II) with an increment of 18°. Experimentally the CTS specimen was loaded at 6 different loading angles through the jig. But, the experimental loading condition of CTS specimen is simulated at six holes in FEA was similar to the work of Borrego et al. [27]. Figure 1 shows the application of uniaxial loads at six holes (F_1 to F_6) in FEA and is given by Eqs. (1)–(3) as below [28]:

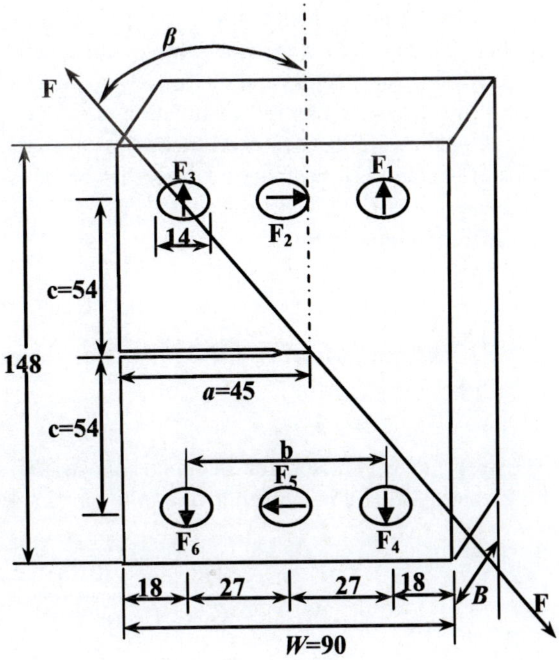

Fig. 1 CTS Specimen dimensions with $W =$ 90 mm used in the analysis. All dimensions are in mm

$$F_1 = F_6 = F\left(\frac{1}{2}\cos\beta + \frac{c}{b}\sin\beta\right) \tag{1}$$

$$F_2 = F_5 = F\sin\beta \tag{2}$$

$$F_3 = F_4 = F\left(\frac{1}{2}\cos\beta - \frac{c}{b}\sin\beta\right) \tag{3}$$

3D finite element analyses (FEA) were carried out on CTS mixed mode (I/II) specimen using FEA software tool, ABAQUS 6.5 [28]. Due to asymmetry loading, full CTS mixed mode (I/II) fracture specimen were considered for FEA. A 20-noded isoparametric 3D solid reduced integration elements (C3D20R in ABAQUS) are used to descritize the CTS mixed mode I/II fracture specimen. Investigators [29, 30] used C3D20R elements in ABAQUS [28] for extraction of stress intensity factors and T-stress. The 3D FEA with various thickness to width ($B/W = 0.3$, 0.4, and 0.5) ratios are done for CTS mixed mode (I/II) fracture specimen. The number of elements used for CTS mixed mode I/II fracture specimen of $B/W = 0.3$ are 8320 and for $B/W = 0.5$ are 9740. Figure 2 shows a finite element mesh of $B/W = 0.3$. It is observed that the number of elements will not be same for different thicknesses and it was increasing as the thickness of the CTS specimen increases so as to avoid coarse mesh, thereby incorporating mesh refinement. The mid nodes of 20-noded 3D solid elements along crack-front were shifted to the quarter-point as shown in Fig. 3 to obtain singularity for the extraction of stress intensity factors. In ABAQUS [28] the stress intensity factors and T-stress along the crack-front were extracted directly from post-processor step. The estimation of K [31] and T-stress [24] is carried out by contour integral or J-integral method as available in ABAQUS [28]. From the work of Kudari [32], the material properties of interstitial free steel (IF) with elastic modulus of 197 GPa, yield strength (σ_y) of 155 Pa, and possion's ratio of 0.3 are considered for elastic FEA.

3 2D Estimation of *K* and *T* in the Mixed Mode (I/II) Cracked Plates

The stress intensity factors for CTS mixed mode (I/II) specimen geometry were given by Richard [33] and it is given by the following expressions:

$$K_{\mathrm{I}} = \frac{F\sqrt{\pi a}\cos\beta}{WB\left(1 - \frac{a}{W}\right)}\sqrt{\frac{0.26 + 2.65\left(\frac{a}{W-a}\right)}{1 + 0.55\left(\frac{a}{W-a}\right) - 0.08\left(\frac{a}{W-a}\right)^2}} \tag{4}$$

$$K_{\mathrm{II}} = \frac{F\sqrt{\pi a}\sin\beta}{WB\left(1 - \frac{a}{W}\right)}\sqrt{\frac{-0.23 + 1.40\left(\frac{a}{W-a}\right)}{1 - 0.67\left(\frac{a}{W-a}\right) - 2.08\left(\frac{a}{W-a}\right)^2}} \tag{5}$$

Fig. 2 FE mesh of $B/W = 0.3$ with 20 node 3D solid elements used in the analysis

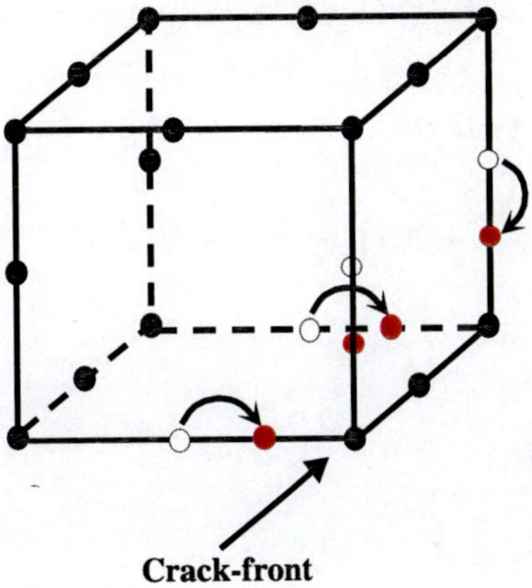

Fig. 3 Singular element used at crack-front for CTS specimen

where F is the applied force, W is the width of the specimen, B is the thickness of the specimen, a is the crack length, and β is the angle of loading direction with respect to the crack plane.

Next, the T can be found by measuring the summation of direct stresses a head at the crack-tip for symmetric properties of Mode-I and antisymmetric properties of Mode-II load, i.e., the stresses along the upper and lower cracked specimen are added to estimate the T-stress. This method provides direct measurement of T-stress from FE results for mixed mode I/II loading and is similar to the earlier Ref. [16]. Form the FE results, the direct estimation of T-stress using this method along the crack-tip is given as [16]:

$$T = \frac{1}{2}\left((\sigma_{xx})_{\theta=-\pi} + (\sigma_{xx})_{\theta=\pi}\right) \tag{6}$$

where σ_{xx} is the stress in x-direction at the crack-tip and θ is the radial position with respect to the crack axis.

4 Results and Discussion

The estimation of 3D stress intensity factors, along the crack-front of the CTS mixed mode I/II fracture specimen was carried out in various load steps, B/W and β using the ABAQUS [28] FE software. The variation of K_{II} versus K_{I} at the center and on the surface of the CTS mixed mode (I/II) fracture specimen for various B/W and β is shown in Fig. 4 typically for a load of 36 kN, which is approximately equal to 0.3 times of yield stress. The results of K_{I} and K_{II} at the center and on the surface of the CTS mixed mode (I/II) fracture specimen for various B/W and β are

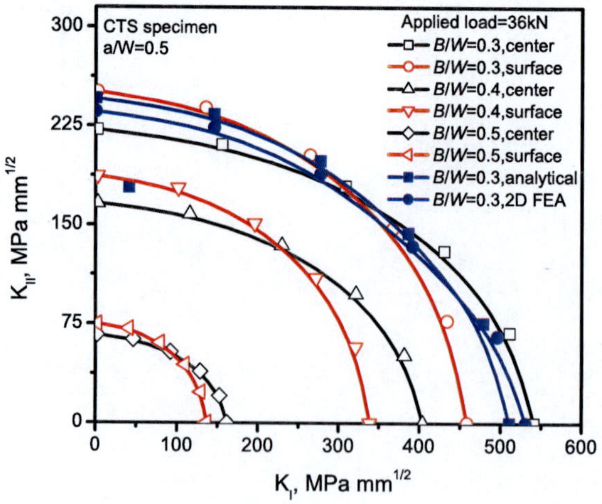

Fig. 4 3D, 2D and analytical comparison of K_{II} versus K_{I} variation for CTS specimen subjected to load of 36 kN

compared with 2D FEA of the CTS mixed mode (I/II) fracture specimen results and analytical formulations cited in the earlier investigation [33]. The extraction of 2D stress intensity factors in FE software for CTS specimen is explained in our earlier work [34, 35]. The nature of variation of K_{II} versus K_{I} at the center and on the surface along the crack-front of the CTS mixed mode I/II fracture specimen for various B/W (0.3, 0.4, 0.5) and β (0°–90°) is in good agreement with the 2D FEA and analytical results. Figure 4 clearly demonstrates that for $\beta = 0°$ (Mode-I) the stress intensity factor at the center is higher compared to the surface of the specimen. But, for $\beta = 90°$ (Mode-II) the stress intensity factor at the surface is higher as compared to the center of the specimen. Figure 4 also shows that for $0° < \beta > 90°$ (mixed mode (I/II)) the stress intensity factor at the center is maximum compared to the surface where Mode-I loading dominates and the stress intensity factor at the surface is maximum compared to the center where Mode-II loading dominates. In mixed mode (I/II) loading, the K_{I} and K_{II} can be handled by a single parameter known as the effective stress intensity factors (K_{eff}) and is given as [27]:

$$K_{\mathrm{eff}} = \sqrt{K_{\mathrm{I}}^2 + K_{\mathrm{II}}^2} \tag{7}$$

Figure 5 shows the 3D FE computed magnitudes of K_{eff} at the center and on the surface of CTS mixed mode (I/II) facture specimen for $B/W = 0.3$, 0.4, and 0.5 are plotted against loading angle (β) for an applied load of 36 kN. Figure 5 also demonstrates that the magnitude of K_{eff} decreases as β increases from 0° to 90°. It is clear from Fig. 5 that for a particular load the K_{eff} in $\beta = 0°$ (Mode-I) is almost 2.45 times more than that of $\beta = 0°$ (Mode-II). The nature of variation of K_{eff} versus β (Fig. 5) is in good agreement with 2D FEA and analytical results [33]. Figure 5 illustrates that the K_{eff} variation at the center and on the surface along the crack-front of the CTS specimen for different B/W ratios depends on the β. This leads to study

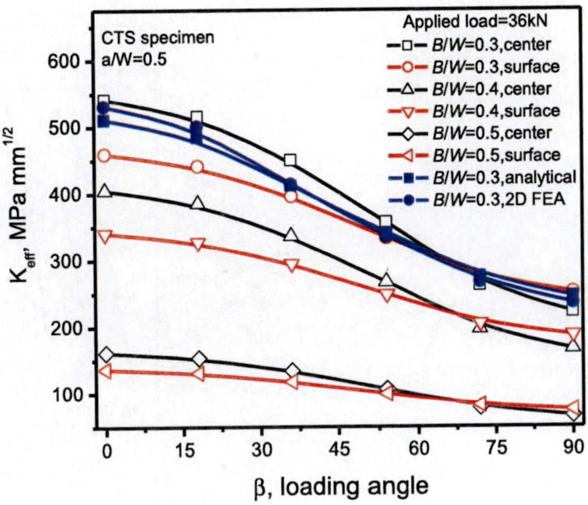

Fig. 5 Variation of K_{eff} versus β typically for 36 kN applied load

of variation of K_{eff} along the crack-front for B/W (0.3, 0.4, and 0.5) and β (0°–90°) of the specimen.

Figures 6 and 7 show the magnitude of 3D FE K_{eff} variation along the crack-front for $\beta = 0°$ and 18° where the Mode-I loads dominate and is superimposed with 2D FE K_{eff} results for a typical load. It is observed that the 3D FE K_{eff} variation is more at the center than on the surface along the crack-front of the CTS specimen and there is a difference in magnitude of K_{eff} between 3D and 2D FE results. The magnitude of K_{eff} difference between 3D and 2D is δK_{eff}. The magnitude of δK_{eff} is minimum at the center than on the surface along the crack-front indicating that the crack initiates at the center for $\beta = 0°$ and 18° in CTS specimen of various thicknesses.

For $\beta = 72°$ and 90°, where the Mode-II loads dominate, the K_{eff} is maximum at the surface than on the center of the CTS mixed mode (I/II) fracture specimen in

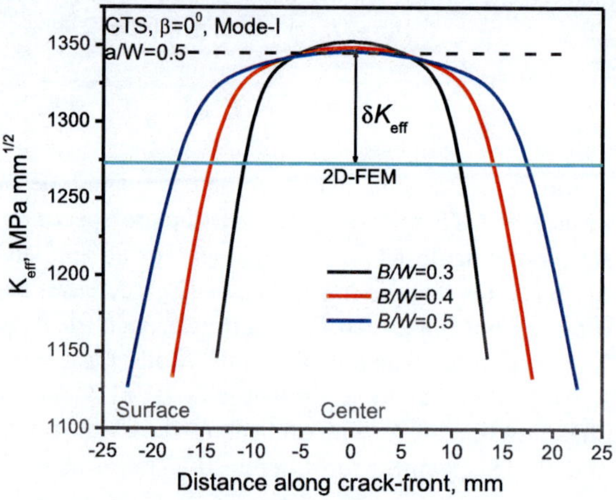

Fig. 6 Variation of K_{eff} versus distance along the crack front for $\beta = 0°$

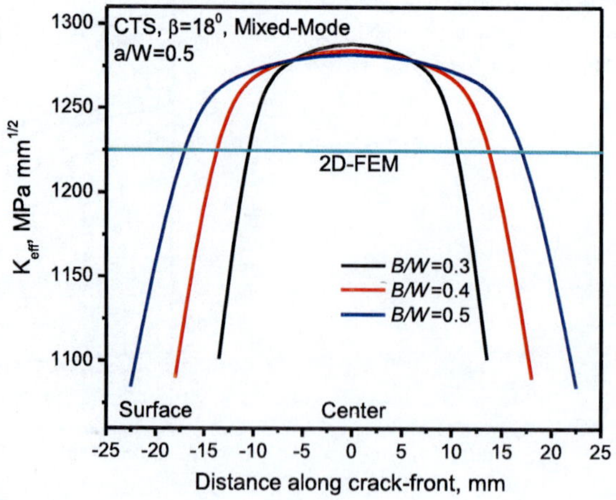

Fig. 7 Variation of K_{eff} versus distance along the crack front for $\beta = 18°$

Figs. 8 and 9. Here, the magnitude of δK_{eff} is minimum on the surface as compared to the center along the crack-front indicating that the crack initiates at the center for $\beta = 72°$ and 90° in CTS specimen of various thicknesses. For the similarly applied load as the β changes from 0° (Mode-I) to 90° (Mode-II), it is observed that δK_{eff} measurement changes from center to the surface of the specimen. This indicates that the CTS mixed mode (I/II) fracture specimen experiences a change of crack-tip constraint from the center toward the surface as the loading angle changes from 0° to 90°.

Next, the estimated 3D FE T-stress variation along the crack-front for CTS mixed mode I/II fracture specimen of various B/W and β were studied. For the validation of T-stress value obtained by 2D FEA (contour integral method) in ABAQUS [28] was

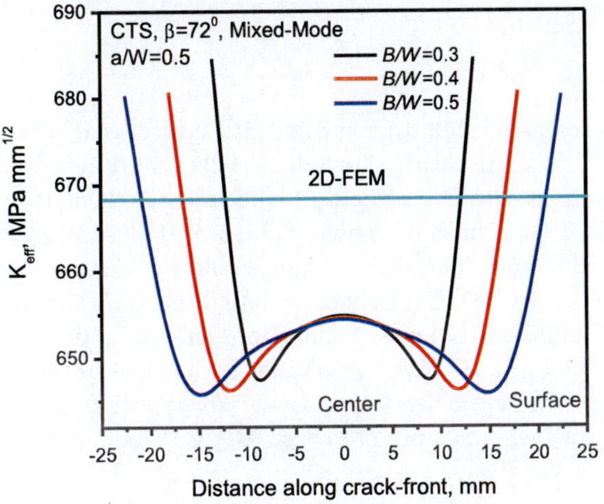

Fig. 8 Variation of K_{eff} versus distance along the crack front for $\beta = 72°$

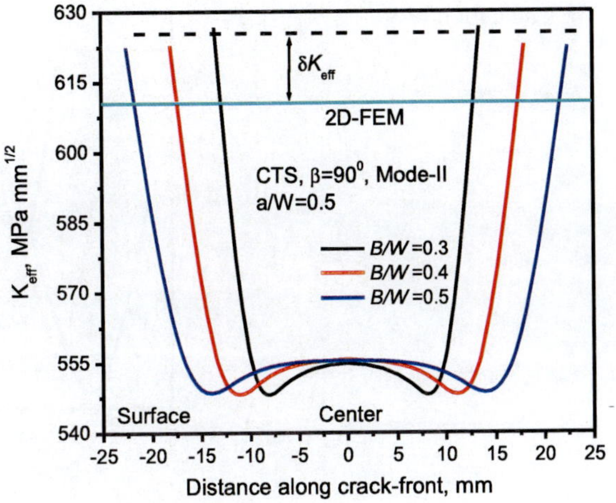

Fig. 9 Variation of K_{eff} versus distance along the crack front for $\beta = 90°$

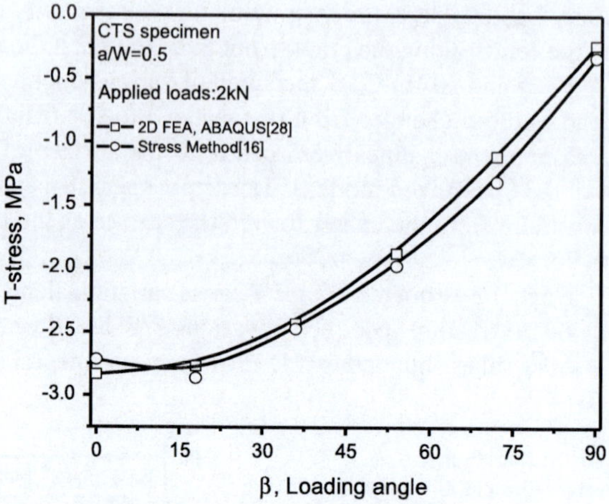

Fig. 10 Variation of T versus β typically for 2 kN applied load

compared with stress method [16] as discussed in earlier Sect. 3 as shown in Fig. 10. This figure clearly shows that the 2D FEA results directly obtained in ABAQUS [28] are in excellent agreement with stress method [16] as explained in session 3 for a CTS specimen of various β. Figure 10 illustrates that the T-stress is negative for $\beta = 0°$ and almost approaching zero for $\beta = 90°$.

Now, 3D FEA have been carried out on CTS mixed mode (I/II) fracture specimen for the extraction of T along the crack-front for B/W (0.3, 0.4, and, 0.5) and β (0°–90°). The 3D FEA T results are normalized with applied stress, $\sigma = F/BW(1 - a/W)$ of CTS specimen for various B/W and β. From Figs. 11 and 12 for $\beta = 0°$ and 18°, the magnitude of normalized T is maximum at the center compared to the surface

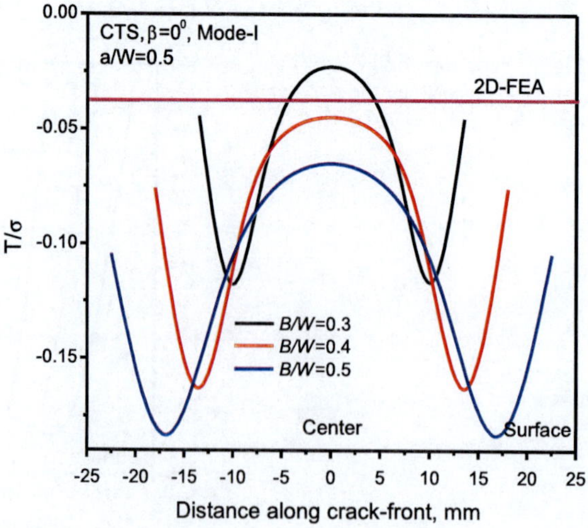

Fig. 11 Variation of T/σ versus distance along the crack front for $\beta = 0°$

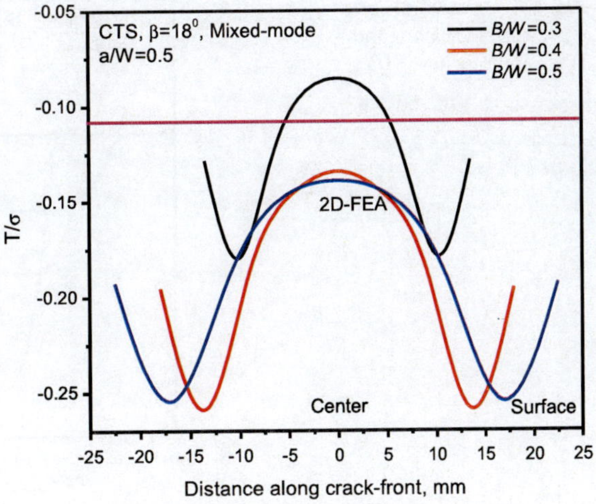

Fig. 12 Variation of T/σ versus distance along the crack front for $\beta = 18°$

of the CTS mixed mode (I/II) fracture specimen. It also observed from these figures that the negative value of normalized T increases as the B/W ratios increases.

From Figs. 13 and 14 for $\beta = 72°$ and 90°, the magnitude of normalized T is maximum at the surface compared to the center along the crack-front of the CTS specimen for various B/W ratios. From Figs. 11, 12, 13 and 14, one has to consider the T value at the center for Mode-I load dominance and on the surface for Mode-II load dominance. It also observed from Figs. 13 and 14 that the negative value of normalized T increases as the B/W ratio increases. For linear elastic condition, Smith et al. [36] clearly revealed that the fracture toughness of a material increases with negative T-stress and decreases with positive T-stress. Hence, according to Figs. 11, 12, 13 and 14, the normalized T-stress FE results are more for $\beta = 90°$ (Mode-II) and

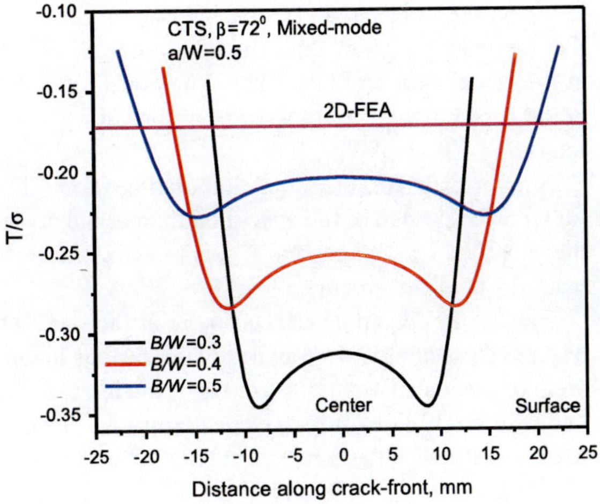

Fig. 13 Variation of T/σ versus distance along the crack front for $\beta = 72°$

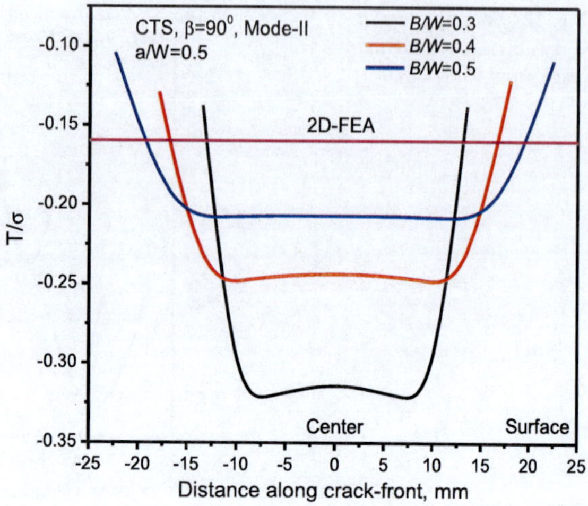

Fig. 14 Variation of T/σ versus distance along the crack front for $\beta = 90°$

less of $\beta = 0°$ (Mode-I) for CTS specimen of various B/W ratios. This indicates that fracture occurs earlier in Mode-II as compared to mixed mode and Mode-I loading for CTS specimen. Figures 11, 12, 13 and 14 depict that the decrease in normalized T will increase in thickness, thereby increases the fracture toughness of the material. But, investigators [37, 38] estimated that as the thickness of the specimen increases the fracture toughness decreases. Consequently from Figs. 11, 12, 13 and 14 clearly demonstrate that the T is not a unique value to measure the fracture toughness of the material. But, T-stress is a constraint parameter which affects the fracture toughness of the material.

Hence, the measurement of fracture toughness of the material in CTS mixed mode (I/II) fracture specimen can be analyzed by considering both K_{eff} and T-stress. From Figs. 6, 7, 8 and 9 and Figs. 11, 12, 13 and 14 for Mode-I load dominating ($\beta = 0°$ and 18°), the 3D CTS FE results indicate that the normalized T and K_{eff} values are lesser on the surface than at the center along the crack-front (thickness direction) of the specimen, in which the constraints (magnitude of both K_{eff} and T) is very high at the center than on the surface. The constraints will be more on the surface than at the center along the crack-front of the CTS specimen for dominance of Mode-II load ($\beta = 72°$ and 90°).

Similarly, for Mode-II load dominating ($\beta = 72°$ and 90°) the constraint is very high on the surface of the specimen than at the center along the crack-front. Since, the magnitude of T-stress and K_{eff} is more on the surface than at the center along the crack-front of the specimen.

Finally, the constraint effects on the surface and at the center along the crack-front of the CTS specimen were analyzed for various loading angle. In this paper, in-plane and out-plane constrains are considered which is similar to the earlier work of Yan and Brocks [39], i.e., in-plane constraint will be considered on the surface and out-plane constraint at the center along the surface of the CTS specimen. For the Mode-I load dominating ($\beta = 0°$ and 18°) CTS specimen, the out-plane constraint is higher than the in-plane constraint along the crack-front, thereby demonstrating the material undergoes failure earlier at the center of the specimen along the thickness than on the surface. The in-plane constraint is higher than the out-plane constraints for Mode-II loading ($\beta = 72°$ and 90°) dominating in CTS mixed mode (I/II) fracture specimen. So, the material failure begins on the surface along the specimen thickness. Also, from the FE results the constraint together with K_{eff} and T-stress is more for $\beta = 0°$ (Mode-I) compared to other loading angle ($\beta > 0°$). This point out that for CTS mixed mode (I/II) fracture specimen subjected to Mode-I loading undergoes earlier fracture than that of mixed mode and Mode-II loading. The constraint will also depend on the *B*/*W* ratio of the specimen and loading angle. The constraint effects will be more for larger thickness and the out-plane constraint changes to in-plane constraint as the loading angle changes from 0° (Mode-I) to 90° (Mode-II).

5 Conclusions

The following conclusions are drawn from the present study:

(i) The variation of stress intensity factors and T-stress along the crack-front will depend on the loading angle and the specimen thickness of the specimen.
(ii) For Mode-I load dominating the magnitude of stress intensity factors and T-stress is more at the center, indicating high constraint at the center than on the surface along the crack-front of the specimen. The constraint is more on the surface for Mode-II load dominating, signifying the amount of stress intensity factors and T-stress is higher on the surface than at the center along the crack-front of CTS specimen.
(iii) The out-plane constraint changes to in-plane constraint for a CTS specimen of various thicknesses, as the loading angle changes from 0° (Mode-I) to 90° (Mode-II). The fracture initiates at center due to high out-plane constraint for $\beta = 0°$. For $\beta = 90°$, fracture will begin on the surface indicating high in-plane constraint. For 0° < β > 90° there may be a transition between out-plane and in-plane constraint, which can be considered as future work under mixed mode I/II loading.

References

1. Larsson SG, Carlsson AJ (1973) Influence of non-singular stress terms and specimen geometry on small-scale yielding at crack tips in elastic-plastic materials. J Mech Phys Solid 21:263–277. https://doi.org/10.1016/0022-5096(73)90024-0
2. Williams ML (1957) On the stress distribution at the base of a stationary cracks. ASME Trans J Appl Mech 24:109–114. ISSN 0021-8936
3. Rice JR (1974) Limitations to the small scale yielding approximation for crack tip plasticity. J Mech Phys Solid 22:17–26. https://doi.org/10.1016/0022-5096(74)90010-6

4. Bilby BA, Cardew GE, Goldthorpe MR, Howard IC (1986) A finite element investigation of the effect of specimen geometry on the fields of stress and strain at the tips of stationary cracks. In: Size effect in fracture. Mech. Eng. Pub. Ltd, London, pp 37–46
5. Betegon C, Hancock JW (1991) Two-parameter characterization of elastic-plastic crack-tip fields. J Appl Mech 58:104–110. https://doi.org/10.1115/1.2897135
6. Du ZZ, Hancock JW (1991) The effect of non-singular stresses on crack-tip constraint. J Mech Phys Solid 39:555–567. https://doi.org/10.1016/0022-5096(91)90041-L
7. O'Dowd NP, Shih CF (1991) Family of crack-tip fields characterized by a triaxiality parameter—I. Structure of fields. J Mech Phys Solid 39:989–1015. https://doi.org/10.1016/0022-5096(91)90049-t
8. Wang YY (1993) On the two-parameter characterization of elastic-plastic crack-front fields in surface-cracked plates. ASTM STP 1171:120–138
9. Cotterell B, Rice JR (1980) On a slightly curved or kinked crack. Int J Fract 16:155–169. https://doi.org/10.1007/BF00012619
10. Leevers PS, Radon JC (1982) Inherent stress biaxiality in various fracture specimen geometries. Int J Fract 19:311–325. https://doi.org/10.1007/BF00012486
11. Cardew GE, Goldthorpe MR, Howard IC, Kfouri AP (1984) On the elastic T term. In: Bilby BA, Miller KJ, Willis JR (eds) Fundamentals of deformation and fracture. Cambridge University Press, Cambridge, pp 465–476
12. Kfouri AP (1986) Some evaluations of elastic T-term using Eshelby's method. Int J Fract 30:301–315. https://doi.org/10.1007/BF00019710
13. Eshelby JD (1975) The calculation of energy release rates. In: Sih GC, van Elst HC, Broek D (eds) Prospect of fracture mechanics. Noordhoff, Leyden, pp 69–84. https://doi.org/10.1007/s10704-006-0044-0
14. Hallback N, Jonsson N (1996) T-stress evaluations of mixed mode I/II fracture specimens and T-effects on mixed mode failure of aluminium. Int J Fract 76:141–168. https://doi.org/10.1007/BF00018534
15. Al-Ani AM, Hancock JW (1991) J-dominance of short cracks in tension and bending. J Mech Phys Solid 39:23–43. https://doi.org/10.1016/0022-5096(91)90029-N
16. Ayatollahi MR, Pavier MJ, Smith DJ (1998) Determination of T-stress from finite element analysis for mode I and mixed mode I/II loading. Int J Fract 91:283–298. https://doi.org/10.1023/A:1007581125618
17. Yang B, Ravi-Chandar K (1999) Evaluation of elastic T-stress by the stress difference method. Eng Fract Mech 64:589–605. https://doi.org/10.1016/S0013-7944(99)00082-X
18. Clatterbuck DM, Chrzan DC, Morris JW Jr (2003) The influence of triaxial stress on the ideal tensile strength of iron. Scripta Mater 49:1007–1011. https://doi.org/10.1016/S1359-6462(03)00490-1
19. Burton WS, Sinclair GB, Solecki JS, Swedlow JL (1984) On the implications for LEFM of the three-dimensional aspects in some crack/surface intersection problems. Int J Fract 25:3–32. https://doi.org/10.1007/BF01152747
20. Nakamura T, Parks DM (1990) Three-dimensional crack front fields in a thin ductile plate. J Mech Phys Solid 38:787–812. https://doi.org/10.1016/0022-5096(90)90040-B
21. Nevalainen M, Dodds RH (1995) Numerical investigation of 3-D constraint effects on brittle fracture in SEN(b) and C(T) specimens. Int J Fract 74:131–161. https://doi.org/10.1007/BF00036262
22. Kwon SW, Sun CT (2000) Characteristics of three-dimensional stress fields in plates with a through-the-thickness crack. Int J Fract 104:291–315. https://doi.org/10.1023/A:1007601918058
23. Nakamura T, Parks DM (1992) Determination of elastic T-stress along three dimensional crack fronts using an interaction integral. Int J Solid Struct 29:1597–1611. https://doi.org/10.1016/0020-7683(92)90011-H
24. Kong XM, Schluter N, Dahl W (1995) Effect of triaxial stress on mixed-mode fracture. Eng Fract Mech 52:379–388. https://doi.org/10.1016/0013-7944(94)00228-A

25. Schutte H, Molla-Abbasi K (2008) On the full set of elastic T-stress terms of internal elliptical cracks under mixed-mode loading condition. Eng Fract Mech 75:1545–1568. https://doi.org/10.1016/j.engfracmech.2007.06.003
26. Liu Q, Qiu J, Ge S, Guo WA (2009) Finite element modeling and analysis of CTS specimen for three-dimensional I-II mixed mode fracture. In: 2009 second international conference on information and computing science, pp 362–365
27. Borrego LP, Antunes FV, Costa JM, Ferreira JM (2006) Mixed-mode fatigue crack growth behaviour in aluminium alloy. Int J Fat 28:618–626. https://doi.org/10.1016/j.ijfatigue.2005.07.047
28. (2004) ABAQUS User's Manual. Version 6.5. Hibbitt, Karlsson & Sorensen, Inc
29. Qu J, Wang X (2006) Solutions of T-stresses for quarter-elliptical corner cracks in finite thickness plates subject to tension and bending. Int J Press Vessel Pip 83:593–606. https://doi.org/10.1016/j.ijpvp.2006.04.003
30. Kodancha KG, Kudari SK (2009) Variation of stress intensity factor and elastic T-stress along the crack-front in finite thickness plates. Fratt ed Inte Strut 8:45–51. https://doi.org/10.3221/IGF-ESIS.08.04
31. Shih CF, Asaro RL (1988) Elastic-plastic analysis of cracks on bimaterial interfaces: Part I—Small scale yielding. J Appl Mech 55:299–316. https://doi.org/10.1115/1.3173676
32. Kudari SK (2004) PhD thesis, IIT, Kharagpur, India
33. Richard HA, Benitz K (1982) A loading device for the creation of mixed mode in fracture mechanics. Int J Fract 22:R55–58. https://doi.org/10.1007/BF00942726
34. Sharanaprabhu CM, Kudari SK (2009) Finite element analysis of minimum plastic zone radius criterion for crack initiation direction under mixed mode loading. Am Inst Phys Conf Proc CP1138:(1):13–22. https://doi.org/10.1063/1.3155125
35. Sharanaprabhu CM, Kudari SK (2010) Analysis of crack-tip plastic zone in a compact tensile shear (CTS) specimen. Fratt Inte Strutt 14:27–35. https://doi.org/10.3221/IGF-ESIS.14.03
36. Smith DJ, Ayatollahi AR, Pavier MJ (2001) The role of T-stress in brittle fracture for linear elastic materials under mixed mode loading. Fatigue Fract Eng Mater Struct 24(2):137–150. https://doi.org/10.1046/j.1460-2695.2001.00377.x
37. Fuchs HO, Stephens RI (1980) Metal fatigue in engineering. Wiley, New York
38. Gdoutos EE (2005) Fracture mechanics an introduction, 2nd edn. Springer, Netherland
39. Yuan H, Brocks W (1998) Quantification of constraint effects in elastic-plastic crack front fields. J Mech Phys Solid 46:219–241. https://doi.org/10.1016/S0022-5096(97)00068-9

Fatigue Behaviour of Polymer Nanocomposites—A Review

Shrinatha R. Katti, M. V. Achutha, and B. K. Sridhara

Abstract Most of the household articles and devices which we use in everyday life are made of natural polymers and synthetic polymers. Existing research also reports that polymer nanocomposites show far better performance versus micro-fillers. Premature failure, of the machine components, occurs well within the endurance limit of the material. This failure due to fatigue is a common phenomenon on many of the applications across industries and hence the prediction and prevention of fatigue failures is critical for safe and economic operation of machines. This work reviews various methods of fatigue characterization of polymer composites. Stress–life and strain–life approaches have been applied by many researchers in this domain. The fatigue characterization leads to the development of life estimation curves by both these methods. There are also works done on other methods to arrive at Stress–life curves. This paper addresses these different methods of characterization and compares them. The analysis of these data for evaluation of various fatigue parameters is also covered in this work.

Keywords Carbon nanotubes · Nanocomposites · Fatigue · S-N curve · Life estimation

1 Introduction

The intensity of loads acting on a material plays a major role in the accurate design of a machine component. The strength of the material when subjected to these loads also is an important parameter. The designer must be knowing the behaviour of the material under various environmental conditions and correlate them with the load-carrying capacity of the component. This ensures that the component performs its duties for which it is designed. Along with this the designer is supposed to know

S. R. Katti (✉)
Dept of ME, NIEIT, Mysuru 570018, Karnataka, India
e-mail: shrinath.katti@nieit.ac.in

M. V. Achutha · B. K. Sridhara
Department of ME, The National Institute of Engineering, Mysuru 570008, Karnataka, India

S. Seetharamu et al. (eds.), *Fatigue, Durability, and Fracture Mechanics*, Lecture Notes in Mechanical Engineering, https://doi.org/10.1007/978-981-15-4779-9_36

the probable failure modes for any component. Fatigue is a common phenomenon leading to failure in structural and mechanical components. It is reported by many researchers that nearly 90% of the total failures which occur in service are due to fatigue. The fatigue failure occurs suddenly without any prior symptoms. Soon after the initiation of crack it propagates until it becomes unstable and suddenly results in failure. This failure due to fatigue is a common phenomenon on many of the applications across industries and hence the prediction and prevention of fatigue failures is critical for safe and economic operation of machines. This paper presents an overview of various fatigue failure analysis methods along with life prediction models. This work compares these methods available [1].

Owing to their broad range of properties and manufacturability, both synthetic and natural polymers play an essential and useful role in everyday life. Inspite of their attractive properties, they have low toughness due to their tendency to be string and brittle but still they are highly deformable. This results in low energy absorption capacity during failure. The combination of the existing usage with specific improvement targets for expanded use makes polymers a prime candidate for composite matrices. Existing research also reports that polymer nanocomposites show far better performance versus micro-fillers [1].

Polypropylene belongs to the category of crystalline polymers. Polypropylene (PP) has wide applications including laboratory equipment, automotive components, packaging and labelling, stationery, plastic parts, and polymer banknotes due to its processability, good balance of physical properties and price. The structure of PP is chemically simple. Hence, PP has been a good choice to be used as matrix material in understanding the properties of composites.

Nanoparticles possess dramatically changed physical properties compared to microparticles. Nanoscale materials have a large surface area for a given volume. Since many important reactions are governed surface properties, a nanostructured material can have substantially different properties from a larger dimensional material of the same composition. Among the several available allotropes of carbon, Carbon nanotube (CNT) which is in cylindrical form is the most fascinating among researchers. Single-walled (SWCNT) and multi-walled CNTs (MWCNTs) are the two types of CNTs. They are formed by rolling the one-atom-thick grapheme sheet into cylinders at a specified angle, variation in which will lead to different CNT structures. CNTs have sp2 bond between carbon atoms. This is the cause for its high tensile strength and Young's modulus [2]. MWCNT will have multiple layers of graphene sheets in the form of concentric cylinders which make them wonderfully strong fibres, which has its strength ten times more than any other fibre which is used as reinforcement. SWCNT is extraordinary electrical and thermal conductor. Researchers have contemplated to produce conductive polymers with low percolation thresholds owing to the conductivity and high aspect ratio of CNTs. In another area, it is thought that their massive thermal conductivity can be exploited to make thermally conductive composites. However, the mechanical enhancement of plastics using carbon nanotubes as reinforcing fillers will be the most researched area of composites [3].

Carbon nanotubes (CNTs) have attracted great interest by the researchers as a reinforcement option due to their remarkable mechanical properties. CNTs have tensile strengths up to 60 GPa, substantially stronger than steel with density less than 17%. Their special mechanical properties are due to strong bonding between the carbon atoms that form carbine plane [4]. The fundamental properties of the composite will be greatly affected by the introduction of the CNT. In spite of having its great potential use as reinforcing polymers, CNTs have not got commercialized yet, due to its vastly large-scale production [5].

PP has a chemically simple structure and hence it is easy to evaluate the effect of adding MWCNT to PP. Carbon nanotubes (CNTs) are promising additives for polymer composites due to their special excellent mechanical, electrical and thermal characteristics. PP/CNT composites produced at laboratory scale showed that low electrical percolation threshold and good reinforcement characteristics may be achieved [6]. CNT/polymer composites combine the good processability of the polymers with the excellent mechanical and other functional properties of the CNTs [2]. The same observations have been reported by Szentes et. al [7] where MFR and flexural modulus have found to be improving when MWCNT is used rather than carbon black. This has been observed to be more impactful when twin screw extruder is used. Coleman [3] opines that CNT will be the best candidate to be used as reinforcement among the family of carbon-based reinforcements. PP/CNT would produce very strong materials which are light at the same time due to the combination of processability of polymers and outstanding properties of the CNTs.

2 Fatigue Testing Methods

Real composite structures experience in-service fatigue loadings. They can rarely be reproduced in laboratory tests, due to two reasons: (i) very few fatigue testing machines are equipped with complex loading conditions as experienced there; and (ii) the available knowledge on in-service fatigue loadings are often insufficient. Therefore, for fatigue testing purposes, representative, standardized load spectra are selected in place of the in-service fatigue loadings. A variety of tests can be done on polymer composites, because huge number of fatigue testing parameters are to be tested. The various categories of tests can be (i) the amplitude controlled tests where stress or strain can be taken as input amplitude (ii) the loading frequency (iii) the direction of loading and (iv) the load ratio R. However, only tension–tension fatigue test with a constant amplitude load has been standardized in both ASTM and ISO standards. The available standards are ASTM D3479/D3479M (2007): 'Standard Test Method for Tension–Tension Fatigue of Polymer Matrix Composite Materials' and EN ISO 13,003:2003: 'Fibre-reinforced plastics—Determination of fatigue properties under cyclic loading conditions'.

Under the category of amplitude controlled tests the constant amplitude test is more commonly used method. This constant amplitude which is the input parameter will be fed into the system based on which the other control variable will be monitored by the system. Practically, many widely used servo-controlled dynamic loading UTM used for the fatigue tests have three modes of loading under the cyclic loading, viz, Load controlled, Displacement controlled and Strain controlled. The first mode accepts the amplitude input of load in kN. Also the mean value of cyclic load has to be entered which decides the load ratio. It implies that at each cycle of loading this load in kN will be applied constantly on either side of the mean value at the specified frequency. The displacement in the specimen will be measured varying with the load applied.

The testing frequency is a parameter which decides the number of cycles of loading per second. This consequently decides the number of loading cycles in a given test duration. While working on the UTM, an important parameter to be taken care of with respect to the frequency is the Data acquisition rate (DAR). As the frequency increases DAR has to increase proportionally, normally it should be at least 40 times the frequency. Although the loading has perfect constant amplitude, if DAR is not appropriate the data will not fetch the true conditions. Frequency greatly influences the fatigue tests on various polymer composites which has been reported by many researchers. The cyclic creep is found to increase at lower stress amplitudes and the fatigue life decreased with reduced cyclic frequency. But on the contrary, at higher stress amplitudes it was observed to be opposite. The fatigue life decreased with higher cyclic frequency [8].

The stress ratio is the ratio of minimum stress to maximum stress in a fatigue test. It is a very important parameter. These two stress levels are evaluated with their algebraic sign. A tension–compression loading will carry a negative stress ratio ($-\infty < R < 0$) while the range $0 < R < 1$ refers to tension–tension loading. Similarly, compression–compression loading is applicable for the range $1 < R < +\infty$. Figure 1 explains this concept in terms of various ratios depicted in a constant-life diagram.

Fully reversed axial fatigue loading is the worst fatigue loading condition for most composite materials. Fully reversed loading with $R = -1$ is the tension–compression

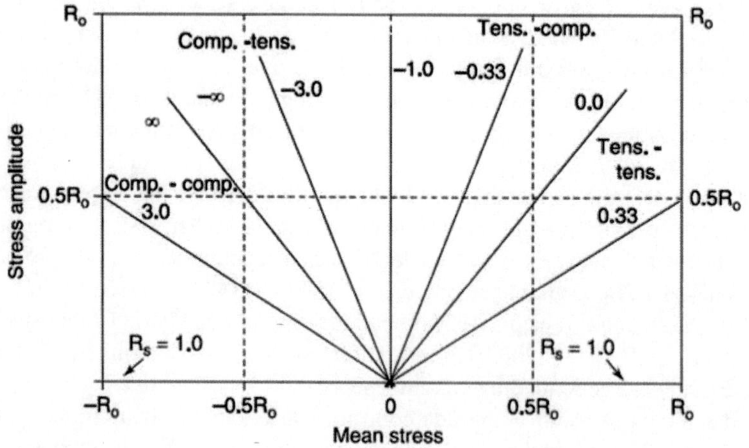

Fig. 1 Constant-life diagram showing lines of constant stress ratio R_S. Reproduced from [8]

loading as explained above. Fatigue lives under this condition are usually shorter than that for the other two types.

3 Fatigue Tests Results

The fatigue tests can be carried out at different loading conditions. Axial and Bending modes are the most commonly used.

3.1 Axial Loading Fatigue

Meneghatti et al. [9] tested the fatigue behaviour of PP composites filled with glass fibre. The applied stress amplitude was in between 36–37 MPa and the test frequencies up to 22 Hz. The number of cycles to failure of the material was observed to increase by 10 times from 10^5 for 10 mm fibre length to 10^6 for 1 mm fibre length. Many researchers have also used the alternate loading amplitude of 45 and 60% of the tensile strength. It has been shown that the elongation rate of the joint would be very slow up to 100 cycles. Beyond this limit the rate of elongation is observed to increase rapidly [10]. It can be noted that the test was carried out in completely reversed mode ($R = -1$) of testing. This observation talks about the deterioration of material after 100 cycles. Till the material structure gets softened, it tolerates the load applied and tries to maintain the elasticity further to which it takes plastic strain and hence the rate of elongation increases rapidly.

The test joint under consideration has been reported to exhibit net tension (N) + bearing (B) mode of failure and shear out (S) + bearing (B) failure mode as indicated in Fig. 2. When completely reversed fatigue load is applied, the tensile load acts at the ends diametrically opposite to the hole.

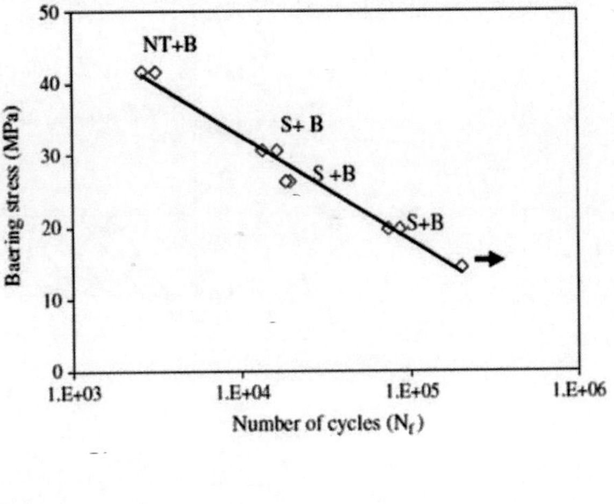

Fig. 2 Bearing stress–life curve under completely reversed loading condition. Reproduced from [11]

As reported by Bureau et al. [12] about the tests on interlaminar fatigue crack propagation for PP-based composites, the moulding conditions have a notable effect. The interlaminar fatigue behaviour will be greatly affected by morphology and microstructure of the composites which change rapidly with the change in the moulding conditions.

3.2 Bending Fatigue Tests

Dube et al. [11] carried out fatigue tests with three-point-bonding on resistance welded PEEK matrix with SS wire coated with TiO as the reinforcement. The loading followed sinusoidal waveforms. The quasi-isotropic specimens are reported to be shown by a linear S-N curve in the region of intermediate fatigue life. At 40% and 35% of the static damage initiation loads, indefinite fatigue life (greater than 1 lakh cycles) was reported.

The crack initiation and progress was monitored by using the travelling microscope which was installed on the test set-up. Figure 3 is plotted including both crack onset points and failure. It can be noted that the crack onset points are lesser compared to the failure points. The failure points can be easily recognized but the onset points are not reported to be noted for all specimens and hence the graph has lesser points. The indefinite fatigue life was defined at a load of 844 N which represents 40% of the static damage initiation load. This load is similar to the observations of many other researchers. At higher cyclic loads, the number of cycles varied linearly with the maximum cyclic load. At load level of 1690 N representing 80% of the damage initiation load, they occurred within a duration of few thousands cycles, indicating shorter fatigue life. This can be attributed to the delamination between the layers of matrix and interlaminar de-bonding between the matrix and reinforcements.

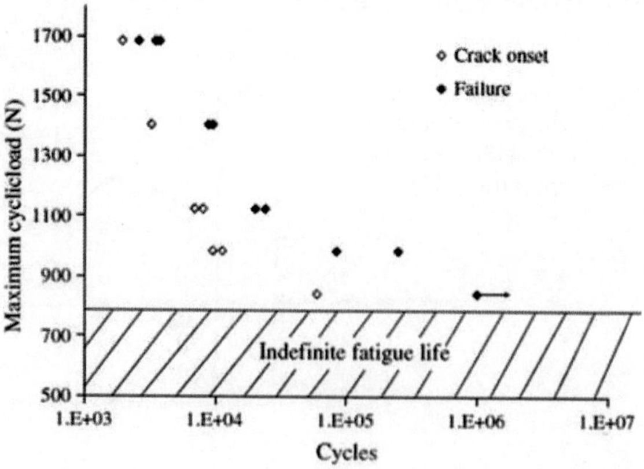

Fig. 3 S-N curve for unidirectional laminate. Reproduced from [11]

4 Conclusions

The researchers have reported extensively on the fatigue behaviour of the polymer composites with glass fibre and carbon fibres as the reinforcements. The testing methods and the consequent results generated give an insight on the development of S-N Curve and its behaviour is dependent on the loading level and frequency. The S-N Curves generated for the axial loading and 3-point bending fatigue loading were studied and compared in the present study. The composites containing PP and MWNT as the constituents were found to be rare especially in the 3-point bending fatique loading. Hence, further research may be carried out in this area. Also dynamic mechanical analysis which is another method of applying dynamic or cyclic loads is used rarely to determine fatigue life. This also can be another area of study.

References

1. Sridhara BK (1996) Fatigue life and reliability evaluation under random loads with Gaussian amplitudes. PhD thesis
2. Prashantha K, Soulestin J, Lacrampe MF, Claes M, Dupin G, Krawezak P (2008) eXPRESS Poly Lett 2:735
3. Coleman JN, Khan U, Blau WJ, Gun'ko YK (2006) Small but strong: a review of the mechanical properties of carbon nanotube–polymer composites. Carbon (Elsevier) 44:1624–1652
4. Novais RM, Simon F, Paiva MC, Covas JA (2012) Composites: Part A 43:2189
5. Sulong AB, Ramli MI, Hau SL, Sahari J, Muhammad N, Suherman H (2013) Composites: Part B, 50:54
6. Leininger W, Wang X, Tangpong XW, McNea M (2012) J Eng Mater Technol 134:01
7. Szentes A, Horvath G, Varga C (2010) Hungarian J Indus Chem 38:67
8. Guedes RM (2011) Creep and fatigue in polymer matrix composites. Woodhead Publishing Limited
9. Meneghatti G, Ricotta M, Lucchetta G, Carnignato S (2014) An hysteresis energy-based synthesis of fully reversed axial fatigue behaviour of different polypropylene composites. Composites: Part B 01–10
10. Subramanian C, Senthilvelan S (2011) Joint performance of the glass fiber reinforced polypropylene leaf spring. Compos Struct 93:759–766
11. Dube M, Hubert P, Yousefpour A, Denault J (2009) Fatigue failure characterization of resistance-welded thermoplastic composites. Int J Fatigue 31:719–725
12. Beureau MN, Perrin F, Denault J, Dickson JJ (2002) Interlaminar fatigue crack propagation in continuous glass fiber/ polypropylene composites. Int J Fatigue 26:99–108

Life Assessment and Extension of RCC Framed Building Subjected to Fatigue Loading Analytical Approach

Shaik Kamal Mohammad Azam, Shaik Abdulla, and Nadeem Pasha

Abstract In the present era, life assessment of building is of paramount importance. In this paper, an attempt has been made in using the concept of fracture mechanics using steel fibers using software such as ANSYS, E-TABS, etc. Before that experimental investigation is carried out using steel fibers, and parameters such as fibers type, aspect ratio, and optimum dosages are covered. In general, fiber-reinforced concrete structures are more durable than normal RCC structures.

Keywords Fatigue · Steel fibers · Aspect ratio · Durability · Fracture mechanics

1 Introduction

Life assessment of RCC buildings is of paramount importance keeping in view the fact that apart from strength parameter more stress is being laid on the durability aspect of structures in the present era. It is also quite evident that steel fibers can be added to concrete because of its post cracking resisting properties due to which the concrete becomes tougher, and more ductility is induced in it. From the literature also it is quite clear that addition of fibers improves various properties of CVC especially those associated with durability, strength, and performance of concrete. Toughness is the important parameter in shrinkage control and crack arrest which also increases with more steel fibers. They are generally used nowadays in concrete for pavements, slabs, and tunnels. Fiber-reinforced concrete is preferred over conventional concrete because of steel fibers ability to resist under extreme loading such as fatigue, impact, and abrasion and minimizes cracking, avoids fractures and spalling of concrete, and thus increases the ductility of concrete.

S. K. M. Azam (✉) · S. Abdulla · N. Pasha
Khaja Bandanawaz College of Engineering, Kalaburagi, India
e-mail: kbnazam9@gmail.com

S. Abdulla
e-mail: shaikmtech@gmail.com

N. Pasha
e-mail: nadeempasha48@yahoo.in

S. Seetharamu et al. (eds.), *Fatigue, Durability, and Fracture Mechanics*, Lecture Notes in Mechanical Engineering, https://doi.org/10.1007/978-981-15-4779-9_37

Fatigue is an important parameter which needs to be studied in detail. It occurs when the structure is subjected to cyclic loading. It is very dangerous, hence to avoid this and resist cracks in concrete, steel fibers are used based *on* ductility of fibers, volume of fiber reinforcement, shape, bond at fiber matrix interface, and aspect ratio of steel fibers. In order to study the effect of fatigue and its effect on structures before carrying out the analysis, a detailed study should also be done on the following parameters such as NLS pushover analysis, Performance Point, Performance Levels (OL), (NP-B), (NP-C), (CP), Plastic Hinge Formation, Evaluation of safety and Ductility Ratio (Figs. 1, 2 and 3).

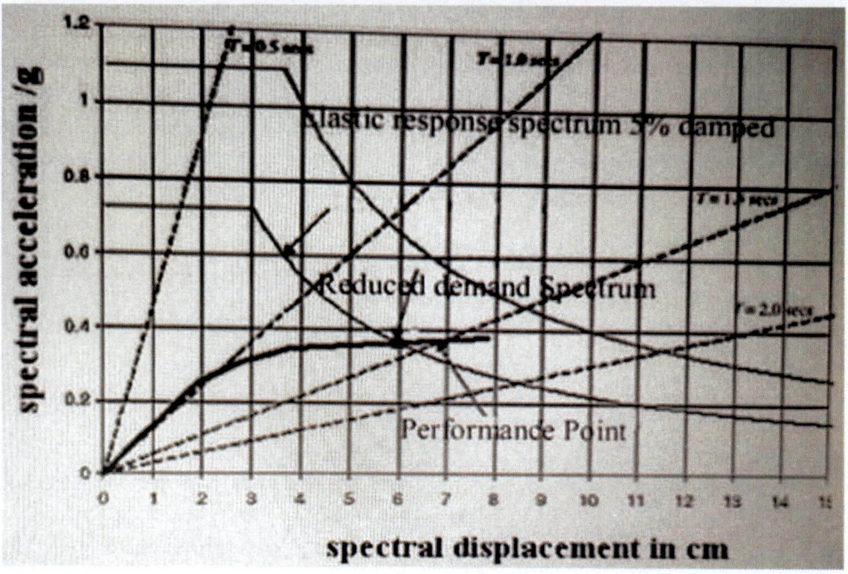

Fig. 1 Determination of performance point

Fig. 2 Hinge property

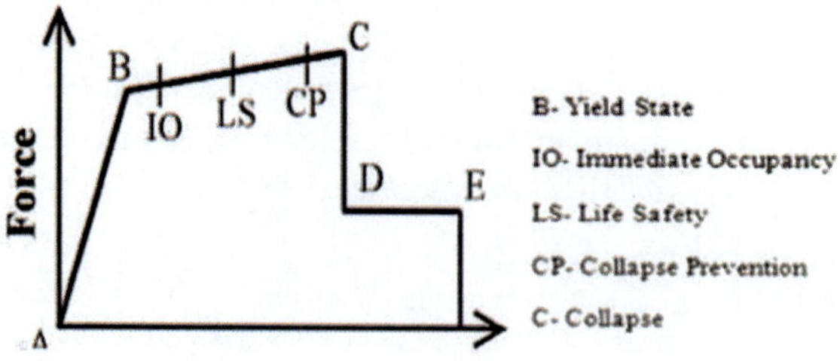

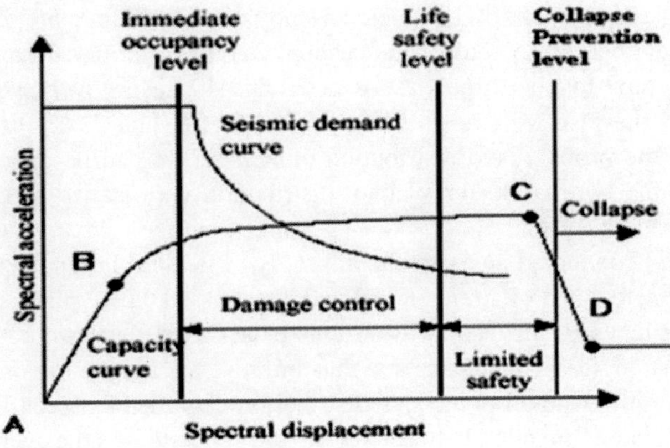

Fig. 3 Different building performance levels

2 Literature Review

Naik et al. [1] conducted a study on "Fatigue behavior plain concrete made with or without fly ash" that presents the state-of-the-art information on fatigue behavior of plain concrete manufactured with or without fly ash.

Normally, endurance or fatigue limit of plain concrete was found to vary between 50 and 60% of its static strength. In compressive mode of loading, concrete containing a class C fly ash showed improved fatigue strength over either concrete contained class F fly ash or no fly ash. However, in flexural mode of loading, inclusion of fly ash in concrete exhibited little effect on the endurance limit of plain concrete.

Rafeeq Ahmed et al. [2] conducted a study on "Influence of steel fibre on the fatigue resistance of concrete by using both straight and hooked fibers were incorporated in the concrete."

The volume fraction was 1% for straight fibers, and both 1 and 2% for hooked fibers. Finally, it was concluded from the study that hooked fibers improved the fatigue strength more than the straight ones; the larger the fiber content, the greater was the enhancement of fatigue strength.

Singh and Kaushik [3] conducted a study on the fatigue strength of Steel Fiber Reinforced Concrete (SFRC) containing fibers of mixed aspect ratio. To obtain the fatigue lives of SFRC at various stress levels, beam specimens of size 500 mm × 100 mm × 100 mm were tested under four-point flexural fatigue loading. The specimens incorporated 1.0% volume fraction of corrugated mixed steel fibers of size 0.6 × 2.0 × 25 mm and 0.6 × 2.0 × 50 mm in different proportions. Fatigue life data obtained has been analyzed in an attempt to determine the relationship among stress level S, number of cycles to failure N, and probability of failure Pf for SFRC.

Mohammadi and Kaushik [4] conducted a study on "The fatigue life distribution of plain and steel fiber reinforced concrete using two types of crimped-type flat steel fibers corresponding to fiber aspect ratios of 20 and 40, respectively, which were used at each of the fiber volume fractions of 0, 1.0, 1.5, and 2.0%. It has been concluded that the probabilistic distribution of fatigue life can be approximately modeled using the two-parameter Weibull distribution with statistical correlation coefficient exceeding 0.90.

Peng et al. [5] conducted an experimental study of flexural fatigue performance of reinforced concrete beams strengthened with prestressed CFRP plates by casting 07 RC beams using CFRP plates. It was observed that CFRP plates were found to be beneficial as it strengthens the beams and thus improve the fatigue life of the structure. The debonding resulted in stress redistribution degraded between CFRP and reinforcing steel bars. The bond between CFRP and concrete was the cyclic loading as a result of which the strength of the beams increases due to the use of CFRP plates. Dong – II, Chang won, and Kyu chai [6] conducted a study in which fracture and fatigue tests were performed in order to investigate the fracture and fatigue behavior of steel-fibre-reinforced concrete (SFRC) structures. 33 SFRC beams were used in the fracture and fatigue tests. From the test results, the effects of the fiber content, fiber aspect ratio, and fatigue behavior were studied. According to the regression technique, some empirical formulae for predicting the fatigue strength of SFRC beams were also suggested. An experimental study by Elisa poveda, Gonzalo, and Ruiz [7] was carried out to evaluate the influence of fiber content on compressive fatigue of SCC reinforced with steel fibers. Five different fiber contents, from plain concrete up to a content of 0.8% in volume, are studied. The addition of fibres proved to be beneficial, and the optimum percentage of steel fibres was found to be (0.6%). It has also been verified that the relationship of the secondary strain rate per cycle with the fatigue life is independent of the amount of fibers, and therefore it depends only on the matrix. Based on these results, a strain-based fatigue failure criterion involving mechanical and time-dependent deformation is proposed. S Goel, S P Singh, and P Singh [8] carried out an experimental investigation on the flexural fatigue strength of Self-Compacting Fibre Reinforced Concrete beams. The fatigue strength of SCFRC containing 0.5%, 1.0%, and 1.5% by volume of steel fibres has been obtained by conducting flexural fatigue tests on approximately 188 beam specimens of size 100 × 100 × 500 mm under third point loading at different stress levels, ranging from 0.90 to 0.70. Approximately, 144 complimentary static flexural tests were also carried out to facilitate fatigue testing. All the static flexural and flexural fatigue tests were conducted on a 100 kN closed-loop, servo-controlled actuator. The results have been represented in the form of S–N diagrams, and to predict the flexural fatigue strength of SCFRC, material coefficients of the fatigue equations have been estimated. Subsequently, two-million cycles of fatigue strength of SCFRC has been found to be higher than that of Normally Vibrated Fibre Reinforced Concrete (NVFRC). Mofreh F Saleh, T Yeow, G MacRae, and A Scott [9] conducted a study for fatigue behaviour of rigid pavements. In this research, steel fibre was added to

Portland cement concrete at 20 kg/m3 and 60 kg/m3 to improve fatigue resistance, which could allow for thinner pavements and hence lower construction costs. Fatigue tests were carried out using constant stress mode. A range of stresses were applied to cover a range of stress ratios from 0.26 to 0.616. It was found that steel fibres improved fatigue resistance. However, high fibre contents showed detrimental effect on fatigue at high stress ratios.

3 Experimental Program

The objective of the experimental investigation is to study the performance of SFRC applied to loading under fatigue for different percentage volume fractions of SF.

In the first stage, test on fresh concrete was performed.

In the second stage, testing of hardened concrete is done. The cube was tested in compression, flexure of prisms, and flexural fatigue strength test or repeated load test of prisms will be carried out. The elastic modulus of concrete at various percentages of steel fibers is determined. In the final stage, ductile nature of concrete is studied in detail by preparing a bare frame model and carrying out lateral load analysis for the bare frame structure by using ETABS software for analysis.

Materials Used

(1) OPC 53 grade cement from a single batch is used for the work of specific gravity 3.14.
(2) Sand of specific gravity 2.65 was used.
(3) Ballast stones will be used for the work of specific gravity 2.71 of size 20 mm downsize.
(4) Conplast was used as superplasticizer.
(5) Dramix (brand) steel fibers are used in aspect ratio 80.

4 Research Methodology

4.1 Compressive Strength Test

M20 Grade concrete is prepared by casting 3 trail mixes at different steel fiber contents. Cubes of (150*150*150) mm sizes are used and tested at 28 days.

4.2 Flexural Strength Test

Prisms of (100*100*500) mm sizes shall be used to cast and this must be tested for flexural strength for both non-fibrous and fibrous specimens.

4.3 Model Analysis

Models of bare frame are prepared for different proportions of steel fibers (09 models) in E-Tabs, and ductility ratio for each was determined.

4.3.1 Analytical Modeling

Lateral load analysis was used for the bare frame structure, and ETABS was used for analysis.

4.3.2 Plan of Proposed Building

See Fig. 4.

Model 1: The building is modeled as bare frame. The LL&FF with 0% fiber and mass of wall 230 mm thick is also considered in all stories.

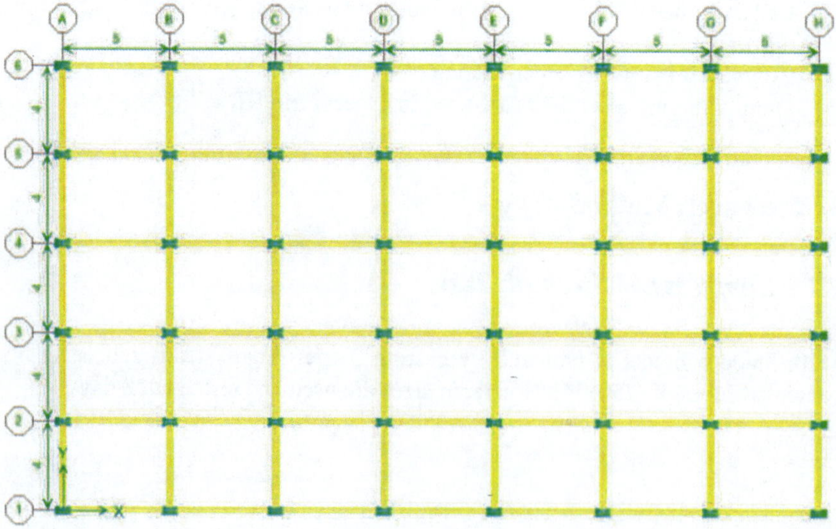

Fig. 4 Plan of building (all dimensions are in m)

Model 2: The building is modeled as bare frame. The LL&FF with 0.25% fiber and mass of wall 230 mm thick is also considered in all stories.

Model 3: The building is modeled as bare frame. The LL&FF with 0.5% fiber and mass of wall 230 mm thick is also considered in all stories.

Model 4: The building is modeled as bare frame. The LL&FF with 0.75% fiber and mass of wall 230 mm thick is also considered in all stories.

Model 5: The building is modeled as bare frame. The LL&FF with 1.0% fiber and mass of wall 230 mm thick is also considered in all stories.

Model 6: The building is modeled as bare frame. The LL&FF with 1.25% fiber and mass of wall 230 mm thick is also considered in all stories.

Model 7: The building is modeled as bare frame. The LL&FF with 1.5% fiber and mass of wall 230 mm thick is also considered in all stories.

Model 8: The building is modeled as bare frame. The LL&FF with 1.75% fiber and mass of wall 230 mm thick is also considered in all stories.

Model 9: The building is modeled as bare frame. The LL&FF with 2.0% fiber and mass of wall 230 mm thick is also considered in all stories.

5 Results and Discussions

See Tables 1, 2, 3, 4, 5, 6, and 7; Figs. 5, 6, 7, 8, 9, 10, 11, 12, 13, 14, 15, 16, 17 and 18.

Table 1 Compressive strength, flexural strength and elastic modulus for various % of steel fibers at 28 days

MIX ID	Percentage (%) of steel fibers	Compressive strength (MPa)	Flexural strength (MPa)	Elastic modulus (MPa)
M1 (CVC)	0	20	22.35	22,360
M2	0.25	26.45	28.50	26,564
M3	0.50	28.75	30.55	27,660
M4	0.75	30.25	32.50	28,350
M5	1.00	31.75	33.45	29,025
M6	1.25	32.45	34.90	29,335
M7	1.50	34.25	36.50	30,111
M8	1.75	30.15	32.50	28,305
M9	2.0	29.75	31.25	28,120

Table 2 Ductility ratio along longitudinal direction

Model	Percentage of fiber added (%)	Time period (S)	EQX (mm)	RSX (Mm)	Yield displacement (U_{Yield}) (mm) Pushover	Ultimate displacement U_{ultimate} (mm) Pushover	Ductility ratio (μ)
Model-01	0	2.389	38.17	29.33	22.30	364.20	16.33
Model-02	0.25	2.192	35.02	26.96	18.80	372.80	19.83
Model-03	0.50	2.148	34.32	26.38	18.00	374.90	20.83
Model-04	0.75	2.12	33.90	26.02	17.60	378.50	21.50
Model-05	1.0	2.097	33.50	25.72	17.20	380.20	22.10
Model-06	1.25	2.086	33.32	25.60	17.00	381.00	22.41
Model-07	1.50	2.059	32.89	25.30	16.60	381.80	23.00
Model-08	1.75	2.137	33.92	26.04	17.60	378.40	21.50
Model-09	2.00	2.130	34.04	26.14	17.70	377.90	21.35

Table 3 Displacement at different % of steel fibers for different models

Story	M-1	M-2	M-3	M-4	M-5	M-6	M-7	M-8	M-9
01	0.79	0.727	0.712	0.704	0.695	0.692	0.683	0.704	0.706
02	4.38	4.023	3.942	3.89	3.849	3.828	3.778	3.897	3.91
03	8.72	8	7.843	7.47	7.656	7.615	7.517	7.753	7.778
04	13.19	12.102	11.859	11.714	11.577	11.516	11.366	11.723	11.762
05	17.6	16.154	15.83	15.636	15.454	15.372	15.172	15.649	15.7
06	21.87	20.066	19.665	19.424	19.197	19.905	18.847	19.439	19.503
07	25.88	23.75	23.275	22.99	22.721	22.6	22.308	23.008	23.084
08	29.54	27.106	26.563	26.23	25.93	25.794	25.459	26.259	26.345
09	32.72	30.02	29.419	29.059	28.719	28.567	28.196	29.02	29.177
10	35.28	32.36	31.721	31.33	30.96	30.802	30.403	31.358	31.461
11	37.1	34.04	33.359	32.951	32.56	32.393	31.972	32.977	33.085
12	38.17	35.023	34.322	33.9	33.5	33.327	32.895	33.928	34.04

Table 4 Displacement as per EQX at different % of steel fibers

% of fiber	Displacement EQX (mm)
0	38.17
0.25	35.02
0.5	34.32
0.75	33.9
1	33.5
1.25	33.32
1.5	32.89
1.75	33.92
2	34.04

Table 5 Displacement as per RSX at different % of steel fibers

% of fiber	Displacement RSX (mm)
0	29.33
0.25	26.96
0.5	26.38
0.75	26.02
1	25.72
1.25	25.6
1.5	25.3
1.75	26.04
2	26.14

Table 6 Yield displacement and ultimate displacement as per push over analysis at different % of steel fibers

% of fiber	Yield displacement (mm)	Ultimate displacement (mm)
0	22.3	364.2
0.25	18.8	372.8
0.5	18	374.9
0.75	17.6	378.5
1	17.2	380.2
1.25	17	381
1.5	16.6	381.8
1.75	17.6	378.4
2	17.7	377.9

Table 7 Ductility ratio as per push over analysis at different % of steel fibers

% of fiber	Ductility ratio
0	16.33
0.25	19.83
0.5	20.83
0.75	21.5
1	22.1
1.25	22.41
1.5	23
1.75	21.5
2	21.35

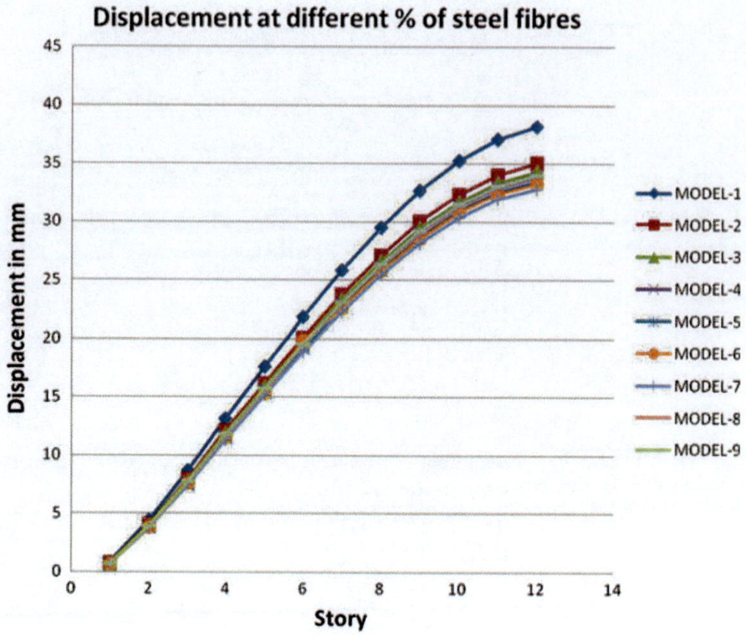

Fig. 5 Displacement versus % of steel fibers

Fig. 6 Displacement versus % of steel fibers as per EQX

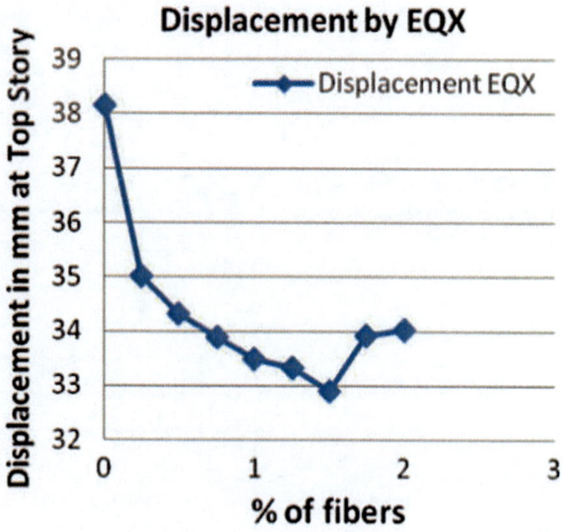

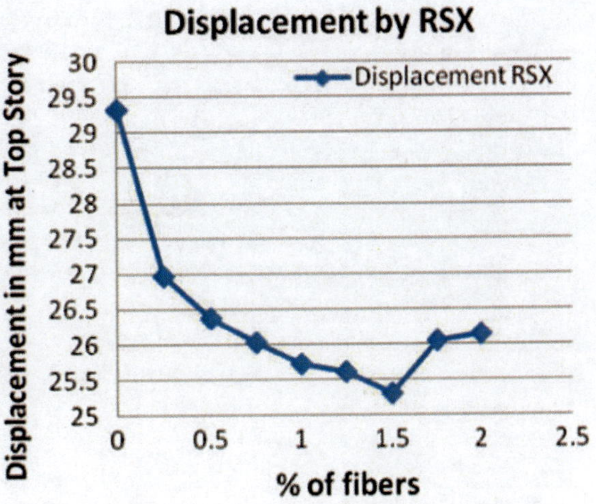

Fig. 7 Displacement versus % of steel fibers as per RSX

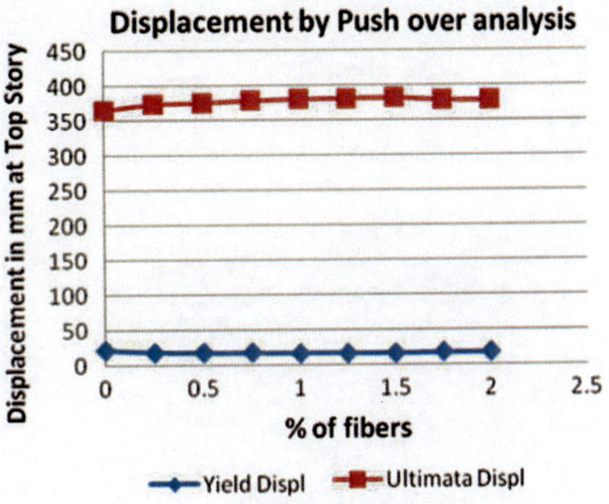

Fig. 8 Yield displacement and ultimate displacement versus % of steel fibers as per push over analysis

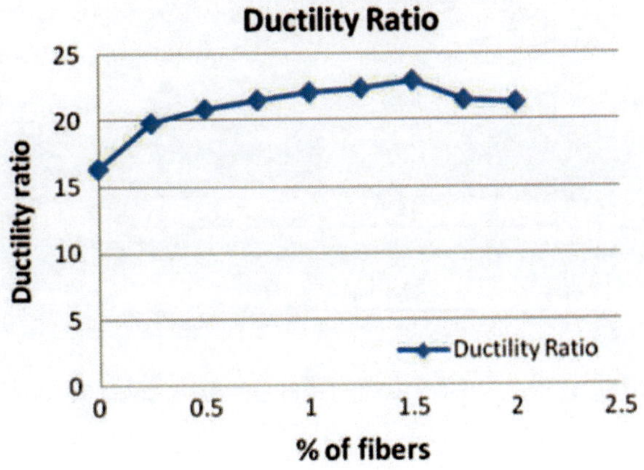

Fig. 9 Ductility ratio versus % of steel fibers as per push over analysis

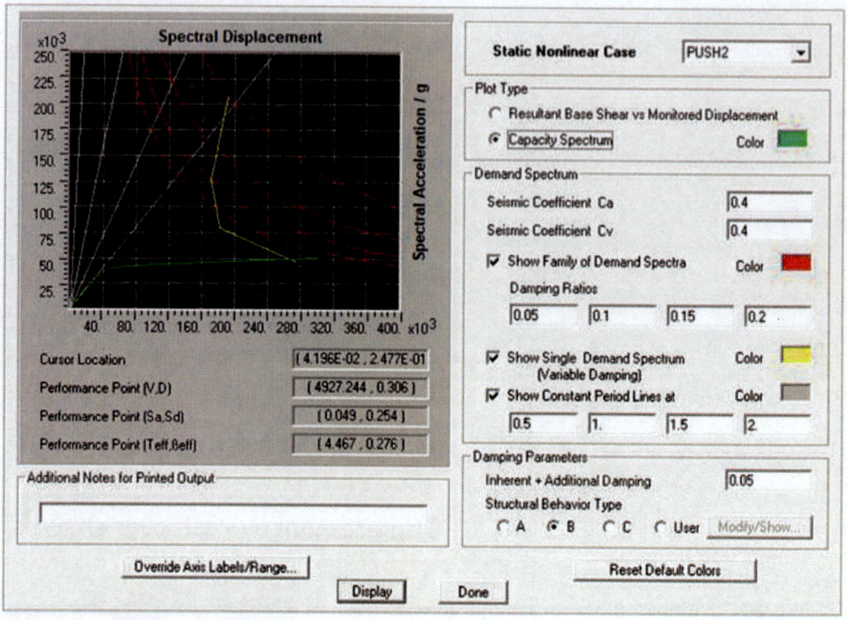

Fig. 10 Performance point as per push over analysis for Model 1

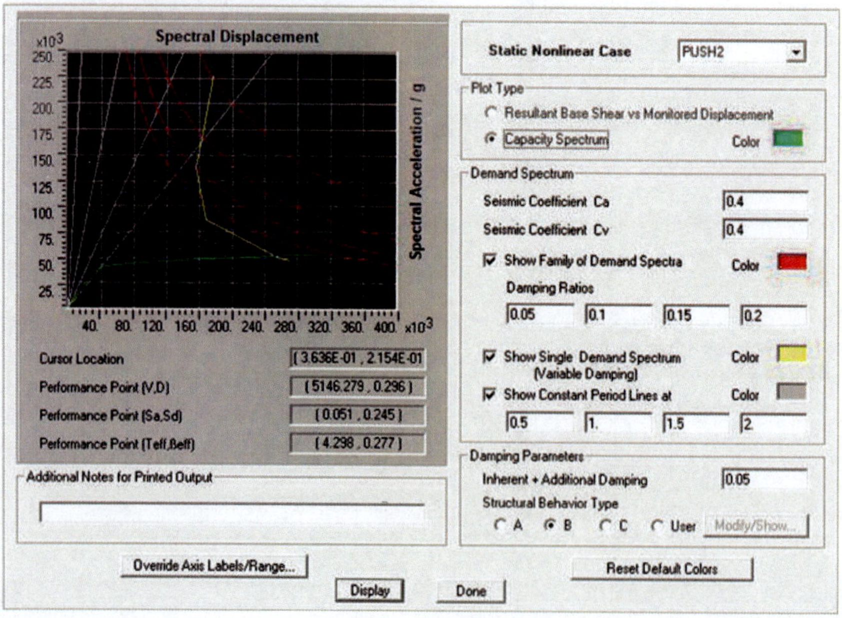

Fig. 11 Performance point as per push over analysis for Model 2

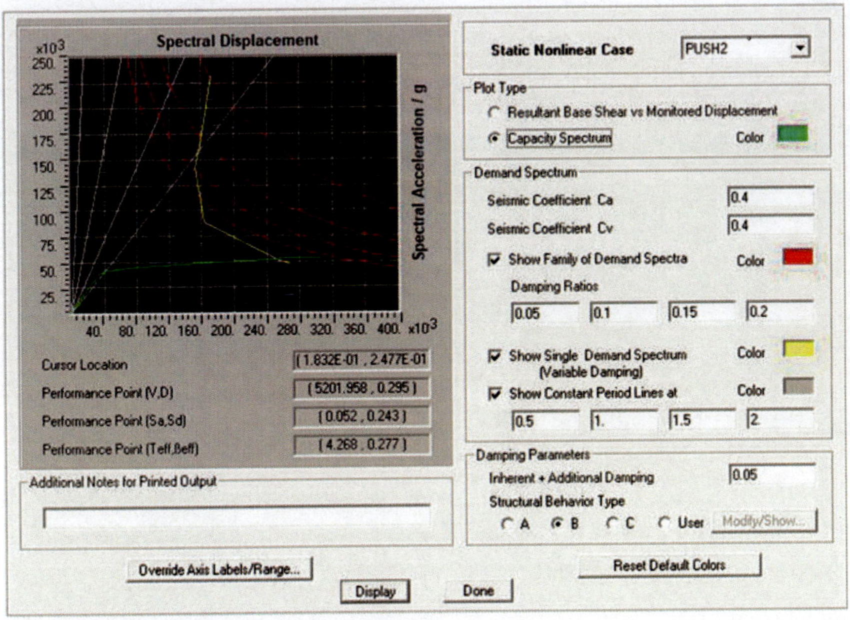

Fig. 12 Performance point as per push over analysis for Model 3

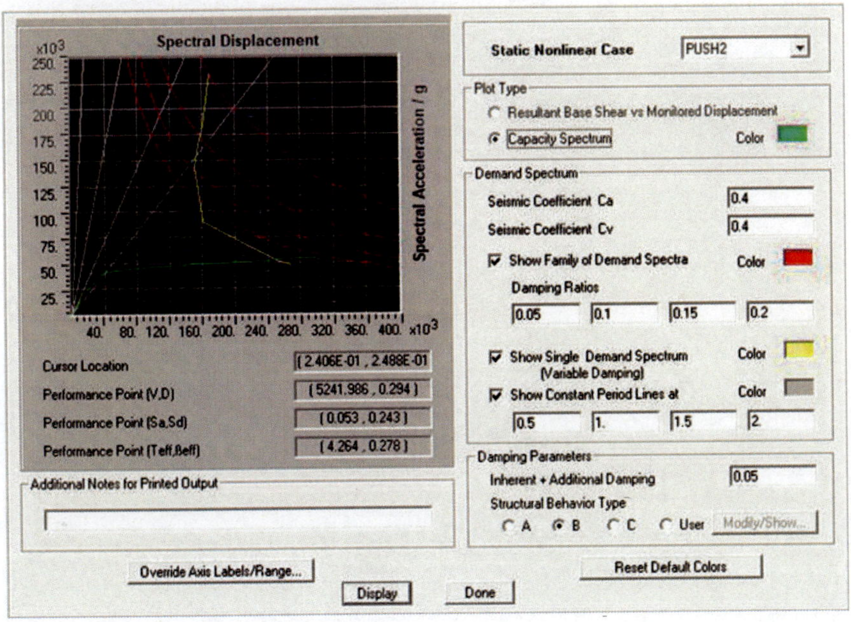

Fig. 13 Performance point as per push over analysis for Model 4

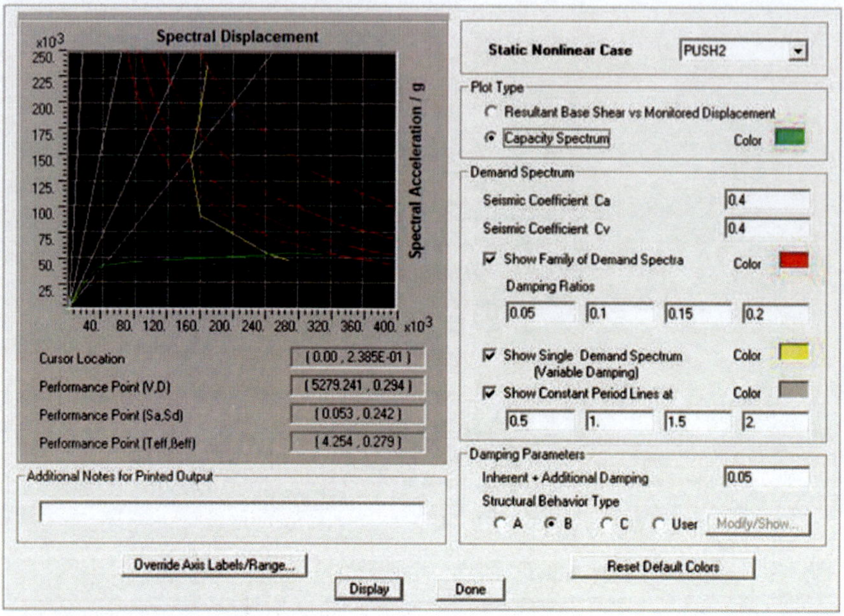

Fig. 14 Performance point as per push over analysis for Model 5

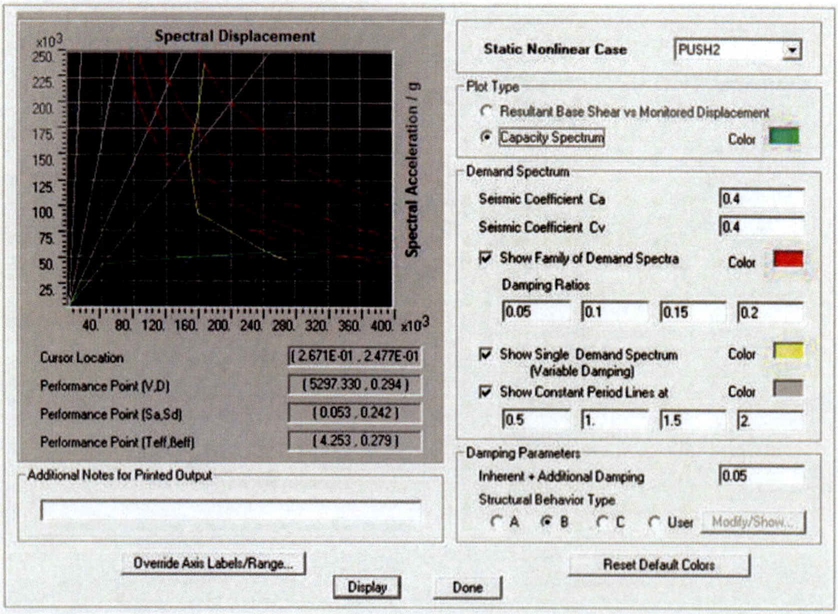

Fig. 15 Performance point as per push over analysis for Model 6

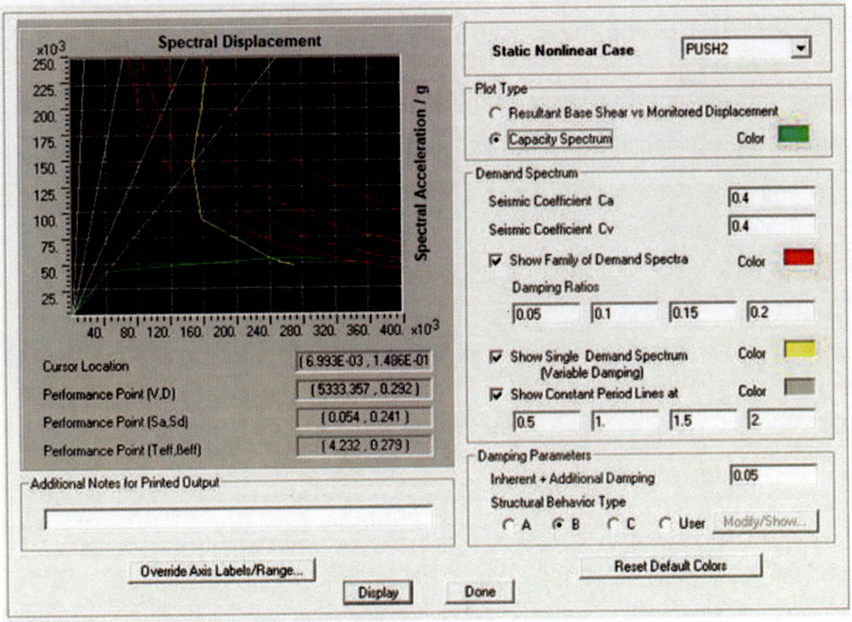

Fig. 16 Performance point as per push over analysis for Model 7

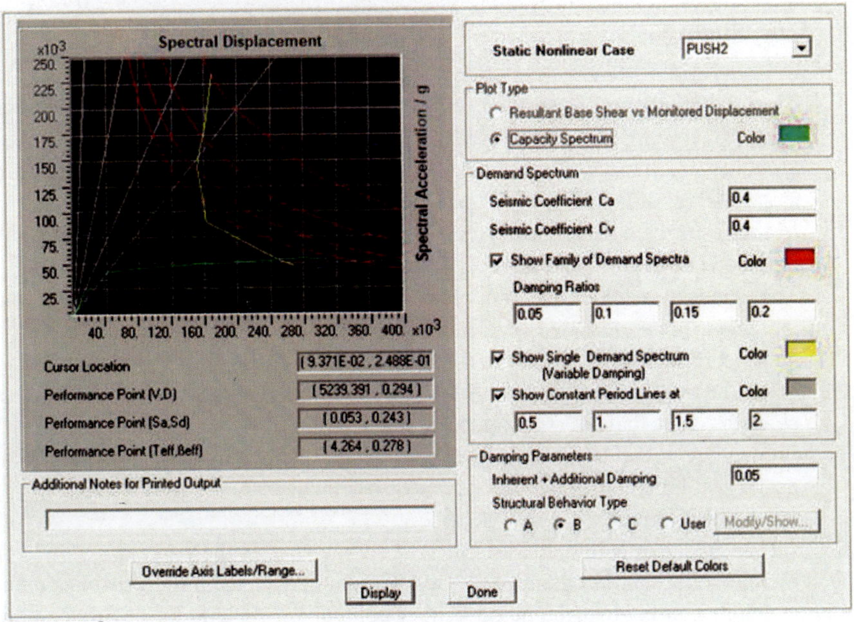

Fig. 17 Performance point as per push over analysis for Model 8

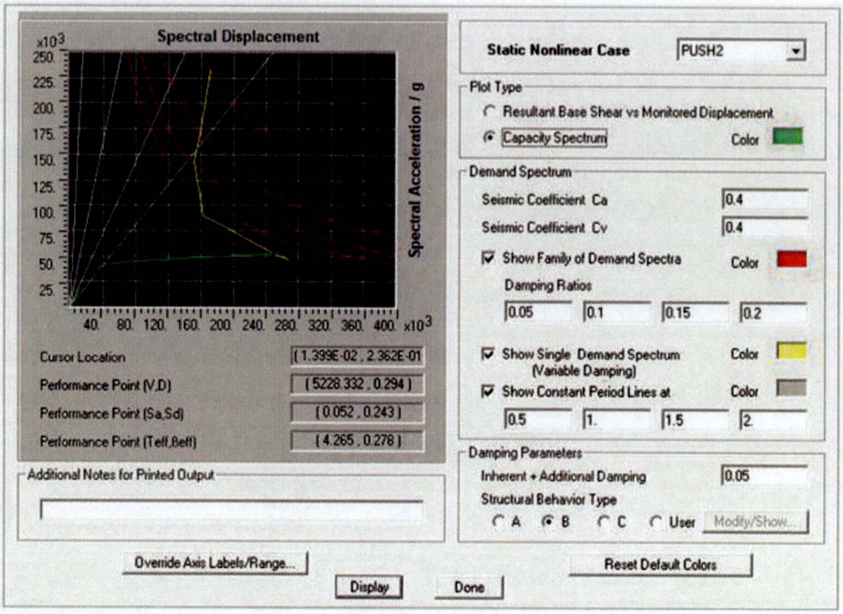

Fig. 18 Performance point as per push over analysis for Model 9

6 Conclusions

1. The fatigue tests are performed on a series of SFRC to investigate the behavior of SFRC varying with the steel fiber contents.
2. Dramix steel fibers are incorporated in the concrete by the same volume fraction at an interval of 0.25% from 0 to 2%.
3. In the result calculated from story-displacement curve, maximum stiffness is obtained at 0.5% of steel fiber.
4. The ductility ratio of the model increases as the fiber content in concrete increases. For model 07, maximum ductility ratio of 23 was achieved which later decreases with increase in fiber content.
5. From the observation, it is evident that SFRC specimens show better stiffness characteristics than RC specimens.
6. From LSA, it can be concluded that as the % of fiber increases displacement decreases. For model 07, minimum displacement of 32.89 mm is observed at 1.5% fiber content.
7. As per linear dynamic analysis, it is observed that as the percentage of fiber increases the displacement decreases. For model 07, minimum displacement of 25.30 mm is observed at 1.5% fiber content which later on decreases with increase in fiber content.
8. According to push over analysis, the yield displacement decreases and ultimate displacement increases with increase in fiber content up to 1.5%.

9. In general, it can be concluded that with addition of steel fibers the strength and durability of RCC structures increases, and the optimum percentage of steel fibers is 1.5%.
10. From the model analysis, the PP increases and then decreases from M 1 to M 7 at 1.5% steel fibers content from model 8.
11. In the present research, the ductile nature and fatigue behavior of RCC structures are assessed and the service life of structures can be improved by using SFRC.

References

1. Naik TR, Singh SS, Ye C. Fatigue behavior plain concrete made with or without flyash. Electric Power Research Institute, Palo Alto, CA. https://www4.uwm.edu/cbu/Papers/1993%20CBU%20Reports/REP-203.pdf
2. Rafeeq Ahmed S, Gupta A, Krishnamoorthy S (2000) Influence of steel fibre on the fatigue resistance of concrete. J Mater Civ Eng 172–179. https://doi.org/10.1061/(ASCE)0899-1561(2000)12:2(172)
3. Singh SP, Kaushik SK (2001) Flexural fatigue analysis of SFRC. Asian J 154–159. https://www.ijert.org/flexural-fatigue-studies-for-sfrc-under-variable-amplitude-loading-for-1-volume-fraction-of-steel-fibres
4. Mohammadi Y, Kaushik SK (2005) The fatigue life distribution of plain and steel fibre reinforced concrete. J Mater Civ Eng 650–658. https://doi.org/10.1061/(ASCE)0899-1561(2005)17:6(650)
5. Peng H, Zhang J, Shang S, Liu Y (2016) Experimental study of flexural fatigue performance of reinforced concrete beams strengthened with prestressed CFRP plates. Elsevier Eng Struct 27:62–72. https://doi.org/10.1016/j.engstruct.2016.08.026
6. Dong-II C, Won-Kyu C. Flexural fracture and fatigue behavior of steel fibre reinforced concrete structures. Elsevier Science Publisher B.V., pp 201–207. https://doi.org/10.1016/0029-5493(94)00946-V
7. Poveda E, Ruiz G. Influence of the fiber content on the compressive low cycle fatigue behavior of self compacting concrete SFRC. Elsevier Science Publisher B.V., pp 9–17. https://doi.org/10.1016/j.ijfatigue.2017.04.005
8. Goel S, Singh SP, Singh P. Flexural fatigue strength and failure probability of self compacting fibre reinforced concrete beams. Elsevier Science Publisher B.V., pp 131–140. https://doi.org/10.1016/j.engstruct.2012.02.035
9. Saleh MF, Yeow T, MacRae G, Scott A. Effect of steel fibre content on the fatigue behaviour of steel fibre reinforced concrete. In: 7th RILEM international conference on cracking in pavements, pp 815–825. https://link.springer.com/chapter/10.1007/978-94-007-4566-7_79

FEA and Experimental Evaluation of Fracture Toughness of LM13 Reinforced With Fused SiO_2

H. S. Harshith and Joel Hemanth

Abstract This paper aims at developing (LM13) a composite made of LM13 reinforced with varying percentage of fused SiO_2 (3, 6, 9, 12%). Also the Fracture Toughness was evaluated both by experimental and FEA. Various metallic and Non-metallic chills were used during the casting process to ensure uniform heat dissipation. The effect of chilling on fracture toughness was also studied. ASTM (E399 1990) standards were used to carry out fracture toughness test. The presence of clusters of particles leads to premature fracture. Upto 9wt% there was increase in fracture toughness, beyond this limit decrease in trend was noticed. Highest fracture toughness was observed at 9 wt% with copper chill and was about 17% higher in comparison to unreinforced matrix alloy. The fracture toughness values as obtained by experimental and FEA are in close agreement with each other and marginal variation of 3–4% has been seen. This is a clear indication of uniform distribution of reinforcement particles in the matrix A- alloy (LM13).

Keywords LM13 · Fused SiO_2 · Chills · Fracture toughness · FEA

1 Introduction

Composite is a mixture of various materials designed for a specific application having excellent mechanical properties; usually, it is a mixture of two or more physically and chemically distinct materials, e.g. metals/ceramics/polymers with a discrete interface between the mixtures, which have a combination of several properties as compared to an unreinforced matrix. They have enhanced properties like high strength to

H. S. Harshith (✉)
Department of Mechanical Engineering, Channabasaveshwara Institute of Technology Gubbi, Tumkur, Karnataka, India
e-mail: hshmech.29@gmail.com

J. Hemanth
Associate Dean, Presidency University, Bengaluru, Karnataka, India
e-mail: joelhemanth@hotmail.com

S. Seetharamu et al. (eds.), *Fatigue, Durability, and Fracture Mechanics*, Lecture Notes in Mechanical Engineering, https://doi.org/10.1007/978-981-15-4779-9_38

weight ratio, higher Young's modulus even at elevated temperature, high toughness, improved fatigue resistance and wear properties with minimum thermal shock and thermal conductivity.

Aluminium-based composite casting is subjected to micro-shrinkage during casting because of its freezing range. The composite materials can be an alternative to conventional materials because of their excellent strength and stiffness. Also composite with desired properties can be obtained by suitably reinforcing the matrix alloy [1, 2]. The remarkable capacity of the composite materials for the particular necessities makes these materials more prominent in many industries, for example, aerospace, automobile (cylinders, chamber liners) and auxiliary segments, bringing about saving in material and energy [3, 4]. The use of MMCs is constrained due to the cost of equipment which are required for developing them and also processing is difficult and laborious. The hot and cold rolling procedure is used to manufacture specific strengthened MMCs of high complexities [5, 6].

The characterizations of mechanical properties of discontinuous MMCs are available from various researchers' reports [7, 8]. With reference to these, the mechanical properties, for example, Young's modulus and quality have been enhanced by 20–40% due to the addition of reinforcements. Notwithstanding, pliability has disintegrated amazingly with the expanding substance of reinforcements [9, 10].

The present work focuses on the preparation and property evaluation of a newly developed chilled MMC with Al-alloy (LM13) as the matrix material and Fused SiO_2 as the reinforcement. From the literature overview, it is comprehended that by the addition of particulates to Al-alloys, improvement in tensile, compression strength hardness, fracture toughness can be obtained. So far no research work has been done to evaluate FT and its validity through FEM for chilled Al-alloy (LM 13)/Fused SiO_2 composite.

2 Experimental Studies

Chills were placed in the mould as shown in Fig 1. The matrix was melted in graphite crucible Fig 2 at a temperature of 640–650 °C and then preheated Fused SiO_2 was added into the matrix. It was followed by constant stirring using mechanical stirrer Fig 3 so that there through mixing of matrix and reinforcement. Figure 4 shows the final casted specimens.

2.1 Fracture toughness test procedure

Close-looped INSTRON material testing machine was used to conduct fracture toughness test. The tests were carried out as per ASTM E 399 1990 standards, containing a Chevron notch at the centre. The specimens were pre-cracked as shown in Fig 5.

Fig. 1 Chills placed in mould

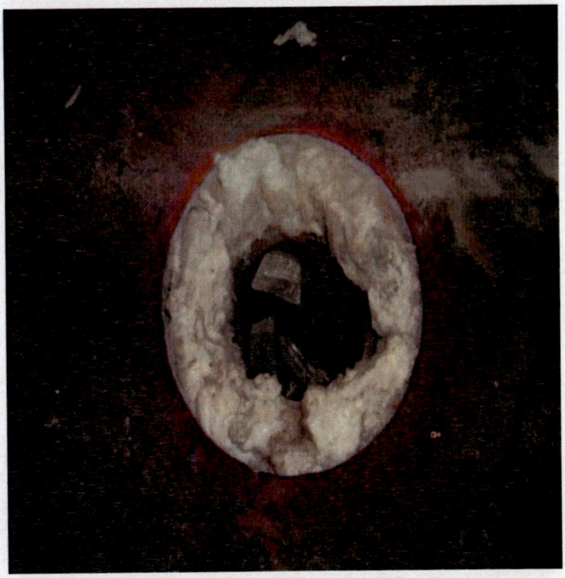

Fig. 2 Aluminium ingots placed in crucible

2.2 *Finite Element Analysis*

The FEA model Fig 6 was made in catiaV5, meshing using Hypermesh Fig 7 and analysis using ansys R15 Fig 8.

Fig. 3 Vortex creation using mechanical stirrer

Fig. 4 Casted specimens

3 Results and Discussions

A 25 mm thick chills were used during the casting of composite specimens. The various chills used and the test results of fracture toughness are listed in Table 1. As observed from the table that copper chill casted specimens have highest fracture

Fig. 5 Fracture specimens

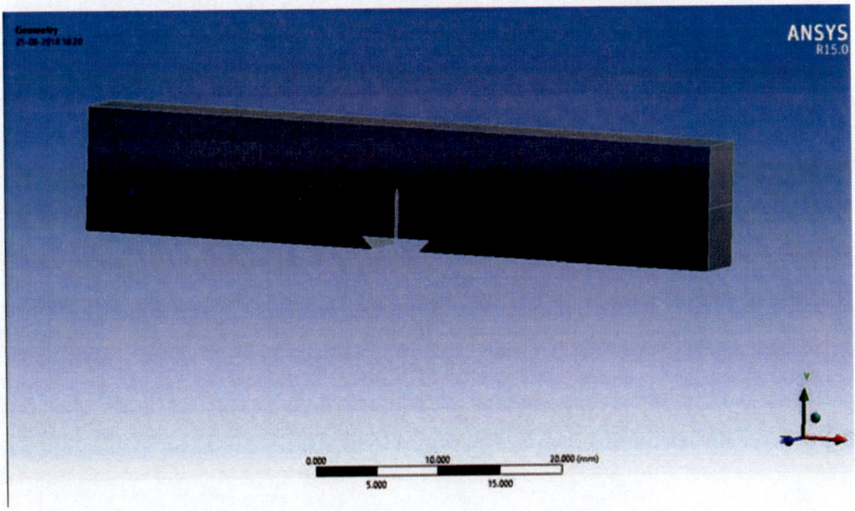

Fig. 6 3D model of fracture specimen

toughness as compared to others. Also from Fig 9 it can be seen that fused SiO_2 contents beyond 9% wt, there is a decrease in fracture toughness.

The possible reasons for this trend are:

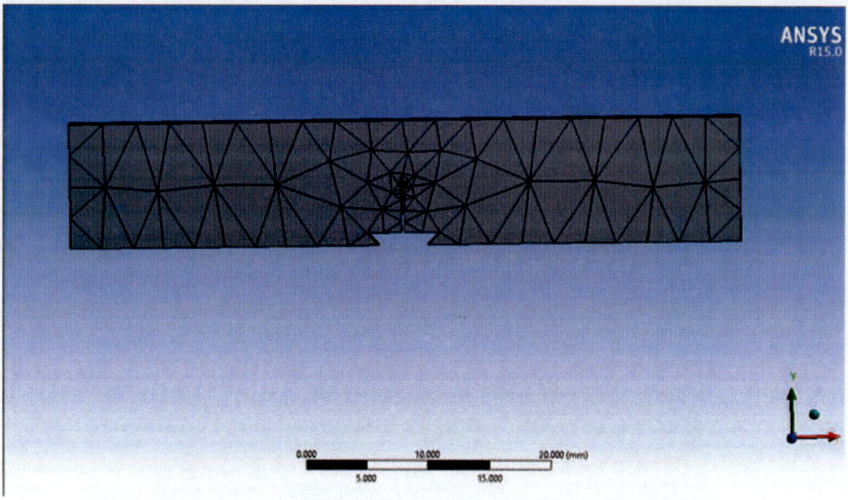

Fig. 7 Meshed 3D model

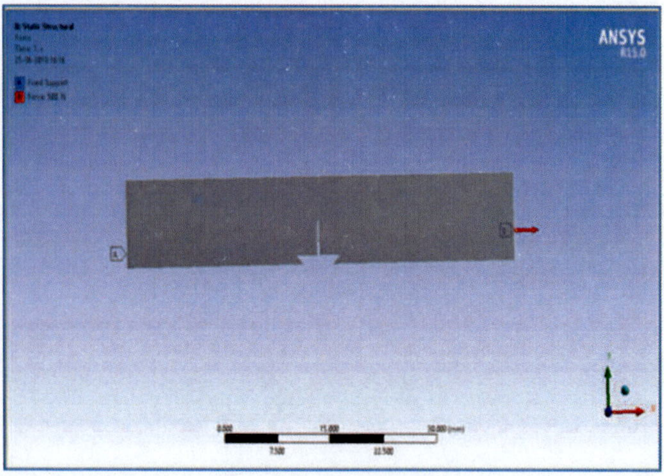

Fig. 8 3D meshed model after applying boundary conditions

- Load transfer between the soft Al-alloy (LM13) matrix and the hard SiO_2 ceramic particle reinforcement.
- Hardening arising from constrained plastic flow and tri-axi-ability in the Al-alloy (LM13) matrix due to the presence of the brittle reinforcements and chilling effect.

Figure 9 shows an increase in the fracture toughness of the MMCs developed which could be primarily attributed to the presence and uniform dispersion of hard

Table 1 Fracture toughness (K_{1C}) test results (in MPa $\sqrt{m}$) Kmax

wt.% Reinforcement → / Type of chill ↓	Fracture Toughness K_{IC} (MPa$\sqrt{m}$)			
	3	6	9	12
Copper	7.840	7.546	9.706	7.227
Mild Steel	5.970	6.592	9.632	6.148
Silicon Carbide	5.250	7.170	9.070	4.830
Graphite	5.187	6.352	8.059	3.959

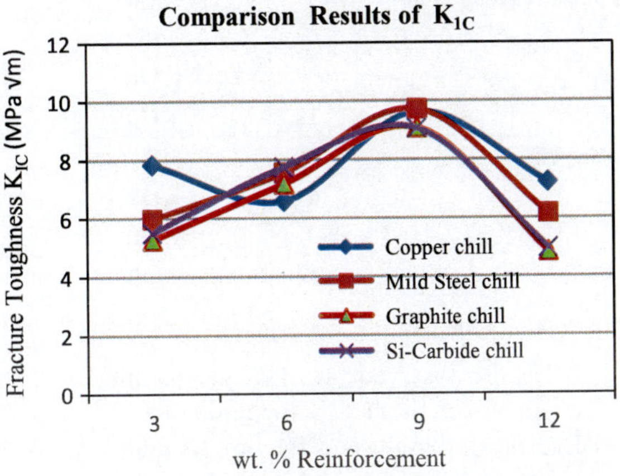

Fig. 9 Plot of K_{1C} V/s wt.% of reinforcement

SiO_2 particulates in Al-alloy (LM13). Also, good bonding and compaction of reinforcement with the matrix alloy. The harder ceramic particulates lead to higher dislocation densities and variation in fracture toughness.

The fracture toughness values as obtained by experimental and FEA as shown in Table 2 are in close agreement with each other and marginal variation of 3–4% has been seen. This is a clear indication of uniform distribution of reinforcement particles in the matrix Al-alloy (LM13).

4 Conclusions

In this research paper, various chills were used for casting out of which copper chill has exhibited excellent strength and fracture toughness.

- Chilled castings of LM13 Fused SiO_2 are going to provide lot of engineering benefits.

Table 2 FEA Results of Fracture toughness test for various wt.% reinforcement and different chills

	Fracture toughness (MPa$\sqrt{m}$)			
Copper chill	wt.%	K_{IC}(Expt)	K_{IC}(FEA)	Error %
	3	7.840	7.760	1.02
	6	6.592	6.532	0.91
	9	9.432	9.265	1.77
	12	5.927	5.628	5.04
Mild Steel chill	3	5.970	5.860	1.84
	6	7.346	7.210	1.85
	9	9.706	9.562	1.48
	12	6.245	6.159	1.37
Silicon Carbide chill	3	5.950	5.850	1.68
	6	6.700	6.625	1.11
	9	9.170	9.150	0.21
	12	4.430	4.321	2.46
Graphite	3	5.587	5.534	0.94
	6	7.952	7.824	1.60
	9	9.120	9.085	0.38
	12	4.559	4.321	5.22

- It was also observed that MMCs containing 9 wt.% have registered highest fracture toughness of about 17% higher in comparison to unreinforced matrix alloy.
- Excellent strength and fracture toughness properties were optimum till 9wt.% reinforcement, beyond this wt.% a decreasing trend was observed. Due to these properties, it sure that LM13 Fused SiO_2 can lead to very good futuristic applications.

References

1. Davidson AM, Regener D (2000) A comparison of aluminum-based metal–matrix composites reinforced with coated and uncoated particulate silicon carbide. Compos Sci Technol 60:865–869
2. Suery M, L_Esperance G (1993) Interfacial reactions and mechanical behavior of aluminum–matrix composites reinforced with ceramic particles. Key Eng Mater 79–80, 33–46
3. McDanels DL (1985) Analysis of stress–strain, fracture, and ductility behavior of aluminum–matrix composites containing discontinuous silicon carbide reinforcement. Metall Trans A 16A:1105–1115
4. Krstic VD (1983) On the fracture of brittle–matrix/ductile–particle composites. Philos Mag A 48:695–708
5. Seah KHW, Hemanth J, Sharma SC (1995) J Mater Sci 30:4986
6. Seah KHW, Hemanth J, Sharma SC (1996) Wear 192:134

7. Seah KHW, Hemanth J, Sharma SC (1998) J Mater Design 19:269
8. Radhakrishna K, Seshan S (1981) Trans AFS 89:158
9. Walther WD (1954) Trans AFS 62:219
10. Ruddle RW (1950) J Inst Mater 77:37

Numerical Analysis and Prediction of FLD For Al Alloy-6063

B. R. Mahesh and J. Satheesh

Abstract Forming limit curves is one of the important parameters to know the formability of a material. In the present work, forming limit diagram of Al-6063 was obtained by using analytical software PAM-STAMP and it is compared by conducting out-of-plane test or Nakazima test. The experimental procedure for determining the forming limit curve requires expensive equipment. Forming limit diagram of Al-6063 is found by conducting out-of-plane test or Nakazima test in PAM-STAMP software. Trials are conducted to obtain actual FLD in terms of major and minor strains with respect to the velocity of the punch. Obtained results are compared with experimental values, there is a good agreement value between numerical and experimental solution.

Keywords Forming limit diagram · Al-6063 · Nakazima test

1 Introduction

Aluminium alloys are widely used in automobile industries due to their superior properties like strength to weight ratio corrosion resistance and superior formability properties. In order to give proper shape to the material [1], formability is one of the criteria, which plays a major role to give shape to the material. To know the formability of a material there are many methods like Erichsen cupping test, Nakazima test, etc. Aluminium alloy like 6063 is a new promising material that has a base material strength of 241 Mpa. Al 6063 is used in automobiles, aerospace, kitchen cabinets however it can be used in some of the automobile [2] parts, so it is important to study forming limit diagram to manufacture certain components.

B. R. Mahesh (✉)
Dayananda Sagar College of Engineering, Banglore, India
e-mail: dsce.mahesh@gmail.com

J. Satheesh
SJB Institute of Technology, Banglore, India
e-mail: satheeshsjbit@gmail.com

S. Seetharamu et al. (eds.), *Fatigue, Durability, and Fracture Mechanics*, Lecture Notes in Mechanical Engineering, https://doi.org/10.1007/978-981-15-4779-9_39

The determination of FLD is a complicated task and researchers on FLD are still a subject of experimental and numerical studies [1]. There are two types of experimental tests that are applied to determine the FLD (1) out-of-plane test (2) Marciniak in-plane formability tests. In the present out-of-plane test is conducted [3]. Out-of-plane test in this method sample sheet metal clamped in circular die and it is deformed by a hemispherical punch [4]. The out-of-plane formability test used more extensively than the Marciniak in-plane test.

Geometries used in Marciniak test are complex; one of the major drawbacks of this test is that positive range of FLD is difficult to obtain [5] full range. FLD Nakazima test is best preferred. Different strain paths are obtained by using the specimens with hemispherical punch and a circular die for varying widths [6].

Nakazima test is the most powerful tool because of the following reason. The geometries of the punch and dies used are not complex [7], FLD can be determined in the full range. In order to evaluate forming limit diagrams in industries as well as in sheet metal testing laboratories Nakazima test is the best method compared to other methods [8].

Forming limit diagram is a useful method for the critical analysis of formability of sheets used in forming processes [9]; FLD is valid only for cases of proportional loading, where the ratio of the principal stresses is constant throughout the forming process [10].

With the use of forming limit diagram in industries, they can evaluate the formability properties of the sheet materials without conducting any experiments to get the required shape for a sheet metal [11]. High strength aluminium alloys like AL-6063 have high strength to weight ratio and good corrosion resistance. Al 6063 possess high percentage of magnesium in the range of about 0.45–0.9% helps in good spring back properties to the sheet metal [12]. Steel sheets of different grades are used in an automobile like Panel Front Header, Panel Skirt, Panel Gutter Deck lid, Panel Body Side Outer, Panel Wheelhouse Outer which have strengths lesser than Al 6063 [13]. These sheets can be replaced by Al 6063 to make the respective structure stronger for the same thickness. It is also evident that the density of the aluminium alloy is very less compared to the different grades of steel used in the car bodies [14].

2 Methodology

2.1 Experimental Materials and Properties

The material chosen for the present studies of formability is Al-6063, because of its high specific strength it can be used to replace certain steel components in automotive industry. Thereby reducing the overall weight of the vehicle and it will help in the fuel economy. Sheet of aluminium alloy 6063 with a blank size of 1 × 1 feet is shown in Fig. 1, and a thickness of 1 mm is used. The thickness of the material is 1 mm chosen based on literature review; in most of the subcomponents and body parts

Fig. 1 Forming limit curve

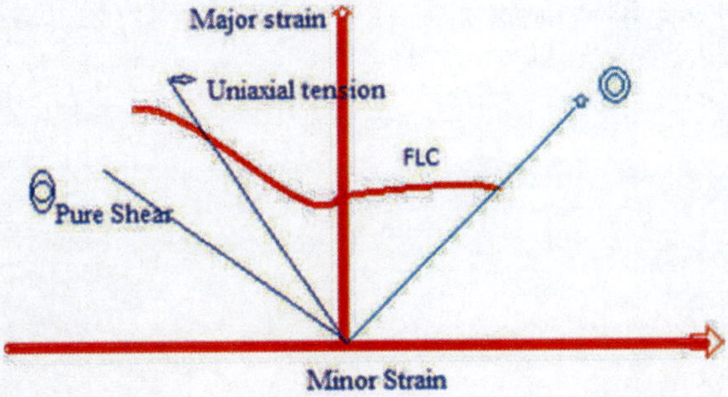

Table 1 Chemical composition of Al 6063

Material	Al	Mg %	Si %	Cr %	Cu %	Fe %	Zn %
Al-6063	95.8–97.7%	2.2–2.8	0.25	0.15–0.35	0–0.1	0–0.4	0–0.1

Table 2 Mechanical properties of Al 6063

Yield strength (MPa)	Ultimate tensile strength (MPa)	Elastic modulus (MPa)	R0	R45	R90	K (MPa)	n	ρ(g/cm^3)
214	241	68947.5	0.433	0.35	0.518	295	0.31	2.69

of automobiles, they use 1 mm thickness. Chemical composition and mechanical properties of aluminium alloy 6063 are given in Tables 1 and Table 2, respectively (Fig. 2).

3 Experimental Set up for Conducting Nakazima Test

3.1 Designing

NX CAD is the software employed for the design of the die, punch and blank. The dimensions of the Nakazima cupping test tool are designed by using ASTM standards obtained from literature review. Clearance is given between the die and the punch to make sure that there is no obstruction. The die and punch are designed using part design, whereas the blank is designed using sheet metal design by giving the sheet metal parameter as 1 mm. Care was taken during the design of the beads to make sure that there is a clearance of 1 mm between the male and the female lock bead to accommodate the sheet metal.

Fig. 2 Blank specimen Al-6063 (1ftx1ft)

3.2 Fabrication

Figures 3 and 4 show the punch and die are designed as per ASTM standards and proper clearance has been maintained between punch and die; clearance depends on the thickness of the sheet used [15]. The thickness of the sheet used is 1 mm so the clearance between punch and die is 1 mm. For Allen screws, provision is made in both top and bottom dies in order to fix the dies properly. Material used for punch and die is hot die steel H13 which has good machinability properties, and hardness range is HRC38–48, which is sufficient to hold the pneumatic pressure during the formation. Chemical composition and mechanical properties of hot die steel are given in Tables 3 and 4.

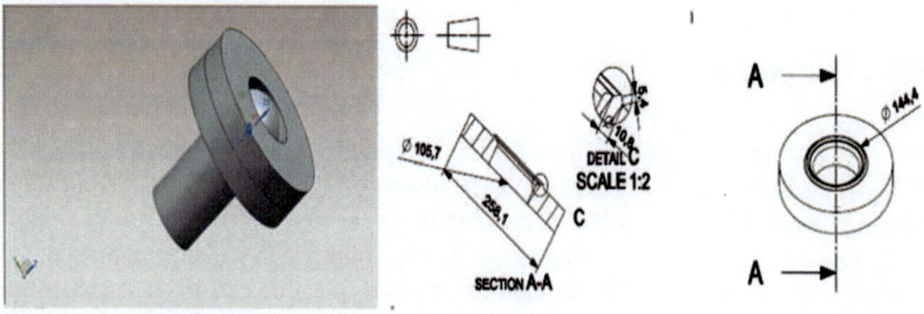

Fig. 3 Punch and die design

Fig. 4 Fabricated punch and die

Table 3 Chemical composition of punch and die (wt%)

Material	C %	Cr %	Ni %	Mn %	Si %	Fe %
Hot die steel	0.07	18.00	9.00	2.00	1.00	69.93

Table 4 Mechanical properties of Hot die steel H13

Yield strength (MPa)	Ultimate tensile strength (MPa)	Elastic modulus (GPa)	n
1380	*1590*	215	0.30

3.3 Numerical Investigation

Forming limit diagram for Al-6063 is obtained using PAM-STAMP software parameters used for Nakazima test is as follows.

Blank Dimensions used is Circular blank of diameter 181.94 mm, Blank holding force is 100 KN, Velocity of punch is 5 mm/second and Clearance between punch and die is 2.03 mm.

The strain path achieved by the Al 6063 blank specimen Figs. 5 and 6 is equal biaxial tension. The region around the lock bead where the blank is held. The blank experiences necking before its failure. The area between the dotted line and the solid line is the marginal zone in the FLD. If the major or minor strain of any element in the blank crosses the dotted lines, the considered element is said to be in the necking region. The major and minor strains can be studied from the readings taken through PAM-STAMP. The thinning percentage of the blank specimen was also found using the analysis software, but they were not found relevant cause aluminium and its alloys are not known to follow the FLD curve as much as steel or other elements do. It can be noticed from the final blank image that the crack (dark blue colour region) is initiated not from the tip of the dome but around it. In the contact force plot, the

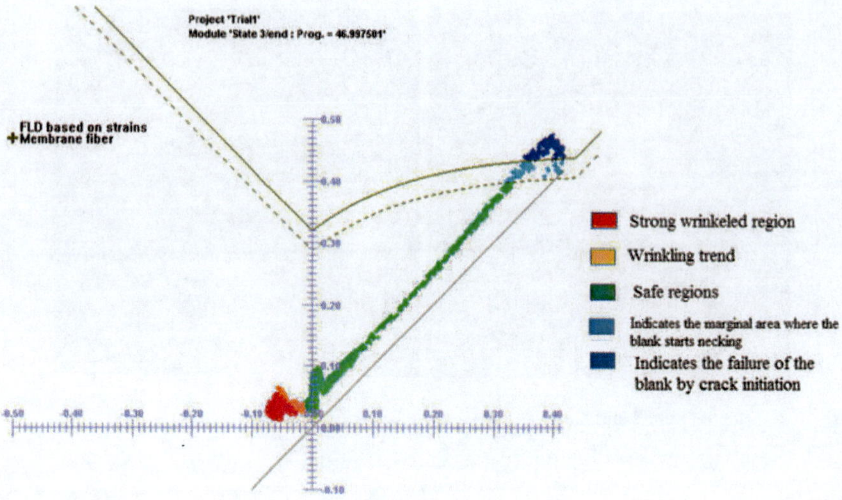

Fig. 5 Forming limit diagram for Al-6063

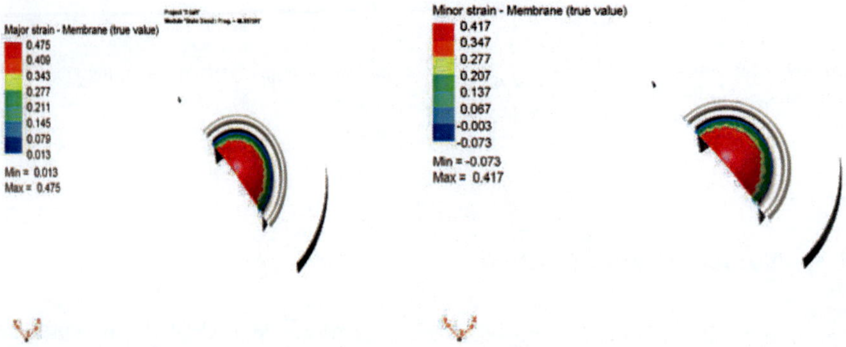

Fig. 6 Major and minor strain of Al-6063

abscissa represents the punch progression, whereas the ordinate represents the force applied by the punch on the blank. From the plotted contact force, we can notice that the force is applied gradually and it is not an impact force. We can also see that at a force of 35.117 kN there is a fracture in the blank.

In the contact force plot load versus displacement shown in Fig. 7 as the force is applied to the blank through punch, material starts deforming linearly up to specific load of 35.117 kN after that it has reached thinning value and material starts cracking at this point of time total displacement of material is about 47 mm.

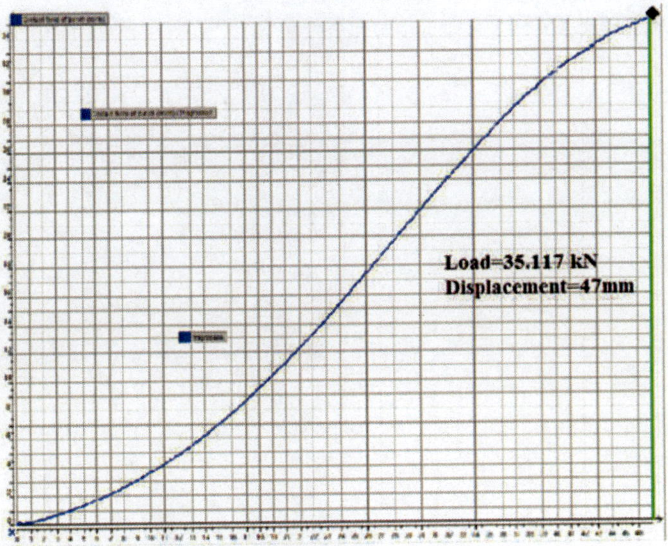

Fig. 7 Contact force plot load versus displacement

3.4 *Experimental Investigation*

After conducting numerical investigation, it is decided to conduct experimental investigation, punch and die designed as per ASTM standards, and the experiment is conducted using hydraulic press shown in Fig. 8.

Blank of thickness 1 mm Al-6063 is fixed between punch and die, and load is applied from hydraulic press through punch. Velocity of the punch is about 5 mm/second which is exactly the same as numerical evaluation velocity. Crack is initiated at a load of 36kN and a displacement of 50 mm is obtained during experiments, which is closely matched with the value of numerical analysis (Figs. 9, 10 and 11).

4 Results and Discussions

Results obtained during numerical analysis are nearly matching with experimental values and the results are tabulated in Table 5.

In order to calculate major and minor strains from the experimental values, small grid circles of 4 mm are drawn on the blank. After conducting the test those circles become ellipse having major and minor diameters, those major and minor diameters are measured using optical microscope, thereby change in length of the diameter is obtained. Formula for major strain/minor strain is Change in length/original length.

Fig. 8 Experimental setup

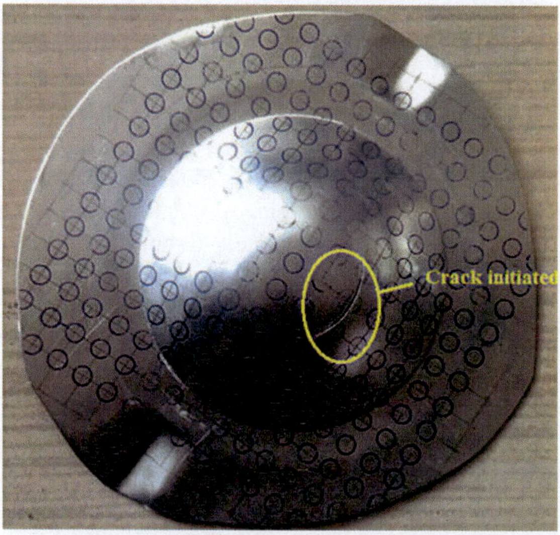

Fig. 9 Blank after Nakazima test

Most of the numerical values obtained through numerical values are matching with experimental values which are shown in Table 6.

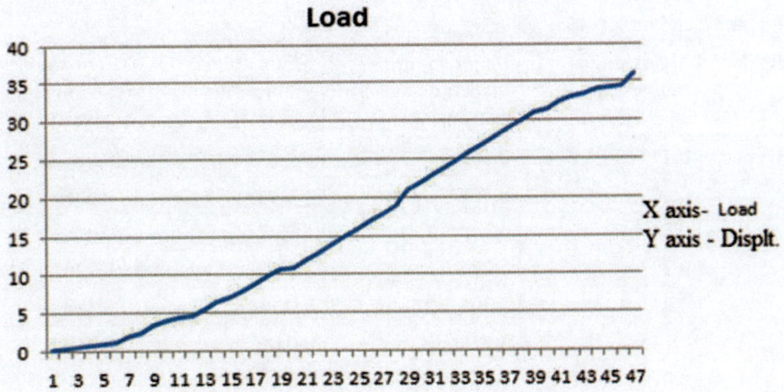

Fig. 10 Contact force plot load versus displacement experiment values

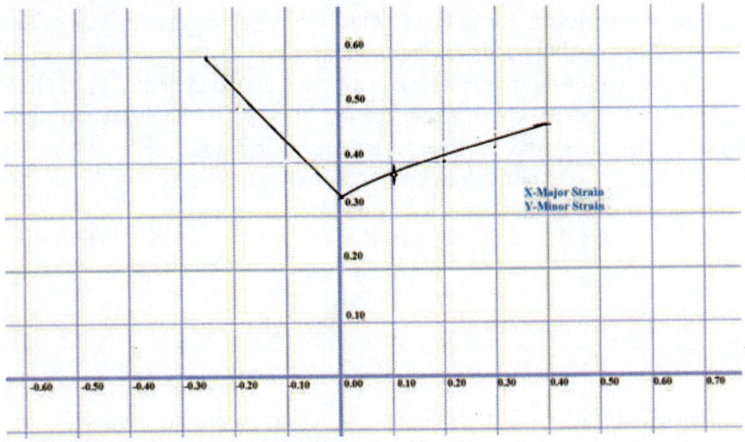

Fig. 11 Forming limit curve from experimental values

Table 5 Al-6063 blank displacement

Sample	Displacement numerical analysis (mm)	Displacement Experimental analysis (mm)	Error(%)
Al-6063	47	50	6

5 Conclusions

(1) Results show that PAM-STAMP FEM software method has a good accuracy for punch load versus displacement. An error of 6% between numerical and experimental values

Table 6 Major and minor strain values

Sample	Major strain (numerical analysis)	Major strain (experimental analysis)	Minor strain (numerical analysis)	Minor strain (experimental analysis)
Al-6063	0.475	0.463	0.417	0.40
	0.409	0.435	0.347	0.335
	0.343	0.348	0.277	0.300
	0.277	0.256	0.207	0.220
	0.211	0.205	0.137	0.100
	0.145	0.125	0.067	0.050
	0.079	0.06	−0.003	−0.25
	0.13	0.10	−0.073	−0.20

(2) Sample of experiment shows that fracture occurs at the bottom pole of the blank due to more thinning occurs at that region.
(3) Contact force plot shows that crack is initiated at a load of 35.117 kN in numerical analysis and 36 kN in experimental analysis. It shows good agreement between the values of experimental and numerical analysis.
(4) Major and minor strain values of numerical values match with the values of major and minor strain conducted through experiments.

References

1. Safdarian R (2016) Stress based forming limit diagram for formability characterization of 6061 aluminium. Trans Nonferrous Met Soc China 26:2433–2441. https://doi.org/10.1016/S1003-6326(16)64350-9
2. Keeler SP, Backofen WA (1963) Plastic instability and fracture in sheets stretched over rigid punches [J]. Trans ASM 56:25–48
3. Goodwin GM (1968) Application of strain analysis to sheet metal forming problems in the press shop [J]. SAE Tech Pap
4. Safdarian R, Jorge RMN, Santos A, Naeini HM, Parente MPL (2015) A comparative study of forming limit diagram prediction of tailor welded blanks [J]. IntJ Mater Form 8:293–304
5. Jain M, Allin J, Lloyd DJ (1999) Fracture limit prediction using ductile fracture criteria for forming of an automotive aluminum sheet[J]. Int J Mech Sci 41:1273–1288
6. Ozturk F, Lee D (2004) Analysis of forming limits using ductile fracture criteria [J]. Mater Process Technol 147:397–404
7. Safdarian R (2015) Forming limit diagram prediction of tailor welded blank by modified M—K model [J]. Mech Res Commun 67:47–57
8. Safdarian KR, Moslemi NH, Torkamany MJ, Liaghat G (2013) Experimental and theoretical investigation of thickness ratio effect on the formability of tailor welded blank [J]. Opt Laser Technol 51:24–31
9. Stoughton TB (2000) A general forming limit criterion for sheet metal forming [J]. Int J Mech Sci 42:1–27

10. Min HE, Fu-guo LI, Zhi-gang WANG (2011) Forming limit stress diagram prediction of aluminum alloy 5052 based on GTN model parameters determined by in situ tensile test [J]. Chin J Aeronaut 24:378–386
11. Müschenborn W, Sonne H (1975) Influence of the strain path on the forming limits of sheet metal [J]. Arch Iron and Steel Ind 46:597–602
12. Panich S, Barlat F, Uthaisangsuk V, Suranuntchai S, Jirathearanat S (2013) Experimental and theoretical formability analysis using strain and stress based forming limit diagram for advanced high strength steels [J]. Mater Des 51:756–766
13. Chen MH, Gao L, Zuo DW, Wang M (2007) Application of the forming limit stress diagram to forming limit prediction for the multi-step forming of auto panels [J]. Mater Process Technol 187–188:173–177
14. Uthaisangsuk V, Prahl U, Bleck W. (2007) Stress based failure criterion for formability characterization of metastable steels [J]. Comput Mater Sci 39:43 – 48
15. Arrieux R (1995) Determination and use of the forming limit stress diagrams in sheet metal forming [J]. J Mater Process Technol 53:47–56

Fault Detection and Health Condition Analysis of Single Stage Gear Box System

Baquer Mohiuddin, S. B. Kivade, and Sangamesh Rajole

Abstract Every machine with the relative motion of parts produces sound and vibration. All the gear boxes usually generate vibrations and respective vibration signatures may be taken as referable characteristics if the condition of the gear is good. During working condition, whenever fault occurs, it may result in serious damage of the gear box. The change in the gear pair meshing could result in changes in vibration signals. The accelerometer mounting on the gear box system is the accurate task for assessment of pair of gear. So the technique of monitoring the condition is very essential to prevent and diagnose the vibration of gear box. Nowadays damage identification and condition monitoring of gear boxes in the industrial machinery have received more attention from the researchers. To analyze the various fault and problems related with gear box failure in a working environment efficiently and accurately, few technologies like material technology, information technology, and processing of signals, etc. bring latest solutions. For the assessment of industrial gear boxes, many investigations are carried out for monitoring the condition of machinery. Signal processing and vibration analysis techniques are well known and much suitable for industrial practices. Since, the signals of vibrations from the gear box are transient and non-stationary in nature. Every technique has some disadvantages and may not be used in all condition, i.e., few failure detection is not possible by simple vibration method. At an early stage, simple analysis by spectral is not very successful to find the injury of gear.

Keywords Damage detection · Gear box · Vibration signal analysis

---

B. Mohiuddin (✉)
Visvesvaraya Technological University, Belagavi 590018, Karnataka, India
e-mail: baqarmohiuddin29@gmail.com

S. B. Kivade
Basavakalyan Engineering College, Basavakalyan, Bidar, Karnataka, India
e-mail: sbkivade@gmail.com

S. Rajole
Department of Metallurgical and Materials Engineering, National Institute of Technology Karnataka, Surathkal, Mangalore 575025, India
e-mail: rajolesangmesh@gmail.com

S. Seetharamu et al. (eds.), *Fatigue, Durability, and Fracture Mechanics*, Lecture Notes in Mechanical Engineering, https://doi.org/10.1007/978-981-15-4779-9_40

1 Introduction

Productivity is a key weapon for producing firms to remain competitive and continuous developing internationally. Enlarged productivity will be achieved through enlarged accessibility. Managing industries into the twenty-first century may be a difficult task. Increasing international competition, quick technological modification, consumer's awareness toward total quality, responsibility, health and safety, environmental issues, and changes in management structure not solely provides several firms with significant opportunities to boost their performance. Gear mechanisms are a vital component in a very sort of mechanical systems. To reveal the severity of the tooth surface, a specific lubricant film thickness is used, which is another method for predicting gear failure [1]. Fault identification in gear has been the subject of thorough investigation and many of the ways supported vibration signal analysis are developed. Commonplace ways embrace crest issue, time-domain averaging, and reception that can be verified to be efficient in fault diagnosing and are presently well described. The intension of exploitation gears in machinery square measures primarily to transmit movement in between shafts whereas keeping the meant angular rate relation along with sleek motion is transferred and gives high efficiency. For the decomposition of signal into number of levels with various frequency bandwidth, Fourier transform is used [2]. Once a defect touching single or additional gear teeth creates on a gear. There are different failure modes like scuffing, pitting, bending fatigue crack, and abrasive wear, which is caused by excessive stress along with an improper lubrication. Fault symptoms have been shown from the signals of vibration of gear [3]. These defects need to be eliminated or diagnosed. The vibrational information output taken from the gear box gives the operation of gear box [4]. Techniques reminiscent of wear and junk analysis and acoustic emissions need accessibility to the shell either to gather samples or to mount the transducers on or close to the shell. The different dynamic responses are taken for comparing the effectiveness of signal processing technique and traditional method [5]. Many of the defects confront within the rotating parts of machinery create a definite vibration outline and hence largely faults will be known. The record of the different responses can be done in vibration monitoring and determination of vibrational data, therefore capable of monitoring rotating machines. The original signal has been breakdown for approximation and detail using discrete wavelet analysis [6].

The philosophy of machine condition monitoring is to observe the state of a machine and to discover any deterioration in condition, to work out the reason for this deterioration, and to predict once failure will be expected. Successful feature in time–frequency data for gear fault identification is taken for gears [7]. Vibration analysis is that the most typically used technology accustomed to monitor the condition of the machine. The frequency of the vibrations also can be mapped or diagrammatic, once sure frequencies are gift. Comparison of the vibration spectra of latest instrumentation versus instrumentation that has been used can offer the knowledge. All the techniques applied for reducing vibration were responsive to measure direction [8]. The monitoring in gear box includes proper examination of

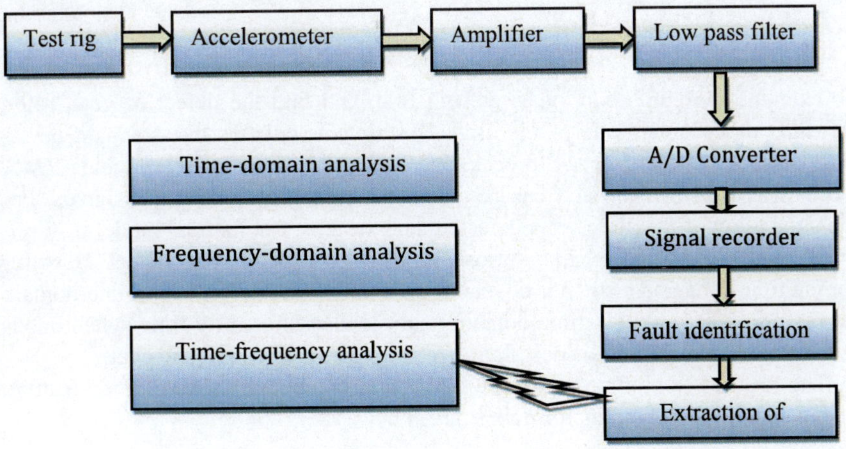

Fig. 1 Vibration signal analysis to diagnose fault

the condition of gear and its modification with respect to time. Gear condition can be examined by considering different physical parameters such as noise and vibration, temperature, wear junk, and contamination of oil. A modification in these parameters known as 'signatures' would, therefore, show the modification within the condition. Identification of the fault is usually carried out in the following phases: knowledge acquisition, extraction of feature, detection and identification of fault as shown in Fig. 1 [9]. Modulated vibration gives the relation between the frequencies and phase between the mesh vibration and the modulation function which is related to the local defect of the gearing [10]. For healthy and defective gear in both time and frequency domains, vibration acceleration levels have been studied [11].

2 Methodology

For diagnosis of mechanical systems, there are number of signal processing methods. To understand the concept of signal processing, knowledge, and investigation are necessary for the selection of suitable signal processing. The development of new method in the damage of gear and its detection by new gear defect determination parameter which depends on energy change in analysis of frequency of vibration signals. Many damages were introduced manually at the gear boxes for proper understanding of impacts on signal analysis and to validate the damage identification [12]. Emission acoustic and vibration signals are the most common data of waveform in monitoring. The time and frequency domain's analysis are the two major categories of stationery data analysis. The methods are able for the improvement of classification level. Experimentation work can be done for determining the robustness of method [13].

2.1 Time-Domain Analysis

To examine the amplitude, phase information, and find the defect of gear, rotor, etc. then time-domain analysis is used. Descriptive statistics like standard deviation, mean, crest factor, peak to peak, and high order statistics that include RMS, kurtosis as well as skewness can be calculated from time waveform signals. For the simulation of dynamics of a defected gear system, a numerical model is developed. Statistical parameter like kurtosis is suitable to identify the defect. However for multiple damaged teeth, it is not enough to localize the defect [14]. Time-domain averaging is the common time-domain analysis also known by time-synchronous averaging. Figure 2 shows a description of the traditional diagnostic chart.

The features of high order statistics are usually known by time-domain feature, which contain RMS and its formula is given by

$$\text{RMS} = \sqrt{1/N \sum_{n=1}^{N} [x(n) - \overline{x}\,]^2}$$

$$\text{Crest Factor} = \frac{\text{Peak}}{\text{RMS}} = \frac{\text{Crest Value}}{\text{RMS Value}} = \frac{\sup|x(n)|}{\sqrt{\frac{1}{N\sum_{n=1}^{N}[x(n)]^2}}}$$

$$\text{Peak} = \text{Crest Value} = 1/2[\max(x(t)) - \min(x(t))]$$

And also kurtosis K is given by

$$\text{Kurtosis (K)} = \frac{1}{\sigma^4}\sum_{i=1}^{N}\frac{(x_i - x)^{-4}}{N} = \frac{\frac{1}{N}\sum_{n=0}^{N-1}[x(t) - \overline{x}]^4}{(\text{RMS})^4},$$

where $\overline{x} = \frac{1}{N}\sum_{n=1}^{N} x(n)$

N Number of samples within the signal,
$x(n)$ Signal of time-domain,

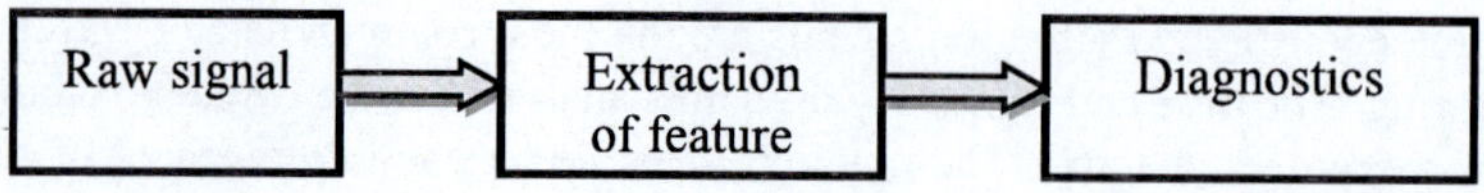

Fig. 2 Gear box diagnostics chart

σ^4 variance square,
$\overline{x}$ Samples of mean value,
X_i Individual sample.

2.2 *Frequency-Domain Analysis*

To represent the domain for analysis of functions of mathematics with frequency then frequency-domain is used to easily isolate and identify few frequency components. Spectrum analysis is most widely used technique using FFT. For better fault predictions, preconditioning of the vibration signal is essential before extraction of features [15]. The spectrum analysis is given by

$$E[x(f)\, x * (f)],$$

where

$x(f)$ is the Fourier transform,
E is expectation,
'*' denotes complex conjugate.

3 Vibration Analysis Techniques and Failure of Gear Box

The major cause for failure of gear is mainly due to design error or dimensional error, error in manufacturing, and error in application. Improper use of gear geometry, non-suitable materials, quality and few other specifications are coming under the design errors. If there is a problem in heat treatment and machining then it is in manufacturing errors. Lastly, application errors are due to difficulties like mounting and installation, maintenance, and cooling. The following Table 1 shows failure part and its causes [16]. Many of the researchers are working on damage identification and vibration analysis technique of gear box. The methods are implemented and tested are time-domain, frequency-domain, time–frequency domain based on short-time Fourier transform, advanced method of signal processing technique, and wavelet transform. The latest process of monitoring the system of gear is mostly recommended for industrial purpose. More systematic experiments with different threshold values, as well as various kinds of faults, should be done before the implementation of method [17].

Table 1 Safety critical failure modes

Failure	Failure mode	Cause	Contributing factors
Fracture of shaft	Fatigue	Unbalance	
		Misalignment	Coupling
		Bent shaft	Bearing failure
	Overloaded	Interference	Incorrect assembly
			Failure of bearing
		Operational	
Gear fracture	Fatigue	Life limit exceeded	
		Surface damage	
	Resonance	Design	
Tooth fracture	Bending fatigue	Life limit exceeded	
		Surface damage	Process related
	Random fracture	Design	Excessive wear
	Overload	Interference	Destructive scoring
		Operational	Process
Overheating	Lubrication	Insufficient oil	Oil line failure
		Loss of oil	Filter bowl failure
	Insufficient cooling	Cooling fan failure	Shaft/gear fracture

4 Results and Discussion

Vibration analysis of gear box system and simulation results are plotted below. The condition monitoring of gear system is based on the analysis of vibration with the meshing of gear tooth of pinion and gear.

Figure 3 shows the graph between the time and acceleration of meshing gears with the particular speed. The gear meshing waveform shows that the high amplitude range of vibration and it is responsible for load transmission. The pinion and gears are connected with the input and output shaft along with bearings. The accelerometer is attached to the bearing and gear housing and it is operating at particular frequency range. The analysis is taken for 20 s of vibration.

The output waveform of the healthy and faulty gear system has shown in Fig. 4. Here the tooth spall has considered as a fault. The frequency impact is high as in the faulty system and it was observed in time–frequency analysis.

Figure 5 shows the power spectra of the gear system with the applied speed of the operating condition. The separate range of the gear, pinion, and meshing of these two has clearly mentioned in the above figure. The fault is not in the pinion and while meshing with pinion its shows the power spectra range.

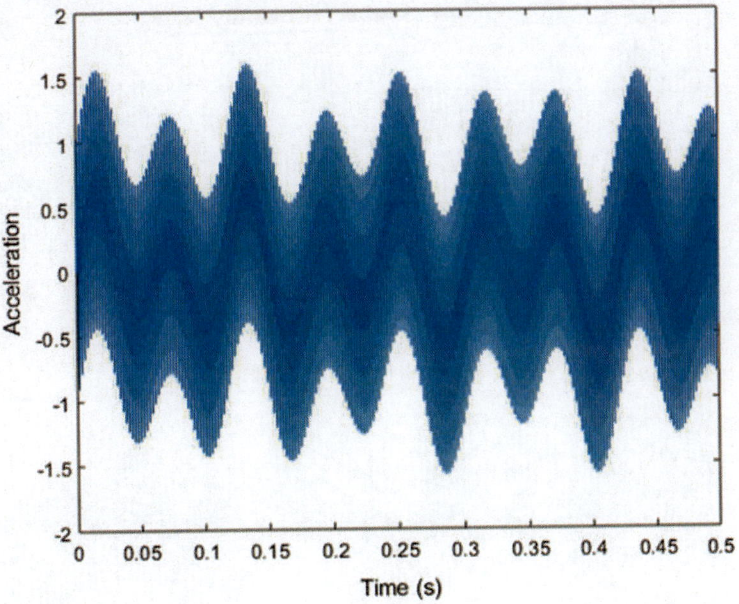

Fig. 3 Time and acceleration

5 Conclusion

To facilitate diagnosis and to reduce cost, vibration analysis technique is used. It is better technique as compared to other monitoring techniques. Skewness has not shown any effective fault but kurtosis is capable to detect the condition of defect. In diagnosis of shaft faults, spectral analysis is useful. To reduce the noise from the signals, time–frequency domain technique is used. Synchronous signal averaging has the ability of reducing complexity of diagnosis of faults of rotating parts like shaft and gear. The experimental results demonstrate that the proposed gear fault diagnosis method is simple, noise tolerant, and efficient.

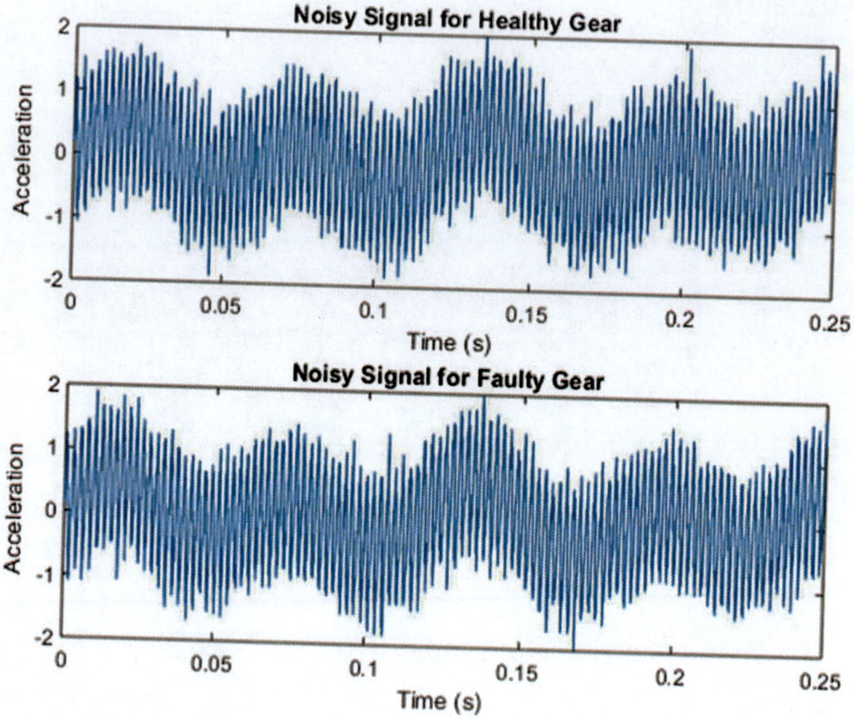

Fig. 4 Noisy and faulty signals

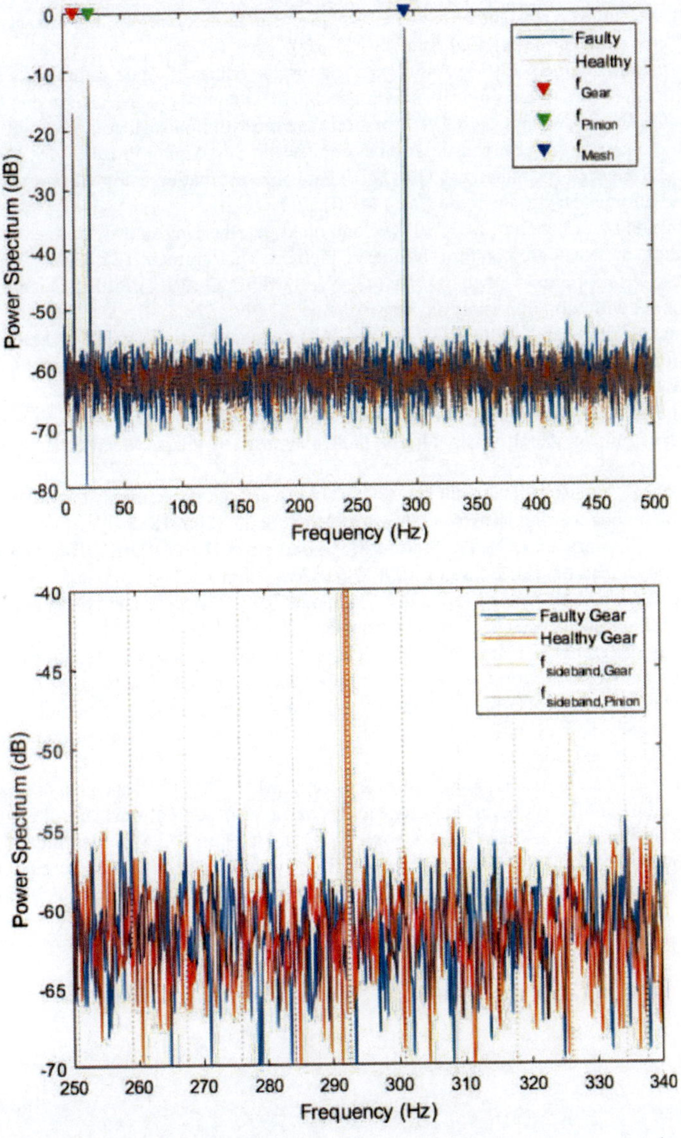

Fig. 5 Power spectrum versus frequency

References

1. Amarnath M, Krishna IP (2013) Detection and diagnosis of surface wear failure in a spur geared system using EEMD based vibration signal analysis. Tribol Int 61:224–234
2. Chen K, Hu J, Peng Z (2017) Analytical framework of gearbox monitoring based on the electro-mechanical coupling mechanism. Energy Procedia 105:3138–3145

3. Liang X, Zuo MJ, Hoseini MR (2015) Vibration signal modeling of a planetary gear set for tooth crack detection. Eng Fail Anal 48:185–200
4. Amarnath M, Lee SK (2015) Assessment of surface contact fatigue failure in a spur geared system based on the tribological and vibration parameter analysis. Measurement 76:32–44
5. Chen Z, Zhai W, Wang K (2019) Vibration feature evolution of locomotive with tooth root crack propagation of gear transmission system. Mech Syst Signal Process 115:29–44
6. Biswal S, George JD, Sabareesh GR (2016) Fault size estimation using vibration signatures in a wind turbine test-rig. Procedia Eng 144:305–311
7. Dhamande LS, Chaudhari MB (2018) Compound gear-bearing fault feature extraction using statistical features based on time-frequency method. Measurement 125:63–77
8. Elasha F, Ruiz-Cárcel C, Mba D, Kiat G, Nze I, Yebra G (2014) Pitting detection in worm gearboxes with vibration analysis. Eng Fail Anal 42:366–376
9. Deepika S, Kashwan KR (2016) Gear assembly fault analysis using dual-tree complex wavelet transform. In: 2016 international conference on energy efficient technologies for sustainability (ICEETS), April, pp 632–636. IEEE
10. Guoji S, McLaughlin S, Yongcheng X, White P (2014) Theoretical and experimental analysis of bispectrum of vibration signals for fault diagnosis of gears. Mech Syst Signal Process 43(1–2):76–89
11. Vernekar K, Kumar H, Gangadharan KV (2014) Gear fault detection using vibration analysis and continuous wavelet transform. Procedia Mater Sci 5:1846–1852
12. Barbeeri N, Barbeiri GDSAV, Martins BM, Barbeiri LDSAV (2017) Damage analysis in automotive gearbox. Procedia Eng 199(2017):1858–1863
13. Wang JJ, Zheng YH, Zhang LB, Duan LX, Zhao R (2017) Virtual sensing for gearbox condition monitoring based on kernel factor analysis. Pet Sci 14(3):539–548
14. Jena DP, Panigrahi SN, Kumar R (2013) Gear fault identification and localization using analytic wavelet transform of vibration signal. Measurement 46(3):1115–1124
15. Bordoloi DJ, Tiwari R (2014) Support vector machine based optimization of multi-fault classification of gears with evolutionary algorithms from time–frequency vibration data. Measurement 55:1–14
16. Wu S, He D (2008) Shaft diagnostics and prognostics development and evaluation using damaged dynamic simulation. In: Proceedings of the institution of mechanical engineers, part O. J Risk Reliab 222(2):219–233. Choy FK, Chen H, Zhou J (2006) Identification of single and multiple teeth damage in a gear transmission system. Tribol Trans 49(3):297–304
17. Syta A, Jonak J, ukasz JedliĹ Ĺ, Litak G (2012) Failure diagnosis of a gear box by recurrences. J Vib Acoust 134(4):041006

A Review on Fracture Toughness Characterization of Aluminium Based Metal Matrix Composites

Doddaswamy Doddaswamy, S. L. Ajit Prasad, and J. Sharana Basavaraja

Abstract Aluminium (Al)-based metal matrix composites have their own applications in automotive components, aerospace structure and many other structural applications because of their low density, good plasticity, high strength, good machinability, excellent resistance to corrosion and sensible thermal and electrical conduction, etc. Many of those applications need higher fracture toughness that is the ability to resist failure because of crack propagation. Fracture properties of composites are essential in assessing the flaw tolerance of the structures. The Characteristics of base metal and the reinforcing materials play an important role in dominant the toughness of composites. The various reinforcements utilized in the processing of metallic-based composites are Silicon Carbide, Aluminium Oxide, Titanium carbide, titanium dioxide and Boron Carbide, etc. Mechanical properties of metal-based composites are greatly affected by the properties of matrix and reinforcement materials, their geometry, composition and also the technique of process these composites. The present paper reviews the early investigation methods and recent methods and practices followed to gauge fracture mechanics parameters like stress intensity factor, J-integral, crack tip opening displacement, etc., for aluminium alloys and aluminium-based composite materials.

Keywords Fracture toughness · Crack propagation · Metal matrix composites · Stress intensity factor · J-integral · CTOD

D. Doddaswamy (✉) · S. L. Ajit Prasad
PES College of Engineering, Mandya, Karnataka, India
e-mail: doddaswamy.1801@gmail.com

S. L. Ajit Prasad
e-mail: palspesce@gmail.com

J. Sharana Basavaraja
BMS College of Engineering, Bangalore, Karnataka, India
e-mail: khinaresh@gmail.com

S. Seetharamu et al. (eds.), *Fatigue, Durability, and Fracture Mechanics*, Lecture Notes in Mechanical Engineering, https://doi.org/10.1007/978-981-15-4779-9_41

1 Introduction

Most aluminium-based MMCs have found applications within the manufacture of various automotive engine parts such as crankshaft and connecting rod [1]. Engineering applications of metal-based composites are increasing in the area of automotive and aerospace industries. However, issues still stay with their poor damage tolerance properties under cyclic stresses. The effect of reinforcement on fracture behaviour of Al MMCs has been extensively investigated. Fracture toughness results are utilized much in the study of fracture mechanics for metallic materials selection to avoid brittle failure, for instance, in atomic energy power plants, aeronautics, offshore, ship building, bridge construction, pressure vessels and pipeline construction. Also composites are used in the rehabilitation process of old bridges and repair of pressure vessels. Material resistance, in terms of strength and fracture resistance, plays a vital role in achieving the desired life of the part.

Fracture toughness values are used to describe the resistance offered by materials against the growth of previously existing cracks by unstable crack development under the particular condition when time influences the loading condition at the crack tip. The fracture toughness to a certain extent describes the plasticity, varying with a change in the volume fraction of reinforcements. Literature showed that, the initial addition of reinforcements increases the toughness values, and decreases slightly with the addition of 10 vol. %. Reinforcement. In addition fracture starting from flaws or geometric discontinuities in materials can be described using the methods of continuum mechanics. Many research works are concentrating on the improvement of damage tolerance properties by controlling the micro structure. The reinforcement type, size, volume fraction and distribution are the vital parameters that might manage the fracture tolerance properties of MMCs [2].

2 Experimental Techniques in Fracture Toughness Testing

It is essential to create a crack in specimens before use in fracture toughness testing. The objective of precracking is to create a crack with a small crack tip radius. Standard servo-hydraulic machines, electromagnetic vibrating machines, and mechanically resonating machines are used to initiate a precrack. All the precracking techniques apart from the fatigue introduce several undesirable effects. Both fluctuating tension loading and completely reversed loading are utilized in hydraulic machines to precrack standard test specimens, whereas rotating beam bending is preferred in the case of circumferentially cracked specimens. The vibrators can even be utilized in conjunction with the servo-hydraulic testing apparatus to simulate the combined action of minor and major cycles acted upon several important parts [3].

2.1 Fracture Toughness Testing of Metals

For the measurement of fracture resistance and subcritical crack growth behaviour, different specimen configurations and procedures to conduct the test have been developed. Altogether cases, a crack should be introduced on the specimen as per standard crack configuration. Therefore, a different configuration of notches with a low notch –tip radius is used. After the introduction of precracking in the specimen, it should be loaded until failure. To calculate the bearing capacity of elements, fracture mechanics have been developed to determine the crack driving force required in the component and to compare with the material's resistance to failure. For fracture toughness testing many standards have been developed worldwide that enables fracture toughness to be obtained for various laboratory specimens. The foremost used testing specimen configurations are the compact tension specimen [1, 4–9] and single edge notched bend [10–14] specimen according to ASTM E-1820 and E-647 standards as given in Fig. 1. As referred in standard the specimen must be thick enough to produce plane-strain conditions near to crack tip. Load,

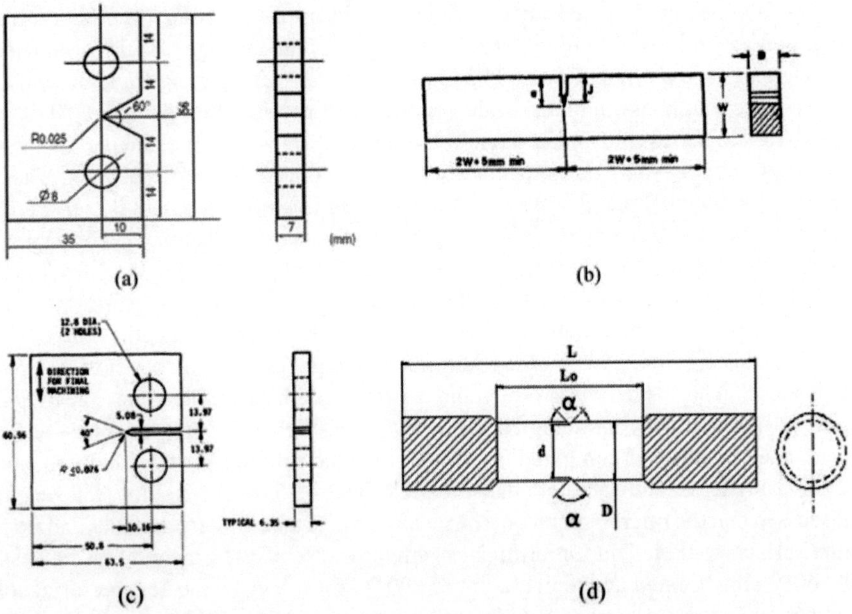

Fig. 1 Specimen configurations: **a** CT specimen with V-notch, **b** SENB specimen, **c** CT specimen with straight notch, **d** circumferentially cracked round bar

displacement curves and crack opening displacement curves are used to calculate the fracture toughness using formulae provided in standards. A condition under the test is conducted, and therefore, the calculation results are recorded and should be documented.

Also the standard testing procedure adopted for fracture toughness of metals is recommended by the American Society for Testing and Material [15]. The 3-point bend specimens are prepared as per ASTM E-399-74 [16]. The testing procedure of the tear test is given in ASTM B871. Normally K_{IC} values are determined by conducting the test using compact tension specimen configuration or by using a single edge notched bend specimen [17]. The ASTM E99 describes the best ways for accurate measurement of K_{IC} of a material having low plasticity and high strength [18].

2.2 *Fracture Toughness of Aluminium Alloys*

Investigation has been carried out to determine the effect of Wire EDM process on fracture property of Al 7075-T6. The Wire EDM creates a heat affected region on the surface nearer the crack that modifies the fracture behaviour [8]. The typical strain energy release rate found was 3186.96 J/m^2. The average CTOD formulated throughout the experiment was 0.00540 mm, at a crack tip opening angle of 40.63°.

Fracture toughness in plane-strain conditions of precipitation-hardened Al 7000 and Al 2000 series alloys and a strain-hardened alloy are been evaluated at room temperature and −320 °F using circumferentially cracked round specimen. These results indicate [19] that the plane-strain condition fracture toughness is considered as an inverse function of the yield strength of the material and at the level of equivalent yield strength the 7000 series alloys is having higher toughness compared to the 2000 series of alloys. The fracture resistance of the alloys increases with an increase in ageing temperature and the maximum value is obtained in the maximum-aged condition (1 h–225 °C). The addition of 2 weight percent of copper to Al alloy indicates that higher hardness values and tensile strength and reduced fracture toughness [20]. The alloy containing Ni showed 20.95% higher fracture toughness at the peak-aged parameters compared to same alloy in the solution-treated condition, and 25.17% higher fracture strength than the peak-aged Al-6Si-0.5 Mg alloy. The results discussed that the microstructure of the material, mechanical characteristics and fracture resistance varies during artificial ageing because of the precipitation process. Al 6082 alloy treated at a temperature of 190 °C for 6 h gives the best performance properties in terms of strength with good fracture toughness [21].

The fracture properties of the extruded Al 2024-T4 and Al 7075-T6 alloys are characterized by a drop tower apparatus [14]. The results formulated that both the crack initiation and the crack propagation toughness increasing with rate of loading. For selected 2000 and 7000 Al alloys, 2024-T4 alloy gives excellent crack initiation toughness, whereas 7075-T6 gives excellent crack growth resistance, which are associated with their different fracture modes. From the point of energy, a crack can grow when the available energy nearer to the crack tip is enough to balance the energy required for crack growth. Therefore, an important consideration in designing of materials which are having high toughness, those are used to promote convenient fracture mechanisms which leads to increase in energy dissipation [22]. Experimental and numerical measures, including the crack-mouth opening displacement and the crack tip opening angle and the *J*-integral, have been widely accepted to show the resistance to crack development in ductile materials. Fracture toughness of Al7075-T6 alloy has been evaluated by utilizing circumferentially cracked round bar specimen having different diameters with different angles of notches. Literatures formulated the effect of specimen diameter and notch angle on the fracture resistance of the Al7075 T6 alloy. The magnitude of fracture toughness is higher for 10 mm diameter and lower for 8 mm diameter [18]. The result shows that the magnitude of fracture toughness increases with an increase in diameter and notch angle.

Fracture toughness (K_{IC}) of Al alloys investigated using 3-point bend specimens, resulting increase in fracture toughness when the initial notch tip radius is greater than 0.5 mm. A review on mechanical and fracture properties, through the experimental procedure has been studied on Al 7075-T6 coated with a process called electroless nickel deposit of 10–20 μm coating thickness [23]. Results were found that, an increase in coating thickness increases the fracture toughness of Al 7075-T6 alloy. Due to the considerable importance in demands of aluminium and its alloys in automobile vehicle components and building industry, it is necessary to enhance its mechanical properties by addition of required alloying elements [24]. Fracture toughness were evaluated using J_{IC} as a parameter for Al alloys and it was measured using single edge bend specimen, indicates fracture energy as 31.37 kJ/m^2 [18].

2.3 *Fracture Toughness of Aluminium Based MMCs*

From the study of literature found that the addition of SiCp and Al_2O_3 as reinforcements with aluminium alloy increases fracture resistance. Fracture toughness of a metal can be identified by the critical stress intensity factor K_{IC} used in linear elastic fracture mechanics (LEFM) to investigate the final load on a structural member containing flaws and cracks. The value of the stress intensity factor for composite material can be experimentally determined to analyze the damage tolerance of composite materials [25]. Literatures indicates that the plane-strain toughness (K_{IC}) of Al alloys reinforced with different reinforcements materials typically lies within the range of 8–38 MPa $\sqrt{m}$, and generally, this property is not as high as compared to the unreinforced alloy. Development of Metal Matrix Composites needs careful microstructure design that can enhance material's fracture resistance even though it is having high strength. Microstructure and constituent material properties are used to investigate the performance in terms of fracture toughness of MMCs by considering the initiation of failure and deformation mechanisms. The fracture characteristics of Al 6061 alloys, reinforced with SiCp processed through powder metallurgy techniques have been investigated. Results indicate [2] that enhancement in matrix strength for a given volume fraction and particle size shows a reduction in both K_{crit} and δ_{crit} of the composite. Investigation has been conducted on aluminium alloy reinforced with SiC [25]. However, the fracture toughness, improved significantly with ageing conditions with the result reported as-cast K_{IC} is 6.63–6.71 MPa $\sqrt{m}$; ageing treatment K_{IC} is 7.57–8.2 MPa $\sqrt{m}$. The fracture toughness of 6061 Al-based composites reinforced with 20% alumina-based discontinuous reinforcement has been investigated over a range of temperature and ageing conditions [26]. Formation of increase in the density of secondary particles is the initial cause of the difference in fracture toughness between these two materials. J -integral method is the method of displacement field measurement by using Digital Image Correlation (DIC) technique [27]. Results reported in the literature show that interface debonding is one of the beneficial failure mechanisms for the improvement of fracture toughness in MMCs. Fracture toughness tests were carried out on two Al alloys 2014 and 2024 with alumina particles of different volume fractions and particle sizes by using a slurry casting technique [28]. The result shows that the fracture occurred through the ductile mechanism. The fracture toughness increased with increasing spacing between particles provided that the particle size was lower than a limiting value, above which unstable crack propagation is observed and the toughness values are lowered.

Aluminium alloy and 6% volume fraction of SiC composites have been synthesized through stir casting at different pouring temperature. The fracture toughness specimens were initially cracked as per the ASTM E399 to provide a crack with small radius and straightness. Finite element (FE) simulations for the proposed 3-point bend (3 PB) geometry was carried out using ANSYS software package (v13) to find intensity factor by applying quarter-point crack tip elements, and the results were validated with experimental outcomes [29]. Fracture characteristic of hybrid composite reinforced with Boron carbide as primary phase and secondary phase as Molybdenum Disulphide with base matrix Al2219 processed using two stage stir casting method has been investigated [30]. Compact tension type specimen that has been used for test and dimensions are as per ASTM E-647. Addition of reinforcement makes strong resistance to open the crack front, which indicates enhancing of fracture toughness up to 27% maximum for Al2219-3%B4C-4%MoS2 hybrid composite. The fracture characteristic of prototype automotive pistons component produced in an aluminium alloy-based composite in industrial conditions has been studied and results are tabulated [31].

Fracture toughness gets increased when the temperature varies in the range of 20–75 °C and maintained a constant temperature up to 250 °C, during this a lower value was observed and recorded. Fracture toughness tests were conducted at temperatures ranging from 25 to 250 °C, accordance with ASTM E399 standard. The effect of thermo-mechanical treatment on the percentage of porosity and mechanical characteristics of SiC reinforced Al 6063 alloy-based composites has been investigated. Secondary processes like cold rolling and solution heat treatment shows reduction in the porosity levels of the materials [17]. The fracture toughness of the composites increased significantly with the secondary processes. The fracture toughness behaviour of Al 6061 alloy reinforced with 20 vol.% Al_2O_3 ceramic microspheres, and processed by liquid metallurgy route, have been investigated [32]. The fracture toughness of the powder metallurgy processed composite was substantially lower than that of the same volume percent processed through liquid metallurgy technique.

A356-SiC composites with different weight percent of 5, 10 and 15% are used to model SENB specimens according to the standard [33]. The majority of reduction in plasticity and also fracture toughness has found with increase in weight by 5%. The effect of heat treatment on the material fracture toughness was also investigated; J-integral criterion was employed [34]. The annealing process shows increase in fracture toughness. Metal matrix composite (MMC) fracture toughness analysis after heat treatment processes, with extrusion of 2124 aluminium alloy with 10% SiC particulates has been carried out [35]. The study of literature indicates that, the most effective technique for increasing fracture toughness of composite is uniform distribution of the particles in matrix materials. The observed K_{IC} values are optimized by using Taguchi technique. The literature reported that the solutionizing temperature will not affect the $K_{1}c$ value but by increasing the ageing time and ageing temperature, the $K_{1}c$ values are observed very high. The aluminium based hybrid metal matrix composite with varying weight percent of corn cob ash (0–4%) is prepared using two step stir casting method [36]. The fracture characteristic of the composites was also determined by using circumferential notch tensile (CNT) specimens. The fracture toughness of the hybrid composites was observed to be greater than the composite without Corn cob ash.

Studies on Al/SiC composites are administered to analyze primary fracture mechanisms that are related with microstructure characteristics, reinforcement properties and deformation behaviour. Investigation of fracture toughness from crack tip opening displacement has been widely used and accepted to indicate material fracture toughness [37]. The methodology is essentially helpful for development of high toughness MMCs through choice of materials and study of microstructures. For thin-walled materials in low constraint conditions, the crack tip opening angle parameter is additionally getting used to indicate stable crack extension, and ASTM D2472 was developed for the CTOA testing procedure [38]. CTOA parameter has been used to describe fracture resistance to stable crack growth for thin-walled structures. Completely different experimental strategies have been developed for the measurement of these parameters to explain fracture toughness of materials. Completely different technique of evaluations of the fracture behaviour in metal matrix composites is reported in literature [34]. Results show that J_{IC} derived K_{IC} values are above the corresponding values obtained by direct application of the linear elastic method. Behaviour of Al2219 with boron carbide particulate metal-based composites has been investigated. The composites were synthesized by reinforcing B_4C particulates using stir casting technique. The performance of the composite was correlated with the Al 2219 alloy to investigate the improvement in fatigue and fracture behaviour with the addition of reinforcement material [6]. Overall, fatigue strength and fracture toughness of the composites material was enhanced to 64% and 33%, respectively, with respect to 4 wt% B_4C particulates reinforcement compared to base alloy.

Fracture toughness K_{IC} and J_{IC} values of the aluminium fly ash composite get reduced than base alloy when produced using liquid metallurgy route [13]. The composites were secondary processed by hot extrusion and evaluated for the mechanical properties. The J_{IC} values of AA2024-fly ash composites material varied between

6 and 15 kJ/m^2 as correlated to 25 kJ/m^2 for the remelted base alloy AA2024. Stress evolution in SiC particle at the time of crack propagation in cast hybrid composites by using Raman Spectroscopy under monotonic loading condition was reported [39]. Experimental results concluded that, cracks propagated by debonding nature of the particle and matrix interface and due to particle fracture, when crack subjected to monotonic loading. Secondary cracks were formed in crack front and subsequent loading causes final fracture. The critical tensile stress for matrix-particle interface debonding and fracture occurred in particles of hybrid MMCs was also determined during the crack opening [40]. It has ascertained that increased addition of Silicon carbide reduces the fracture toughness. Analysis show that the best results for fracture toughness is at 25% of SiC. Further increase in volume percent shows reduction in stress concentration [41].

3 Conclusion

Review of literature on fracture toughness characterization of alluminium alloys and aluminium-based MMCs gives the effect of secondary process like heat treatment and extrusion process and micro structural parameter of the reinforcement on fracture toughness. The compact tension specimen and SENB specimens are most commonly used in evaluating the fracture resistance of the metallic materials. ASTM will give the procedure and configuration of the specimen for fracture toughness characterization. The materials subjected to heat treatment experiences enhancement in fracture toughness value. The importance of study of the precracked specimen under different standards provided by ASTM was also reviewed. The facility of precracking using wire EDM machining also presents the effect of machining on fracture characteristics of aluminium alloy.

Several authors have reported investigation of fracture toughness of MMCs which are manufactured using stir casting method. Little work has been presented on MMCs which are produced through powder metallurgy technique. There are some advantages of producing MMCs through powder metallurgy process are, more volume and weight percent of reinforcement can be added compared to stir casting method and uniform distribution of the particles can be achieved. Literature study illustrates that one can achieve maximum fracture toughness with uniform distribution of reinforcement particles. The fracture toughness of aluminium-based composites is studied for SiC and Al_2O_3, B4C, fly ash as a reinforcement, there is a need to investigate the effect of titanium dioxide, silicon nitrade, etc., as reinforcing materials. It has been observed that the increase in silicon carbide percentage beyond 25% resulting in reduction of fracture toughness. Analysis indicates that the 25% addition of SiC gives best result for fracture toughness. Further increase in SiC shows reduction in fracture toughness.

Thermal ageing treatment of aluminium alloys and aluminium MMCs increase the resistance of fracture in the material. Authors reported that the material aged at a temperature of 190 °C for 6 h increases the toughness. The percentage of porosity and mechanical behaviour of Al 6063 with SiC reinforcement composites was influenced by thermo-mechanical treatment that have been investigated. A higher level of reduction in porosity was observed in composites by applying cold rolling and solution heat treatment processes. The fracture toughness of the composites gets increased considerably with the application of the cold rolling process and solution heat treatment. Literature shows that the critical Stress Intensity Factor (SIF), K in tensile, i.e. mode I was calculated to formulate quantitative and qualitative data of the material. Fracture toughness K_{IC} of aluminium alloys determined experimentally under 3-point bend specimens, it shows higher values, when the initial crack tip radius is greater than 0.5 mm. The results showed that the fracture toughness enhanced as the time and temperature condition of over-ageing treatment increased. The fracture strength of the alloys materials were found to increase with increasing ageing temperature and found to be greater in the peak-aged condition. The addition of 2% Cu to the Al alloy shows substantial reduction of the ductility and fracture toughness.

The results illustrated that the mechanical, microstructural and fracture characteristics changes when the materials treated in artificial ageing because of the precipitation strengthening process. Al 6082 alloy at 190 °C for 6 hour ageing gives the best performance in properties including strength and good fracture toughness.

References

1. Kumai S, Tanaka T, Zhu H, Sato A (2004) Tear toughness of permanent mold cast and DC cast A356 aluminum alloys. Mater Trans 45(5): 1706–1713
2. Prantik M (2012) Alloy designation, processing, and use of AA6XXX series aluminium alloys. Int Scholar Res Netw ISRN Metallurgy 15: 165082. https://doi.org/10.5402/2012/165082
3. Zhu X-K, Joyce JA (2012) Review of fracture toughness (G, K, J, CTOD, CTOA) testing and standardization. Eng Fract Mech 85: 1–46
4. Zhang T, Wang S, Wang W (2019) A comparative study on fracture toughness calculation models in spherical indentation tests (SITs) for ductile metals. Int J Mech Sci 160: 114–128
5. Wang A, Wang Y, Zhang C, Zhang T, Liao H (2019) On the estimation and modeling of fracture toughness in structural ceramics in a simple way. Theor Appl Fract Mech 103: 102273
6. Lutz BEP, Goodwin CJ (1996) catastrophic failure mode assessment of the international space station Alpha. NASA contractor report 4720
7. Venkateswara Rao KT, Ritchie RO (2013) Mechanical properties of Al–Li alloys Part 1 fracture toughness and microstructure. Mater Sci Technol 5(9): 882–895
8. Hossain A, Kurny ASW (2013) Effect of ageing temperature on the mechanical properties of Al-6Si-0.5 Mg cast alloys with Cu additions treated by T6 heat treatment. Univ J Mater Sci 1(1): 1–5
9. Verrilli MJ, Salem JA (2001) Mechanical testing: standardization, international homologization: fracture strength: an overview. Encycl Mater Sci Technol
10. A textbook of (1973) Plane strain fracture toughness (K_{IC}) Data handbook, for metals: army materials and mechanics research center
11. Collins DA, Barkley EL, Lach TG, Byun TS (2019) Effects of thermal aging on the fracture toughness of cast stainless steel CF8. Int J Pressure Vessels Piping 173: 45–54
12. Zhu X-K (2015) Advances in fracture toughness test methods for ductile materials in low-constraint conditions. Procedia Eng 130: 784–802

13. Bhandakkar A, Prasad RC, Sastry SML (2014) Fracture toughness of AA2024 aluminum fly ash metal matrix composites. Int J Compos Mater 4(2): 108–124. https://doi.org/10.5923/j.cmaterials.20140402.10
14. Xing M, Wang Y, Jiang Z (2013) Dynamic fracture behaviors of selected aluminum alloys under three-point bending. Defence Technol 9: 193e200
15. Milan MT, Bowen P (2004) Tensile and fracture toughness properties of SiCp reinforced al alloys: effects of particle size, particle volume fraction, and matrix strength. ASM Int JMEPEG 13: 775–783. doi:https://doi.org/10.1361/10599490421358
16. Ranganatha S, Srinivasan MN (1982) Fracture toughness of cast aluminium alloys. Bull Master Sci 4(1): 29–35
17. García-Romero AM, Egizabal P, Irisarri AM (2010) Fracture and fatigue behaviour of aluminium matrix composite automotive pistons. Appl Compos Mater 17: 15–30. https://doi.org/10.1007/s10443-009-9117-3
18. Wei T, Carr DG, Budzakoska E, Payten W, Harrison RP, Ripley MI (2006) Assessment of the fracture toughness of 6061 aluminium by the small punch test and finite element analysis. Mater Forum 30
19. Madyira DM (2015) IAENG: effect of wire EDM on microstructure and fracture toughness of 7075-T6511 AluminumAlloy. In: Proceedings of the world congress on engineering 2015 vol II. WCE, London, U.K
20. Hossain A, Kurny ASW (2014) Effect of Cu and Ni on the mechanical properties and fracture behavior of Al–Si–Mg cast alloys. Chem Met Alloys 7: 154–159
21. Mrówka-Nowotnik G, Sieniawski J, Nowotnik A (2006) Tensile properties and fracture toughness of heat treated 6082 alloy. J Achieve Mater Manufact Eng 17(1–2)
22. Li Y, Zhou (2019) Effect of competing mechanisms on fracture toughness of metals with ductile grain structures. Eng Fract Mech 205: 14–27
23. Mohan Kumar S, Pramod R, Shashi Kumar ME, Govindaraju HK (2014) Evaluation of fracture toughness and mechanical properties of aluminum alloy 7075, T6 with Nickel Coating: 12th global congress on manufacturing and management, GCMM 2014. Procedia Eng 97: 178–185
24. Carman CM, Armiento DF, Markus H (1965) Plane-strain fracture toughness of high-strength aluminum alloys. J Basic Eng 87(4): 904–916. https://doi.org/10.1115/1.3650841
25. Li Y, Zhou M (2016) Prediction of fracture toughness scatter of composite materials. Comput Mater Sci 116: 44–51
26. Prosenjit D, Jayaganthan R, Singh V (2013) Experimental finding of initiation fracture toughness and FEM simulation of fracture behaviour of UFG 7075 Al Alloy. Adv Mater Lett 4(9): 668–681
27. Alaneme KK, Aluko AO (2012) Fracture toughness (K1C) and tensile properties of as-cast andage-hardened aluminium (6063)–silicon carbide particulate composites. Scientia Iranica A 19(4): 992–996
28. Davidson DL (1989) Fracture surface roughness as a gauge of fracture toughness: aluminium particulate SiC composites. J Mater Sci 24: 681–687
29. Kamat SV, Hirth JP (1991) Fracture of alumina particulate reinforced alumina alloy matrix composites. Bull Mater Sci 14(5): 1197–1203
30. Rajaravi C, Lakshminarayanan PR (2015) Experimental and finite element analysis of fracture toughness on Al/SiCp MMCs in different conditions. Int J Eng Manage Res 5(5): 320–324
31. Niranjan DB, Shiva Shankar GS, Shankar LB (2018) Study on fracture behaviour of Hybrid Aluminium Composite. MATEC Web of Conferences, vol 144, 02012
32. Alaneme KK (2012) Influence of thermo-mechanical treatment on the tensile behaviour and CNT evaluated fracture toughness of borax premixed SiCp reinforced Al 6063 composites. Int J Mech Mater Eng 7(1): 96–100
33. Park BG, Crosky AG, Hellier AK (2008) Fracture toughness of microsphere Al_2O_3–Al particulate metal matrix composites. Composites: Part B 39: 1270–1279
34. Srinivasa Rao KSRK, Kamaluddin S (2017) Machining and fracture characteristics of SiC reinforced A356 alloy composites. Int J Adv Trends Eng Sci Technol (IJATEST) 2(4)

35. Perez Ipiñaa JE, Yawnyb AA, Stukeb R, Gonzalez Oliverb C (2000) Fracture toughness in metal matrix composites. Mater Res 3(3): 74–78
36. Pillai PN, Lekshminarayanan PR (2008) Evaluation of fracture toughness of metal matrix composites. Mater Sci Res India 5(2): 331–337
37. Fatile OB, Akinruli JI, Amori AA (2014) Microstructure and mechanical behaviour of stir-cast Al-Mg-Sl alloy matrix hybrid composite reinforced with corn Cob Ash and Silicon Carbide. Int J Eng Technol Innov 4(4): 251–259
38. Yan L, Leon P, Cyril W (2019) 3D multi scale modeling of fracture in metal matrix composites. J Mater Res 34(13)
39. (2000) A textbook of mechanical properties of engineered materials. Mater Re 3(3): 74–78
40. Mileiko ST (2017) Fracture-toughness/ notch sensitivity correlation for metal and ceramic based fibrous composites. Compos part B 116: 1–6
41. Asif Iqbal AKM, Chen S, Arai Y, Wakako A (2015) Study of stress evolution in SiC particles during crack propagation in cast hybrid metal matrix composites using Raman spectroscopy. Eng Failure Anal 52: 109–115

Evaluation of Fracture Toughness of Precracked Steel Specimen Using Split Hopkinson Pressure Bar

Sonal Chibire, Nitesh P. Yelve, and Vivek M. Chavan

Abstract The mechanical properties of the materials depend on strain rate. The mechanical properties of material respond variously at quasi static conditions as well as high strain rate conditions. The Split-Hopkinson Pressure Bar is mostly used for high strain rate testing in range of 10^2–10^4 s^{-1} of strain rate. The SHPB is used to perform dynamic three-point bend test to measure fracture toughness. The strain rate of the pre-cracked specimen is measured using the strain gauges attached to the incident bar and transmitted bar. Additionally, a simulation study is carried out in which three dimensional models of a modified SHPB and a specimen are prepared and analysed using ANSYS©. The transient dynamic analysis technique is used for simulating the high strain-rate condition for the pre-cracked specimen. The results obtained through the simulation are compared with those obtained through the experiments. From the output of the simulation, the values of load and displacement are obtained. These values from the simulation are substituted to the analytical formula for calculation of the fracture toughness of pre-cracked specimen.

Keywords Split hopkinson pressure bar · Load point displacement · Strain rate · Finite element analysis · Fracture toughness

1 Introduction

The mechanical behaviour changes differently at different strain-rate loadings as well as temperatures. Thus, description of these change in the material behaviour is necessary under the high strain rate loading in numerical simulation. This description of material involves response of stress–strain as a function of high strain-rates, variation of stress states, high temperatures, damage accumulation and modes of failure [1].

S. Chibire (✉) · N. P. Yelve
Department of Mechanical Engineering, Fr. C. Rodrigues Institute of Technology, Vashi, Navi Mumbai, India
e-mail: Sonal.chibire22@gmail.com

V. M. Chavan
Bhabha Atomic Research Center, Trombay, Mumbai, India

S. Seetharamu et al. (eds.), *Fatigue, Durability, and Fracture Mechanics*, Lecture Notes in Mechanical Engineering, https://doi.org/10.1007/978-981-15-4779-9_42

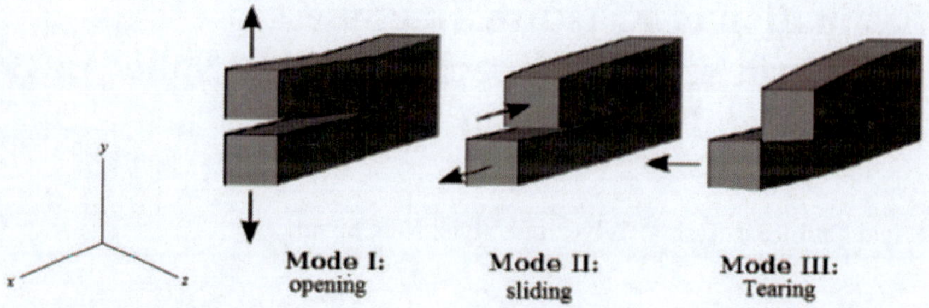

Fig. 1 Modes of crack extension

The classification of the modes of fracture, proposed by Irwin, as represented in Fig. 1. In the mode I (also called as opening mode of fracture), the tensile forces are applied on the specimen in such a way that, the crack surfaces of specimen are pulled separately in the direction of Y. The deformed surfaces of specimen are symmetric with respect to the planes which is normal to Y and Z-axis. In the mode II (also called as sliding mode of fracture), the shear forces which is parallel to the crack surfaces of specimen is applied on the specimen which slide in the direction of X-axis one over the other. The deformed surfaces of specimen are symmetric with the plane which is normal to the z-axis and skew symmetric with the plane which is normal to the Y-axis. In the mode III (also called as tearing mode of fracture), the shear forces parallel to the crack front as well as crack surfaces of the specimen are applied on the specimen which slide in the direction of Z-axis one over the other. The deformed surfaces are skew-symmetric with the plane which normal to the Z and Y-axis [2].

The following equation is modified by Irwin from Griffith's equation:

$$K_C = \beta \sigma_f \sqrt{\pi a} \tag{1}$$

Its unit is MPa $\sqrt{}$m, where 'β' is a geometry factor, 'σ_f' is the stress at the failure and its unit is 'MPa', and 'a' is the crack length of specimen. In Fracture Handbook, there are various equations for different geometries. The stress intensity and fracture toughness depend on the loading conditions on the material and crack geometry of specimen [2].

Han et al. [3] evaluated three methods of SIF calculation, i.e. displacement or stress extrapolation method, node displacement method and J-integral method. However, the accuracy of SIF calculation of specimen was influenced by an element type and mesh density, in finite element analysis. Rubio-González et al. [4] investigated the effect of damage due to fatigue on quasi-static and dynamic fracture toughness of steel (AISI 4140-T) and aluminium (6061-T6) specimens. In this analysis, the dynamic fracture toughness of the samples was evaluated. Kim et al. [5] evaluated the intrinsic static or dynamic fracture toughness of casting steel (Al 7175-T74 and 17-4PH) from the apparent static or dynamic fracture toughness of notched specimen by using an instrumented impact and three-point bending testers. As notch radius increases above the critical nose radius, the fracture toughness of specimen also increases. Sahraoui et al. [6] conducted an experiment using three-point bend specimen using an instrumented Charpy pendulum with a laser displacement measurement. The dynamic fracture toughness at high impact velocities is measured. Tosal et al. [7] compared the curves of ductile to brittle transition of a structural steel using static and dynamic conditions, and the crack growth of specimen was evaluated. Qiu et al. [8] investigated the fracture toughness of steel which is influenced by pre-strain under static and dynamic loading conditions. As strain rate increases, the fracture toughness decreases. Foster et al. [9] determined the dynamic crack initiation toughness of 4340 steel. The fracture toughness of metals under quasi-static loading conditions is determined. Xian-Zhu et al. [10] obtained the analysis of fracture toughness (K, G, J, CTOA, CTOD). In this, the mechanical properties of metal are reviewed.

If the rate of the growth of crack is calculated, an engineer can inspect appropriately and before failure occurs, the part can repair or replace. Thus, the life of part increases. Finite Element analysis is used for a pre-cracked three-point bend specimen in the present paper. ANSYS© is used to evaluate the Fracture Toughness of specimen. In this analysis, pre-cracked three-point bend specimen is considered. Width of specimen is 10 mm and the length of crack is 5 mm.

In the present paper, the finite element method is applicable for simulation of the high strain rate condition for the pre-cracked specimen using transient dynamic analysis technique. The results obtained through the simulation are further compared with those obtained through the experiment.

The experimental setup is explained in Sect. 2 and simulation work is given in Sect. 3. The results and discussion on them are given in Sect. 4. The concluding remarks on the present study are given in Sects. 5.

2 Experimental Setup

The Split-Hopkinson pressure bar generates only one-dimensional wave propagation [11]. It consists of four main components, i.e. a gas gun, a striker bar (SB), an incident bar (IB) and a transmission bar (TB) (Fig. 2). The striker bar is held in the gas gun chamber. The material and area of cross section of incident bar, transmission bar and striker bar are same. During the test, a striker, incident and transmission bar must be remained elastic. The pre-cracked specimen is held between the incident and transmission bar. The striker bar is struck by gas pressure on the incident bar. The compressed elastic wave is generated in the incident bar due to impact. The plastic deformation of pre-cracked specimen occurs due to propagation of repeated elastic wave propagation. The part of the repeated wave transmits to the transmission bar and the other part of the repeated wave reflects to the incident bar. Each of the wave is sensed by the strain gauges which is connected to both the bars. Strain gauges are placed in the form of half Wheatstone bridge for measuring the axial strain only. The generated elastic strain in both the bars are used to evaluate the graph of stress and strain in the specimen [12].

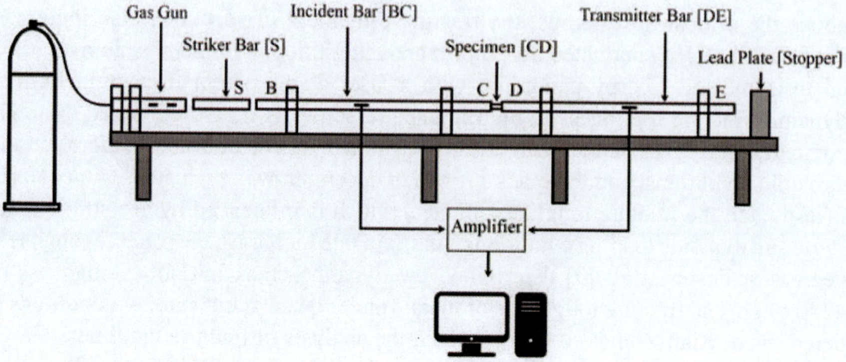

Fig. 2 Split Hopkinson bar setup

3 Simulation

Simulation of impact experiments is increasingly being used by automobile manufacturers, defence and manufacturers of telecom and electronic equipment [13]. Ansys LS Dyna has been used to test crash worthiness of cars; it has not been validated to simulate stress wave theory. An explicit finite element analysis is carried out for Split Hopkinson pressure bar loading of a three-point bend specimen.

Figure 3 gives the schematic and dimensions of the experimental apparatus required to be modelled. The incident and transmitted bar has a length of 1.3 m. The projectile is 0.5 m long. Both the bars are made up of high strength steel and have a common diameter of 13 mm.

3.1 Material Properties

As the material of striker, incident and transmitted bars was very rigid as compared to that of specimen, these were modelled as Linear Elastic and material properties of Structural Steel were used, which are given in Table 1.

As mentioned earlier, the bilinear isotropic property is used for the specimen. Since, the experimental shows the material of specimen to be rate sensitive, exhibiting strain hardening in the plastic regime, a Rate Sensitive Non-linear Inelastic material model was selected for the specimen, which are given in Table 2.

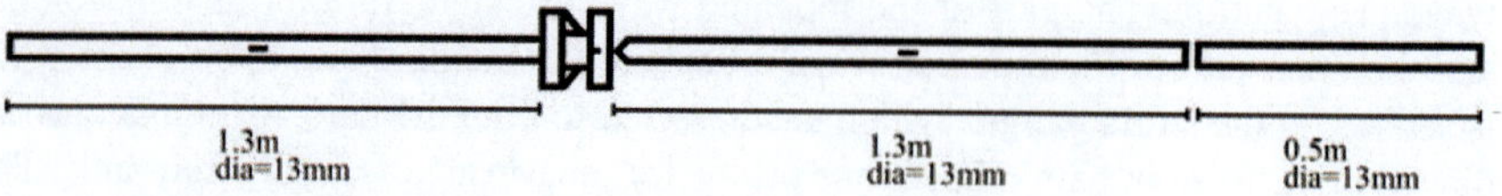

Fig. 3 Schematic configuration of split Hopkinson bar setup

Table 1 Material properties of linear elastic structural steel

Property	Value	Unit
Modulus of elasticity	200	GPa
Density	7800	kg/mm^3
Poisson's ratio	0.30	–

Table 2 Material properties of plastic structural steel

Property	Value	Unit
Youngs modulus	250	MPa
Density	$7.83e^{-6}$	kg/mm^3
Poisson's ratio	0.33	–
Tangent modulus	$860e^{6}$	Pa

3.2 Element Type

For modelling of solid structures, two main types of elements are the tetrahedrons (SOLID168), which are suitable for free meshing, and hexahedrons/bricks (SOLID164) that are used for mapped meshing. Since the solid modelling was done with a view to be map meshed later, SOLID164 (eight node brick) element was selected as the element type for modelling the bi-material specimen and the striker and incidence bars [3].

3.3 Element Size

Hourglass energy must be constant at all the time of simulation. Due to hourglass energy, every time the result will change and correct results will not be found. It is clear by energy graph (Fig. 4) that, hourglass energy is constant at all the time of simulation [14].

Plastic strain should be uniformly distributed to get accurate results. So, it is required to change the size of mesh and check the results. It was given 1 mm mesh size for specimen and 10 mm for bars. It was showing that 10 elements of specimen were interacting with 1 element of bars. Hence, it is found that, the size of specimen and bars must be same for getting the results uniformly as the applied load is uniform.

3.4 Loading and Boundary Condition

The specimen displacement was constrained in *X*-direction of the global coordinate system to achieve a three-point bending configuration to obtain only axial strain of

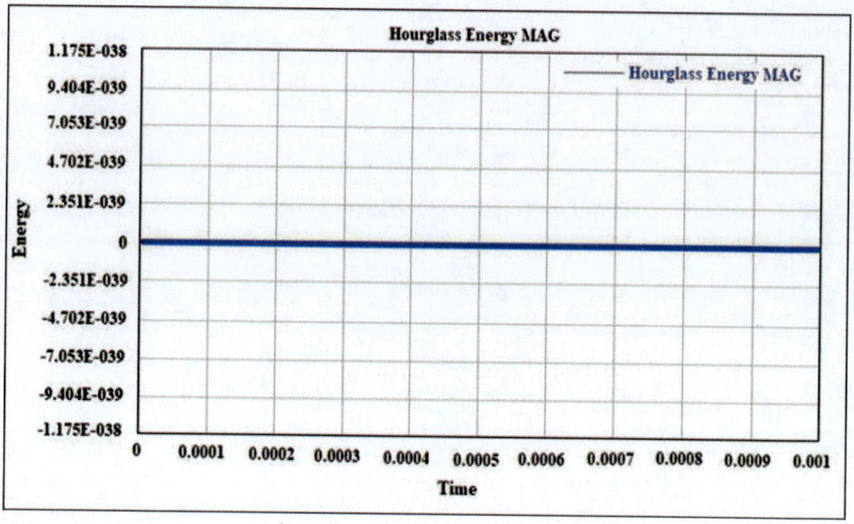

Fig. 4 Hourglass energy

incident bar and transmitted bar. An initial axial velocity of 15 m/s was used for the striker bar in the *X* direction of the global coordinate system.

4 Results and Discussion

4.1 Experimental Result

The pre-cracked specimen has been tested in this study, and the specimen broke after being loaded. The split Hopkinson bar technique is used for measuring the dynamic load applied to the specimen [14]. The longitudinal repeated compressive pulses are propagated along the incident bar due to impact of striker bar. The part of the compressed elastic wave is transmitted to the specimen, thus fracture occurs and the part of other elastic wave is reflected to the incident bar as a tensile wave [14]. The result obtained, that is, incident pulse, transmitted pulse and reflected pulse are plotted on a graph of strain versus time. Figure 5 shows typical signals obtained from the bars and the specimen.

4.2 Simulation Results

Transient dynamic analysis technique has been used for the simulation of strain as a function of time and measurement of load point displacement. ANSYS© has been

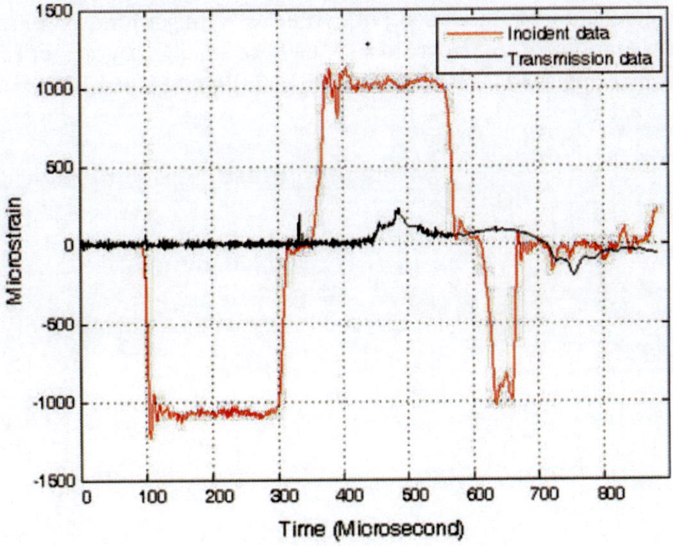

Fig. 5 Signal obtained from the bars and the specimen

used to perform simulation. Strain gauges are used to obtain the strain rate as shown in Fig. 6.

For determination of quasi-static fracture toughness, the curve of load versus load line displacement is required. Strain-rate obtained from experiment is compared

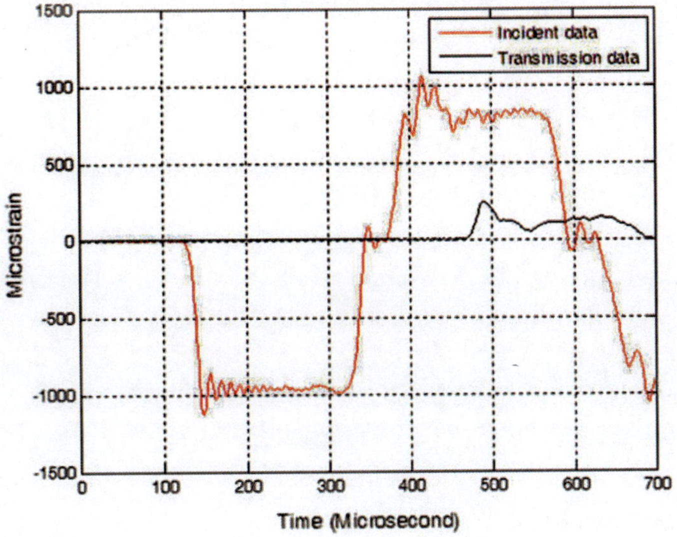

Fig. 6 Graph of strain at crack tip versus time

with the simulation for validation [14]. After validating with experimental result, if both the strain rates are in accordance, then we can plot a graph of Load versus Deformation using MATLAB as shown in Fig. 7 (Figs. 8, 9 and 10).

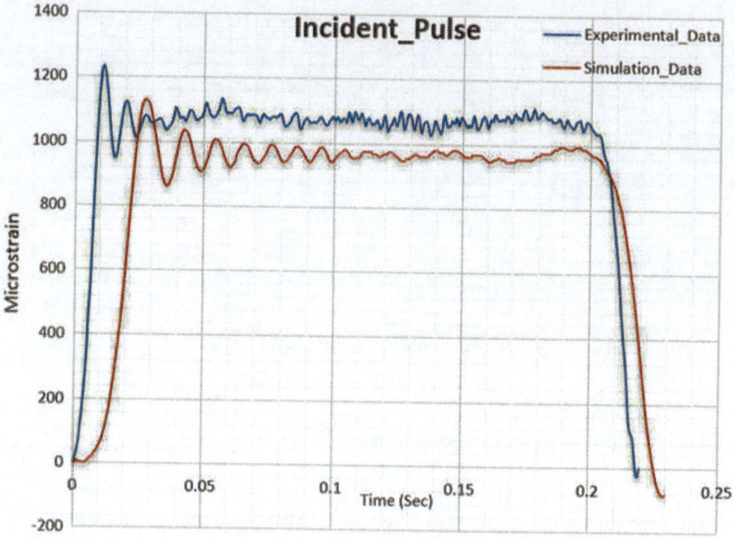

Fig. 7 Incident pulse

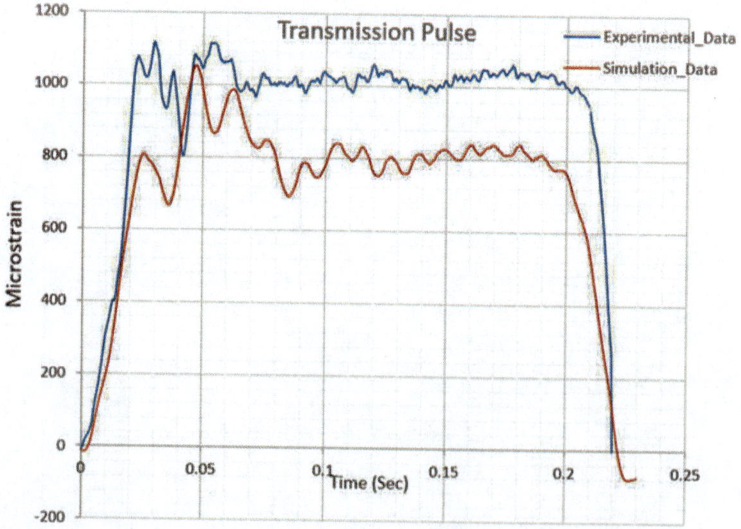

Fig. 8 Transmission pulse

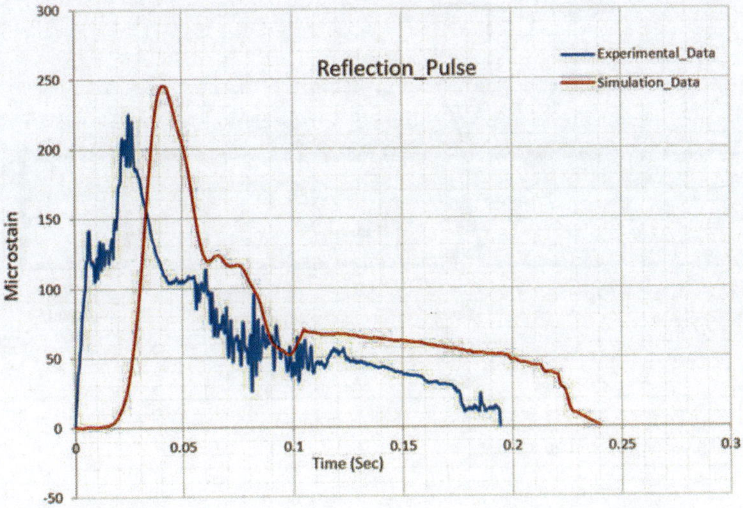

Fig. 9 Reflection pulse

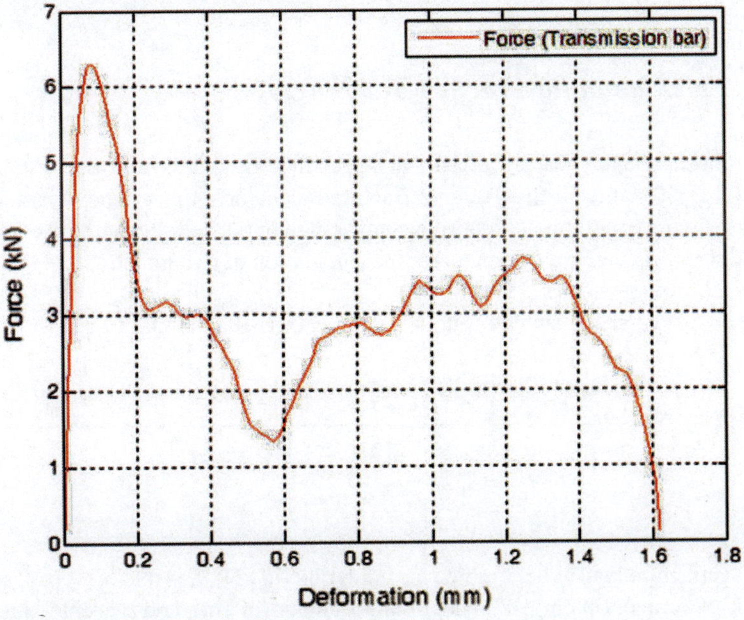

Fig. 10 Graph of load versus displacement

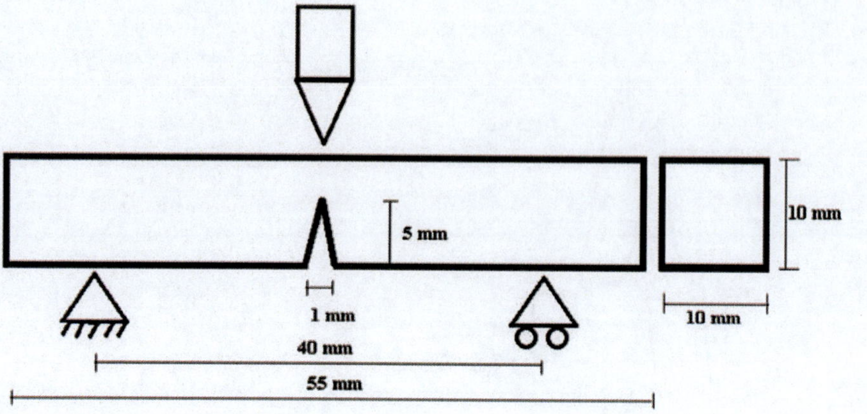

Fig. 11 Three-point bend specimen

From the above graph, we get force of 6.2 kN on transmission bar to propagate the crack. This force is used for the further calculation to calculate fracture toughness.

4.3 Calculation for Fracture Toughness

The procedures to measure quasi-static fracture toughness are well established (ASTM E399) whereas the dynamic fracture mechanic's theory and dynamic fracture toughness experimental procedures are still under development [8]. Researchers have developed analytical formula for the calculation of dynamic fracture toughness at a crack initiation time [15] (Fig. 11).

For the bend specimen calculate K_Q in units of (MPa) as follows [15].

$$K_Q = \frac{P_Q * S}{B * W^{3/2}} \frac{3\left(\frac{a}{w}\right)^{1/2} * \left[1.99 - \left(\frac{a}{w}\right)\left(1 - \frac{a}{w}\right)\left\{2.15 - 3.93\left(\frac{a}{w}\right) + 2.7\left(\frac{a}{w}\right)^2\right\}\right]}{2\left[1 + 2\left(\frac{a}{w}\right)\right]\left(1 - \frac{a}{w}\right)^{3/2}}, \quad (2)$$

where,

P_Q is fracture load as determined in simulation (in kN), B is thickness of specimen (in cm), S is span (in cm), W is depth of specimen (in cm), and a is crack length of specimen (in cm), By simulation, we get P_Q is obtained as 6.2 kN, which gives K_Q as 20.86 MPa $\sqrt{m}$. The obtained value of K_Q is in accordance with the results given in reference paper [16].

5 Conclusion

The mechanical behaviour of metals and non-metals are strain-rate dependent. Thus, the study of materials under high strain rates is required for optimal design. Split-Hopkinson pressure bar is one of the testing techniques which works on one dimensional wave propagation theory to measure high strain rates. This test method is widely accepted for stain rate range of 10^2–10^4 s^{-1}. To meet this requirement, the SHPB facility is designed and developed by Refueling Technology Division and Mechanical Metallurgy Division, BARC. The set-up is installed at Material Dynamics lab. Engg. Hall No. 3. The Split Hopkinson Bar apparatus is also modeled using ANSYS©. The simulated impact experiment in ANSYS© depicts stress waves generated with pulse width and magnitude in conformity with simulation and experimental data.

Acknowledgements Authors are thankful to Mr. Sudhanshu Sharma, SO-E, BARC, for his help in carrying out the experiments at BARC.

References

1. Anderson TL (1995) Fracture mechanics: fundamentals and applications. CRC Press, Boston
2. Sarath Kumar Nagoju MK, Gopinath V (2013) Computation of stress intensity factor and critical crack length of ASTM A36 steel using fracture mechanics. Int J Eng Res Technol (IJERT) 2(9)
3. Han Q, Yin Y, Wang D (2015) Determination of stress intensity factor for mode I fatigue crack based on finite element analysis. Eng Fract Mech 138:118–126
4. Rubio-González C, Gallardo-González JA, Mesmacque G, Sanchez-Santanaa U (2008) Dynamic fracture toughness of pre-fatigued materials. Int J Fatigue 30:1056–1064
5. Kim JH (2004) Evaluation of static and dynamic fracture toughness using apparent fracture toughness of notched specimen. Mater Sci Eng A 387–389:381–384
6. Sahraoui S (2009) Measurement of the dynamic fracture toughness with notched PMMA specimen under impact loading. Polym Testing 28:780–783
7. Tosal L (2000) Comparison of the static and dynamic fracture behavior of an AE-460 structural steel. Eng Fract Mech 66:537–549
8. Qiu H (2003) A model for the dynamic fracture toughness of ductile structural steel. Eng Fract Mech 70:589–598
9. Foster JT (2011) Dynamic crack initiation toughness of 4340 steel at constant loading rates. Eng Fract Mech 78:1264–1276
10. Zhu XK (2012) Review of fracture toughness (G, K, J, CTOD, CTOA) testing and standardization. Eng Fract Mech 85:1–46
11. Sharma S, Chavan VM Split-Hopkinson pressure bar: an experimental technique for high strain rate tests
12. Junaid A (2008) LS Dyna as a valid tool for analysing stress wave propagation under impact loading. J Eng Appl Sci 27(1):53–61
13. Lindholm US (1971) High strain rate tests. In: Bunshah R (ed) Chapter 3, Appendix A in techniques of metals research, vol 5, Pt. 1. Wiley
14. Verma SK (1998) An experimental cum numerical technique to determine dynamic interlaminar fracture toughness. Eng Fract Mech 60(5–6):583–596

15. ASTM E 399-90 Standard test method for plane-strain fracture toughness of metallic materials. American Society for Testing and Materials, 100 Barr Harbor Dr., West Conshohocken, PA 19428
16. Nath SK (2006) Effect of microstructure and notches on the fracture toughness of medium carbon steel. J Naval Architect Mar Eng

Surface Modification, Characterization and Optimization of Hybrid Bio Composites

G. Venkatesha Prasanna

Abstract The objective of the work was to well understanding of the impact of exterior treatment on the bio reinforcement and optimization of mechanical strengths of the Bio fibril composites. Reseacher during this research provides better viewpoint over outcome of chemical treatment on the surface of Luffa-Tamarind fibrils by the NaOH and benzene diazonium chloride treatment for the tractile, flexural and compressive properties of crossover composites. The outcome of % of raw and chemically treated fiber content with and without presence of moisture, unwanted materials/impurities on mechanical properties was studied. The mechanical properties were ideally expanded at 40% wt. of surface customized Luffa-Tamarind fibers loading and underneath or above 40% wt. it exhibited a poorer mechanical properties. The alkali and benzene diazonium chloride treatment consumes the voids spaces of fiber, diminishes the unwanted materials, aligns uneven fibers, provides rough surface topography, improves the aspect ratio and strengthening efficiency.

Keywords Luffa-Tamarind fibers · Surface treatment and mechanical tests · Morphology

1 Introduction

In over-all artificial fibers explicitly Kevlar, carbon, glass, are broadly utilized in resin compounds because they are highly solid and firm. Aside from the numerous advantages possed by man-made fibers in diverse applications, these man made fibers have few restrictions. So as to overcome this emergency, more appeal was raised towards the production of natural/renewable fibers based composites to make the world 'green fibers' [1, 2]. Bio fibers have been used as an substitute fibers to synthetic fibers as well as fortification for polymer matrix composites. Cellulose, hemicelluloses, pectin, lignin and waxes are contained in bio fibers [3–5]. The water retaining property contained by the natural fibers confines their utilization in making

G. Venkatesha Prasanna (✉)
Mechanical Engineering Department, CVR College of Engineering, Hyderabad 501510, India
e-mail: gvpcvrmech@gmail.com

S. Seetharamu et al. (eds.), *Fatigue, Durability, and Fracture Mechanics*, Lecture Notes in Mechanical Engineering, https://doi.org/10.1007/978-981-15-4779-9_43

of amalgams. As a result of hydrophobic nature of Polymer matrix, which may cause fiber and matrix incongruent and brings about frail coupling among resin and fibrils. Fundamental reason for the peripheral of the fibrils was for lessening the moisture ingestion nature of the bio fibers and furthermore to increase the compatibility with polymer matrix.

Just a little level of the propelled composites were created by Thermoplastics. Yet, thermoset plastics add to bigger part of 70% in the manufacture of composites. As a outcome of their inbuilt structure, these matrix materials became the most imperative resins. Resin epoxy was a multipurpose as well as widely used matrix material for varied applications such as advanced composites due to their superior properties [6]. Be that as it may, the utilization of thermoset materials was regularly controlled as a result of low toughness property. Then blending technique can be employed proficiently and viably to defeat the few substandard quality of two resins. Compatible matrix combinations give the different efficient substance from the inferior predominant separate resins [6]. Accordingly, when the mass measures of epoxies are utilized, typically it is made to blend with tough resins. Vinyl ester was the resilient element utilized alongside epoxy matrix. Vinyl ester was the extra result of an epoxy resin [6]. To improve the exhibition in superior products epoxy resin was blended with vinyl ester for the increase of tensile property and stiffness [3, 7, 8]. Hydroxyl bunches enhance the polarity of Vinyl ester and make easy the adhesion and pigment wetting properties, which drove the wetness from the fibrils. In this exploration epoxy/vinyl ester (80/20% w/w) resin mix was organized. Composite specimens with the 10, 20, 30, 40 and 50% wt. content of unprocessed, NaOH processed and benzene diazonium chloride processed Luffa-Tamarind fibrils fortified epoxy/vinyl ester blend were manufactured.

2 Experimental

2.1 Materials

The two resins namely: 1. epoxy araldite LY 556 as resin, hardener HY 951 2. Vinyl ester.

The two Fibers namely: 1. Luffa fiber 2. Tamarind fiber were utilized to create the composites.

2.2 Chemical Alteration of Bio Fibrils

Because of presence of aquaphilic hydroxyl bunches in the bio fibrils, face little issues during the support of natural fibers into the polymers. This hydrophilic nature hinder the viable bonding of fibers with polymers. Besides pectin, grease, oil materials covers the active faces of the fibrils and also acts as obstruction during coupling with the resin. For the adequate coupling, fibrils outside should be transformed using diverse exterior alteration methods and constituents.

2.3 Chemical (Alkali) Treatment

The bio fiber's hydrophilic property, ruins the successful bonding of fibers with matrix blend. Furthermore, unwanted materials like pectin, grease, oil materials encase the active surfaces of the fibrils and behave as a obstruction for the coupling of fibrils with the resin. In order to increase the exterior adhesion, fibrils exterior should be reformed using diverse exterior alteration methods and constituents. Chemical treatments provide more reactive bunches on the bio fibrils exterior, which expands the proficient coupling with the matrix blend. Exterior of the fibers required to be altered by benzene diazonium chloride processing and alkali treatment using concentration of 10 wt% NaOH was conducted out in this work. Later the alkali treatment the fibrils were flooded in refined solution for nearly one to eradicate the left behind alkali. Further washing was finished with the refined solution having little quantity of acetic acid. Wet fibrils were dried in the oven.

2.4 Development of Benzene Diazonium Chloride

Concentrated HCl of 8 cm^3 was added to a bubbling tube containing 3 cm^3 of phenyl amine (aniline) and 10 cm^3 of water, the blend was shaken until the amine liquefy, and then the solvent was cooled to 5 °C by cooling it in an ice bath. In the wake up of cooling a solution of sodium nitrite (3 g in 8 cm^3 of water) was added to the pre-cooled blend. The temperature of the blend was kept up underneath 10 °C in the course of enclosure of sodium nitrite [5].

2.5 Benzene Diazonium Treatment of Tamarind-Luffa Fibers

Luffa-Tamarind fibers were sliced to 10 mm length, afterward washing with superior water, afterwards dried dehydrated in the oven for one day at 70 °C. The fibers cured then flooded in 6% alkali solution taken through glass beaker. Newly made diazo solvent should be added gradually to the mix at steady vibration. After that the fibrils were withdrawn outside, and dressed with soap solution then using superior solution and ultimately dried outside for 48 h [8].

2.6 Fabrication of Blended Hybrid Bio Composites

A lean layer of hard wax was poured inside the mould cavity for simple discharge and pull back of manufactured composite specimens. After the wax was cured, a slim layer of poly vinyl alcohol (PVA) was put. Bio fiber crossover composites were manufactured employing the hand lay-up method. Then the hybrid bio fibers of the unprocessed & treated Luffa-Tamarind fibers were fortified to the matrix blend to get the essential bio-composite specimens for the investigation of performance and mechanical qualities. Then air bubbles were evacuated vigilantly by using the roller with delicate rolling. With the goal of complete cure, composite samples were post cured for 2 h by keeping the specimens in hot roaster at temperature of 80 °C. After the complete curing, raw and treated samples of bio-composite specimens were examined.

3 Results and Discussion

3.1 Tensile Properties

Tensile test samples, with dimensions 150 mm × 15 mm × 3 mm were evaluated according to ASTM principles. The effect of fibril amount and modification of fibrils by the two chemical treatments using: NaOH processing and benzene diazonium processing on the flexile property of bio composites were exhibited in the Fig. 1. In general, chemically processed bio fibers fortified composites have higher tensile property than the unprocessed fibril compounds. In case of unrefined bio composites, the repellency and insufficient compatibility among hydrophobic matrix and hydrophilic bio fibrils which thwarts and unfavorably effects the interfacial bonding between fiber and matrix blend and which brings about fading of the fortification limit of the fibril due to the insufficient stress convey from the resin to fibrils [5, 9–11]. The tensile quality of unprocessed, NaOH processed and benzene diazonium treated Luffa-Tamarind fibers fortified composites revealed most extreme and ideal at 40% of fiber loading than 10, 20, 30 and 50% fiber load. The most raised tensile strength was

Fig. 1 **a** Luffa fiber 1 **b** Tamarind fiber

seen at benzene diazonium processed 40% Luffa-Tamarind fibrils fortified composite as a result of ideal amount of fibrils, finer fiber distribution in resin material, interfacial bonding and enough load transport arises among the fibrils and resin [5, 8, 12]. But, an considerable tensile strength was seen in NaOH processed Luffa-Tamarind fiber compounds off course critical raise was seen in the flexile strength of benzene diazonium chloride-processed fibril compounds. Benzene diazonium chloride processed composite point outs greater diminishing of the aquaphilic nature of Luffa-Tamarind fibers because of the mix of the hydroxyl gathering of fibrils with benzene diazonium chloride [9, 10]. Surface alteration by chemical processing of fiber diminished the aquatic assimilation property of the fibers and additionally yields a rough surface topography on the fibers [10]. Consequently the coupling among the fibrils and resin reveal increment in fortification limit of fibers [12, 13]. The amount of load transport among the fibrils and resin in a compound was determined by the exterior coupling, fibril angle, arrangement of fibrils and fibrils dispersion in the resin amalgamation [7, 10, 11].

3.2 Compressive Strength

The composite specimens were fabricated and tested for Compressive strength properties as per the ASTM D specifications. The deviation in the values of compressive strength property with respect to fiber amount with and without surface alteration was appeared in the Fig. 2. It was revealed from the test that benzene diazonium chloride processed bio composites conveyed prevalent and ideal qualities for compressive strength than the alkali treated and unprocessed bio composites [11, 14–16]. The superlative condition was a direct result of ideal fiber stacking and chemical treatment, which advanced the boosted interfacial bonding among the fibers and matrix blend effected in plenty stress transfer and performance [6,

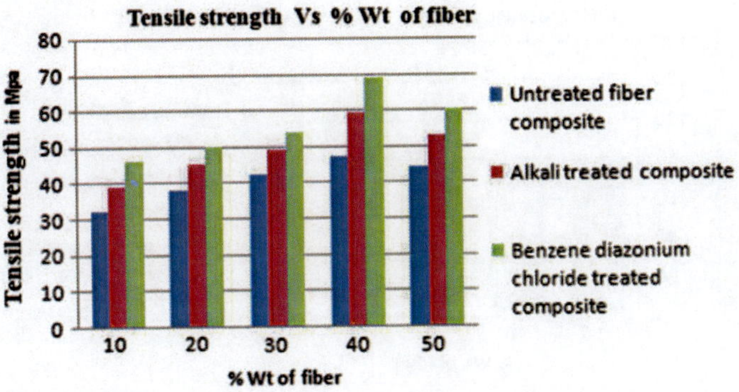

Fig. 2 Tensile strength of untreated, alkali treated and Benzene diazonium chloride treated composites

12, 16]. Earlier works reveal alteration of the fiber surface by different chemical treatments increased the aspect ratio and amplify the surface roughness [6, 17]. The surface treatment through of NaoH, also gives the passage for the clearing of uninvited tainted materials like waxes, oils, impurities, hemicelluloses, moisture and lignin from the natural fibers, give up the upper gain of (alpha) cellulose in natural fibers and also wipes the fiber surface but benzene diazonium chloride treatment furthermore cleans the fiber surface [1, 7, 8]. Chemical treatment affords more active area for the fiber surface required for moistening in the polymer in order to have good bonding [4, 10, 18] and results in enhanced mechanical properties.

Correlation of the strength values of NaOH processed and benzene diazonium chloride processed Luffa-Tamarind fibrils exposes that benzene diazonium chloride processed fibrils have uniform and legitimate arrangement of fibers in the matrix blend effected in superior strength values [6, 9, 11]. The treatment of fiber also enhanced exterior bind amid the Luffa-Tamarind fibrils and the resin [6, 10, 13, 16]. Thus most noteworthy compressive proprty was seen at 40% fibril stack.

3.3 Flexural Strength

The specimens of composite with dimensions 150 mm × 15 mm × 3 mm were produced and examined for flexural properties according to ASTM D 5943-96 requirements. Difference in the values of flexural property as a function of fibril load, with and without alkali treatment was presented in the Fig. 3. After the test it was realized that the flexural property of the Luffa-Tamarind fibers reinforced epoxy/vinyl ester blended composite demonstrated advanced and ideal incentive at 40% fiber loading. This is a direct outcome of the decrease in the hydrophilic property of Luffa-Tamarind fibers because of the pairing of the hydroxyl gathering of fibril with benzene diazonium chloride. Benzene diazonium chloride process evacuates the

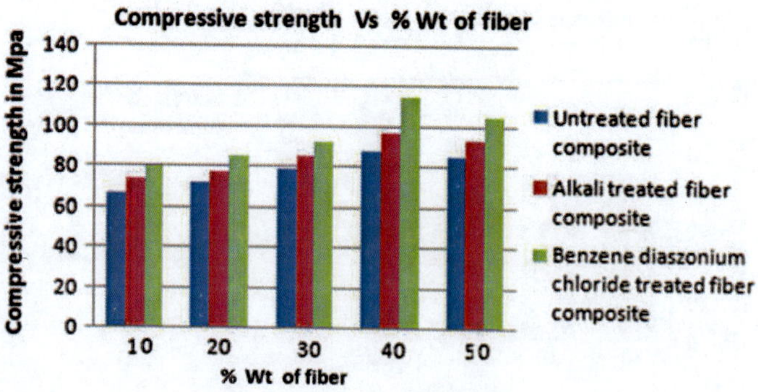

Fig. 3 Compressive strength of unprocessed, NaOH processed and benzene diazonium chloride processed composites

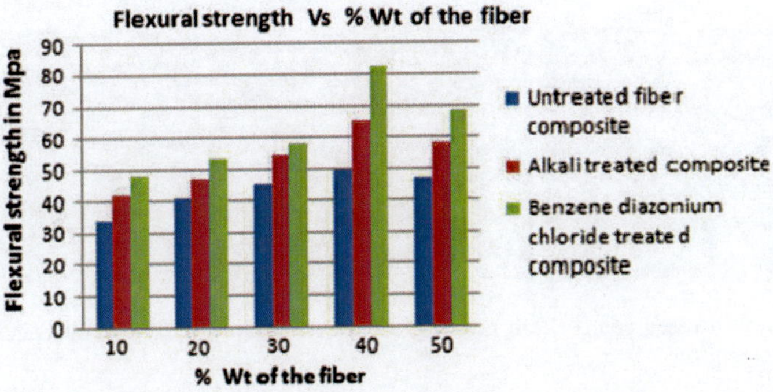

Fig. 4 Flexural strength of unprocessed, NaOH processed and benzene diazonium chloride processed amalgams

voids of fibrils and line up the uneven fiber [3, 7, 10]. Furthermore, benzene diazonium chloride treatment trim down the fibrils area and thereby increases the aspect ratio, subsequently builds the perspective proportion, which expands the powerful fiber surface region for adequate interfacial adhesion with the matrix blend [11, 14, 16] (Fig. 4).

3.4 Moisture Content Test

The Fig. 5 exposes the percentage retaining of wetness quantity by unprocessed, NaOH processed and benzene diazonium chloride processed amalgams. The bio fibrils composites fabricated from the aquaphilic quality of the fibril exhibit hydrophilic qualities [2, 9, 14]. But the NaOH responsive to hydroxyl bunches were breaks and transfer out from the fibrils exterior during chemical treatment. Hydrophilic hydroxyl bunches present in the natural fibers were shrink by the alkali and benzene diazonium chloride process, hence surface of the fiber was tailored and builds the fiber dampness obstruction property, however which was more in benzene

Fig. 5 Moisture content absorption of unprocessed, NaOH processed and benzene diazonium chloride processed amalgams

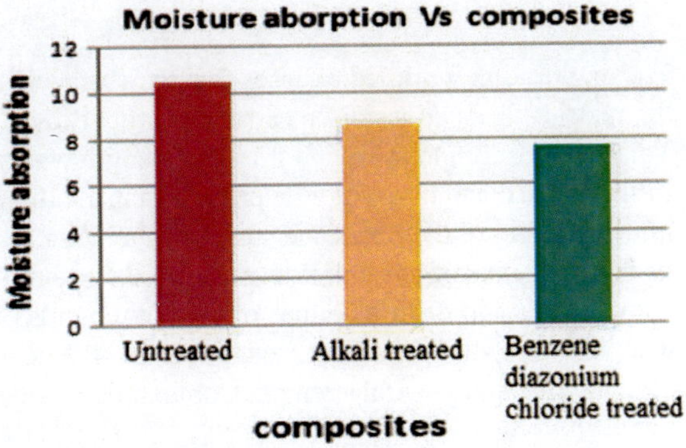

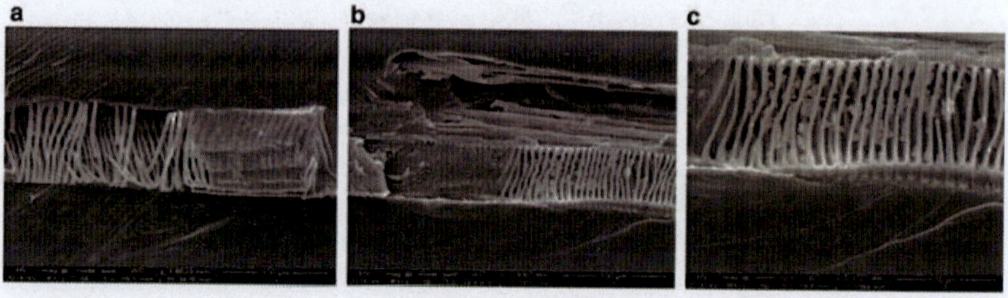

Fig. 6 **a** Unprocessed composite **b** NaOH processed composite **c** Benzene diazonium chloride processed composite

diazonium chloride treatment [3, 9, 12]. From the outline it was revealed that retention rate of dampness amount by natural fiber composite, was higher, when corelated to NaOH processed and benzene diazonium chloride processed amalgams [4, 19].

3.5 Morphology

Figure 5 presents the pictures of the Luffa-Tamarind fibers, before and after chemical action modified fibrils posses lesser lignin content, because of the limited removal of wax, grease, moisture and other impurities [4, 11, 18, 20]. It was observed that surface treatment has expelled the peripheral debris from the fibrils, effected in coarse peripheral fibril and increment in the compelling region accessible for interfacial bonding [16, 19, 21]. This coarse surface texture gives opportunity for enhancement in coupling response due to the appearance hydroxyl gatherings to the matrix, along these lines expanding the fibers spreading in the matrix blend and fiber–matrix sticking together [5, 13, 14, 16, 22]. Anyway the benzene diazonium chloride treatment creates significantly ever more better surface and better fiber–matrix adhesion compared to alkali treatment (Fig. 6).

4 Conclusion

The exploration work led exposes that the surface variation extensively accelerates the matrix-fibrils adhesion, mechanical properties of the natural fibril compounds. Thus tensile, compressive and flexural strength was realized to have improved more for Benzene diazonium chloride process than alkali treatment of the fibril because of more cleaning of the fibers and superior fibers/matrix blend interface was perceived in benzene diazonium chloride process. Results obviously show that the flexile, compressive and flexural values of the compounds expanded with increase in fiber heaping up to 40% and underneath or past 40% it indicated a decline in a flexile, compressive and flexural strength. Comparatively huge quantity of fibril, recline one over the other in spite of being mixed with resin at 50% fiber amount, deficient wetting of fibers in the matrix blend and substandard fibers/matrix blend interfacial bonding. However at relatively less amount of fiber prompted declined stress transfer ability. The rate ingestion of dampness amount by unrefined fiber composite was more than the NaOH processed composite and Benzene diazonium chloride processed composite. But the chemical action of the fibrils, and at optimum amount of fibril (40%) it was seen as conceivable to elevate the properties of the composites with the ejection of impurities/unwanted materials and disposal or evacuation of dampness and builds the fortifying effectiveness of the fiber in the composites with the improvement of load convey from the resin to load bearing fibrils.

References

1. Kaith BS, Singha AS, Sanjeev K, Susheel K (2008) Mercerization of flax fibre improves the mechanical properties of fibre-reinforced composites. Inter J Polym Mater 57(1):54–72
2. Venkatesha Prasanna G (2020) Influence of fiber length and chemical treatment on the mechanical properties of blended composites. Springer
3. Edeerozey AM, Akil HM, Azhar AB, Ariffin MZ (2007) Chemical modification of kenaf fibres. Mater Lett 61:2023–2025
4. Venkatesha Prasanna G, Jha NK, Harish C (2018) Chemical treatment and analysis of mechanical properties of hybrid biocomposites. Int J Mech Prod Eng Res Dev. ISSN (P): 2249-6890; ISSN (E): 2249-8001, Special Issue: 189–193
5. Venkatesha Prasanna G, Jha NK (2019) Optimisation & mechanical testing of hybrid bio composites. Mater Today, 1395–1404. Elsevier
6. Raghavendra Rao H, Ashok kumar M, Ramachandra Reddy G (2011) Hybrid composites: effect of fibers on mechanical properties. Int J Macromol Sci 1(1):9–14
7. Yu T, Ren J, Li S, Yuan H, Li Y (2010) Effect of fibre surface-treatments on the properties of poly (lactic acid)/ramie compo-sites. Compos Part A: Appl Sci Manuf 41:499–505
8. Venkatesha Prasanna G (2018) Modification, optimization of mechanical properties of bio fibers blended composites. Int J Eng Technol 7(2.33):799–801
9. Bessadok A, Marais S, Roudesli S, Lixon C, Metayor M (2008) Influence of chemical modifications on water sorption and mechanical properties of Agave fibre. Compos Part A 39:29–45
10. Venkatesha Prasanna G, Venkata Subbaiah K, Varada Rajulu A (2012) Chemical resistance, impact, flexural, compressive properties and optimization of fibers of natural fibers reinforced blend composites. Sch J Eng Res 1(6):85–89
11. Zhang XR, Zhao P, Pei XQ, Wang QH, Jia Q (2007) Flexural strength and tribological properties of rare earth treated short CFs/polyimide composites. Ex Poly Lett 1(10):667–672
12. Venkatesha Prasanna G, Venkata Subbaiah (2013) Modification, flexural, impact, compressive properties and chemical resistance of natural fibers reinforced blend composites. Malays Polym J 8(1):38–44
13. Morrison RT, Boyd RN (1989) Organic chemistry. Prentice-Hall, London
14. Mohanty AK, Misra M, Drzal LT (2001) Surface modifications of natural fibres and performance of resulting bio-composites. Compos Interfaces 8(5):313–343
15. Zhang XR, Pei XQ, Wang QH (2007) The effect of fiber oxidation on the friction and wear behaviors of short-cut CFs/polyimide composites. Ex Poly Lett (5):318–25
16. Venkatesha Prasanna G, Venkata Subbaiah K (2013) Hardness, tensile properties and morphology of blend hybrid biocomposites. Sch J Eng Res 2(1):21–29

17. Zhiwei X, Yudong H, Yuanjun S, Chunhua Z, Li L (2007) Surface characteristics of rare earth treated CFs and interfacial properties of composites. J Rare Earths 25:462–468
18. Akin DE, Gamble GR, Morrison WH, Rigsby L (1996) Chemical and structure analysis of fibre and core tissues from flax. J Sci Food Agric 72:155–165
19. Sreekal MS, Thomas S (2003) Effect of fibre surface modification on water-sorption characteristics of oil palm fibres. Compos Sci Technol 63:861–869
20. Myrtha K, Holia O, Anung S (2007) Physical and mechanical properties of natural fibers filled polypropylene composites and its recycle. J Biol Sci 7:393–396
21. Qianqian S, Xianhua C (2008) Effect of rare earths surface treatment on tribological properties of CFs reinforced PTFE composite under oil lubricated condition. J Rare Earths, 2, 6(4):584–592
22. Varada Rajulu A, Ganga Devi, Babu Rao G (2003) Miscibility studies of epoxy/unsaturated polyester resin blend in chloroform by viscosity, ultrasonic velocity and refractive index methods. J App Polym Sci (89):2970–2972
23. KaliaS, Kaith BS, Kaur I (2009) Pretreatments of natural fibres and their application as reinforcing material in polymer composites. Rev Polym Eng Sci 49:1253–1272
24. Su F, Zhang Z, Wang K, Jiang W, Liu W (2005) Tribological and mechanical properties of the composites made of carbon fabrics modified with various methods. Compos Part A, 36:1601–1607
25. Seena J, Koshy P, Thomas S (2005) The role of interfacial interactions on the mechanical properties of banana fibre reinforced phenol formaldehyde composites. Compos Interfaces 12:581–600
26. Venkatesha Prasanna G, Venkata Subbaiah K (2013) Modification of tensile, compressive properties & chemical resistance of hybrid biocomposites. Int J Nanomater Biostruct. 3(1):9–12. ISSN 2277-3851
27. Alamgi MK, Monimul MH, Islam RM, Bledzki AK (2010) BioResources 5:1618–1625
28. Punyamurth R, Sampathkumar D, Bennehalli B, Badyankal P, Vekateshappa SC (2014) Surface modification of abaca fiber by benzene diazonium chloride treatment and its influence on tensile properties of abaca fiber reinforced polypropylene composites. Ciencia Technol Dos Mater 26(2):142–149
29. Anon A (1992) A guide to unusual natural fibres: pineapple leaf fibre (PALF). Textiles 21(3)
30. Venkatesha Prasanna G. Tirunagari Jayadeep G, Nikhitha Poornabhodha (2019) Chemical treatment, influence of fiber content and optimization of hybrid natural fibers reinforced composites. Smart Innov Syst Technol (SIST). Springer. ISSN: 2190-3018

Impact of Fiber Length and Chemical Alteration on the Mechanical Properties of Blended Composites

Gowdagiri Venkatesha Prasanna

Abstract The effort of instigator in this recent work was to investigate the effect and impact of fiber length and surface treatment on the mechanical properties of biofibers fortified composites. Biofibers are hydrophilic and resin is aquaphobic, which does the fibers and resin contrary and effects in deprived interfacial binding among the resin blend and fibers. Fundamental inspiration driving this chemical action was to shrink their wetness retention property of the fibers and, further more, to expand the compatibility with matrix blend. In the current research work, hybrid biofibers composites were fabricated by blending 10% vinyl ester matrix with 90% epoxy with the reinforcement of Bagasse–Luffa fibers into the resins blend. The superiority and optimal values of tensile, compressive, and flexural properties were observed for 2-cm fiber length, benzene diazonium chloride-processed composites than the unprocessed fiber composites, 5% NaOH-treated, 10% NaOH-treated hybrid biofibers composites, and other chemically customized composites.

Keywords Bagasse–luffa fibers · Surface treatment · Chemical resistance test and mechanical tests

1 Introduction

Unreal strands namely carbon, Kevlar, glass, and aramid have been most broadly utilized in polymer matrix composites because they are highly sturdy and stiff. Aside from the many advantages acquired by synthetic fibers in diverse directions and applications, these synthetic fibers have many limitations in terms of their biodegradability, initial processing costs, recyclability, energy utilization, machine abrasion, wear, and health hazards, etc. In order to overcome this crisis, more demand was raised toward the fabrication of natural/renewable fiber-based composites to make the world 'green fibers' [1, 2]. Biofibers have been utilized as a substitute fiber to synthetic fibers as well as fortification for polymer matrix composites. Natural fibers generally contain

G. V. Prasanna (✉)
Mechanical Engineering Department, CVR College of Engineering, Hyderabad 501510, India
e-mail: gvpcvrmech@gmail.com

S. Seetharamu et al. (eds.), *Fatigue, Durability, and Fracture Mechanics*, Lecture Notes in Mechanical Engineering, https://doi.org/10.1007/978-981-15-4779-9_44

cellulose, hemicelluloses, pectin, lignin, and waxes [3–5]. The water absorption character possessed by the natural fibers controls the utilization of fibers in the fabrication of composites. Because of the aquaphobic property of polymer matrix may cause resin and fiber contrary and effects in frail bonding between the resin and fiber. The reason for this chemical action was to curtail the water retention quality of the biofibers which enhances the rapport with resin.

The matrix blend comprise various advantages in terms of the product end use applications and enhancing resin's utilization, performance, and properties [4, 6, 7]. Epoxy is a multipurpose and widely used matrix material for various applications such as advanced composites due to good pairing abilities. Thermoset resin applications were limited due to loxer toughness. Further blending strategy can be utilized productively to overcome the poorer properties of both the polymers. Miscible polymer blends produce a recently improved material from less predominant separate materials [8]. Vinyl ester is a toughening material which can be used together with epoxy materials. Vinyl ester was the extra resulting product of epoxy. The blending of epoxy with flexible polymers and elastomers improved the hardness [2, 4]. Subsequently, an appropriate polymer was required for the enhancement of stiffness and resistance against impact load. Vinyl ester was largely utilized for thermosetting material due to its low price and immense mix of qualities like resistance against chemicals. The extremity of vinyl ester was boosted by hydroxyl groups and make easy the adhesion and pigment soaking properties, promote grip, and color wetting properties, which gave the route to the damping of the fibers in matrix. For this reason a framework mix of matrix mix of vinyl ester/epoxy (90/10% w/w) resins was developed. Composite specimens with the 1, 2 and 3 cm length of unrefined, NaOH- and benzene diazonium chloride-processed Bagasse–Luffa fibers with a blend of epoxy/vinyl ester were fabricated.

2 Experimentation

2.1 *Resins and Fibers*

Two resins, namely: 1. Epoxy Araldite LY 556 as resin, hardener HY 951, 2. Vinyl ester.

Two Fibers, namely: 1. Luffa fiber, 2. Bagasse fiber were utilized to fabricate the composites (Fig. 1).

2.2 *Chemical (Alkali) Treatment*

The biofiber's hydrophilic feature, damage the successful bonding of fibers with a matrix blend. Furthermore, unwanted materials like pectin, grease, oil, and waxy

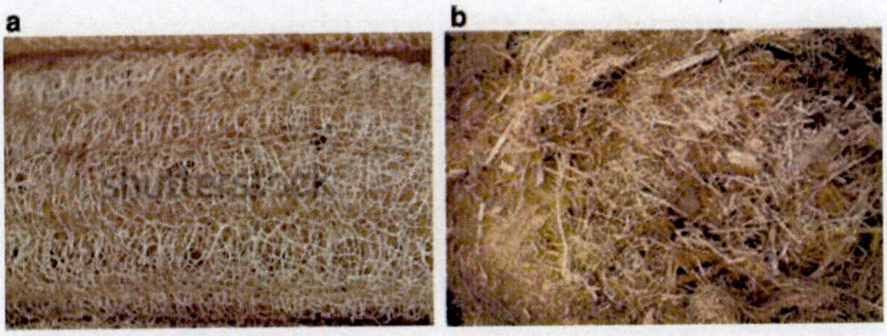

Fig. 1 **a** Luffa fiber. **b** Bagasse fiber

substances cover the reactive functional groups of the fiber and act as an obstruction for the interlock of fibers with the matrix. For the progress of the interfacial adhesion/bonding, surface of the fibers should be tailored with different chemical treatments. Chemical treatments provide more reactive bunches on the biofibers and this improves proficient coupling with the matrix blend. Surface of the fibers required to be modified by alkali action using different concentrations of NaOH: 5 and 10 wt% was conducted out in this work. After the alkali treatment, the reinforcing materials were immerged in the refined solution for 24 h to eliminate the remaining alkali. Further cleaning would be finished using the refined solution having a less quantity of acetic acid. In an oven, moist fibers were get cured for 3 h at 70 °C.

2.3 *Development of Benzene Diazonium Chloride*

Concentrated hydrochloric acid of 8 cm^3 was poured to a bubbling tube having phenylamine (aniline) of 3 and 10 cm^3 of H_2O, then the blend was shaked for amine liquefaction. Afterward, the solvent was chilled to 5 °C by using a frost bath. In the wake of cooling, a solvent of sodium was added to the pre-cooled merge. The temperature of the blend was kept underneath 10 °C, while purifying sodium nitrite [9].

2.4 *Benzene Diazonium Treatment of Bagasse–Luffa Fibers*

Bagasse–luffa fibrils were sliced to 10 mm length, then were cleaned using a refined solution. Afterward, this got cured by heating in the oven for 24 h at 70 °C. The fibers were cured and then immerged in the solution having 6% NaOH by curing the glass beaker at about 5 °C for 10 min. Recently prepared diazo solution was poured gradually to the blend by steady vibrating. After that, the fibrils were taken outside, cleaned with soap solution pursued by refined solution, and ultimately cured outside for 48 h [10].

2.5 Development of Blended Hybrid Biocomposites

The mold cavity was coated with a thin layer of hard wax for trouble-free discharge and pulled back of manufactured composite specimens. After the wax was cured, a lean layer of the aqueous solution of polyvinyl alcohol (PVA) was applied. Then the hybrid biofibers of the untreated and processed Bagasse–Luffa fibers were fortified to the matrix merge to get the required biocomposite specimens for the exploration of performance and mechanical properties. Then air bubbles were removed from cautiously by using the roller with delicate rolling. With the goal of a complete cure, composite samples were post-cured at a temperature of 80 °C for 2 h by keeping the specimens in a hot oven. After the complete curing, unprocessed and treated samples of biocomposite specimens were examined.

3 Results and Discussion

3.1 Tensile Strength Properties

Composite samples for the tensile strength test, prepared with 150 mm × 15 mm × 3 mm, were assessed as per ASTM standards and were investigated utilizing Universal testing machine with a cross-head speed of 50 mm/min. Due to the hydrophilic nature of natural fibers, inconsistency was noticed between natural fibers and matrix, which prompted an improper and lacking wetting of fibers in the matrix blend, which resulted in low-grade properties for the untreated composites. The impact of alteration of fiber's surface by the chemical treatment, alkali treatment, and quantity of reinforcement on the tensile property of bioconglomerates is presented in Fig. 2. In untreated biofiber composites, the repellency between hydrophobic matrix and hydrophilic natural fibrils would antagonistically influence the interfacial bonding between fiber's surface and matrix blend, and bring about decline of the load-bearing capability of the fibers due to the inadequate stress convey from the resin to fibers. Among a variety of surface variations performed, 2-cm-length benzene diazonium chloride-treated Bagasse–Luffa fibers blended composite displayed superior and optimum value for tensile strength than 1-cm and 3-cm fiber length 5% NaOH-treated, 10% NaOH-treated composite, and 1-cm, 2-cm, and 3-cm fiber length unprocessed biocomposites [7, 9]. However, an substantial tensile strength was noticed in NaOH-treated Bagasse–Luffa fibers composites, while noteworthy incease in the tensile property of benzene diazonium chloride-processed composites was noted. Benzene diazonium chloride-treated composite demonstrates more drop of the aquaphilic character of Bagasse–Luffa fibers due to the mixing of the hydroxyl bunches of fiber with benzene diazonium chloride and surface of the reinforcement turns out to be progressively uniform and coarse because of the annihilation of smaller scale gaps and thus the load conveys capability amid the cells increments [5, 11, 12]. Most significant tensile strength was seen for 2-cm length

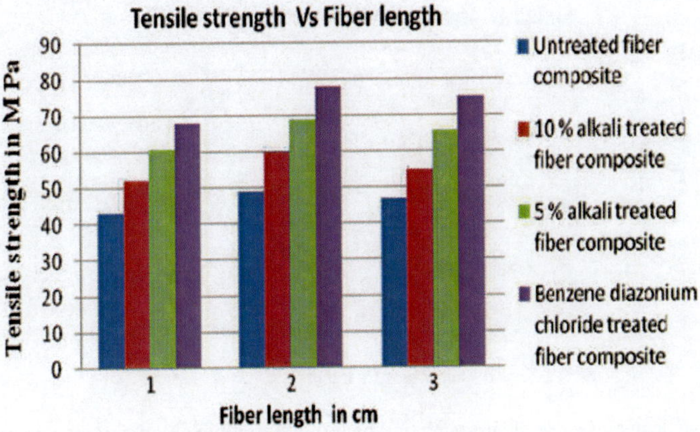

Fig. 2 Shows the tensile strength of untreated, alkali treatment, enzene diazonium chloride-treated biocomposites

benzene diazonium chloride-treated composite, because which improved the aspect ratio and surface roughness gave the channel to upgraded reinforcement distribution in resin and prompted increment in the interfacial adhesive coupling amid the resin and Bagasse–Luffa fibers and also affected in the upgraded morphological qualities [1, 2, 4]. The investigation clearly indicates that a sturdy interface with reinforcement cell boundaries and Coupling reaction will take place among cellulose of reinforcement and benzene diazonium chloride and results in the development of a diazo cellulose compound, represents the enormous increment in tensile strength values [3, 5]. The fibers surface treatment by means of both benzene diazonium chloride, NaOH, gives the route for the evacuation of undesirable materials like waxes, oils, impurities, hemicellulose and lignin from the natural fibers in addition to the diminution of moisture retention, yields the higher level of (alpha) cellulose in natural fibers and, furthermore, cleans the fiber surface [2, 13]. Be that as it may, where as at 3-cm fiber length treated composite, even benzene diazonium chloride-treated composite, 5% NaOH-treated, 10% NaOH-treated, untreated biofiber composite presented inferior values [7, 14, 15] on account of more fiber length.

3.2 *Flexural Properties*

The specimens with dimensions 150 mm × 15 mm × 3 mm were fabricated and investigated for flexural properties as per the ASTM D 5943-96 standards. The deviation of flexural strength as a function of reinforcement length, with and without chemical action was noted and shown in Fig. 3. From the test it was noticed that the flexural property of the 2-cm length benzene diazonium chloride-treated Bagasse–Luffa fibers blended composite demonstrated elevated and optimum value than the 1-cm

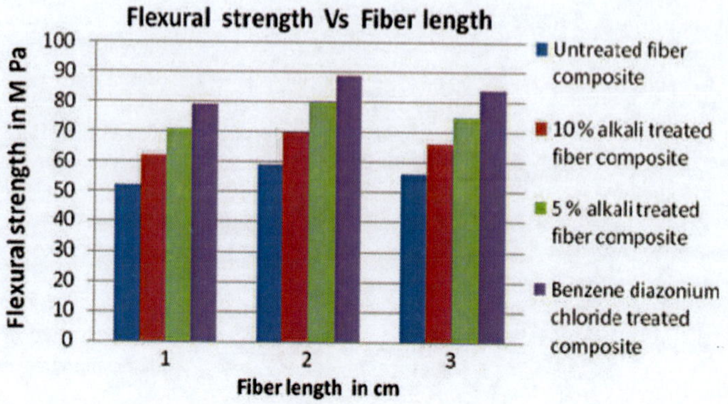

Fig. 3 Reveals the effect of alkali treatment, benzene diazonium chloride treatment and length of the fiber on the flexural property of biocomposites

and 3-cm length fiber, untreated, 5% NaOH-treated, 10% NaOH-treated composites [3, 9, 15, 16]. Due to the further alkali concentration at (10% NaOH), excess delignification of the fiber occurs, which brought about debilitating or harming of the fiber surface and effected in diminished flexural strength, than the (5% NaOH) treated composites and is indicated in the Fig. 3 [2, 17, 18]. Better fiber-matrix interaction and interfacial bonding was observed for 5% NaOH-treated composite compared with 10% NaOH-treated biocomposites [4, 11, 13, 18]. Alkali treatment mostly drive out the undesirable and debased materials like waxes, oil, hemicellulose and lignin, which yields a higher level for (alpha) cellulose in natural fibers and also cleanse the fiber's surface [5, 13, 16, 19]. But superior and noteworthy improvement of aspect ratio, fiber distribution in matrix blend, and stress convey capacity was viewed in benzene diazonium chloride-treated composites than in alkali-treated biofibers composites and untreated composites.

3.3 *Compressive Strength*

Specimens were manufactured and examined for Compressive strength properties as per ASTM D specifications. It was also seen that Benzene diazonium chloride-treated biocomposites disclosed predominant and ideal character for compressive strength than the alkali-treated and untreated biocomposites [3, 15]. The most favorable condition was an instantaneous outcome of factors like perfect fiber length, and chemical treatment, which was a direct result of variables like ideal fiber length, the better exterior coupling among resin and fibers resulted in ample stress transfer and performance [2, 14, 15]. Past works uncovers that fiber exterior modification by chemical treatment increases surface coarseness with the and lessening in dampness [12, 19, 20]. However, inferior compressive strength for the composite was detected at a reinforcement length of 3-cm treated composite. For reinforcement length of 3 cm,

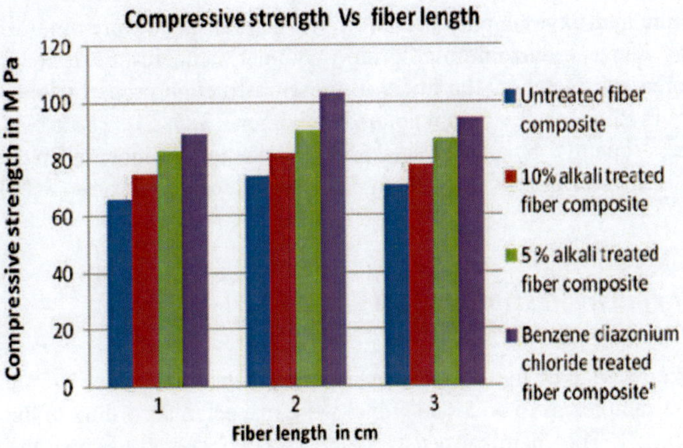

Fig. 4 Represents the compressive strength of untreated, alkali treatment, benzene diazonium chloride treated biocomposites

untreated and treated composites, there was a decline in the compressive property because of inadequate fiber–matrix interfacial bonding and micro-crack formation at the interface prompted deficient stress transfer. But benzene diazonium chloride treatment occupies the gaps of fiber with further aligning irregular fiber, more and better Compressive strength values as compared to alkali-treated composites [4, 16, 19] (Fig. 4).

3.4 Moisture Content Test

Figure 5 shows the incorporation rate of wetness amount by unprocessed reinforced composite, NaOH-processed and Benzene diazonium chloride-processed composites. The natural fibers composites show evidence of hydrophilic properties [2, 10].

Fig. 5 Moisture content absorption of unprocessed, NaOH-processed and benzene diazonium chloride-processed composites

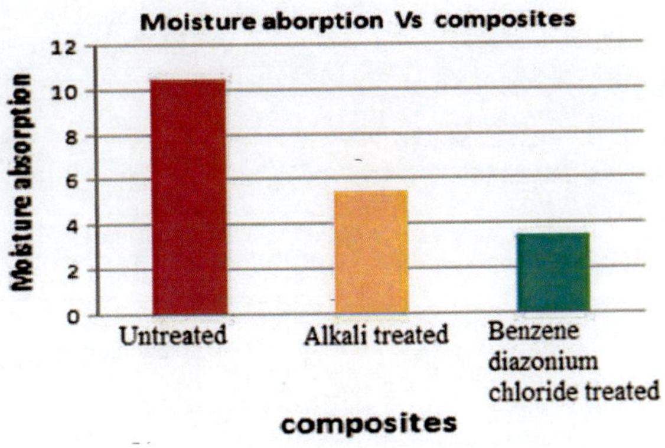

Hydrophilic hydroxyl bunches present in the natural fibers were diminished by the alkali and Benzene diazonium chloride treatment, henceforth exterior of the fiber was tailored and increases the fibers wetness obstruction property, however which was more in Benzene diazonium chloride treatment [3, 5, 14]. The assimilation rate of dampness quantity by unprocessed fiber composite, was more when contrasted to alkali-treated and Benzene diazonium chloride-treated composites [4, 21].

3.5 Chemical Resistance Test

To gain knowledge of the chemical resistance of the composites in each case, the composite samples of $(5 \times 5 \times 3)$ mm^3 were inspected according to the ASTM D 543-87 [5, 22]. The chemical resistance test was conducted to know the impact of underneath referenced chemicals on unprocessed and treated fibers fortified composites. Then % weight loss or % weight gain values for unprocessed and processed biocomposites submersed in the mentioned chemicals was shown in Table 1. Put on in weight was observed for almost all the chemicals with the exception of toluene and carbon tetrachloride and which was indicated by in the table. The table obviously point out that processed biocomposites also have a weight drop in carbon tetrachloride [4, 9, 13]. The explanation for the weight reduction was a direct outcome of the assault of chlorinated hydrocarbons on the cross-connected vinyl ester-epoxy matrix blend system. The weight gain after the test demonstrated that the composites were

Table 1 shows the effect of chemicals on the weight of unprocessed and surface modified Biocomposites

Sl. no.	Chemicals used for the test	Weight loss (−) or gain (+) in %	
		Untreated composites	Treated composites
1	8% Acetic acid	11.85	12.31
2	40% Nitric acid	12.01	12.86
3	10% Hydrochloric acid	13.03	13.89
4	10% Sodium hydroxide	12.14	12.87
5	10% Sodium carbonate	10.07	10.68
6	10% Ammonium hydroxide	9.7	10.56
7	Benzene	2.03	2.73
8	Toluene	−1.74	−0.89
9	Carbon tetrachloride	−0.97	−0.78
10	Water	2.48	2.75

swelled with the gel creation rather than dissolving in chemical reagents [2, 5–7]. It was seen after the test that composites were also opaque to water, unaffected by almost all the chemicals excluding toluene and carbon tetrachloride.

3.6 Dielectric Strength

The investigation of dielectric strength of raw, NaOH-processed, and benzene diazonium chloride-processed composites, directed according to ASTM D-149 standards. The composite specimens with measurements of 120 mm × 120 mm × 3 mm were formed. The voltage was perceived for five points to all specimens, then the normal value was noted during the exploration. The test was directed at 50 Hz frequency and room temperature. Digital micrometer of 0.001 mm least count was employed to discover the thickness of the composites at the breakdown point. It was substantial that the dielectric strength of hybrid biofiber composites increments with an increase in the length of fiber from 1 cm to 2 cm but reduced at 3-cm length fiber treated and untreated composite [2, 4, 7, 15, 19]. At 3-cm length fiber composite, a decline in dielectric strength was noticed because of inappropriate fiber circulation in the matrix blend and meager interfacial bonding amid the fibers and matrix blend [5, 9] (Fig. 6).

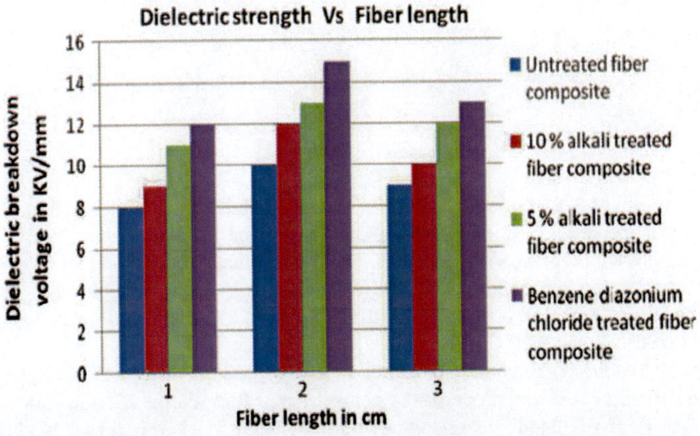

Fig. 6 Shows the dielectric strength of unprocessed, NaOH-processed and benzene diazonium chloride-processed composites

4 Conclusion

Chemical treatment carried out in the current work illustrates that, which extensively improves the fiber–matrix adhesion, enlarged the strength, and thus the performance of the natural fiber composites. The assimilation rate of dampness amount by unrefined fiber composite was more when contrasted with the alkali-treated and benzene diazonium chloride -treated composites. The tensile, compressive, and flexural strength esteems were improved for both alkali- and benzene diazonium chloride processed composites. But the benzene diazonium chloride treatment consumes the void gaps of fiber and adjusts uneven fibers looked superior strength values as compared to alkali-treated composites. The results show that the tensile, compressive and flexural strength of 2-cm length benzene diazonium chloride-treated fiber composites have exhibited have displayed prevalent and ideal values but 3-cm length fiber untreated and treated composite illustrated a substandard tensile, compressive and flexural strength, because of the improper fiber allocation in the matrix blend and deficient interfacial bonding between the fibers and matrix blend which additionally prompted diminished stress transferability. But after the chemical treatment or surface modification of fibers, it was exposed that improvement in the properties of the composites because of removal of impurities, hemicellulose, oils, waxes, pectin grease, and lignin from the surfaces of the fibers, and yields rough surface topography, which improves the aspect ratio of fibers and expands the strengthening output of the fiber with the improvement of stress convey from the matrix to load-bearing fibers.

References

1. Kaith BS, Singha AS, Sanjeev K, Susheel K (2008) Mercerization of flax fibre improves the mechanical properties of fibre-reinforced composites. Int J Polym Mater 57(1): 54–72
2. Prasanna GV, Rohit K (2019) Surface modification, characterization and optimization of hybrid bio composites. Springer (2019)
3. Edeerozey AM, Akil HM, Azhar AB, Ariffin MZ (2007) Chemical modification of kenaf fibres. Mater Lett 61: 2023–2025
4. Zhiwei X, Yudong H, Yuanjun S, Chunhua Z, Li L (2007) Surface characteristics of rare earth treated CFs and interfacial properties of composites. J Rare Earths 25: 462–468
5. Prasanna GV (2018) Modification, optimization of mechanical properties of bio fibers blended composites. Int J Eng Technol 7(2.33): 799–801
6. Prasanna GV, Subbaiah KV, Rajulu AV (2012) Chemical resistance, impact, flexural, compressive properties and optimization of fibers of natural fibers reinforced blend composites. Scholarly J Eng Res 1(6): 85–89
7. Prasanna GV, Jayadeep GT, Poornabhodha N (2019) Chemical treatment, influence of fiber content and optimization of hybrid natural fibers reinforced composites. Springer, The Smart Innovation, Systems & Technologies (SIST)
8. Raghavendra Rao H, Ashok Kumar M, Ramachandra Reddy G (2011) Hybrid composites: effect of fibers on mechanical properties. Int J Macromol Sci 1(1): 9–14
9. Alamgi MK, Monimul MH, Islam RM, Bledzki AK (2010). BioResources 5: 1618–1625

10. Akin DE, Gamble GR, Morrison WH, Rigsby L (1996) Chemical and structure analysis of fibre and core tissues from flax. J Sci Food Agric 72: 155–165
11. Kalia S, Kaith BS, Kaur I (2009) Pretreatments of natural fibres and their application as reinforcing material in polymer composites. Rev Polym Eng Sci 49: 1253–1272
12. Mohanty AK, Misra M, Drzal LT (2001) Surface modifications of natural fibres and performance of resulting bio-composites. Compos Interfaces 8(5): 313–343
13. Yu T, Ren J, Li S, Yuan H, Li Y (2010) Effect of fibre surface-treatments on the properties of poly (lactic acid)/ramie compo-sites. Compos Part A Appl Sci Manuf 41: 499–505
14. Bessadok A, Marais S, Roudesli S, Lixon C, Metayor M (2008) Influence of chemical modifications on water sorption and mechanical properties of Agave fibre. Composites Part A 39: 29–45
15. Myrtha K, Holia O, Anung S (2007) Physical and mechanical properties of natural fibers filled polypropylene composites and its recycle. J Biol Sci 7: 393–396
16. Prasanna GV, Subbaiah KV (2013) Hardness, tensile properties and morphology of blend hybrid biocomposites. Scholar J Eng Res 2(1): 21–29
17. Seena J, Koshy P, Thomas S (2005) The role of interfacial interactions on the mechanical properties of banana fibre reinforced phenol formaldehyde composites. Compos Interfaces 12: 581–600
18. Prasanna VG, Subbaiah KV (2013) Modification of Tensile, Compressive properties & Chemical Resistance of hybrid biocomposites. Int J Nanomater Biostruct 3(1): 9–12
19. Rajulu AV, Devi G, Rao GB (2003) Miscibility studies of epoxy/unsaturated polyester resin blend in chloroform by viscosity, ultrasonic velocity and refractive index methods. J App Polym Sci 89: 2970–2972
20. Zhang XR, Zhao P, Pei XQ, Wang QH, Jia Q (2007) Flexural strength and tribological properties of rare earth treated short CFs/polyimide composites. Ex Poly Lett 1(10): 667–672
21. Sreekal MS, Thomas S (2003) Effect of fibre surface modification on water-sorption characteristics of oil palm fibres. Compos Sci Technol 63: 861–869
22. Qianqian S, Xianhua C (2008) Effect of rare earths surface treatment on tribological properties of CFs reinforced PTFE composite under oil lubricated condition. J Rare Earths 6(4): 584–592
23. Zhang XR, Pei XQ, Wang QH (2007) The effect of fiber oxidation on the friction and wear behaviors of short-cut CFs/polyimide composites. Ex Poly Lett 5: 318–325
24. Su F, Zhang Z, Wang K, Jiang W, Liu W (2005) Tribological and mechanical properties of the composites made of carbon fabrics modified with various methods. Compos A 36: 1601–1607
25. Prasanna GV, Jha NK, Harish C (2018) Chemical treatment and analysis of mechanical properties of hybrid biocomposites. Int J Mech Prod Eng Res Dev: 189–193
26. Prasanna GV, Subbaiah KV (2013) Modification, flexural, impact, compressive properties and chemical resistance of natural fibers reinforced blend composites. Malaysian Polym J 8(1): 38–44
27. Prasanna GV, Jha NK (2019) Optimisation and mechanical testing of hybrid bio composites. Mater Today: 1395–1404
28. Punyamurth R, Sampathkumar D, Bennehalli B, Badyankal P, Vekateshappa SC (2014) Surface modification of abaca fiber by benzene diazonium chloride treatment and its influence on tensile properties of abaca fiber reinforced polypropylene composites. Ciencia Technologia dos Mater 26(2): 142–149
29. Anon A (1992) A guide to unusual natural fibres: pineapple leaf fibre (PALF). Textiles 21(3)
30. Morrison RT, Boyd RN (1989) Organic chemistry. Prentice-Hall, London

Experimental Study to Understand the Effect of Residual Stress and Applied Load on Creep Relaxation

Anilkumar Shirahatti

Abstract In structural integrity analysis of components operating at high temperature, it is important to understand whether the presence of residual stresses lead to failure [1]. Accurate prediction of the creep crack initiation is needed in structural integrity assessments of components. General assessment of structures uses the experimental data obtained from laboratory test specimens subjected to either displacement or load tests but in actual operating condition, components are subjected to both inherent residual stresses and applied load. In the current research work, two test rigs are designed and tests are performed to understand the effects of EFU, long-range residual stress and external applied load on creep behaviour of 316H stainless steel. Results obtained show that, for the same total initial reference stress, the time for crack to grow is lower in the case of mixed loading conditions compared to load-controlled tests. The longer crack growth times are a consequence of the relaxation and redistribution of the residual loads in the structure. The initiation time is also a function of the elastic follow-up.

Keywords Crack growth · Residual stress · Creep · Elastic follow-up

1 Introduction

A nuclear power station has a large amount of welds, viz. nearly 2000 welds are present in large bore steam pipe work [2, 3]. Welded joints are particularly vulnerable to plant degradation and material ageing. Welded engineering components operating at high temperature will be subjected to the combinations of residual stress and applied load. When components are subjected to the applied load, it is usually termed as to induce primary stresses or load-controlled stress, while residual stresses present in the components are often considered as secondary stresses or a displacement-controlled stress. When components are working under high temperature condition, residual stresses are expected to relax with respect to time [4, 5]. The interaction will

A. Shirahatti (✉)
Department of Mechanical Engineering, Jain College of Engineering, Belagavi, India
e-mail: anilshirahatti@gmail.com

S. Seetharamu et al. (eds.), *Fatigue, Durability, and Fracture Mechanics*, Lecture Notes in Mechanical Engineering, https://doi.org/10.1007/978-981-15-4779-9_45

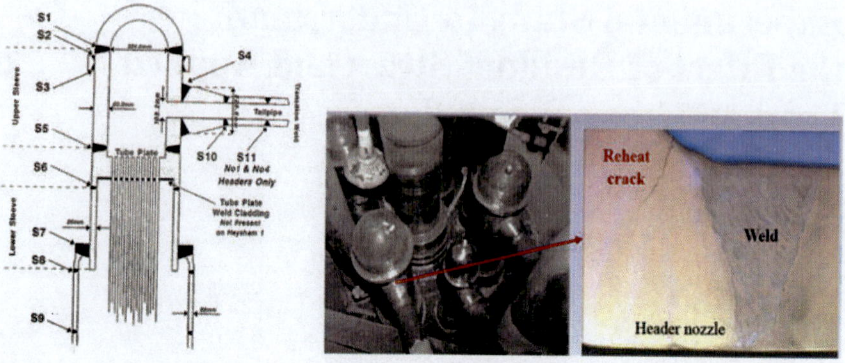

Fig. 1 Superheater header: type 316H austenitic stainless steel thick section weldments [5]

become more complex, if there is a combination of both residual stress and applied load. Weld residual stresses are created by misfit strains and as such essentially represent displacement-controlled boundary conditions. Usually, welded components are then subjected to an externally applied load. The subsequent deformation of the welded sample can be viewed as being associated with combined boundary conditions, i.e. combined displacement and load-controlled situations. Figure 1 shows thick section weldments of 316H austenitic stainless steel in the superheater header. Crack initiation was observed in welded components operating at high temperature due to the presence of combined weld residual stress and applied stress [6].

In the life assessment of structural components, residual stress plays an important role. Residual stresses may arise in the components or structures as a consequence of the manufacturing process adopted and final fabrication [7, 8]. If residual stress is tensile in nature, then it can combine with applied loads and lead to failure of the component at a load seems to safer from design point of view. Thus, for accurate assessments of components, a quantitative understanding of behaviour of residual stresses with applied load is needed.

In the case of creep crack initiation and growth, many experiments are carried out using standard laboratory-scale creep test specimens tested under load-controlled [9, 10] and occasionally under displacement-controlled conditions [11]. However, structures are subjected to stresses that arise from a combination of residual and applied stresses. The relaxation of residual stress in one location is compensated by changes in residual stress distribution in other locations so as to retain equilibrium, i.e. components are often subjected to combined displacement- and load-controlled conditions. Furthermore, it is now known [2, 12–18] that, depending on the stiffness of the structure, relaxation of the residual stress can be associated with elastic follow-up.

The purpose of this paper is to illustrate a three-bar test rig that was designed [5, 7] to introduce the combinations of residual and applied stresses into a compact tension C(T) specimen. The residual stresses can be induced in a controlled manner and can be characterized easily. Tests are performed to understand the effect of residual stresses and applied load on creep crack initiation and relaxation.

2 Test Rig Design

The concept of test rig is illustrated in Fig. 2. The test rig was designed such that in a controlled manner residual stresses can be induced and will also have further provision to apply external load to the complete test rig. The middle bar in test rig consists of C(T) specimen and other part of bar with stiffness X_{sp} and X_m respectively. There are two bars having a stiffness X_{out} each. An initial tensile residual stress is induced in middle bar using initial misfit Y, which is further balanced by combine compressive forces acting outer bars. Once known, residual force is induced and equilibrium is reached, an external load is applied to the complete test rig. More details about the design are given in [5, 7].

Stiffness of the specimen relative to the stiffness of the other parts of middle bar and outer bars will decide the value of elastic follow-up of the structure, Z. The overall EFU value, Z can be found using the Eq. (1).

$$Z = Z_s Z_{eff} \tag{1}$$

where

$$Z_s = \left(\frac{1+\beta}{\beta}\right),\ Z_{eff} = \left(\frac{1+\alpha_{eff}}{\alpha_{eff}}\right),\ \frac{1}{X_{eff}} = \frac{1}{X_{sp}} + \frac{1}{X_m},\ \alpha_{eff} = \frac{2X_{out}}{X_{eff}},\ \beta = \frac{X_m}{X_{sp}}$$

A detailed derivation of Eq. (1) is explained in [7]. As per R5 Standards [4], the value of overall EFU Z greater than 5 represents load-controlled behaviour, while a value near to 1 gives displacement-controlled behaviour. Hence, it was decided to design and manufacture two test rigs, which fulfill R5 Standard criteria. Figure 3 shows the detailed schematic diagram of test rig and more details about design and manufacturing are given in [7].

Thermocouples are connected to measure the specimen temperature, room temperature and strain gauge temperature. A potential drop system is used to measure the crack initiation and growth, while capacitance gauge mounted on C(T) specimen

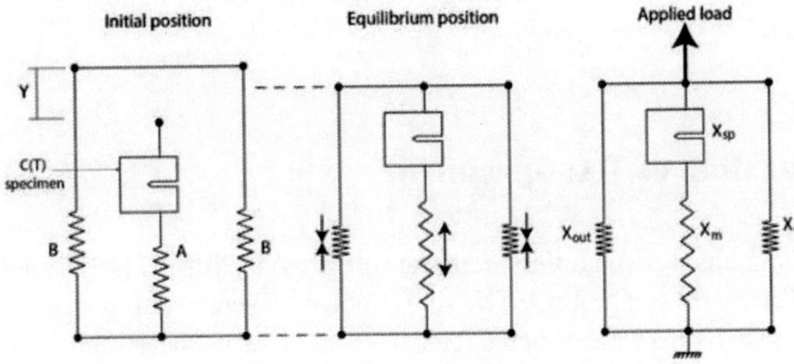

Fig. 2 Concept of three-bar structure

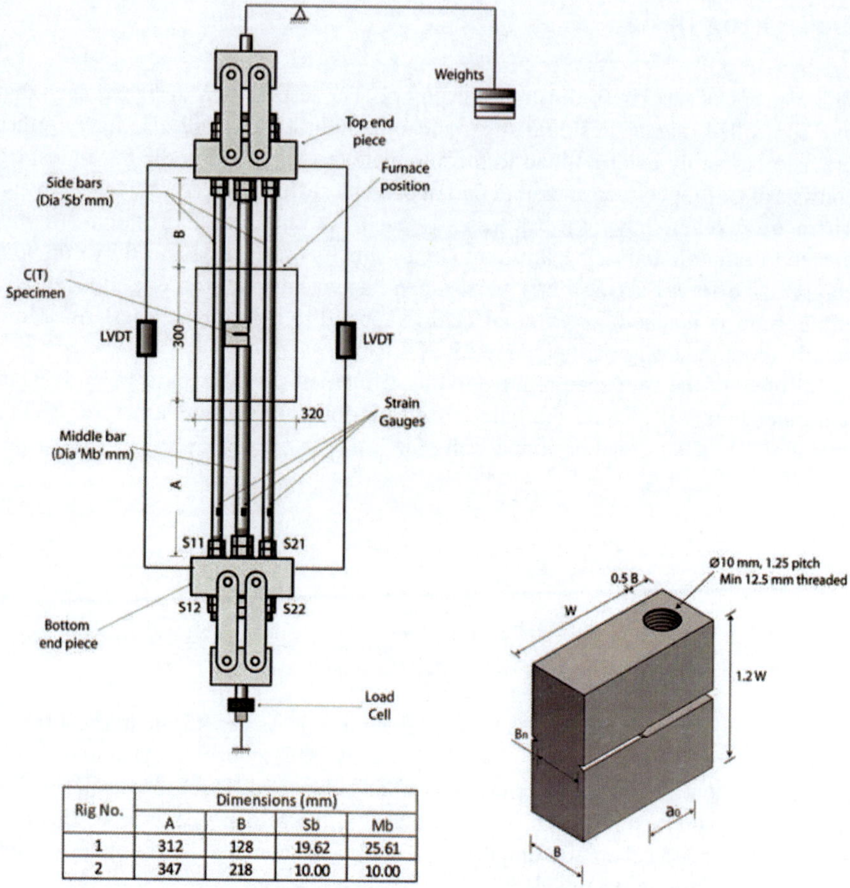

Rig No.	Dimensions (mm)			
	A	B	Sb	Mb
1	312	128	19.62	25.61
2	347	218	10.00	10.00

Fig. 3 Three-bar test rig

is used to measure load line displacement. Two LVDTs are mounted at side to measure the overall displacement of the test rig. The external load is applied via a lever arm arrangement.

3 Preparation of Test Specimen

The test specimens were manufactured according to ASTM 1457 [19] from ex-service Type 316H stainless steel supplied by EDF Energy. A screw-fitting arrangement (see Fig. 3) rather than pins were adopted to provide an accurate measurement of stiffness. A 2-mm pre-crack was introduced using 0.1-mm EDM wire such that crack length to width ratio of 0.5 is maintained. Tables 1 and 2 give details of C(T) specimen and material composition of the test specimens.

Table 1 Details of C(T) specimens

Rig No.	Test ID	W (mm)	B (mm)	B_n (mm)	a_0 (mm)
Load control	LC-01	38.07	19.04	15.36	19.36
Low EFU: Z = 2.15	MX-01	38.05	19.03	15.38	19.59
High EFU: Z = 7.34	MX-02	37.83	19.04	15.36	19.59

4 Introduction of Residual Stress in the Structure

First, in the C(T) specimen, middle bars were connected to top and bottom end pieces. The outer bars were then connected to top end piece and they have provision to move freely through the clearance holes provided at the bottom end piece. Instruments, viz. strain gauges, LVDTs, thermocouples, were connected to data logger and furnace was switched on to achieve a temperature of 550 °C for the C(T) specimen. When C(T) specimen reached a stable temperature as per ASTM standards, nuts S21 and S11 (shown in Fig. 3) were screwed down, so that the both the end pieces were forced apart. Under the equilibrium conditions, this results in tensile load in middle bar and balancing compressive forces in outer bars. Force acting in bars can be easily found by strain gauges mounted on them. Finally, when the desired amount of residual force was introduced into the structure, nuts S22 and S12 on side bars were fixed and the entire assembly was then subjected to an applied load.

5 Creep Crack Tests

In total, three creep crack growth tests were performed and more details are given in Table 3. LC-01 test is the load-controlled test performed on conventional lever arm test rig. Additionally, one test each was performed on three-bar test rig. The initial total reference stress on three-bar test rig was selected such a way that it matches with the load control test. However, the combination of applied and initial residual stresses was different depending on the ease of inducing residual stress and to avoid any twisting or bending. During the tests, the applied load was monitored using the load cell shown in Fig. 3. The combined residual and applied forces in the middle and outer bars were measured using the strain gauges.

Table 3 provides the details of the test conditions together with calculated reference stresses. The tests were continued to permit sufficient changes in potential drop to be measured. The tests were then stopped, specimen were broken open and final crack lengths measured.

The reference stress in Table 3 was determined from [4, 20].

$$\sigma_{\mathrm{ref}} = \frac{P}{W B_n n_L} \tag{2}$$

Table 2 Chemical composition of the specimen made of 316H stainless steel

Chemical	P	Cr	Mo	C	S	N	Si	Al	V	W	Mn	Ti	Cu	W	Co
Weight (%)	0.02	17.1	2.38	0.04	0.014	11	0.29	<0.005	0.02	0.042	1.49	0.013	0.09	0.042	0.09

Table 3 Summary of tests

Details/Specimens	LC-01	MX-01 (Z = 2.15)	MX-02 (Z = 7.34)
Initial condition (Start of Tests)			
Initial residual force (kN)	–	8.91	6.36
Initial residual reference stress (MPa)	–	183	130
Applied load (kN)	12.46 kN	2.79	5.43
Applied reference stress (MPa)	240 MPa	57	110
Total initial reference stress (MPa)	240 MPa	240 (223)	240
Final condition (End of Tests)			
Total load (kN)	12.46	5.27	10.97
Total reference stress (MPa)	240 MPa	108	224
Final crack growth (mm)	0.882	0.030	0.169
Test duration (hrs)	1508	4882.4	4229

where n_L is a normalized limit load function given by

$$n_L = \sqrt{(1+\gamma)\left(1+\gamma(a/W)^2\right)} - (1+\gamma(a/W)) \text{ with } \gamma = 2/\sqrt{3}$$

6 Results and Discussion

In tests MX-01 and MX-02, a residual force of 8.91 kN and 6.36 kN was introduced into the C(T) specimen, respectively. In both the tests, residual force acting in the middle bar was approximately equal to the summation of the compressive forces acting in side bars. The load versus load line displacement behaviour of all specimens is shown in Fig. 4. Initial residual forces acting on test specimens MX-02 and MX-01 are represented by points A2 and B2, respectively. The curve A2–A3 and B2–B3 represents the application of applied loads, while A3–A4 and B3–B4 corresponds to the relaxation of load acting on the specimen with respect to time. Figure 5 shows the load line displacement for all the tests.

The variation of forces in all bars with respective to time for both rigs is shown in Figs. 6 and 7. It is evident that when external load was applied in case of MX-01, C(T) specimen on test rig 1 with Z = 2.15, relaxation of load from 11.7 to 10.87 kN took place. Also, there was an additional plastic deformation that took place in specimen when compared to conventional load control test. The MX-02 C(T) specimen test on test rig 2 with Z = 7.34 showed less additional plastic deformation during external loading. This phenomenon is due to the higher EFU value in rig 2.

During the creep process, it was found that the total load was relaxed in both the specimens and as expected the rate of relaxation is high for MX-01 than MX-02.

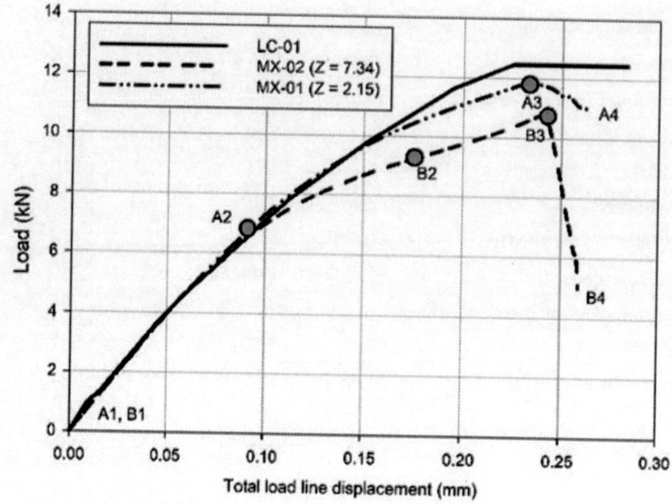

Fig. 4 Load versus total load line displacement for C(T) specimens

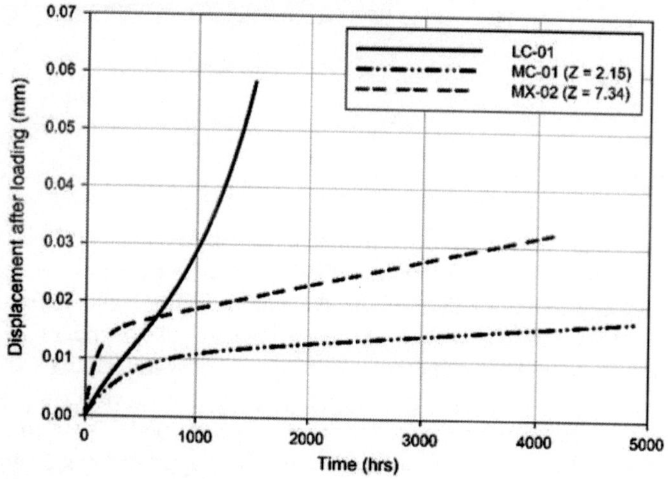

Fig. 5 Comparison of displacements after loading

In case of MX-01, after 4883 h of creep, the initial induced residual stress was relaxed from 183 MPa (8.91 kN) to 51 MPa (2.48 kN). While in case of specimen MX-02, after 4229 h, the initial induced residual stress was relaxed from 130 MPa (6.36 kN) to 113 MPa (5.54 kN). It is observed that the stiffness of the C(T) specimen was decreasing due to creep and crack growth and hence with respect to time a greater portion of the applied load was carried by the outer bars. The slope of relaxation curves A3–A4 and B3–B4 in Fig. 4 is dependent on the relative stiffness of C(T) specimen and other components of the test rig, and which in turn corresponds to

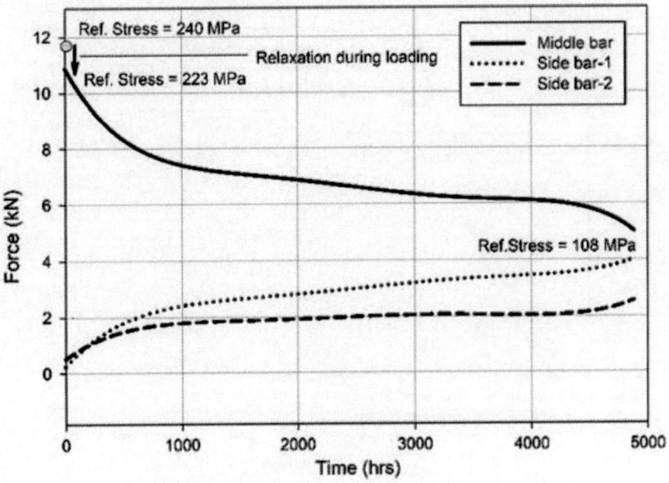

Fig. 6 Variation of force in all bars with time (MX-01)

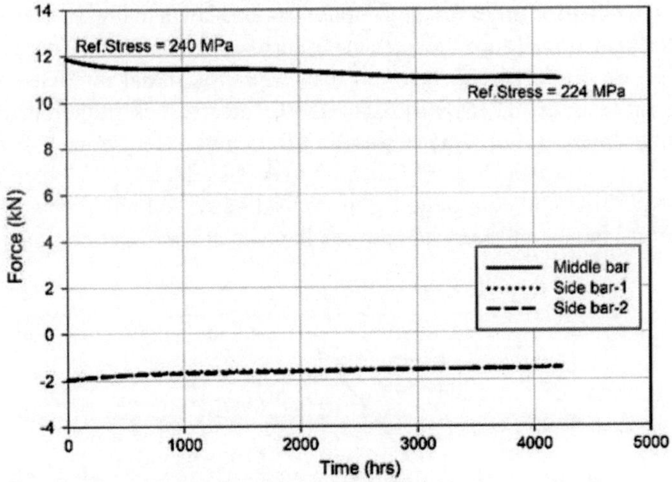

Fig. 7 Variation of force in all bars with time (MX-02)

EFU values associated with test rigs. It can be concluded that rig with higher EFU considers the residual load as load control, while the test rig with lower EFU considers it as displacement control.

Figures 5 and 8 show the measure crack load line displacement and crack growth, respectively, for LC-01, MX-01 and MX-02. It is evident that for the same total initial reference stress, the creep load line displacement was greater for MX-02 than MX-01. Also, crack initiation of 0.03 mm occurred much early in case of MX-02 than MX-01, but both were much later than LC-01.

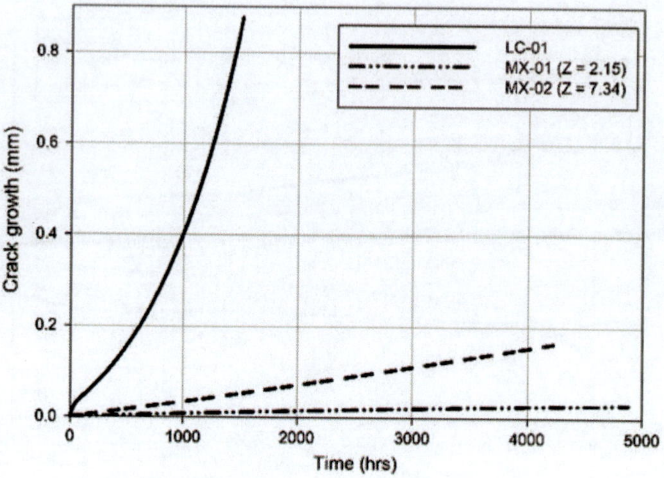

Fig. 8 Comparison of crack growth

The crack initiation times for the various tests are shown in Fig. 9, where, together with the tests performed and explained in the current paper additional tests performed in [6, 7, 9] are also reported. It reveal that, while the initial reference stress was the same for tests LC-01, MX-01 and MX-02, creep crack initiation times were significantly different. For a crack length of 0.03 mm in load control case, it took 52.46 h, while for low and high EFU tests, it took 4882.4 h and 873.9 h, respectively. Rig with low EFU will contribute for more load relaxation and redistribution than high EFU and thus it will leads to increase in creep crack initiation time. Figure 10

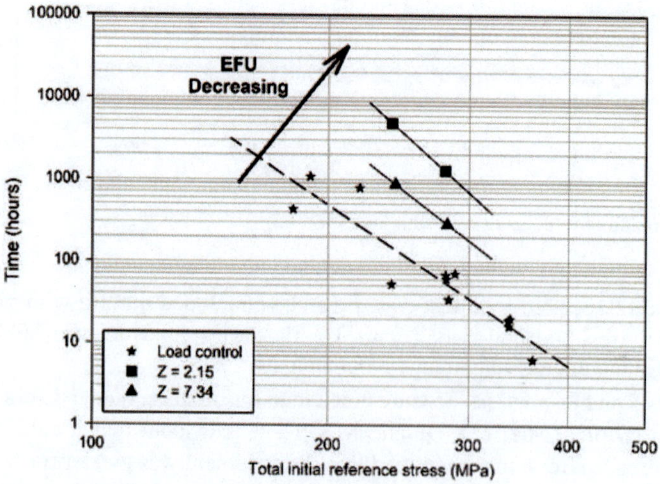

Fig. 9 Comparison of time for cracks to grow 0.03 mm

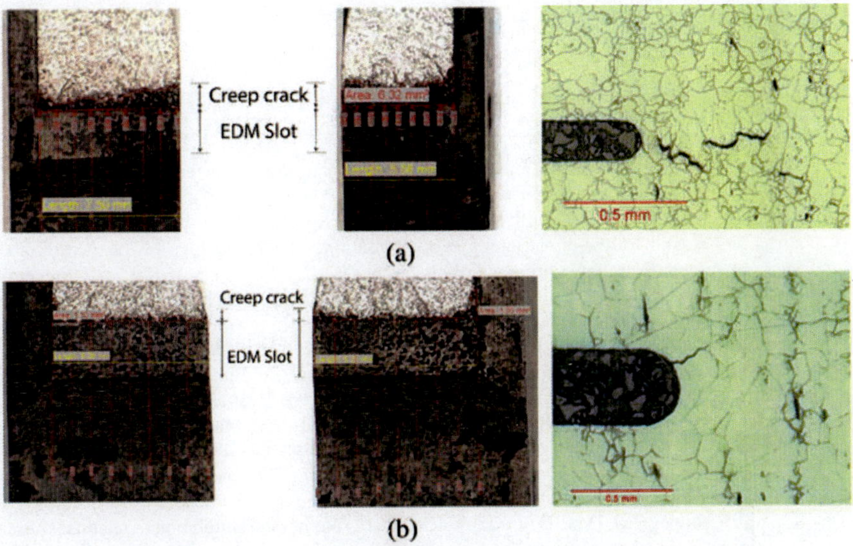

Fig. 10 Comparison of Crack growth and Micrograph of specimen **a** LC-01, **b** MX-02

shows the crack growth and micrographs of the specimen LC-01 and MX-02. All the creep tests appeared to fail in an intergranular crack growth and to consist mainly of wedge cracking. More detailed micrographic analysis is given in [7].

7 Conclusion

Structural integrity analysis of components is based on conventional laboratory load-controlled or displacement-controlled tests, but in actual working condition, they are subjected to combination of residual stress and applied load. This in turn represents mixed load conditions. In the current paper, two test rigs with EFU near 2 and 7 were developed. The rigs are designed on the concept of three-bar structure, in which residual stresses can be introduced in a controlled manner. The results obtained shows that even though the initial reference stresses were the same for mixed and load-controlled test, the initiation times were longer in mixed boundary condition test. Also, it is found that with initiation times were high for lower values of elastic follow-up test rig. The longer initiation times were due to the relaxation and redistribution of the residual and applied loads in the structure.

Acknowledgements The author gratefully acknowledges EDF Energy, United Kingdom for the financial support for this work and research facility provided by the University of Bristol, United Kingdom. The author also acknowledges Dr. David Dean and Mr. Mike Spindler of Structural Integrity Group at EDF Energy, for lending their expertise and advice.

References

1. Shirahatti AM, Smith DJ (2015) The effects of long-range residual stress, elastic follow-up and applied load on creep crack incubation and material toughness. J Strain Anal Eng Des, JSA Golden Issue, July 2015, pp 1–15
2. Robinson EL (1955) Steam piping design to minimize creep concentration. Trans ASME 77:1147–1162
3. Shirahatti AM, Truman CE, Smith DJ (2013) Experiments to determine the influence of residual stress and elastic follow-up on creep crack initiation. In: Proceedings of the ASME 2013 pressure vessels & piping conference, Paris, France
4. British Energy Generation Ltd. (2003) Assessment procedure for the high temperature response of structures. R5 Standards, Issue 3
5. Shirahatti AM, Smith DJ (2019) A novel creep test rig to study the effects of long-range residual stress, elastic follow-up and applied load. J Braz Soc Mech Sci Eng 41(230):1–12
6. Bouchard PJ, Withers PJ, McDonald SA, Heenan RK (2004) Quantification of creep cavitation damage around a crack in a stainless steel pressure vessel. Acta Mater 52(1):23–34
7. Shirahatti AM (2014) A novel test rig to study the effects of elastic follow-up, long range residual stress and applied load on creep crack initiation. PhD thesis, University of Bristol, United Kingdom
8. Kapadia P, Davies CM, Dean DW, Nikbin KM (2012) Numerical simulation of residual stresses induced in compact tension specimens using electron beam welding. In: Proceedings of the ASME 2012 pressure vessels & piping conference, Toronto, Ontario, Canada
9. Dean DW, Gladwin DN (2007) Creep crack growth behavior of Type 316H steels and proposed modifications to standard testing and analysis methods. Int J Press Vessels Pip 84:378–395
10. Fookes AJ (2003) Assessment of crack growth in steels at high temperature. PhD thesis, Bristol University, United Kingdom
11. Gladwin DN, Allport L, January (2010) The constant static displacement controlled creep crack growth response of 316H stainless steel header material. Technical report, British Energy Generation Ltd., Report No. E/EAN/BDBB/0024/AGR/06
12. Hadidi-Moud S, Smith DJ (2008) Use of elastic follow-up in integrity assessment of structures. In: Proceedings of the ASME 2008 pressure vessels & piping conference, Chicago, Illinois, USA
13. Aird CJ, Mahmoudi AH, Mirzaee Sisan A, Truman CE, Smith DJ (2006) Generating well defined residual stresses in laboratory specimens. In: Proceedings of the ASME 2006 pressure vessels & piping conference, Vancouver, BC, Canada
14. Shirahatti AM, Wang Y, Truman CE, Smith DJ (2013) A new method of introducing long range residual stresses to study creep crack initiation. In: 13th international conference on fracture, Beijing, China
15. Hossain S, Truman CE, Smith DJ (2011) Generation of residual stress and plastic strain in a fracture mechanics specimen to study the formation of creep damage in type 316 stainless steel. Fatigue Fract Eng Mater Struct 34:654–666
16. Wang YQ, Coules HE, Truman CE, Smith DJ (2018) Effect of elastic follow-up and ageing on the creep of an austenitic stainless steel. Int J Solids Struct 135:219–232
17. Kapadia P, Davies CM, Dean DW, Nikbin KM (2015) Assessment of creep crack growth due to assessment of creep crack growth due to secondary and combined loading. Transactions, SMiRT-23, Manchester, United Kingdom
18. Khayatzadeh S, Tanner D, Truman CE, Flewitt P, Smith DJ (2017) Influence of thermal ageing on the creep of a P92 martensitic steel. Mater Sci Eng A 708:544–555
19. American Society for Testing and Materials (2000) Standard test method for measurement of creep crack growth rates in metals. ASTM E1457-00
20. R6 Revision 4 Amendment 8 (2010) Assessment of the integrity of structures containing defects. BEGL procedure